Energy and Environment Research in China

More information about this series at http://www.springer.com/series/11888

Ye Yao · Yuebin Yu

Modeling and Control in Air-conditioning Systems

上海交通大學出版社
SHANGHAI JIAO TONG UNIVERSITY PRESS

Springer

Ye Yao
Shanghai Jiao Tong University
Shanghai
China

Yuebin Yu
University of Nebraska–Lincoln
Lincoln
USA

ISSN 2197-0238 ISSN 2197-0246 (electronic)
Energy and Environment Research in China
ISBN 978-3-662-53311-6 ISBN 978-3-662-53313-0 (eBook)
DOI 10.1007/978-3-662-53313-0

Jointly published with Shanghai Jiao Tong University Press, Shanghai, China

Library of Congress Control Number: 2016948282

Printed on acid-free paper

This Springer imprint is published by Springer Nature
The registered company is Springer-Verlag GmbH Germany
The registered company address is: Heidelberger Platz 3, 14197 Berlin, Germany

Preface

With the global warming and the rapid improvement of people's living standards, energy consumption by air-conditioning (AC) systems in buildings is on the rise. According to the US Energy Information Administration (EIA) and the US Department of Energy, the consumption of electrical energy by HVAC (heating, ventilation, and air-conditioning) systems in the residential, commercial, and industrial sectors corresponds to 18.62 %, 16.20 %, and 2.34 % of the total electrical energy consumed in the USA, respectively (totalizing 37.16 %). In China, building sector accounted for 23.4 % and 28 % of total energy use in 2011 and 2012, respectively, and about half of total building energy is consumed by HVAC systems. Thus, energy conservation in HVAC systems will play an important role in search of solutions to meet the growing global energy demand. Any technological measures for HVAC systems' energy consumption reduction require effective models based on which the high-performance HVAC systems and optimal control schemes for highly efficient operations can be designed.

This book mainly concerns about modeling and control in air-conditioning systems. Some advanced modeling methods including state-space method, graph-theory method, and structure-matrix method, as well as combined forecasting method, are employed for the modeling of air-conditioning systems. The virtual sensor calibration and virtual sensing methods (which will be very useful for the real system control) are illustrated together with the case study. The model-based predictive control and the state-space feedback control are introduced to the air-conditioning systems for a better local control, and the air-side synergic control scheme and the global optimization strategy with the decomposition-coordination method are developed aiming at energy conservation of the entire system. Lastly, control strategies for VAV systems including the total air volume control and the trim-and-response static pressure control are investigated with practice. The book comprises ten chapters that are summarized as below:

Chapter 1 (written by Dr. Ye Yao and Dr. Yuebin Yu) introduces background of the topic related to this book, gives a literature overview about modeling approaches in HVAC field, and presents proposed methods to be used in this book.

Chapter 2 (written by Dr. Ye Yao) illustrates in detail the modeling process for HVAC components and system with the state-space modeling method.

Chapter 3 (written by Dr. Ye Yao) presents simulation results on transient responses of HVAC components with the state-space models under different perturbations and initial conditions.

Chapter 4 (written by Dr. Ye Yao and Dr. Yuebin Yu) is related to development of graph-theory approach for modeling HVAC components and system, and introduces the structure-matrix analysis method to study control characteristics of HVAC state-space models.

Chapter 5 (written by Dr. Yuebin Yu and Dr. Ye Yao) deals with the virtual sensor calibration and virtual sensing methods.

Chapter 6 (written by Dr. Yuebin Yu and Dr. Ye Yao) is about control design based on the state-space model.

Chapter 7 (written by Dr. Ye Yao) is about forecasting models for air-conditioning load prediction. The two original forecasting models based on the combined principle are introduced.

Chapter 8 (written by Dr. Ye Yao) deals with energy models for HAVC components based on which the energy analysis program is developed and used for the energy analysis on variable-air-volume (VAV) air-conditioning systems.

Chapter 9 (written by Dr. Ye Yao and Dr. Yuebin Yu) is about optimal control of HVAC system aiming at energy conservation.

Chapter 10 (written by Dr. Ye Yao and Dr. Yuebin Yu) mainly deals with modular modeling, control strategies, and sequences as well as test script for VAV system.

Acknowledgement

The study work related to the book has been financially supported by several National Nature Science Foundations (No. 50708057; No. 51110105012).

Shanghai, China Ye Yao
June 2016

Contents

About the Authors

Dr. Ye Yao is an Associate Professor at the School of Mechanical Engineering, Shanghai Jiao Tong University, China. He received his Ph.D. from Shanghai Jiao Tong University (SJTU), China. He was promoted as Associate Professor of SJTU in December 2008. From September 1, 2009 to September 1, 2010, he performed his research work in Ray W. Herrick Lab at Purdue University (PU), USA. He was awarded as Excellent Reserve Youth Talent and SMC Excellent Young Faculty by SJTU, respectively, in the year 2009 and 2015, and got Shanghai Pujiang Scholars Talent Program in the year 2012. His current research interests mainly include (a) HVAC modeling and optimal control for energy conservation; (b) Heat and mass transfer enhancement assisted by ultrasound. He has successfully published about 100 academic publications and two academic books (first author) and owns 30 Chinese patents. He is now the peer reviewer of many international academic journals such as 'International Journal of Heat and Mass Transfer', 'International Journal of Thermal Sciences', 'International Journal of Refrigeration', 'Energy', 'Building and Environment', 'Energy and Buildings', and 'Applied Energy'.

Dr. Yuebin Yu is an Assistant Professor in the Durham School of Architectural Engineering and Construction at University of Nebraska-Lincoln, USA. He received his Ph.D. degree in Building Performance and Diagnostics from Carnegie Mellon University, Pittsburgh, PA, USA. He devotes his research efforts to the fields including (a) smart building technology, including automated continuous commissioning and advanced controls, automated fault detection and diagnosis, virtual sensing and virtual calibration; (b) active utilization of renewable energy for heating, ventilation and air-conditioning, including low-grade energy, solar and geothermal thermal energy, active phase change material, bionic building enclosure; and (c) built environment modeling and evaluation. At UNL, he maintains a state-of-the-art laboratory with well-instrumented facilities and advanced web-based AFDD platform for smart buildings and advanced building envelope studies. He is an active and voting member in the Technical Committee TC7.5 for Smart Building

Systems and serves as the sub-committee chair of Fault Detection and Diagnostics in ASHRAE. He participated in the revision of ASHRAE Handbooks on Fault Detection and Diagnostics and Energy Estimating and Modeling Methods. He has published about 50 academic publications.

Abbreviations

ACH	Air change rate
AFDD	Automated fault detection and diagnostics
AHU	Air handling unit
CLFTOT	Glass total cooling load factor
COP	Coefficient of Performance
DDC	Direct digital control
DX	Direct expansion
EEV	Electronic expansion valve
ERV	Enthalpy recovery wheel
FIR	Finite impulse response
FP	First principle
FPS	Fraction of possible sunshine
HVAC	Heating ventilation and air-conditioning
HW	Hammerstein–Weiner structure
IAHU	Integrated air handling unit system
IAQ	Indoor air quality
inv	Inverse function
LTI	Linear time-invariant
MAT	Mixed air temperature
MD	Measured disturbance
MIMO	Multiple input and multiple output
MO	Measured output
MPC	Model predictive control
MSHGF	Maximum solar heat gain factor
MV	Measured input
NL	Nonlinear
OA	Outside air
OAT	Outdoor air temperature
OAD	Outdoor air damper
PCA	Principal component analysis
PID	Proportional, integral, and derivative

PMV	Predicted mean vote
RAT	Return air temperature
RTU	Rooftop units
rpm	Round per minute
SAT	Supply air temperature
SC	Shading coefficient
SCFM	Supply air flowrate
SCOP	System coefficient of performance
SISO	Single input, single output
SLF	Solar load factor
Th	Threshold
UD	Unmeasured disturbance
UO	Unmeasured output
VAHU	Virtual air handling unit
VFD	Variable frequency drive

Nomenclature

a	Heat transfer coefficient (W/(m^2·°C)
A	Area (m^2)
b	Width (m); or filter coefficient
B	Thermal energy consumption (J)
B_m	Equivalent friction coefficient of motor (kg.m^2/s)
c	Mass specific heat (J/(kg. °C))
C	Percentage of opening degree of VAV terminal valve
d	Diameter, m
$D_{w.a}$	Mass diffusivity of water vapor in the air (m^2/s)
e	Fin spacing of heat exchanger (m); or error
G	Air/water flow rate (kg/s); or volume flow rate at VAV terminal device (m^3/s) (or m^3/h)
G_{contam}	Contaminant generating rate (kg/s)
h	Enthalpy (J/kg)
H	Lift, m H_2O
H_{stage}	Heating stage
H_t	Packing height of cooling tower (m)
i_a	Input current of motor (A)
I	Radiant intensity (W/m^2)
J	Moment of inertia of rotor (kg.m^2); or cost function
K_m	Torsional torque motor constant (N.m.rev/A)
K_v	Motor back emf constant (V.s/rev)
l	Length of coiled tube or air duct (m)
L_a	Motor inductance, Heng
M	Mass (kg)
$n(n_o)$	Motor actual (nominal) speed, (r/min)
n_e	Longitudinal average number of tube columns of evaporator
n_r	Introduction ratio
N	Power (W); or function of normal distribution
P	Pressure (Pa); or the size of the measurements
Pr	Prandtl number

q	Latent heat of condensation (J/kg); or irreversible loss rate (W); or solar radiation (W/m^2); or heat flux (W/m^2)
Q	Amount of heat exchange; or cooling capacity; or cooling load (W)
Qn	Thermal load in the working zone of air-conditioned space (W)
r	Radius (m); or cooling load factor
R	Heat conduction resistance (°C/W)
R_a	Electric resistance of motor, Ω
R_r	Gas constant of gaseous refrigerant (J/kg · K)
S	Finned tube center spacing of surface of heat exchanger, (m); or standard deviation
$T[t]$	Temperature (K [°C]); or T-distribution
u	Velocity, m/s
Δu	Rate of control inputs
U_m	Input electric voltage (V)
v	Variable
V	Vector of variable (v); or total air volume
w	Vector of disturbance
W	Air humidity (kg/(kg dry air))
W_n	Moisture load rate in working zone of air-conditioned space (kg/s)
Z	Compression factor of refrigerant gas
x	Reading of the measurand; or independent variable

Greek Letters

α	Confidence internal
α_ρ	Coefficient of absorption of radiant heat
β	Outdoor intake ratio
β_{IAQ}	Outside air ratio for acceptable IAQ in normal operation mode
λ	Heat conductivity coefficient (W/(m.°C)); or eigenvalue
λ_m	Mass transfer coefficient (kg/(m^2/s))
μ	Dynamic viscosity coefficient (kg/(m.s))
η	Efficiency
ς	Contaminants concentration (kg/kg)
v	Kinematic viscosity coefficient (m^2/s)
ε	Transmissivity
ε_a	Air volume ratio in heat exchanger
ρ	Density (kg/m^3)
δ	Exhaust air ratio from the interior zone
δ_c	Fin thickness of heat exchanger (m)
τ	Time (s)
ω_m	Angular speed of motor rotor (rad/s)
θ	Damper opening angle (rad)
θ_m	Angular displacement of motor rotor (rad)
γ	Circulation air ratio

ϕ_m	Power factor of the motor
ψ_e	Heat flux density of evaporator (W/m^2)
φ	Interior zone supply airflow rate ratio
κ_s	Adiabatic compression index of compressor
σ_r	Surface tension of the refrigerant boils bubbles (N/m)
σ^2	Variance of the population
$\sigma_{\bar{X}}^2$	Variance of the samples
$\chi_{f/p,G}$	Fan/pump flow dimensionless coefficient
$\chi_{f/p,P}$	Fan/pump outlet pressure dimensionless coefficient
χ_n	Motor rotor and fan/pump vane angle ratio
ξ	Interior zone relief air ratio
ltg	Lighting
L	Exit

Subscripts

a	Air or air windward section of heat exchanger
as	Saturated moist air
as–an	Air exchange between air supply and working zone
an–ar	Air exchange between working and air-return zone
b	Benchmark
bd	Building
c	Fin outer ring of heat exchanger or condenser; or cold deck; or cooling
cc	Cooling coil
cg	Shell wall of condenser
cl	Calibrated
col	Cloth
com	Compressor
co	Between refrigerant and shell wall of condenser
coil	Water-to-air heat exchanger
cp	Coolant liquid pump
cr	Refrigerant in condenser
cv	Control volume
cw	Coolant liquid of condenser
d	Cross section of straight duct or longitudinal length
da	Between air and internal surface of duct
dg	Wall of straight duct
di	Internal surface of duct
dis	Discharge
dsg	Design condition
dsc	Desiccant dehumidifier
e	Evaporating; or exterior region; or inlet
ef	Evaporator refrigerant side
eg	Shell wall of evaporator

env	Environment
ei	Inner ring of shell tube of evaporator
eco	Economizer
eo	Between refrigerant and shell wall of evaporator
ep	Chilled fluid pump
eqt	Equipment
er	Refrigerant in evaporator
erv	Energy recovery wheel
ew	Coolant liquid of evaporator
ex	Exhaust fan
E	Inlet
f/p	Fan/pump
fg	Moisture
g	Shell wall of heat exchanger
ga	Between air and exterior surface of heat exchanger
gain	Heat gain
gas	Gas
G	Glass
gw	Between water and interior surface of heat exchanger
gb	Exterior surface of heat exchanger; or saturated air layer
GB	Gray box
hc	Heating coil
H	Heating
i	Inner surface of finned tube of heat exchanger; or inner surface of room wall; or interior region
id	Induction
inf	Infiltration
IVC	Improved virtual calibration
lat	Latent
max	Maximum
meas	Measured
mfr	Manufacturer
min	Minimum
mix	Mixed air
mrt	Mean radiant
n	Working zone of air-conditioned room
o	Initial value; or exterior surface; or outdoor
out	Outside air
p	Pipe
pc	Peaking cooling
ph	Peaking heating
pg	Pipe wall
pi	Inner surface of water pipe
pw	Between water and inner surface of pipe
r	Refrigerant; air-return zone; or room air

rew	Outer wall of room
reg	Regeneration fan
rh	Reheat coil
riw	Inner wall of room
rm	Circulating refrigerant in chiller
rmois	Moisture source
rq	Heat source
r, *g*	Gaseous refrigerant
r,*f*	Liquid refrigerant
r-surf	Solar radiation corrected to the plane of window
run	Operation
s	Supply air zone of air-conditioned room; or discharge air; or sensible
set	Set point
shc	Subcooled degree of liquid refrigerant leaving condenser
shr	Superheat degree of gaseous refrigerant leaving evaporator
si, *sp*	Counter for sensors
signal	Signal from a sensor
sol	Solar radiant intensity
st	Status
sup	Supply
t	Between air and water film in cooling tower
tb	Saturated air layer on water film of fill of cooling water tower
ter	Terminal
thm	Thermal
tot	Total
ts	Transmission due to solar effect
val	Validated
vc	Virtual calibration
vir	Virtual
w	Water
wa	Between air and water film
wl	Wall

Chapter 1
Introduction

1.1 Background

With the development of economy and growth of people's living standards, air-conditioning systems are increasingly popular for the improvement of thermal environment indoors, which results in ever-increasing energy consumption of buildings. The need for rational energy use is a global concern, and challenges are energy consumption reduction without losing efficiency and quality of life. According to the US Energy Information Administration (EIA) [1] and the US Department of Energy [2], the consumption of electrical energy by HVAC (heating, ventilation and air-conditioning) systems in the residential, commercial, and industrial sectors corresponds to 18.62, 16.20, and 2.34 % of the total electrical energy consumed in the USA, respectively (totalizing 37.16 %). In China, building sector accounted for 23.4 and 28 % of total energy use in 2011 and 2012, respectively, and about half of the total building energy is consumed by HVAC systems [3]. Thus, energy conservation in HVAC systems will play an important role in the search for solutions to meet the growing global energy demand.

As well known, any technological measures for HVAC systems' energy consumption reduction require effective models based on which the relevant software tools, such as DeST [4], HVACSIM [5], and CONTAMW 2.0 [6], have been developed and assist us to design high performance HVAC systems and optimal control schemes for highly efficient operations. Generally, three types of approaches are used for modeling HVAC systems. The first type is known as physics-based (or white box) approach which uses detailed physics-based equations to model HVAC components, sub-systems, and systems to predict the whole systems' behaviors and performances. The second is data-driven (or black box) approach, viz. that the system performance data is collected under a specific test or a relationship is found between the input and output variables using some mathematical techniques, e.g., statistical regression and artificial neural network (ANN). This type of models needs on-site measurements over a period of time. In the third type, known as hybrid

Y. Yao and Y. Yu, *Modeling and Control in Air-conditioning Systems*,
Energy and Environment Research in China, DOI 10.1007/978-3-662-53313-0_1

approach, the basic structure of the model is formed with the physics-based methods and the model parameters are determined by using the parameter estimation algorithms on the measured data of the system. In order to build the physics-based models and to determine their parameters, the detailed knowledge about the system and its working process is needed. Physics-based models have good generalization capability but suffer from poor accuracy, while the data-driven models have relatively high accuracy but suffer from generalization beyond the training domain. Hybrid models combine the advantages of the two, providing good generalization capability as compared to the data-driven models and better accuracy as compared to the physics-based models. In the following sections, the state-of-the-art modeling in the field of HVAC is summarized in terms of modeling approach.

1.2 Modeling Approaches in HVAC Field

1.2.1 Physics-Based Modeling Approach

Physics-based approach is mostly applied to the modeling of HVAC components. HVAC components include primary and secondary. The primary system mainly includes chillers, boiler, cooling towers, and liquid distribution system; and the secondary system includes air-handling equipment, air-distribution system, and liquid distribution system between the primary system and the building interior. The distribution components are pumps, fans, dampers, valves, ducts, and pipes. They affect the energy flow in buildings by consuming electrical energy which drives pumps and fans and transferring thermal energy to (or from) the working fluid in all distribution components [7].

1.2.1.1 Chiller Model

A chiller unit basically consists of four individual components, i.e., evaporator, condenser, compressor, and expansion valve, which can be modeled, separately. The chiller works on the basis of vapor compression cycle in which the phase change of liquid refrigerant in evaporator takes away heat from the air-conditioned space; then, compressor increases the pressure of gas refrigerant making it superheated and releases it into the condenser where the gas refrigerant is condensed and the condensation heat is rejected to the water or air; afterward, the expansion valve reduces the pressure by releasing the refrigerant in the evaporator in a cool state.

Sami et al. [8] used a lumped-parameter approach to build a set of component models including shell-and-tube condenser and evaporator, air-cooled condenser, direct expansion (DX) evaporator, capillary tube, and thermostatic expansion valve. The heat exchangers were modeled using a drift-flux model that consists of separating the vapor and liquid phases and coupling the mass and energy balances of the individual phases through the evaporation or condensation. The model was then expanded by Sami and his cooperators [9, 10] and used to predict system's performance.

Lei et al. [11] developed a lumped-parameter dynamic model of a water-cooling refrigeration system based on the component models by using mass and energy balance principles. With the dynamic model, the effects of control inputs such as compressor operational frequency and TEV opening fraction on the output performance of the system were investigated.

To simulate the quick start-ups of centrifugal chillers, Schalbart et al. [12] established a vapor compression cycle model in which the two-phase flow models were used for modeling the condenser and evaporator of chiller. Meanwhile, the mechanical inertia of the fluid was considered, and all properties were derived from the internal specific energy and density and determined locally by the integration of mass and energy equations applied to finite volumes of the components: evaporator, condenser, and pipes.

The moving-boundary (MB) formulation is characterized by phase boundaries that move with time within the heat exchanger. Subject to the same assumptions as with the finite-volume (FV) approach, the MB formulation begins with dividing the heat exchanger volume into variable control volumes that encompass each phase region existing in the heat exchanger [13, 14]. Nyers and Stoyan [15] built a model of evaporator with the MB formulation using finite differencing within each phase. This model was used to predict the evaporator's behavior under step jump, exponential saturation, and periodic oscillation of the temperature and flow rate of the secondary fluid, compressor speed, condenser pressure, and throttle coefficient. The MB formulation was adopted as well by He et al. [16] in the development of a system model of a basic vapor compression refrigeration system for the purpose of studying the effect of multivariable feedback control and by other researchers for the other purposes [17–19]. Bendapudi et al. [20] discussed the FV and MB approaches applied to shell-and-tube heat exchangers. Detailed model formulations of both the FV and the MB approach for shell-and-tube heat exchanger modeling were provided, and stability was demonstrated as components and within a complete centrifugal chiller system model. They concluded that the FV formulation would be more robust through start-up and all load-change transients, but execute slower, while the MB method could handle all load-change transients, but start-up stability would be more sensitive to compressor and expansion valve formulations.

1.2.1.2 Cooling Tower Model

Cooling towers are widely used to remove heat from industrial processes and from HVAC systems. Heat rejection in cooling towers is accomplished by heat and mass transfer between hot water droplets and ambient air. Physical models of cooling tower are usually developed using Merkel's method and Effectiveness-NTU method [21]. In the Merkel's method, the water loss of evaporation was neglected and the Lewis number was assumed to be one in order to simplify the analysis. Although the Merkel's model has been the basis for most modern cooling tower analysis, it does not accurately represent the physics of heat and mass transfer process in the cooling tower fill [22]. The physical model based on the effectiveness-NTU method is relatively accurate compared to that based on the Merkel's method, but its reliability still depends on the accuracy of geometric information of cooling towers.

The physical models of cooling tower are also developed with the numerical method. Tan and Deng [23] derived a numerical model for the reversibly used water-cooling tower. The model could be used to make a detailed numerical analysis on the air and water states at any horizontal plane along the tower height. Fisenko et al. [24] presented a mathematical model for the performance of a cooling tower. The model consisted of two interdependent boundary-value problems. The first boundary-value problem described evaporative cooling of water drops in the spray zone of a cooling tower; the second boundary-value problem described film cooling in the pack. In the following years, Fisenko et al. [25] developed a new mathematical model of a mechanical draft cooling tower. The model represented a boundary-value problem for a system of ordinary differential equations, describing a change in the droplets velocity, its radii, and temperature, and also a change in the temperature and density of the water vapor in a mist air in a cooling tower. In addition, the model allowed one to calculate contributions of various physical parameters to the processes of heat and mass transfer between water droplets and damp air and the influence of atmospheric conditions on the thermal efficiency of the tower.

1.2.1.3 Air-Cooling/Heating Coil Model

In a HVAC system, the cooling/heating coil handles the supply air to anticipated conditions. The heat exchanger model can be obtained by the energy and mass balance on the water and air side of the coil. The physics-based approach for heat exchanger modeling results in a large number of governing equations. Solutions to the equations can be grouped into three categories: ① numerical solutions, ② lumped-parameter solutions, and ③ analytical solutions.

For the numerical solution, air-cooling coil is physically divided into numerous segments, and the inlet and outlet variables of each segment are calculated in turn [26, 27]. With this approach, the distributions of air temperature and humidity in the coil can be obtained, and the coil's thermal performances can be accurately evaluated, but it requires large computational cost.

The use of a lumped-parameter model can be, however, more preferred than the use of a numerical solution due to its lower computational costs. In a lumped-parameter model, the enthalpy difference between air and cooling medium is treated as the driving force of simultaneous heat and mass transfer. Such a treatment was firstly proposed by Threlkeld [28], based on the assumption of unit Lewis Factor, and has been popularly adopted by many other researchers [29–31].

Analytical solutions may be able to evaluate the thermal performance of different types of heat exchangers in an accurate manner but less computational cost.

Bielski and Malinowski [32] used the analytical method to solve a set of partial differential governing equations describing the transient temperature field in a parallel flow three-fluid heat exchanger by using Laplace transform. They compared the results obtained from their analytical solution with that from the numerical and semi-analytical solutions developed in other studies to validate their analytical solution. Yin and Jensen [33] developed a dynamic response model of a heat exchanger by using the integral method. In their model, the temperature distribution in the single-phase fluid was expressed by a combination of the initial and final temperature distributions as well as a determined time function. The model was used to investigate transient temperature responses of the single-phase fluid and the heat exchanger's shell wall when subjected to a step change in temperature or mass flow rate of inlet fluid. Ren and Yang [34] derived an analytical solution model for the coupled heat and mass transfer process in parallel/counter indirect evaporative coolers under real operating conditions. The analytical solution model was validated by comparing its results with those from numerical integrations. Xia et al. [35] obtained analytical solutions for evaluating the thermal performances of both chilled water wet cooling coils and DX wet cooling coils, respectively, under both unit and non-unit Lewis Factors. With the analytical solutions, the distributions of air temperature and humidity ratio along air flow direction in a wet cooling coil could be predicted, and the differences in the thermal performances of the cooling coils under both unit and non-unit Lewis Factors could be identified. To solve the nonlinear problem existing in the dynamic model of heat exchanger, Yao et.al. [36] developed a linearized model for air-cooling coil with small disturbances that was solved by means of Laplace transform. The model could be used to study the dynamic influences of inlet variables on the cooling capacities of cooling coil under different initial conditions.

In recent years, the state-space model for air-cooling coil was put forward by Yao et al. [37]. The state-space model is featured by its convenience in describing the dynamic characteristics of a multiple-input-multiple-output (MIMO) system. The detailed information about the state-space modeling will be introduced in the following chapter (Chap. 2).

1.2.1.4 Fan and Pump Model

Fans/pumps are the main power equipment in HVAC system, which consume over 30 % of total energy consumption of a central air-conditioning system [38]. The power of a fan or pump depends on the flow rate, pressure difference between inlet and outlet and efficiency of the fan or pump. Carrado et al. [39] established a dimensionless model for centrifugal ventilator and studied the influence of blade angle and rotate speed on the dimensionless parameters including fluid mass flow rate, pressure head, and input power. The model could assist us to design a controller for the control of air flow rate. Wang [40] developed a model for variable frequency pump in which the influence of electric frequency on the pump's performance was considered.

1.2.1.5 Duct/Pipe Model

As transmission channels of fluid in HVAC system, the drag loss and energy loss of duct and pipe are the two important factors concerned in the physics-based model development. Shao et al. [41] established a CFD (computational fluid dynamics) model for duct and analyzed the resistance loss of duct system in HVAC system. Fisk et al. [42] established a duct system model in large commercial buildings. The duct system model included physical characterization, air leakage, and heat conduction gains. Rokni and Gatski [43] built a mathematical model of turbulent convective heat transfer in duct and investigated the turbulent convective heat transfer in fully developed duct flows. Sugiyama et al. [44] adopted algebraic heat flux models to study turbulent heat flux distributions in a square duct with one roughened wall.

1.2.1.6 Room Model

The air-conditioning system is usually designed with respect to the thermal characteristic of the air-conditioned room, and the system control scheme is developed based on the requirement of indoor thermal environment. The physical model for air-conditioned room can be obtained according to energy and mass balance of air within a zone. Heat is transferred to the zone through the supply air, conduction through walls and windows, infiltration and supply air, and internal and external gains [45]. Heat transfer to a zone is commonly modeled by using heat balance method and thermal network zone method [46].

In heat balance method, the energy flow is modeled using the first law of energy conservation. For a zone, generally, each heat transfer element (wall, window, ceiling, floor, etc.) and zone air will result in a heat balance equation. These equations are simultaneously solved to find the unknown temperature of zone and surface of each heat transfer element using the matrix algebra techniques. The heat balance method has been adopted for studying dynamic responses of air temperature indoors. As early as the 1980s, Metha et al. [47] established a dynamic response model for the indoor air temperatures with the assumption that the air in the room was well mixed and the heat capacity of furniture was included in the air. Borresen et al. [48] discussed several simplified dynamic room models in which the thermal interaction between room air and surrounding walls were taken into account in different ways, e.g., the influence of the walls on the indoor air temperatures was neglected; the convective heat transfer between the air and the walls was considered, and the walls' thermal response time was used for the dynamic simulation. The study concluded that the choice of the simplified level employed depended on how closely the long-term responses and steady-state values fit the actual room response. The dynamic room temperature model developed by Tashtoush et al. [49] was obtained by energy balance on the room air, two walls, and the ceiling. The dynamic room model was employed for the control analysis on a HVAC system.

In thermal network zone model, the building is divided into a network of nodes with interconnecting paths through which the energy flows. The implementation of this method varies based on the selection of nodes on which energy balance is applied. Zone models exist with different levels of complexity: from simple 'well-mixed' models with one air node representing the whole air volume in the room to complex CFD models solving the equations of conservation of mass, momentum, and energy [50]. Chen and Peng [51] firstly studied the indoor dynamic temperature distributions by using a fixed-flow-field CFD model in which the room thermal response and indoor air distribution were computed considering the outdoor air temperature, solar radiation, indoor heat sources, and other thermal boundary conditions. Although the CFD results have been shown to be accurate, the calculation is too time-consuming. A multi-zone model simplifies the CFD model by dividing a room, which is otherwise modeled in CFD by thousands of finite volumes, into several air zones each of which is assumed to be well-mixed as one air node. Wu et al. [52] developed a nodal room model with four temperature nodes at different heights above the floor. The model could be used to predict vertical temperature distribution in a typical office room with floor heating and displacement ventilation. Yao et al. [53] developed a state-space room model in which the room was divided into three zones in terms of the air momentum and the air flow rate based on the degradation of fluid mechanics equations, i.e., supply air zone, working zone, and return air zone. Compared to the single-zone model, the simplified multi-zone room model is much closer to real situations while saving much computational time compared to the CFD model. So, it may be the best choice for studying the room temperature distributions and dynamic thermal characteristics of indoor air.

1.2.2 Data-Driven Modeling Approach

Physics-based models provide good generalization capability but lack of accuracy compared with the data-driven ones. In addition, the calibrations of physics-based models are confronted with great challenge due to the identification of large number of parameters. Major methods used for data-driven modeling of HVAC systems include frequency-domain, data mining, fuzzy logic, and statistical method [46].

1.2.2.1 Frequency-Domain Models

Due to the heavy thermal inertia existing in the HVAC system, many processes such as dynamics of exit air temperature of air-cooling/heating coils are changing slowly with time delay. Such processes can be modeled using the first- and the second-order formula with dead time as follows [54, 55]:

$$G(s) = \frac{Y(s)}{U(s)} = \frac{K}{\tau s + 1} \mathrm{e}^{-Ls} \tag{1.1}$$

$$G(s) = \frac{Y(s)}{U(s)} = \frac{1}{a \cdot s^2 + b \cdot s + c} e^{-Ls} \quad (1.2)$$

where, K is static gain; L is apparent dead time of the process; τ is time constant; a, b, and c are unknown coefficients.

The first- and second-order models are developed for SISO systems and can be extended to MIMO systems. The first-order process model has been employed for developing advanced PID autotuner of AHU and describing dynamics of flow meter [55] and cooling coils [56]. First- and second-order frequency-domain models with dead time have simple structure and a small number of parameters to be determined from the measured data. The identification techniques are often applied to identify these parameters.

1.2.2.2 Data Mining Models

The data mining and machine learning algorithms such as ANN and support vector machine (SVM) are often applied to complicated and nonlinear system like HVAC systems. The network is trained by a supervised learning algorithm, and the SVM-based approach projects the nonlinearly separable data into higher dimensional feature space through a mapping function in which it can be separated linearly.

The ANN method is often employed to estimate the performance of HVAC systems and components. For example, Hosoz and Ertunc [57] predicted the performance of an automobile air-conditioning system using ANN. They developed a multilayer feed-forward networks (MLFFN) for predicting the performance parameters such as compressor power, heat rejection rate in the condenser, refrigerant mass flow rate, compressor discharge temperature with reference to compressor speed, cooling capacity, and condensing temperature. Ertunc and Hosoz [58] compared the performance predictions of an evaporative condenser using ANN techniques. They predicted the condenser heat rejection rate, exit refrigerant temperature of condenser as well as dry- and wet-bulb temperatures of the leaving air stream with respect to the following parameters: inlet air temperature and humidity, air flow rate, refrigerant flow rate, water flow rate, absolute pressure, and temperature of the refrigerant at the inlet of the condenser.

The SVM method has been applied to load prediction in building HVAC systems. In the study by Ding et al. [59], the SVM was used to establish a load forecasting model based on measured data of cooling load over a period. The global optima of SVM penalty parameter, intensive loss function, and kernel function were found by using the ant colony optimization (ACO). To improve the load forecasting capacity, the SVM model was often combined with the other data analysis algorithms, e.g., hybrid SVM combined with autoregressive integrated moving average (ARIMA), hybrid SVM combined with kernel principal component analysis (KPCA), and hybrid SVM combined with particle swarm optimization (SAPSO) algorithm.

Unlike the ANN, the SVM is resistant to over-fitting the data and has better performance than simple ANN. SVM finds the global optimum solution in the data and provides the best fit for the data. In order to build the models using data mining algorithms, large amount of training and testing data is needed. No physical interpretation of the developed model is possible and the performance degrades when conditions deviate from training and testing conditions. These algorithms are suitable for offline model development, and online implementation of these algorithms is often cumbersome [46].

1.2.2.3 Fuzzy Logic Models

The fuzzy logical model is developed based on the expert knowledge, and it is implemented through the if-then-else statements or rules which are written in the form of a table or database. Fuzzy logical model is developed usually for the control of HVAC system. For example, Becker et al. [60] designed a fuzzy controller for temperature and relative humidity in refrigeration system by considering their thermodynamic coupling. He et al. [61] established a multiple model predictive control (MMPC) strategy based on Takagi-Sugeno (T-S) fuzzy models for temperature control of air-handling unit (AHU) in HVAC systems. The overall HVAC control system was constructed by a hierarchical two-level structure. The higher level is a fuzzy partition based on AHU operating range to schedule the fuzzy weights of local models in lower level, while the lower level is composed of a set of T-S models based on the relation of manipulated inputs and system outputs correspond to the higher level. There are also other new fuzzy logical models combined with some data mining algorithms, such as artificial neural fuzzy interface system (ANFIS) model [62], gradient autotuned Takagi-Sugeno fuzzy forward model [63], and genetic algorithm-based adaptive fuzzy logic model [64]. Models developed with fuzzy logic require experiences and comprehensive knowledge about the objects as well as large amount of field data for training. However, these may not be readily available for many HVAC components and thus presents a difficulty in modeling these components using fuzzy logical method.

1.2.2.4 Statistical Models

The statistical models used in the HVAC fields mainly include autoregressive exogenous (ARX), autoregressive moving average exogenous (ARMAX), ARIMA, Box-Jenkins (BJ), and output error (OE). The mathematical expression for the generalized structure of statistical black box models in a simple input–output relationship is given below:

$$\alpha(q^{-1})y(\tau) = \frac{\beta_1(q^{-1})}{\theta_1(q^{-1})}u(\tau) + \frac{\beta_2(q^{-1})}{\theta_2(q^{-1})}w(\tau) \tag{1.3}$$

where $\alpha, \beta_1, \beta_2, \theta_1, \theta_2$ are polynomials; q^{-1} is back shift operator; $u(\tau), y(\tau)$, and $w(\tau)$ are input, output, and noise at the time point τ.

The model ARMAX is superior to ARX as it incorporates the time series of error in the model structure which is essential for capturing the dynamics of the error and better control performance. ARIMA is a generalization of ARMAX, modeling the stationary and non-stationary data into a single step, and consists of autoregressive, integrated, and moving average parts [46]. Mustafaraj et al. [65] investigated Box-Jenkins (BJ), autoregressive with external inputs (ARX), autoregressive moving average with external inputs (ARMAX), and OE models to identify the thermal behavior of an office room in a modern commercial building in London. Their study manifested that these numerical models could all be potentially used for improving the performance of the thermal environment control system. Other examples of statistical model applications in HVAC areas include the following: predicting the room temperature variations for both short-term and long-term periods with ARMAX method [45], and load forecasting in air-conditioned non-residential buildings with the ARX and the ARIMA method [66]. Since the processes in an HVAC system depend on their previous values, a time series regression model (i.e., ARX, ARMAX, and ARIMA) captures these correlations by including the process variables from the previous sampling times, which will result in a favorably accurate model of the process dynamics.

1.2.3 Hybrid Modeling Approach

Hybrid models use physics-based models as the model structure, and their model parameters are estimated from the measured data. These models provide physical meaning and are superior to data-driven models in terms of generalization capability. Hybrid models choose appropriate parameters to capture regular patterns of the system which cannot be well modeled with physical equations. The determination of model parameters requires knowledge of both the physical phenomena and the data from the process.

Wang et al. [67] built a gray box model for predicting the performance of chilled water-cooling coils in a static state. The model was built based on the heat transfer mechanism and the energy balance principle. The key model parameters that represent the lumped geometric terms were identified by catalog information with the linear or nonlinear least-squares methods. Ghiaus et al. [68] used a gray box model to identify parameters in the partially known models of AHU elements including air-cooling coil, electric heater, and humidifier. In another study by Yao et al. [69], a dynamic response model of wet cooling coils was developed based on the first law of energy and mass conservations, and the coefficients of heat and mass transfer in the dynamic model were determined by test data with ANN method.

Hybrid models provide better accuracy than physics-based models and better generalization capability compared to data-driven models, but they are also difficult to develop. In order to develop hybrid models, both the knowledge of underlying physical phenomenon and input–output data of the system are required. For some HVAC sub-systems, the underlying physical phenomenon could be very complicated to model, and for the other systems, the input–output data may not be readily available making it difficult to develop the hybrid models. The parameters for hybrid models also need to be updated timely when the operating conditions deviate from the training data in order to ensure higher accuracy.

1.3 Proposed Methods

1.3.1 State-Space Modeling

State-space model can be obtained directly from the input and output data measurements [70]. However, it is often derived from a series of physics-based equations in many studies. As early as the 1990s, Jiang [71] applied the state-space method to the study of heat transfer characteristics of building wall. Afterward, the state-space method has been employed by limited researchers to develop dynamic models for HVAC components [37, 72, 73], indoor environments [50, 74], and distributed air-conditioning systems [75, 76]. In this book, the state-space method will be used to develop dynamic models of all components in HVAC system, and these state-space models can be used to study dynamic characteristics of HVAC equipment and promote applications of some advanced control models in the HVAC field.

The potential benefits of state-space modeling method mainly include the following:

(1) Unlike the ANN, ARMAX, and ARX models, it can be of physics-based model which allows us to understand the essences of dynamic relationship between the input perturbations and the output response variables of the system.
(2) It is flexible and easy to simulate the transient response of MIMO systems.
(3) It has uniform representation and hence enhances the integration of system modeling.
(4) It provides not only information of system outputs, but also that of the system states which may be used as feedback signals to improve control performances.
(5) It is an important basis for the development of MPC (model-based predictive controller) that may have good application prospect in the HVAC fields [77, 78].
(6) Lastly, state-space model is expressed in the form of matrix that is convenient for computer calculation.

1.3.2 Graph-Theory Modeling

Graph theory [79], which considers sub-systems and interactions, has been applied in various areas. Garg et al. [80] developed a deterministic quantitative model based on graph theoretical methodology to compare various technical and economical features of wind, hydro, and thermal power plants and also used to evaluate and rank the power plants in ascending or descending order in accordance with the value of their suitability index. Prabhakaran et al. [81] developed an integrated systems model for the structure of the composite product system. The composite product was first modeled with the help of graph theory, then by variable adjacency matrix and then by a multinomial known as permanent function. Grekas and Frangopoulos [82] developed automatic synthesis of mathematical models by using graph theory for optimization of thermal energy systems. The topology of the graph was stored in the computer memory, and the computer model of the respective system could be constructed automatically by object-oriented programming (OOP). Singh and Agrawal [83] built a graph-based model of manufacturing system. The matrix models and the variable permanent function models have been developed to carry out decomposition, characterization, and the total analysis. Wang et al. [84] employed the graph-theory-based method to describe the flexible flow circuits of different liquefaction processes of LNG spiral wound heat exchanger. The graph-theory description helps to solve the distributed-parameter model in a proper way.

The literature about the applications of graph theory in the HVAC field is very limited although the topology of a HVAC system may make the theory particularly applicable and beneficial. To develop an integrated systems model for the structural analysis of air-conditioning system and integrate different components of the air-conditioning system and takes into account the effect of interaction between these components for the better understanding of the structure of the air-conditioning system, the graph-theory approach is employed in this book for modeling HVAC components and system. The potential benefits of graph-theory modeling in HVAC system mainly include the following:

(1) Recognizing different components or sub-systems constituting the air-conditioning system and identifying interactions among these different components;
(2) Makes it possible to automatically develop the model of the whole system by using models of components from a library and information about the topology of the system (i.e., the components comprising the system and their interconnections);
(3) Development of matrix models and the variable permanent function models to carry out decomposition, characterization, and overall analysis.

1.3.3 Combined Forecasting Modeling

Accurate prediction of air-conditioning load is the critical basis for energy-saving operations of central air-conditioning system including adjusting the starting time of cooling to meet start-up loads, minimizing or limiting the electric on-peak demand, optimizing costs and energy use for cool storage systems [85]. Although there have developed many forecasting models for air-conditioning load prediction, such as ARIMA model, exponentially weighted moving average model (EWMA), multiple linear regression model (MLR), ANN model, and gray model (GM) [86–90], any individual forecasting model has its own limitations. For example, the ARIMA and EWMA model is based on time series analysis method, and they fit only for short-term forecasting; the ANN model needs large amount of data to train, and it is valid only within the scope of the training data; the MLR model may result in big forecasting errors because nonlinear relationships always exist in actual situations; the GM uses the operations of accumulated generation to build differential equations for prediction, and it may be more suitable for long-term forecasting. To overcome shortcomings and make full use of advantages of individual forecasting model, the combined forecasting method is introduced in this book. The theory of the combined forecasting method is based on a certain linear combination of various results from different forecast models. The fitting capacity of the combined forecasting model is greatly improved, and the forecasted result will present a higher precision [91]. Formulations have been developed in the literature [92] for the optimal combined forecasting method whose deviation reaches the minimum and is less than that of each individual forecasting method.

1.3.4 Decomposition–Coordination Algorithm for Global Optimization Model

Control functions of building automation systems (BASs) can be divided into two categories, local control functions and supervisory control (or energy management) functions. Local control functions are the basic control and automation that allow the building services systems to operate properly, which are established based on the dynamic models of controlled objects (e.g., state-space model). The supervisory control functions determine optimum operation points of equipment in HVAC system aiming at the minimum energy input or operating cost to provide the satisfactory indoor comfort and healthy environment. Aiming at energy saving of central air-conditioning systems, many researchers have done a lot of work on improving energy efficiency of local systems. For example, the optimal controls of chilled water system were intensively focused on by some studies [93–96], while that of the air-distribution sub-systems were particularly paid attention to by the other studies [97–99]. As well known, local optimal operations will not always lead to global optimization results. A central air-conditioning system normally consists of a primary chilled water plant and numerous air-handling and air-distribution sub-systems. When the whole HVAC system is taken into account, there ought to exist a trade-off of energy consumption among different equipment in this system. It is easily understood that increasing the fan speed of the cooling tower increases fan power, but it improves chillers' efficiency and reduces chillers' power; increasing the chilled water set point temperature reduces chiller power but increases pump power, and so does increasing the supply air set point to chiller power and fan power [100]. Therefore, it is necessary to consider the optimization of HVAC systems as a whole, and global optimization model is suggested for the supervisory control of whole system.

Searching optimization method is equally important in realizing global optimization control strategies. Many optimization algorithms have been developed in the past two decades, including direct search methods [101], gradient-based and nonlinear programming methods [102], and ANN [103]. In addition, the new concepts of genetic and evolutionary algorithm have been introduced as well to tackle HVAC optimization problems [104, 105]. These algorithms have their own features and roles in solving certain optimization problems, but they have their own limitations in real applications. The direct search methods may be less computationally efficient although they are reliable and easy to implement; the gradient-based and nonlinear programming methods are only applicable for solving the optimization problems that can be directly presented by a continuous function; the ANN relies too much on the input data for training and is unable to reliably extrapolate beyond the calibration range. Although genetic can quickly evaluate a large solution area and work quite well for problems with high degree of complexity, it has the biggest disadvantage due to its nature of evolution that may easily evolve away from a bad solution and lead to a large number of iterations for arriving at near-optimal solution [106]. Central air-conditioning systems are often large complicated nonlinear systems that consist of many strongly coupled sub-systems. There will encounter big challenge in solving the global optimization problem in which a large number of equipment are involved and numerous decision variables are to be determined. The high-dimensional problems will easily result in a computation collapse (which is also called 'high-dimensional disaster') no matter what kind of optimization algorithm (mentioned above) is adopted.

In this book, a global optimization model for central air-conditioning system is developed. The global optimization problem is formulated based on the energy models of components in HVAC system, including chiller, water-to-air surface heat exchanger, cooling water tower, and inverter pump/fan. To solve the 'high-dimensional disaster' problem of global optimization model for some super-large HVAC system, the decomposition–coordination algorithm, which is considered to be an effective way to tackle large-scale optimization problem [107–110], is employed for the optimization model solving.

1.3.5 Virtual Calibration for HVAC Sensors

Due to the large quantity and complexity of devices in central HVAC systems, this process can barely be handled manually; instead, it is fulfilled automatically with an enhanced BAS that electronically integrates the mechanical devices through sensing, computing, data processing, and actuating. However, due to the consideration on initial cost, building systems are generally under-sensed with near-zero sensor redundancy. Physical variables of our interest in HVAC systems and buildings may be measured with only one sensor or even not measured. For example, the outdoor intake ratio in AHUs is seldom acquired. Meanwhile, many sensors in building HVAC systems are improperly installed, wrongly placed, damaged, or gradually failed in the adverse working environment [111]. Readings from these sensors or transmitters could be inaccurate or totally wrong. Because of zero redundancy in typical building systems, it becomes difficult to tell the reliability and accuracy of measurements. Using erroneous data or wrong information could lead to a significant energy penalty or even direct failure of control and operation algorithms.

Sensor errors generally comprise precision degradation, reading bias, drifts, noise, or sensor failure. Conventional approaches for correcting the errors and improving the accuracy of measurements from various sensors and meters in real buildings can be categorized as (1) sensor calibration [112, 113] and (2) statistics-driven data fusion [114–116]. The essence of a physical sensor calibration is a well-designed comparison against a standard instrument in a predefined environment to bring the working sensor back to its normal condition. A sensor calibration is the fundamental method of correcting suspicious sensors. Generally, all sensors in a dynamic system should be checked regularly against standard instruments to ensure measurements' quality. Based on some rule-of-thumb, for example, for temperature measurements, sensors should be calibrated every 12 months; for pressure gauges, it is desired for every six months. Beside the sensor calibration, statistic data fusion methods may also be applied to obtain the representative value of physical variables. With a data-driven method, different data or information sources (for instance, direct measurements from physical sensors and indirect measurements from models) are integrated in a data fusing process to obtain the accurate, complete, or dependable information. The main procedure of data-driven methods consists of various filtering algorithms and statistical processes.

A sensor calibration is more preferable over a data fusion method since the former works frontend on a sensor itself for maintaining the quality of direct measurements. Meanwhile, a calibration is the most effective method in reducing systematic errors and eliminating failure of sensors. Despite the necessity, a sensor calibration is barely carried out regularly on various sensors in building HVAC systems unless significant measurement errors or malfunctions are identified. Main challenges to conduct a regular calibration on sensors are as follows:

(1) *Time and monetary cost*. A complete calibration process of an individual sensor includes multiple steps, from removing a working sensor from a system, conducting a calibration, to reinstalling it back; any of the steps could be time-consuming and expensive.
(2) *Disruption to a normal operation*. Removing and reinstalling a sensor will more or less disrupt the normal operation of HVAC systems. Missing measurements from the removed sensor also need to be covered temporarily to resume the operation during the process.

(3) *Access to various sensors.* Due to the space and installation constraints, it could be impractical or very costly to remove some sensors (e.g., a flow rate meter in a pipeline, a temperature sensor hiding behind the ceiling) from its working environment.
(4) *Large quantity of sensors.* Building HVAC systems have a large sensor network to acquire different types of information (e.g., temperature, humidity, flow rate, CO_2, etc.) and from different levels on the operation of the system. This factor further amplifies the difficulties listed above.

In addition to these challenges, there is one more limitation directly associated with a conventional calibration. A physical sensor after calibration may not have a favorable working environment, as that in the calibration, to work properly and provide close measurement to the real value. For example, Yu et al. [117] found that the commonly preinstalled supply air temperature sensor in compact rooftop air conditioners cannot accurately measure the real temperature of supply air. Due to the compact size, poor air distribution, and intensive thermal radiation of gas heating chamber, errors associated with the sensor could be up to 19.2 °C and be erratic.

In addition to acquiring improved accuracy and resiliency against errors, an ideal calibration process should be conducted as in situ, hence avoiding the differences in the medium and changes of working environment and the associated effects on the measurements. Some studies have recently been conducted in the area of automated virtual in situ calibration of sensors [118–124]. Terms, such as blind calibration [122] and self-calibration [121], are used synonymously as virtual in situ calibration [111, 117] in the studies. A blind calibration was considered by Balzano and Nowak for sensor networks without a dense deployment [122]. It was assumed that the sensor calibration function can be depicted with a linear model; therefore, a calibration problem was transformed to obtain the unknown gains and offsets. Slightly over-sampling was assumed for general applications in order to solve the linear system of equations. A virtual standard concept was proposed by Dulev et al. to calibrate measuring devices [120]. A calibration problem in a sensor network was treated as a general parameter estimation problem in a study conducted by Whitehouse and Culler [118]. Measuring devices for localization are calibrated as a whole with parameters identified that can eventually optimize the overall system response. The average measurement errors were reduced from 74.6 to 10.1 % after the implementation of the method. An iterative registration and fusion approach was proposed for calibrating multi-3-D sensors [114]. The foundation of their calibration approach was to minimize the squared distance error through the least-squares from the sensors' data. A collaborative calibration scheme was proposed by Bychkovskiy et al. for calibrating sensors with dense deployment [119]. The redundancy was utilized to calibrate one sensor against the others. The relative calibration relationships, as temporal correlations between pairs of co-located sensors, were first determined. After that, heuristic optimization was applied to maximize the consistency of the pairwise functions among sensors. Wireless thermistors were tested to evaluate the proposed method. A self-calibration method, formulated as an inference problem on a graphical model, was investigated for calibrating wireless localization sensors [121]. Nonparametric belief propagation was then applied to obtain the solution for the problem.

Recently, Yu et al. [111, 117] proposed a model-based virtual calibration technology for the measurement of supply air temperature in packaged air-conditioning units. The measurement error, which is up to 19.2 °C and erratic, can be improved with an uncertainty in ±0.7 °C. The model was later improved to ensure high robustness over a wider range of operating and fault conditions. The book will arrange a chapter to present an innovative virtual in situ calibration algorithm developed by Yu et al. [111, 117], which is noninvasive and real time and can be potentially automated to handle the aforementioned challenges and limitation of a conventional calibration.

1.3.6 Model-Based Predictive Control (MPC)

Functions of building control are typically fulfilled in a two-level structure: a supervisory level and a local level. The supervisory-level controller comprises a group of logics and typically resides in a central station. It takes actions based on preset conditions and/or rules, or the commands from the operators, such as heating/cooling switch-over, components sequencing, event scheduling, etc. However, this type of rule-based responsive control, with local on/off and proportional–integral–differential (PID) controllers, is expensive in the long run since they operate at a non-optimal efficiency [125]. In addition, neither of the two control objectives, thermal comfort and energy savings, can be explicitly expressed in the conventional control laws [126]. An advanced building control system needs to possess some kind of predictive capability with building dynamics and deterministic and stochastic disturbances considered and can evaluate different objectives rather than decoupled simple references. One of such enabling technologies is termed *model-based predictive control* (MPC) [127]. It includes a building system model, a cost function, a set of constraints, and an optimization solver and runs in a receding horizon manner. From a general point of view, an MPC is a strategy, rather than a control law, which uses a combination of control methods linked by common ideas [128]. Due to its advantages in handling constraints, cross-coupling issues, and multiple objectives in a nonlinear multiple-input-multiple-output (MIMO) system, MPCs have been greatly utilized in the manufacturing process industry since its introduction in the 1970s and are regarded as one of the most mature advanced controls [129].

Research on MPC in building systems has just recently started to thrive [130–133]. Upon the literature review, we find the majority of these studies are in the area of general MPCs, which usually couple a simulator with an optimization solver for the applications. With this approach, the mathematical structure of the simulator is invisible to the outside solver. Stochastic global search strategies, such as genetic algorithm, simulated annealing, particle swarm optimization, are applied to identify the optimums. This approach is generally associated with very high computational cost and not suitable for online implementation. For example, Zhang and Hanby [130] applied genetic algorithm to a system with different energy sources and thermal storage. They found that while significant improvements in system operation were possible, a search for 13 hourly commands took over 24 h to complete. Because of the computational cost associated with general MPCs, an offline MPC was recommended followed by the logistic regression on the data to extract some simple rules for real implementation. Another approach of formulating an MPC is to use an explicit linear model of a physical system and make the gradients visible to the optimization solver. This allows the application of powerful tools, such as the linear quadratic regulator theory, Kalman filtering theory, to explore the potentials. Study of utilizing the good features of linear MPC on multivariate HVAC systems is gradually gaining more interest recently. For instance, Yuan and Perez [131] applied MPCs to a variable air volume system to achieve an acceptable indoor air quality. The ventilation condition in the six zones with MPC was found to be improved compared to the traditional control. Freire et al. [133] utilized an MPC for thermal comfort control in a single-zone one-actuator air-conditioning building. Two case studies in terms of an MPC and different metabolic rate and clothing index were simulated to validate the proposed method. Hazyuk et al. [134] applied linear programming method in the canonical form to the predictive control of intermittent heated buildings. The mathematic summation of energy consumption and thermal comfort was treated as the cost function and constraints, respectively. Heat flux from the radiator was used as the control input.

Despite the findings from those studies, very few studies considered both thermal comfort and energy savings explicitly in the formulation of MPCs. In addition, the method and application of a linear classical MPC for MIMO nonlinear buildings systems have not been fully explored. The book will arrange a chapter to investigate the methodologies and potentials of utilizing linear classical MPCs for nonlinear systems, considering both thermal comfort and energy cost.

1.4 Organization of This Book

As shown in Fig. 1.1, the book totally comprises ten chapters. And relationships among these chapters are presented in Fig. 1.2. The main contents of each chapter are summarized as below:

This chapter introduces background of the topic related to this book, gives a literature overview about modeling approaches in HVAC field including the physics-based, the data-driven, and the hybrid. Afterward, the proposed methods to be used in this book are mentioned, and their potential benefits are summarized.

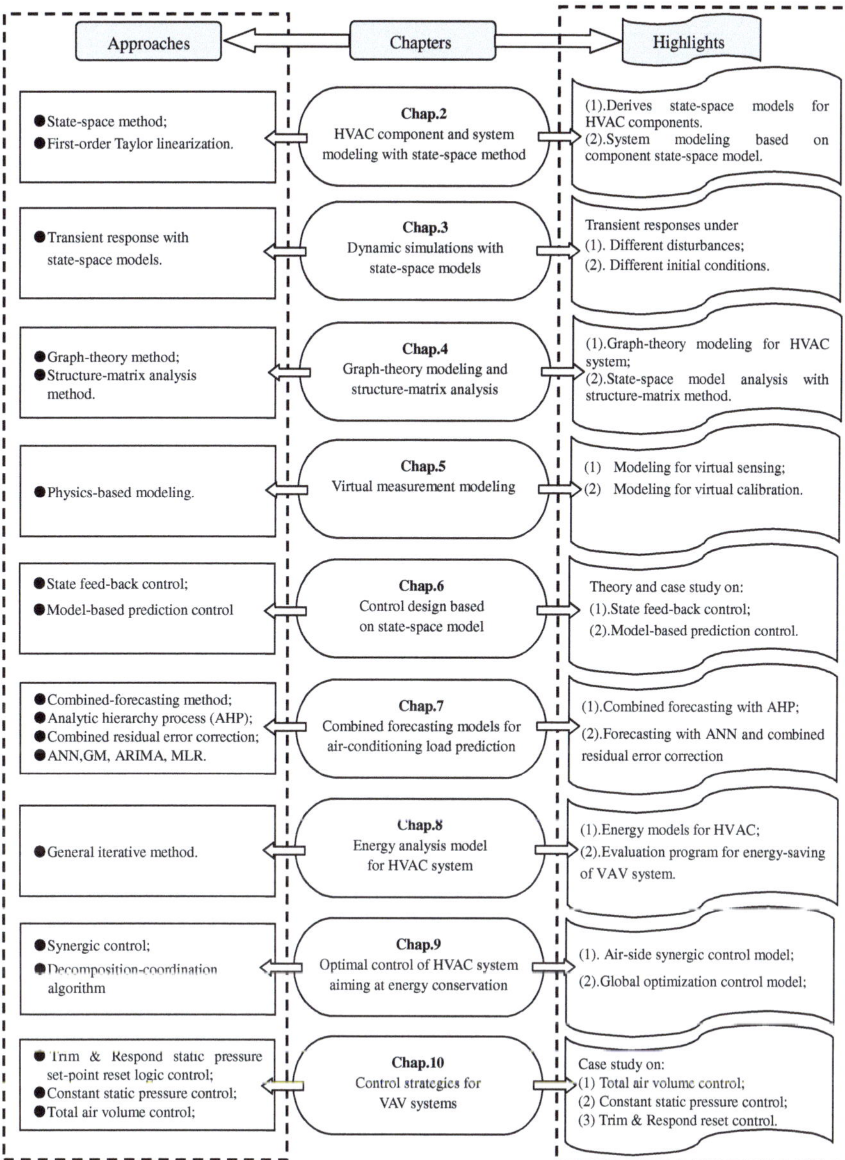

Fig. 1.1 Basic structure of this book

Chapter 2 illustrates in detail the state-space modeling process for HVAC components including water-to-air surface heat exchanger, chiller, cooling tower, air-conditioned room, and duct (pipe) as well as fan (pump). Then, the HVAC system modeling based on the component state-space models is given. Meanwhile, experimental validations for all the state-space models are presented.

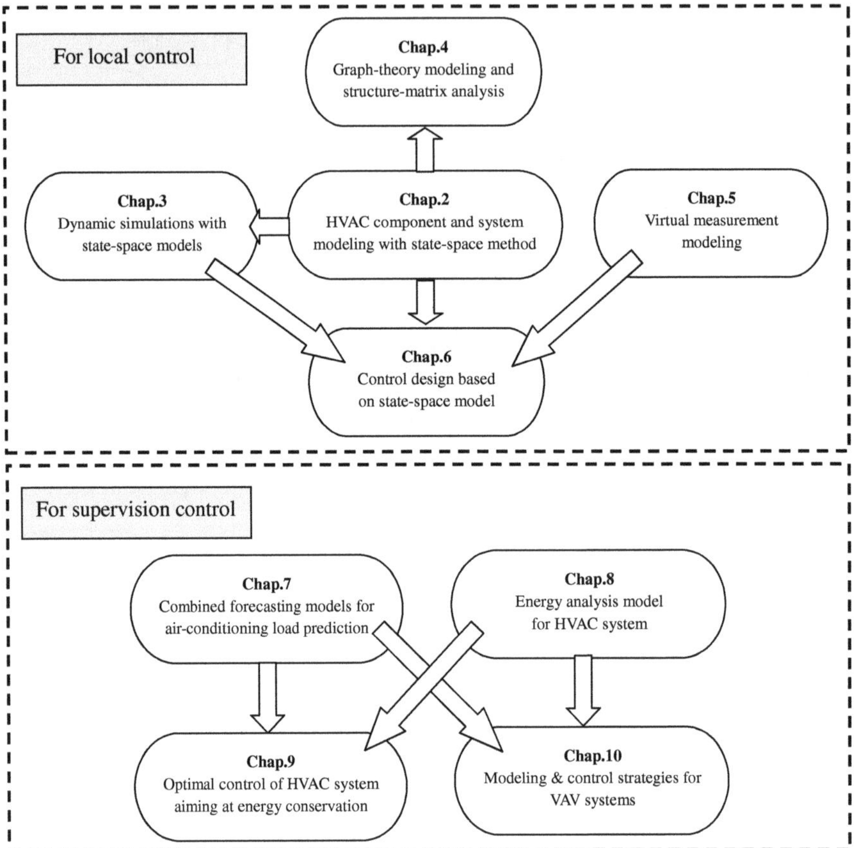

Fig. 1.2 Relationships among chapters in this book

Chapter 3 discusses dynamic characteristics of HVAC components with the state-space models under different perturbations and initial conditions. These components include water-to-air surface heat exchanger, chiller, cooling water tower, air-conditioned room, and air duct as well as water pipe.

Chapter 4 is related to development of graph-theory approach for the HVAC components and system modeling. Meanwhile, the structure-matrix analysis method is introduced to study model's control characteristics including reachability, controllability, and observability.

Chapter 5 deals with the virtual calibration method for sensors used in HVAC system and virtual sensing that has potential applications in the control and fault diagnostic of HVAC systems. A modeling-based in situ virtual calibration to the supply air temperature sensor in packaged rooftop units (RTUs) is given, and a virtual sensor for supply airflow measurement is presented as the case study.

Chapter 6 proposes a dual-loop MPC and feedback linearization to decouple the building physical model from equipment physical models.

The Wiener–Hammerstein N-L-N approach is employed to solve the modeling difficulty, complex control variables, and discrete constraints in the application of advanced control. Two examples are given for the proposed MPC applications in the HVAC field. Afterward, the state-space feedback control is introduced, and the control system design for water-to-air heat exchanger and refrigeration system are presented.

Chapter 7 is about forecasting models for air-conditioning load prediction. The two original forecasting models based on the combined principle are introduced: One is combined forecasting model based on analytic hierarchy process (AHP), and the other is forecasting model based on neural network and combined residual error correction. Case studies are presented for the proposed forecasting models.

Chapter 8 deals with energy models for HAVC components including chiller, boiler, pump, fan, cooling tower, and water-to-air heat exchanger. Then, the energy analysis program is developed based on theses energy models and used for studying energy performance of variable air volume (VAV) air-conditioning systems.

Chapter 9 is about optimal control of HVAC system aiming at energy conservation. The air-side synergic control model and the global optimal control model for HVAC systems are presented, and the decomposition–coordination algorithm is employed for solving the global optimization model with high dimension.

Chapter 10 deals with the development of component model library of VAV system, control strategies including total air volume control and trim-and-respond logic control, control sequences, and test script for VAV system.

References

1. US Energy Information Administration (EIA).: Electric Power Monthly, Table 5.1, 11 March 2011
2. US Department of Energy (DOE).: 2010 Buildings Energy Data Book, Sections 2.1.5 and 3.1.4 (2011)
3. Building energy conservation research center of Tsinghua University, Annual Report on China Building Energy Efficiency (2013)
4. Tsinghua university architecture technology science. DeST User Manual (2004)
5. Park, C.: HVACSIM+ User's Guide Update. National Institute of Standards and Technology (2008)
6. Dols, W.S., Walton, G.N.: CONTAMW 2.0 User Manual. National Institute of Standards and Technology (2002)
7. ASHRAE.: ASHRAE Handbook—Fundamentals. Atlanta (USA), GA (2009)
8. Sami, S.M., Duong, T., Mercadier, Y., Galanis, N.: Prediction of the transient response of heat pumps. ASHRAE Trans. **93**, 471–478 (1987)
9. Sami, S.M., Comeau, M.A.: Development of a simulation model for predicting dynamic behavior of heat pumps with non-azeotropic refrigerant mixtures. Int. J. Energy Res. **16**(4), 443–450 (1992)
10. Sami, S.M., Dahmani, A.: Numerical prediction of dynamic performance of vapor compression heat pump using new HFC alternatives to HCFC-22. J. Appl. Thermal Eng. **16**(8/9), 691–705 (1996)

11. Lei, Z., Zaheeruddin, M.: Dynamic simulation and analysis of a water chiller refrigeration system. Appl. Therm. Eng. **25**(14–15), 2258–2271 (2005)
12. Schalbart, P., Haberschill, P.: Simulation of the behaviour of a centrifugal chiller during quick start-up. Int. J. Refrig. **36**(1), 222–236 (2013)
13. Grald, E.W., MacArthur, J.W.A.: Moving-boundary formulation for modeling time-dependent two-phase flows. Int. J. Heat Fluid Flow **13**(3), 266–272 (1992)
14. Willatzen, M., Pettit, N.B.O.L., Ploug-Sørensen, L.: A general dynamic simulation model for evaporators and condensers in refrigeration. Part I: moving-boundary formulation of two-phase flows with heat exchange. Int. J. Refrig. **21**(5), 398–403 (1998)
15. Nyers, J., Stoyan, G.: A dynamical model adequate for controlling the evaporator of a heat pump. Int. J. Refrig. **17**(2), 101–108 (1994)
16. He, X., Liu, S., Asada, H.H.: Modeling of vapor compression cycles for multivariable feedback control of HVAC systems. ASME J. Dyn. Syst. Measur. Control **119**(2), 183–191 (1997)
17. McKinley, T.L., Alleyne, A.G.: An advanced nonlinear switched heat exchanger model for vapor compression cycles using the moving-boundary method. Int. J. Refrig. **31**(7), 1253–1264 (2008)
18. Cecchinato, L., Mancini, F.: An intrinsically mass conservative switched evaporator model adopting the moving-boundary method. Int. J. Refrig. **35**(2), 349–364 (2012)
19. Bell, I.H., Quoilin, S., Georges, E., Braun, J.E., Groll, E.A., Horton, W.T., Lemort, V.: A generalized moving-boundary algorithm to predict the heat transfer rate of counterflow heat exchangers for any phase configuration. Appl. Therm. Eng. **79**(3), 192–201 (2015)
20. Bendapudi, S., Braun, J.E., Groll, E.A.: A comparison of moving-boundary and finite-volume formulations for transients in centrifugal chillers. Int. J. Refrig. **31**(8), 1437–1452 (2008)
21. Jin, G.Y., Cai, W.J., Lu, L., Lee, E.L., Chiang, A.: A simplified modeling of mechanical cooling tower for control and optimization of HVAC systems. Energy Convers. Manag. **48** (2), 355–365 (2007)
22. Kloppers, C., Kroger, G.: Cooling tower performance evaluation: Merkel, Poppe, and e-NTU methods of analysis. J. Eng. Gas Turbines Power **127**(1), 1–7 (2005)
23. Tan, K., Deng, S.: A numerical analysis of heat and mass transfer inside a reversibly used water cooling tower. Build. Environ. **38**(1), 91–97 (2003)
24. Fisenko, S.P., Petruchik, A.I., Solodukhin, A.D.: Evaporative cooling of water in a natural draft cooling tower. Int. J. Heat Mass Transf. **45**(23), 4683–4694 (2002)
25. Fisenko, S.P., Brin, A.A., Petruchik, A.I.: Evaporative cooling of water in a mechanical draft cooling tower. Int. J. Heat Mass Transf. **47**(1), 165–177 (2004)
26. Mirth, D.R., Ramadhyani, S., Hittle, D.C.: Thermal performance of chilled-water cooling coils operating at low water velocities. ASHRAE Trans. **99**(1), 43–53 (1993)
27. Yu, X., Wen, J., Smith, T.F.: A model for the dynamic response of a cooling coil. Energy Build. **37**(12), 1278–1289 (2005)
28. Threlkeld, J.L.: Thermal Environmental Engineering, 2nd edn. Prentice-Hall, Engledood Cliffs (1970)
29. Mullen, C.E., Bridges, B.D., Porter, K.J., Hahn, G.W., Bullard, C.W.: Development and validation of a room air-conditioning simulation model. ASHRAE Trans. **104**(1), 389–397 (1998)
30. Deng, S.M.: A dynamic mathematical model of a direct expansion (DX) water cooled air-conditioning plant. Build. Environ. **35**(7), 603–613 (2000)
31. Wang, J., Hihara, E.: Prediction of air coil performance under partially wet and totally wet cooling conditions using equivalent dry-bulb temperature method. Int. J. Refrig. **26**(3), 293–301 (2003)
32. Bielski, S., Malinowski, L.: An analytical method for determining transient temperature field in a parallel-flow three-fluid heat exchanger. Int. Commun. Heat Mass Transfer **32**(8), 1034–1044 (2005)

33. Yin, J.: M. Jensen K. Analytic model for transient heat exchanger response. Int. J. Heat Mass Transf. **46**(17), 3255–3264 (2003)
34. Ren, C., Yang, H.: An analytical model for the heat and mass transfer processes in indirect evaporative cooling with parallel/counter flow configurations. Int. J. Heat Mass Transf. **49** (3–4), 617–627 (2006)
35. Xia, L., Chan, M.Y., Deng, S.M., Xu, X.G.: Analytical solutions for evaluating the thermal performances of wet air cooling coils under both unit and non-unit Lewis factors. Energy Convers. Manag. **51**(10), 2079–2086 (2010)
36. Yao, Y., Lian, Z., Hou, Z.: Thermal analysis of cooling coils based on a dynamic model. Appl. Therm. Eng. **24**(7), 1037–1050 (2004)
37. Yao, Y., Huang, M., Mo, J., Dai, S.: State-space model for transient behavior of water-to-air surface heat exchanger. Int. J. Heat Mass Transfer **66**(9), 173–192 (2013)
38. Eppelheimer, D.M.: Variable flow—the quest for system energy efficiency. ASHRAE Trans. **102**(2), 673–678 (1996)
39. Carrdo, V., Mazza, A.: Axial fan. IEA annex 17 Report, Politecnico Ditorino, Italy (1991)
40. Wang, S.W.: Dynamic simulation of a building central chilling system and evaluation of EMCS on-line control strategies. Build. Environ. **33**(1), 1–20 (1998)
41. Shao, L., Riffat, S.B.: Accuracy of CFD for predicting pressure losses in HVAC duct fittings. Appl. Energy **51**(3), 233–248 (1995)
42. Fisk, W.J., Delp, W., Diamond, R.: Duct systems in large commercial buildings: physical characterization, air leakage, and heat conduction gains. Energy Build. **32**(1), 109–119 (2000)
43. Rokni, M., Gatski, T.B.: Predicting turbulent convective heat transfer in fully developed duct flows. Int. J. Heat Fluid Flow **22**(4), 381–392 (2001)
44. Sugiyama, H., Akiyama, M., Nemoto, Y., Gessner, F.B.: Calculation of turbulent heat flux distributions in a square duct with one roughened wall by means of algebraic heat flux models. Int. J. Heat Fluid Flow **23**(1), 13–21 (2002)
45. Wu, S., Sun, J.Q.: A physics-based linear parametric model of room temperature in office buildings. Build. Environ. **50**(1), 1–9 (2012)
46. Abdul, A., Farrokh, J.S.: Review of modeling methods for HVAC systems. Appl. Therm. Eng. **67**(1–2), 507–519 (2014)
47. Metha, D.P., Woods, J.E.: An experimental validation of a rational model of dynamic responses of building. ASHRAE Trans. **86**(1), 546–558 (1980)
48. Borresen, B.A.: Thermal room models for control analysis. ASHRAE Trans. **87**(2), 251–260 (1981)
49. Tashtoush, B., Molhim, M., Al-Rousan, M.: Dynamic model of an HVAC system for control analysis. Energy **30**(10), 1729–1745 (2005)
50. Riederer, P., Marchio, D., Visier, J.C., Husaunndee, A., Lahrech, R.: Room thermal modeling adapted to the test of HVAC control systems. Build. Environ. **37**(8–9), 777–790 (2002)
51. Chen, Q., Peng, X., Paassen, A.H.C.V.: Prediction of room thermal response by CFD technique with conjugate heat transfer and radiation models. ASHRAE Trans. **101**(part 1), 50–60 (1995)
52. Wu, X., Olesen, B., Fang, W., Zhao, L.: A nodal model to predict vertical temperature distribution in a room with floor heating and displacement ventilation. Build. Environ. **59**(1), 626–634 (2013)
53. Yao, Y., Yang, K., Huang, M.: A state-space model for dynamic response of indoor air temperature and relative humidity. Build. Environ. **64**(6), 26–37 (2013)
54. Huang, G., Wang, S., Xu, X.: Robust model predictive control of VAV air-handling units concerning uncertainties and constraints. HVAC R. Res. **16**(1), 15–33 (2010)
55. Huang, G.: Model predictive control of VAV zone thermal systems concerning bi-linearity and gain nonlinearity. Control Eng. Pract. **19**(7), 700–710 (2011)
56. Wang, S., Zhou, Q., Xiao, F.: A system-level fault detection and diagnosis strategy for HVAC systems involving sensor faults. Energy Build. **42**(4), 477–490 (2010)

57. Hosoz, M., Ertunc, H.M.: Artificial neural network analysis of an automobile air conditioning system. Energy Convers. Manag. **47**(11–12), 1574–1587 (2006)
58. Ertunc, H.M., Hosoz, M.: Comparative analysis of an evaporative condenser using artificial neural network and adaptive neuro-fuzzy inference system. Int. J. Refrig. **31**(8), 1426–1436 (2008)
59. Ding, L., Lv, J., Li, X., Li, L.: Support vector regression and ant colony optimization for HVAC cooling load prediction. In: International Symposium on Computer Communication Control and Automation (3CA), vol. 1 IEEE, Tainan, Taiwan, pp. 537–541 (2010)
60. Becker, M., Oestreich, D., Hasse, H., Litz, L.: Fuzzy control for temperature and humidity in refrigeration systems. IEEE Trans., FM-4-2: 1607–1611 (1994)
61. He, M., Cai, W.J., Li, S.Y.: Multiple fuzzy model-based temperature predictive control for HVAC systems. Inf. Sci. **169**(1), 155–174 (2005)
62. Soyguder, S., Alli, H.: An expert system for the humidity and temperature control in HVAC systems using ANFIS and optimization with fuzzy modeling approach. Energy Build. **41**(3), 814–822 (2009)
63. Homoda, R.Z., Saharia, K.S.M., Almuribb, H.A.F., Nagia, F.H.: Gradient auto-tuned Takagi-Sugeno fuzzy forward control of a HVAC system using predicted mean vote index. Energy Build. **49**(6), 254–267 (2012)
64. Khan, M.W., Choudhry, M.A., Zeeshan, M., Ali, A.: Adaptive fuzzy multivariable controller design based on genetic algorithm for an air handling unit. Energy **81**(3), 477–488 (2015)
65. Mustafaraj, G., Chen, J., Lowry, G.: Development of room temperature and relative humidity linear parametric models for an open office using BMS data. Energy Build. **42**(3), 348–356 (2010)
66. Penya, Y.K., Borges, C.E., Agote, D., Fernandez, I.: Short-term load forecasting in air-conditioned non-residential Buildings. In: 20th IEEE International Symposium on Industrial Electronics (ISIE), Gdansk, Poland, 1359–1364 27–30 June 2011
67. Wang, Y.W., Cai, W.J., Soh, Y.C., Li, S.J., Lu, L., Xie, L.: A simplified modeling of cooling coils for control and optimization of HVAC systems. Energy Convers. Manag. **45**(18–19), 2915–2930 (2004)
68. Ghiaus, C., Chicinas, A., Inard, C.: Grey-box identification of air-handling unit elements. Control Eng. Pract. **15**(4), 421–433 (2007)
69. Yao, Y., Liu, S.: Transfer function model for dynamic response of wet cooling coils. Energy Convers. Manag. **49**(12), 3612–3621 (2008)
70. Ferkl, L., Siroky, J.: Ceiling radiant cooling: comparison of ARMAX and sub-space identification modelling methods. Build. Environ. **45**(2), 205–212 (2010)
71. Jiang, Y.: State space method for analysis of the thermal behavior of rooms and calculation of air-conditioning load. ASHRAE Trans. **101**(1), 122–132 (1995)
72. He, X.D., Liu, S., Asada, H.H., Itoh, H.: Multivariable control of vapor compression systems. HVAC&R Res. **4**(3), 205–230 (1998)
73. Yao, Y., Huang, M., Yang, K.: State-space model for dynamic behavior of vapor compression liquid chiller. Int. J. Refrig. **36**(8), 2128–2147 (2013)
74. Li, M., Wu, C.L., Zhao, S.Q., Yang, Y.: State-space model for airborne particles in multizone indoor environments. Atmos. Environ. **42**(21), 5340–5349 (2008)
75. Qi, Q., Deng, S.: Multivariable control-oriented modeling of a direct expansion (DX) air conditioning (A/C) system. Int. J. Refrig. **31**(5), 841–849 (2008)
76. Kumar, M., Kar, I.N., Ray, A.: State space based modeling and performance evaluation of an air-conditioning system. HVAC R Res. **14**(5), 797–816 (2008)
77. Moroşan, P.D., Bourdais, R., Dumur, D., Buisson, J.: Building temperature regulation using a distributed model predictive control. Energy Build. **42**(9), 1445–1452 (2010)
78. Prívara, S., Široký, J., Cigler, J.: Model predictive control of a building heating system: the first experience. Energy Build. **43**(2–3), 564–572 (2011)
79. Balakrishnan, V.K.: Graph Theory. Shaum's Outline Series, New York (1997)

80. Garg, R.K., Agrawal, V.P., Gupta, V.K.: Selection of power plants by evaluation and comparison using graph theoretic methodology. Electr. Power Energy Syst. **28**(3), 429–435 (2006)
81. Prabhakaran, R.T.D., Agrawal, V.P.: Structural modeling and analysis of composite product system: a graph theoretic approach. J. Compos. Mater. **40**(22), 1987–2007 (2006)
82. Grekas, D.N., Frangopoulos, C.A.: Automatic synthesis of mathematical models using graph theory for optimisation of thermal energy systems. Energy Convers. Manag. **48**(12), 2818–2826 (2007)
83. Singh, V., Agrawal, V.P.: Structural modelling and integrative analysis of manufacturing systems using graph theoretic approach. J. Manuf. Technol. Manage. **19**(7), 844–870 (2008)
84. Wang, T., Ding, G., Duan, Z., Ren, T., Chen, J., Pu, H.: A distributed-parameter model for LNG spiral wound heat exchanger based on graph theory. Appl. Therm. Eng. **81**(1), 102–113 (2015)
85. Braun, J.E.: Reducing energy costs and peak electrical demands through optimal control of building thermal storage. ASHRAE Trans. **96**(part 1), 587–601 (1990)
86. Kawashima, M., Dorgan, C.E., Mitchell, J.W.: Hourly thermal load prediction for the next 24 hours by ARIMA, EWMA, LR, and an artificial neural network. ASHRAE Trans. **101**(part 1), 186–200 (1995)
87. Hwang, R.C., Huang, H.C., Chen, Y.J., Hsieh, J.G., Chen, H.C., Chuang, C.W.: Selection of influencing factors for power load forecasting by grey relation analysis. In: Second National Conference on Grey Theory and Applications, Taiwan, pp. 109–113 (1997)
88. Neto, A.H., Fiorelli, F.A.S.: Comparison between detailed model simulation and artificial neural network for forecasting building energy consumption. Energy Build. **40**(12), 2169–2176 (2008)
89. Zhou, Q., Wang, S., Xu, X., Xiao, F.: A grey-box model of next-day building thermal load prediction for energy-efficient control. Int. J. Energy Res. **32**(15), 1418–1431 (2008)
90. Guo, J.J., Wu, J.Y., Wang, R.Z.: A new approach to energy consumption prediction of domestic heat pump water heater based on grey system theory. Energy Build. **43**(6), 1273–1279 (2011)
91. Chen, H.: The Validity of the Theory and Its Application of Combination Forecast Methods. Science Press, Beijing (China) (2008)
92. Granger, C.W.J., Ramanathan, R.: Improved methods of combining forecasts. J. Forecast. **3** (3), 197–204 (1984)
93. Braun, J.E., Klein, S.A., Mitchell, J.W., Beckman, W.A.: Methodologies for optimal control of chilled water systems without storage. ASHRAE Transact. **95**(1), 652–662 (1989)
94. Olson, R.T., Liebman, S.: Optimization of a chilled water plant using sequential quadratic programming. Eng. Optim. **15**(1), 171–191 (1990)
95. Ahn, B.C., Mitchell, J.W.: Optimal control development for chilled water plants using a quadratic representation. Energy Build. **33**(4), 371–378 (2001)
96. Chang, Y.C.: Application of genetic algorithm to the optimal chilled water supply temperature calculation of air-conditioning systems for saving energy. Energy Res. **31**(8), 796–810 (2007)
97. Engdahl, F., Johansson, D.: Optimal supply air temperature with respect to energy use in a variable air volume system. Energy Build. **36**(3), 205–218 (2004)
98. Xu, X., Wang, S., Sun, Z., Xiao, F.: A model-based optimal ventilation control strategy of multi-zone VAV air-conditioning systems. Appl. Therm. Eng. **29**(1), 91–104 (2009)
99. Mossolly, M., Ghali, K., Ghaddar, N.: Optimal control strategy for a multi-zone air conditioning system using a genetic algorithm. Energy **34**(1), 58–66 (2009)
100. Sun, J., Reddy, A.: Optimal control of building HVAC and R systems using complete simulation-based sequential quadratic programming (CSB-SQP). Build. Environ. **40**(5), 657–669 (2005)
101. Hanby, V.I., Wright, J.A.: HVAC optimization studies: component modeling methodology. Build. Serv. Eng. Res. Technol. **10**(1), 35–39 (1989)

102. Zaheer-Uddin, M., Zheng, G.R.: Optimal control of time-scheduled heating, ventilating and air conditioning processes in buildings. Energy Convers. Manag. **41**(1), 49–60 (2000)
103. Chow, T.T., Zhang, G.Q., Lin, Z., Song, C.L.: Global optimization of absorption chiller system by generic algorithm and neural network. Energy Build. **34**(1), 103–109 (2000)
104. Huang, L.: Using genetic algorithms to optimize controller parameters for HVAC systems. Energy Build. **26**(2), 277–282 (1997)
105. Fong, K.F., Hanby, V.I., Chow, T.T.: System optimization for HVAC energy management using the robust evolutionary algorithm. Appl. Therm. Eng. **29**(11–12), 2327–2334 (2009)
106. Kido, T., Takag, K., Nakanishi, M.: Analysis and comparisons of genetic algorithm, simulated annealing, TABU search, and combination algorithm. Informatics **18**(4), 399–410 (1994)
107. Arturo, M.C., Lorenz, T.B.: A stable elemental decomposition for dynamic process optimization. J. Comput. Appl. Math. **120**(1), 41–57 (2000)
108. Thi, H.A.L., Pham, D.T., Thoai, N.V.: Combination between global and local methods for solving an optimization problem over the efficient set. Eur. J. Oper. Res. **142**(2), 258–270 (2002)
109. Huang, Y.J., Reklaitis, G.V., Venkatasubramanian, V.: Model decomposition based method for solving general dynamic optimization problems. Comput. Chem. Eng. **26**(6), 863–873 (2002)
110. Wang, Y.J., Ying, L.: Global optimization for special reverse convex programming. Comput. Math Appl. **55**(6), 1154–1163 (2008)
111. Yu, D., Li, H., Ni, L., Yu, Y.: An improved virtual calibration of a supply air temperature sensor in rooftop air conditioning units. HVAC & R Res. J. **17**(5), 798–812 (2011)
112. United Nations.: Role of measurement and calibration in the manufacture of products for the global market (2006)
113. Li, Z., Huang, G.: Preventive approach to determine sensor importance and maintenance requirements. Autom. Constr. **31**(3), 307–312 (2013)
114. Huang, Y., Qian, X., Chen, S.: Multi-sensor calibration through iterative registration and fusion. Comput. Aided Des. **41**(2), 240–255 (2009)
115. Djuric, N., Huang, G., Novakovic, V.: Data fusion heat pump performance estimation. Energy Build. **43**(6), 621–630 (2011)
116. Mitchell, H.B.: Data Fusion: Concepts and Ideas, 2nd ed. Springer (2012)
117. Yu, D., Li, H., Yu, Y., Xiong, J.: Virtual calibration of a supply air temperature sensor in rooftop air conditioning units. HVAC&R Res. **17**(1), 31–50 (2011)
118. Whitehouse, K., Culler, D.: Calibration as parameter estimation in sensor networks. In: Proceedings of the 1st ACM International Workshop on Wireless Sensor Networks and Applications, pp. 59–67 (2002)
119. Bychkovskiy, V., Megerian, S., Estrin, D., Potkonjak, M.: A collaborative approach to in-place sensor calibration. Lect. Notes Comput. Sci. **263**(2), 301–316 (2003)
120. Dulev, V., Ermishin, S., Khoteev, N., Lopatin, A., Shabanov, P., Menshikov, A., Startsev, A.: Automated measuring system designed to calibrate measuring devices using virtual standards technology. In: 2004 IEEE Autotestcon, pp. 158–164
121. Fisher, J.W., Moses, R.L., Willsky, A.S.: Nonparametric Belief Propagation for Self-calibration in Sensor Networks. IPSN' 04, Berkeley, California, 26–27 April 2004
122. Balzano, L., Nowak, R.: Blind calibration of sensor networks. In: Proceedings of IPSN'07. Cambridge, Massachusetts, 25–27 April 2007
123. Yu, D., Li, H., Yang, M.: A virtual supply airflow rate meter for rooftop air-conditioning units. Build. Environ. **46**(6), 1292–1302 (2011)
124. Tan, R., Xing, G., Liu, X., Yao, J., Yuan, Z.: Adaptive calibration for fusion-based cyber-physical system. ACM Trans. Embed. Comput. Syst. **11**(4), 80, (1–25) (2012)
125. Mosh, B., Rashidi, F.: Self-tuning based fuzzy PID controllers: application to control of nonlinear HVAC systems. In: Proceedings of 5th International conference: Intelligent Data Engineering and Automated Learning. Exeter, UK (2004)

126. Yu, Y.: Model-based multivariate control of conditioning systems for office buildings (PhD. thesis). Carnegie Mellon University (2012)
127. Rawlings, J.B., Mayne, D.Q.: Model Predictive Control: Theory and Design. Nob Hill Publishing (2009)
128. Camacho, E.F., Bordons, C.: Model Predictive Control in the Process Industry. Springer, Berlin (1995)
129. Qin, S.J., Badgwell, T.A.: A survey of industrial model predictive control technology. Control Eng. Pract. **11**(7), 733–764 (2003)
130. Zhang, Y., Hanby, V.I.: Model-based control of renewable energy systems in buildings. HVAC&R Res. **12**(3a), 739–760 (2006)
131. Yuan, S., Perez, R.: Multiple-zone ventilation and temperature control of a single-duct VAV system using model predictive strategy. Energy Build. **38**(10), 1248–1261 (2006)
132. Wang, S., Ma, Z.: Supervisory and optimal control of building HVAC systems: a review. HVAC&R Res. **14**(1), 3–32 (2008)
133. Freire, R.Z., Oliveira, G.H.C., Mendes, N.: Predictive controllers for thermal comfort optimization and energy savings. Energy Build. **40**(7), 1352–1365 (2008)
134. Hazyuk, I., Ghiaus, C., Penhouet, D.: Optimal temperature control of intermittently heated buildings using model predictive control: part II—control algorithm. Build. Environ. **51**(5), 388–394 (2012)

Chapter 2
Component Modeling with State-Space Method

2.1 Basic Knowledge About State-Space Modeling Method

A state-space representation, also known as the 'time-domain approach', is a mathematical model of a physical system as a set of input, output, and state variables related to first-order differential equations. To abstract from the number of inputs, outputs, and states, these variables are expressed as vectors. Additionally, if the dynamical system is linear, time-invariant, and finite-dimensional, then the differential and algebraic equations may be written in matrix form [1]. The state-space representation provides a convenient and compact way to model and analyze systems with multiple inputs and outputs. With p inputs and q outputs, we can encode all the information about a system instead of $q \times p$ Laplace transforms. Unlike the frequency-domain approach, the use of the state-space representation is not limited to systems with linear components and zero initial conditions [2].

The most general state-space representation of a linear system with m inputs, n outputs, and p state variables is written in the following form [3]:

$$\dot{X}(\tau) = A \cdot X(\tau) + B \cdot U(\tau) \tag{2.1}$$

$$Y(\tau) = C \cdot X(\tau) + D \cdot U(\tau) \tag{2.2}$$

where

- $U(\tau)$ is input variable vector, $U(\tau) = [u_1(\tau), u_2(\tau), \ldots, u_m(\tau)]^T$;
- $Y(\tau)$ is output variable vector, $Y(\tau) = [y_1(\tau), y_2(\tau), \cdots, y_n(\tau)]^T$;
- $X(\tau)$ is state variable vector, $X(\tau) = [x_1(\tau), x_2(\tau), \cdots, x_p(\tau)]^T$;
- A is the state (or system) matrix, $\dim[A] = p \times p$;
- B is the input matrix, $\dim[B] = p \times m$;
- C is the output matrix, $\dim[C] = n \times p$;
- D is the feed-forward matrix, $\dim[D] = n \times m$;

Y. Yao and Y. Yu, *Modeling and Control in Air-conditioning Systems*,
Energy and Environment Research in China, DOI 10.1007/978-3-662-53313-0_2

The internal state variables are the smallest possible subset of system variables that can represent the entire state of the system at any given time. The minimum number of state variables required to represent a given system is usually equal to the order of the system's defining differential equation. If the system is represented in transfer function form, the minimum number of state variables is equal to the order of the transfer function's denominator after it has been reduced to a proper fraction [4].

In this chapter, the state-space method is firstly employed to develop dynamic models of the main components in a general HVAC system including water-to-air heat exchanger, refrigeration system, cooling tower, air-conditioned space, air duct, water pipe, and valves (or dampers). Afterward, the method of component models' integration is introduced and an example of HVAC system modeling based on the component models is given.

2.2 Modeling for HVAC Components

2.2.1 Water-to-Air Heat Exchanger

Water-to-air heat exchanger is used for air cooling with or without accompanying dehumidification. They work under dry conditions (without dehumidification) when the external surface temperature is lower than the air dew-point temperature or otherwise under wet conditions (with dehumidification). For general comfort conditioning, cooling, and dehumidifying, the extended-surface (finned) heat exchanger design is the most popular. In finned heat exchanger, the external surface of the tubes is primary and the fin surface is secondary. The primary surface generally consists of rows of round tubes or pipes that may be staggered or placed in line with respect to the airflow. The inside surface of the tubes is usually smooth and plain, but some heat exchanger designs have various forms of internal fins or turbulence promoters to enhance the performance. The individual tube passes in a heat exchanger are usually interconnected by return bends to form the serpentine arrangement of multi-pass tube circuits [5]. Heat exchangers for water, aqueous glycol, or halocarbon refrigerants usually have aluminum fins on copper tubes, or copper fins on copper tubes, or aluminum fins on aluminum tubes. The outside diameters of core tube are commonly 8, 10, 12.5, 16, 20, and 25 mm, with fins spaced 1.4–6.4 apart. Tube spacing ranges from 15 to 75 mm, depending on the width of individual fins and on the other performance considerations.

Figure 2.1 shows the physical map of a water-to-air heat exchanger which is taken as the modeling object. The objective of the modeling is to study the dynamic characteristics of the water-to-air heat exchanger when subjected to perturbations of different inlet variables including entering air temperature and humidity, entering water temperature, and flow rate of the two fluids.

Fig. 2.1 Physical map of a water-to-air heat exchanger

2.2.1.1 Model Development [6]

1) Assumptions and basic equations

Figure 2.2 shows the schematic diagram of water-to-air heat exchanger for modeling. Further, the following assumptions are necessary to be made for the convenience of modeling.

(1) Moisture air is treated as a mixture of ideal gases. The specific heat and density of air are considered as constants in the process of heat and mass transfer;
(2) Air-side heat transfer coefficient includes the additional thermal resistance due to the presence of partially or completely wet extended surfaces;
(3) Under wet-condition mode, the air on the surface of coils and fins is considered as saturated state;
(4) Temperature and humidity of the air change linearly from the inlet to the outlet of the heat exchanger. So does temperature of the water [7].

Thus, a series of equations can be established according to the law of energy and mass conservation.

(1) *Mass and energy equation for water passing through coils*:

$$G_{\mathrm{w,E}} = G_{\mathrm{w,L}} = G_{\mathrm{w}} \tag{2.3}$$

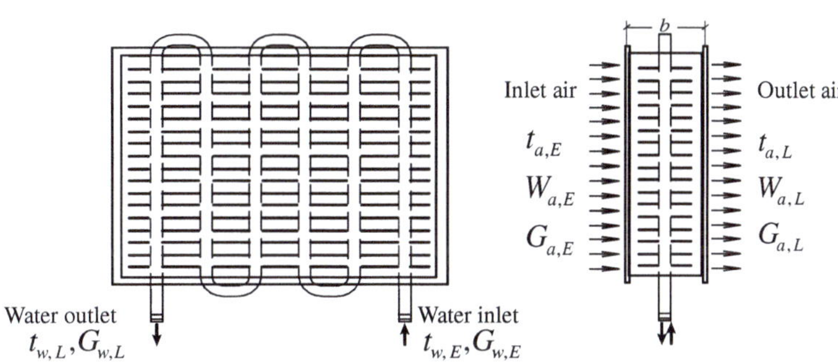

Fig. 2.2 Schematic diagram of water-to-air heat exchanger for modeling

$$\frac{1}{2}\rho_{\mathrm{w}}c_{\mathrm{w}}A_{\mathrm{w}}l\frac{\mathbf{d}(t_{\mathrm{w,L}}+t_{\mathrm{w,E}})}{\mathbf{d}\tau}=G_{\mathrm{w,E}}c_{\mathrm{w}}(t_{\mathrm{w,E}}-t_{\mathrm{w,L}})+a_{\mathrm{gw}}A_{\mathrm{i}}\left(t_{\mathrm{g}}-\frac{t_{\mathrm{w,E}}+t_{\mathrm{w,L}}}{2}\right)\tag{2.4}$$

(2) *Mass and energy equation for air passing through heat exchanger*:

$$G_{\mathrm{a,E}}=G_{\mathrm{a,L}}=G_{\mathrm{a}}\tag{2.5}$$

- Dry-condition process:

$$W_{\mathrm{a,E}}=W_{\mathrm{a,L}}\tag{2.6a}$$

$$\frac{1}{2}\varepsilon_{\mathrm{a}}\rho_{\mathrm{a}}c_{\mathrm{a}}A_{\mathrm{a}}b\frac{\mathbf{d}(t_{\mathrm{a,L}}+t_{\mathrm{a,E}})}{\mathbf{d}\tau}=G_{\mathrm{a,E}}c_{\mathrm{a}}(t_{\mathrm{a,E}}-t_{\mathrm{a,L}})+a_{\mathrm{ga}}A_{\mathrm{o}}\left(t_{\mathrm{m}}-\frac{t_{\mathrm{a,E}}+t_{\mathrm{a,L}}}{2}\right)\tag{2.6b}$$

- Wet-condition process:

$$\frac{1}{2}\varepsilon_{\mathrm{a}}\rho_{\mathrm{a}}bA_{\mathrm{a}}\frac{\mathbf{d}(W_{\mathrm{a,L}}+W_{\mathrm{a,E}})}{\mathbf{d}\tau}=G_{\mathrm{a,E}}(W_{\mathrm{a,E}}-W_{\mathrm{a,L}})+\lambda_{\mathrm{m}}A_{\mathrm{o}}\left(W_{\mathrm{gb}}-\frac{W_{\mathrm{a,E}}+W_{\mathrm{a,L}}}{2}\right)\tag{2.7a}$$

$$\frac{1}{2}\varepsilon_{\mathrm{a}}\rho_{\mathrm{a}}bA_{\mathrm{a}}\frac{\mathbf{d}(h_{\mathrm{a,L}}+h_{\mathrm{a,E}})}{\mathbf{d}\tau}=G_{\mathrm{a,E}}(h_{\mathrm{a,E}}-h_{\mathrm{a,L}})+a_{\mathrm{ga}}A_{\mathrm{o}}\left(t_{\mathrm{m}}-\frac{t_{\mathrm{a,E}}+t_{\mathrm{a,L}}}{2}\right)+q_{\mathrm{r}}A_{\mathrm{o}}\lambda_{\mathrm{m}}\left(W_{\mathrm{gb}}-\frac{W_{\mathrm{a,E}}+W_{\mathrm{a,L}}}{2}\right)\tag{2.7b}$$

(3) *Energy equation for coils and fins*

- Dry-condition process:

$$M_{\mathrm{g}}c_{\mathrm{g}}\frac{\mathbf{d}t_{\mathrm{g}}}{\mathbf{d}\tau}=a_{\mathrm{gw}}A_{\mathrm{i}}\left(\frac{t_{\mathrm{w,E}}+t_{\mathrm{w,L}}}{2}-t_{\mathrm{g}}\right)+a_{\mathrm{ga}}A_{\mathrm{o}}\left(\frac{t_{\mathrm{a,E}}+t_{\mathrm{a,L}}}{2}-t_{\mathrm{m}}\right)\tag{2.8a}$$

- Wet-condition process:

$$M_{\mathrm{g}}c_{\mathrm{g}}\frac{\mathbf{d}t_{\mathrm{g}}}{\mathbf{d}\tau}=a_{\mathrm{gw}}A_{\mathrm{i}}\left(\frac{t_{\mathrm{w,E}}+t_{\mathrm{w,L}}}{2}-t_{\mathrm{g}}\right)+a_{\mathrm{ga}}A_{\mathrm{o}}\left(\frac{t_{\mathrm{a,E}}+t_{\mathrm{a,L}}}{2}-t_{\mathrm{m}}\right)+q_{\mathrm{r}}A_{\mathrm{o}}\lambda_{\mathrm{m}}\left(\frac{W_{\mathrm{w,E}}+W_{\mathrm{w,L}}}{2}-W_{\mathrm{gb}}\right)\tag{2.8b}$$

In Eqs. (2.3) through (2.8), the symbol '*G*' stands for flow rate, kg/s; '*c*' for mass specific heat, J/(kg °C); '*A*' for area, m^2; '*h*' for enthalpy, J/kg; '*l*' for length of coiled tube, m; '*b*' for width of heat exchanger, m; '*t*' for temperature, °C; '*W*' for air humidity, kg/(kg dry air); 'M' for mass, kg; 'q_r' for latent heat of condensation of water vapor, J/kg; 'a_{gw}' for heat transfer coefficient on the water side, W/(m^2 °C); 'a_{ga}' for heat transfer coefficient on the air side, W/(m^2 °C); 'λ_m' for mass transfer coefficient, kg/ (m^2/s); 'τ' for time, s; 'ρ' for density, kg/m^3; and 'ε_a' for air volume ratio in heat exchanger. The subscript '*a*' stands for air; '*w*' for water; 'g' for shell wall of heat exchanger; 'E' for inlet; 'L' for outlet; 'm' fins of coil; '*gb*' for saturated air near external surface of heat exchanger; 'o' for air-side surface of heat exchanger; and '*i*' for water-side surface of heat exchanger.

2) Parameter determination and linearization

Two kinds of variables are considered in the equations: the fundamental and the lumped. The fundamental variables include the temperature and humidity of inlet and exit air ($t_{a,E}$, $W_{a,E}$, $t_{a,L}$, $W_{a,L}$), the temperature of inlet and exit water ($t_{w,E}$, $t_{w,L}$), and the air and water flow rates (G_a and G_w). These variables can be considered as the summation of initial value (θ_o) and a small increment ($\Delta\theta$):

$$\theta = \theta_o + \Delta\theta \tag{2.9}$$

where θ denotes, $t_{a,E}$, $W_{a,E}$, $t_{a,L}$, $W_{a,L}$, $t_{w,E}$, $t_{w,L}$, G_a, and G_w, respectively.

The lumped variables include the coefficients of heat and mass transfer between fluids and heat exchanger (i.e., a_{gw}, a_{ga} and λ_m), which are usually expressed as an equation dependent on the fluid flow rate (G_a or u_a, G_w or u_w), respectively, as below:

$$Nu_w = C_1 Re_w^{n_1} \tag{2.10}$$

$$Nu_a = C_2 Re_a^{n_2} \tag{2.11}$$

$$Nu_d = C_3 Re_a^{n_3} \tag{2.12}$$

where $C_1, C_2, C_3, n_1, n_2, n_3$ are empirical constants that can be determined with experimental data; $Nu_w = \frac{2a_{gw}r_i}{\lambda_w}$; $Re_w = \frac{2u_w r_i}{\nu_w}$; $Re_a = \frac{2u_a r_d}{\nu_a}$; $Nu_d = \frac{2\lambda_m r_d}{\rho_a D_{w,a}}$, $Re_a = \frac{2u_a r_d}{\nu_a} = \frac{2G_a r_d}{\nu_a \rho_a A_a}$; $r_d = \frac{(S-2r_i)(e-\delta_c)}{(S-2r_i)+(e-\delta_c)}$. The symbol '$\lambda_w$' is heat conductivity coefficient of water, W/ (m °C); 'r_i' is inner radius of coil, m; 'u_w' and 'u_a' are velocities of water and air, respectively, m/s; 'ν_w' and 'ν_a' are kinematic viscosity coefficients of water and air, respectively, m^2/s; '$D_{w,a}$' is mass diffusivity of water vapor in the air, *m*^{2}*/s;* 'S' is finned tube center spacing of surface of heat exchanger, *m;* '*e*' is fin spacing of heat exchanger, *m;* and 'δ_c' is fin thickness of heat exchanger, m.

Thus, the lumped variables can be linearized by using the first-order Taylor series:

$$\sigma = \sigma_o + \left(\frac{\partial\sigma}{\partial\theta'}\right)_o \Delta\theta' \tag{2.13}$$

where σ stands for a_{gw}, a_{ga} and λ_m, respectively; θ' for u_a (G_a) or u_w (G_w); and Δ for increment of variables.

The efficiency of sensible heat exchange, η_s, is defined as follows:

$$\eta_s = \frac{\left(t_m - \frac{t_{a,E} + t_{a,L}}{2}\right)}{\left(t_g - \frac{t_{a,E} + t_{a,L}}{2}\right)} \tag{2.14}$$

where t_m is the average surface temperature of fins. η_s is mainly related to the structure of finned heat exchanger. For a specific water-to-air heat exchanger, η_s can be considered as a constant, normally as 0.7–0.8.

According to Eq. (2.14), t_m can be expressed as follows:

$$t_m = \eta_s t_g - (1 - \eta_s)\frac{t_{a,E} + t_{a,L}}{2} \tag{2.15}$$

The humidity of saturated air, W_{gb}, on the fins can be expressed by Eq. (2.16).

$$W_{gb} = a_1 t_m^3 + a_2 t_m^2 + a_3 t_m + a_4 \tag{2.16}$$

where the coefficients, a_1, a_2, a_3 and a_4, are obtained as 0.0008, −0.0190, 0.7287, and 1.6910, respectively, through data fitting of the thermodynamic properties of moist air. The maximum relative error of Eq. (2.16) compared with the data in ASHRAE Handbook [8] is less than 3.5 % when the temperature changes from 5 to 40 °C.

By using the first-order Taylor series, Eq. (2.16) can be converted into a linear one as below:

$$W_{gb} = (W_{gb})_o + \left(3a_1 t_m^2 + 2a_2 t_m + a_3\right)_o \Delta t_m \tag{2.17}$$

Let $\beta_1 = \left(3a_1 t_m^2 + 2a_2 t_m + a_3\right)_o$, Eq. (2.17) can be written as:

$$W_{gb} = (W_{gb})_o + \beta_1 \Delta t_m \tag{2.18}$$

The enthalpy of moisture air, h_a, can be calculated by:

$$h_a = c_a t_a + (2501 + 1.84 t_a) \cdot W_a \tag{2.19}$$

Let $\beta_2 = (2501 + 1.84 t_a)$, Eq. (2.19) can be simplified as follows:

$$h_a = c_a t_a + \beta_2 W_a \tag{2.20}$$

Since the value of $1.84 t_a$ is very small compared with 2501, β_2 can be considered as a constant dependent on the initial air temperature.

The latent heat of condensation of water vapor (q_r: J/kg) can be calculated by Eq. (2.21).

$$q_r = (2501 - 2.35t_a) \times 10^3 \tag{2.21}$$

Since the air temperature (t_a) changes in a small range, q_r can be similarly equal to a constant as well, i.e., q_r = 2,441,250.

3) State-space representation

Through parameter linearization and removing the high-order item (product of two increment items), Eqs. (2.3) through (2.8) can be transformed as follows:

For dry-condition process

$$T_w \frac{d\Delta t_{w,L}}{d\tau} = X_1 \Delta t_{w,L} + X_2 \Delta t_g + X_3 \Delta t_{w,E} + X_4 \Delta G_{w,E} + \xi_{\Delta t_{w,L,dry}} \tag{2.22}$$

$$T_a \frac{d\Delta t_{a,L}}{d\tau} = Y_{dry,1} \Delta t_{a,L} + Y_{dry,2} \Delta t_g + Y_{dry,3} \Delta t_{a,E} + Y_{dry,4} \Delta G_{a,E} + \xi_{\Delta t_{a,L,dry}} \tag{2.23}$$

$$\begin{aligned} T_g \frac{d\Delta t_g}{d\tau} &= Z_{dry,1} \Delta t_{w,L} + Z_{dry,2} \Delta t_{a,L} + Z_{dry,3} \Delta t_g + Z_{dry,4} \Delta t_{a,E} \\ &\quad + Z_{dry,5} \Delta t_{w,E} + Z_{dry,6} \Delta G_{w,E} + Z_{dry,7} \Delta G_{a,E} \end{aligned} \tag{2.24}$$

$$\Delta G_{w,L} = \Delta G_{w,E} \tag{2.25}$$

$$\Delta G_{a,L} = \Delta G_{a,E} \tag{2.26}$$

$$\Delta W_{a,L} = \Delta W_{a,E} \tag{2.27}$$

For wet-condition process

$$T_w \frac{d\Delta t_{w,L}}{d\iota} = X_1 \Delta t_{w,L} + X_2 \Delta t_g + X_3 \Delta t_{w,E} + X_4 \Delta G_{w,E} + \xi_{\Delta t_{w,L,wet}} \tag{2.28}$$

$$\begin{aligned} T_a \frac{d\Delta t_{a,L}}{d\tau} &= Y_{wet,1} \Delta t_{a,L} + Y_{wet,2} \Delta W_{a,L} + Y_{wet,3} \Delta t_g + Y_{wet,4} \Delta t_{a,E} \\ &\quad + Y_{wet,5} \Delta G_{a,E} + Y_{wet,6} \Delta W_{a,E} + \xi_{\Delta t_{a,L,wet}} \end{aligned} \tag{2.29}$$

$$\begin{aligned} T_m \frac{d\Delta W_{a,L}}{d\tau} &= M_{wet,1} \Delta t_{a,L} + M_{wet,2} \Delta W_{a,L} + M_{wet,3} \Delta t_g + M_{wet,4} \Delta t_{a,E} \\ &\quad + M_{wet,5} \Delta G_{a,E} + M_{wet,6} \Delta W_{a,E} + \xi_{\Delta W_{a,L,wet}} \end{aligned} \tag{2.30}$$

$$\begin{aligned} T_g \frac{d\Delta t_g}{d\tau} &= Z_{wet,1} \Delta t_{w,L} + Z_{wet,2} \Delta t_{a,L} + Z_{wet,4} \Delta t_g + Z_{wet,5} \Delta t_{a,E} + Z_{wet,6} \Delta t_{w,E} \\ &\quad + Z_{wet,7} \Delta G_{w,E} + Z_{wet,8} \Delta G_{a,E} + Z_{wet,9} \Delta W_{a,E} \end{aligned} \tag{2.31}$$

$$\Delta G_{w,L} = \Delta G_{w,E} \tag{2.32}$$

$$\Delta G_{\mathrm{a,L}} = \Delta G_{\mathrm{a,E}} \tag{2.33}$$

All coefficients in Eqs. (2.22) through (2.31) are listed in Table 2.1. Combining Eqs. (2.22) through (2.33), the state-space equation for water-to-air heat exchanger can be expressed by Eq. (2.34):

$$\dot{x}_{\mathrm{coil}} = A_{\mathrm{coil}} \cdot x_{\mathrm{coil}} + B_{\mathrm{coil}} \cdot u_{\mathrm{coil}} + \xi_{\mathrm{coil}} \tag{2.34}$$

And the corresponding output equation is written as follows:

$$y_{\mathrm{coil}} = C_{\mathrm{coil}} \cdot x_{\mathrm{coil}} + D_{\mathrm{coil}} \cdot u_{\mathrm{coil}} \tag{2.35}$$

To transform into the standard form of state-space representation, Eq. (2.34) can be written as follows:

$$\dot{x}_{\mathrm{coil}} = A_{\mathrm{coil}}(x_{\mathrm{coil}} + A_{\mathrm{coil}}^{-1}\xi_{\mathrm{coil}}) + B_{\mathrm{coil}}u_{\mathrm{coil}} \tag{2.36}$$

Let $X_{\mathrm{coil}} = x_{\mathrm{coil}} + A_{\mathrm{coil}}^{-1}\xi_{\mathrm{coil}}$, we have $x_{\mathrm{coil}} = X_{\mathrm{coil}} - A_{\mathrm{coil}}^{-1}\xi_{\mathrm{coil}}$. Thus, the standard state-space model for water-to-air heat exchanger can be obtained as follows:

$$\dot{X}_{\mathrm{coil}} = A_{\mathrm{coil}}X_{\mathrm{coil}} + B_{\mathrm{coil}}u_{\mathrm{coil}} \tag{2.37}$$

$$y_{\mathrm{coil}} = c_{\mathrm{coil}}X_{\mathrm{coil}} + D_{\mathrm{coil}}u_{\mathrm{coil}} - C_{\mathrm{coil}}A_{\mathrm{coil}}^{-1}\xi_{\mathrm{coil}} \tag{2.38}$$

For the dry-condition process:

$$x_{\mathrm{coil}} = [\Delta t_{\mathrm{w,L}}, \Delta t_{\mathrm{a,L}}, \Delta t_{\mathrm{g}}]^T;\ y_{\mathrm{coil}} = [\Delta t_{\mathrm{w,L}}, \Delta G_{\mathrm{w,L}}, \Delta t_{\mathrm{a,L}}, \Delta W_{\mathrm{a,L}}, \Delta G_{\mathrm{a,L}}]^T;$$

$$u_{\mathrm{coil}} = [\Delta t_{\mathrm{w,E}}, \Delta G_{\mathrm{w,E}}, \Delta t_{\mathrm{a,E}}, \Delta W_{\mathrm{a,E}}, \Delta G_{\mathrm{a,E}}]^T;\ \xi = \left[\xi_{\Delta t_{\mathrm{w,L,dry}}}, \xi_{\Delta t_{\mathrm{a,L,dry}}}, 0\right]^T;$$

$$A_{\mathrm{coil}} = \begin{bmatrix} \frac{X_1}{T_{\mathrm{w}}} & 0 & \frac{X_2}{T_{\mathrm{w}}} \\ 0 & \frac{Y_{\mathrm{dry},1}}{T_{\mathrm{a}}} & \frac{Y_{\mathrm{dry},2}}{T_{\mathrm{a}}} \\ \frac{Z_{\mathrm{dry},1}}{T_{\mathrm{g}}} & \frac{Z_{\mathrm{dry},2}}{T_{\mathrm{g}}} & \frac{Z_{\mathrm{dry},3}}{T_{\mathrm{g}}} \end{bmatrix};\ B_{\mathrm{coil}} = \begin{bmatrix} \frac{X_3}{T_{\mathrm{w}}} & \frac{X_4}{T_{\mathrm{w}}} & 0 & 0 & 0 \\ 0 & 0 & \frac{Y_{\mathrm{dry},3}}{T_{\mathrm{a}}} & 0 & \frac{Y_{\mathrm{dry},4}}{T_{\mathrm{a}}} \\ \frac{Z_{\mathrm{dry},5}}{T_{\mathrm{g}}} & \frac{Z_{\mathrm{dry},6}}{T_{\mathrm{g}}} & \frac{Z_{\mathrm{dry},4}}{T_{\mathrm{g}}} & 0 & \frac{Z_{\mathrm{dry},7}}{T_{\mathrm{g}}} \end{bmatrix};$$

$$C_{\mathrm{coil}} = \begin{bmatrix} 1 & 0 & 0 \\ 0 & 0 & 0 \\ 0 & 1 & 0 \\ 0 & 0 & 0 \\ 0 & 0 & 0 \end{bmatrix};\ D_{\mathrm{coil}} = \begin{bmatrix} 0 & 0 & 0 & 0 & 0 \\ 0 & 1 & 0 & 0 & 0 \\ 0 & 0 & 0 & 0 & 0 \\ 0 & 0 & 0 & 1 & 0 \\ 0 & 0 & 0 & 0 & 1 \end{bmatrix}.$$

Table 2.1 Coefficients in Eqs. (2.22) through (2.31)

Equation No.	Coefficient expression
Eq. (2.22)	$T_w = \frac{1}{2}\rho_w c_w A_w l; X_{dry,1} = -c_w(G_{w,E})_o - \frac{A_i}{2}(a_{gw})_o;$ $X_{dry,2} = A_i(a_{gw})_o; X_{dry,3} = c_w(G_{w,E})_o - \frac{A_i}{2}(a_{gw})_o;$ $X_{dry,4} = c_w(t_{w,E}-t_{w,L})_o + A_i\left(\frac{\partial a_{gw}}{\partial G_{w,E}}\right)_o\left(t_g - \frac{t_{w,L}+t_{w,E}}{2}\right)_o;$ $\xi_{\Delta t_{w,L,dry}} = -\frac{1}{2}\rho_w c_w A_w l\frac{\partial \Delta t_{w,E}}{\partial \tau}$
Eq. (2.23)	$T_a = \frac{1}{2}\varepsilon_a\rho_a c_a A_a b_a; Y_{dry,1} = -c_a(G_{a,E}) - \frac{A_o\eta_s}{2}(a_{ga})_o;$ $Y_{dry,2} = A_o\eta_s(a_{ga})_o; Y_{dry,3} = c_a(G_{a,E})_o - \frac{A_o\eta_s}{2}(a_{ga})_o;$ $Y_{dry,4} = c_a(t_{a,E}-t_{a,L})_o + \left(\frac{\partial a_{ga}}{\partial G_{a,E}}\right)_o\eta_s A_o\left(t_g - \frac{t_{a,E}+t_{a,L}}{2}\right);$ $\xi_{\Delta t_{a,L,dry}} = \frac{1}{2}\varepsilon_a\rho_a c_a A_a b\frac{\partial\Delta t_{a,E}}{\partial\tau}$
Eq. (2.24)	$T_g = c_g M_g; Z_{dry,1} = Z_{dry,5} = \frac{A_i}{2}(a_{g,w})_o;$ $Z_{dry,2} = Z_{dry,4} = \frac{A_o\eta_s}{2}(a_{ga})_o; Z_{dry,3} = -A_i(a_{gw})_o - A_o\eta_s(a_{ga})_o;$ $Z_{dry,6} = Z_{wet,7} = \frac{A_i}{2}\left(\frac{\partial a_{gw}}{\partial G_{w,E}}\right)_o(t_{w,L}+t_{w,E}-2t_g)_o;$ $Z_{dry,7} = \frac{A_o\eta_s}{2}\left(\frac{\partial a_{ga}}{\partial G_{a,E}}\right)_o\left[\frac{1}{2}(t_{a,L}+t_{a,E}) - (t_g)_o\right].$
Eq. (2.28)	$T_w = \frac{1}{2}\rho_w c_w A_w l\ X_{wet,1} = -c_w(G_{w,E})_o - \frac{A_i}{2}(a_{gw})_o;$ $X_{wet,2} = A_i(a_{gw})_o; X_{wet,3} = c_w(G_{w,E})_o - \frac{A_i}{2}(a_{gw})_o;$ $X_{wet,4} = c_w(t_{w,E}-t_{w,L})_o + A_i\left(\frac{\partial a_{gw}}{\partial G_{w,E}}\right)_o\left(t_g - \frac{t_{w,L}-t_{w,E}}{2}\right)_o;$ $\xi_{\Delta t_{w,L,wet}} = -\frac{1}{2}\varepsilon_a\rho_w c_w A_w\frac{\partial\Delta t_{w,E}}{\partial\tau}.$
Eq. (2.29)	$T_a = \frac{1}{2}\varepsilon_a\rho_a c_a A_a b;$ $Y_{wet,1} = -c_a(G_{a,E}) - \frac{A_o}{2}\left[(a_{ga})_o\eta_s + \beta_1(q_r-\beta_2)(1-\eta_s)(\lambda_m)_o\right];$ $Y_{wet,2} = Y_{wet,6} = \frac{A_o(\beta_2-q_r)}{2}(\lambda_m);$ $Y_{wet,3} = A_o\eta_s(a_{ga})_o + (q_r-\beta_2)A_o\beta_1\eta_s(\lambda_m)_o;$ $Y_{wet,4} = c_a(G_{a,E})_o - \frac{A_o}{2}\left[2(a_{ga})_o - (a_{ga})_o\eta_s + \beta_1(q_r-\beta_2)(1-\eta_s)(\lambda_m)_o\right].$ $Y_{wet,5} = c_a(t_{a,E}-t_{a,L})_o - \frac{A_o}{2}\left[\left(\frac{\partial a_{ga}}{\partial G_{a,E}}\right)_o\eta_s + \beta_1(q_r-\beta_2)(1-\eta_s)\left(\frac{\partial\lambda_m}{\partial G_{a,E}}\right)_o\right](t_{a,E}+t_{a,L})_o$ $+A_o(t_g)_o\left[\left(\frac{\partial a_{ga}}{\partial G_{a,E}}\right)_o\eta_s + \beta_1(q_r-\beta_2)\eta_s\left(\frac{\partial\lambda_m}{\partial G_{a,E}}\right)_o\right] + \frac{A_o(\beta_2-q_r)}{2}\left(\frac{\partial\lambda_m}{\partial G_{a,E}}\right)(W_{a,E}+W_{a,L})_o;$ $\xi_{\Delta t_{w,L,wet}} = -\frac{1}{2}\varepsilon_a\rho_a A_a c_a\frac{\partial\Delta t_{a,E}}{\partial\tau}.$
Eq. (2.30)	$T_m = \frac{1}{2}\varepsilon_a\rho_a A_a b; M_{wet,1} = M_{wet,4} = -\frac{\beta_1(1-\eta_s)A_o}{2}(\lambda_m)_o;$ $M_{wet,2} = -(G_{a,E})_o - \frac{A_o}{2}(\lambda_m)_o; M_{wet,3} = \beta_1\eta_s A_o(\lambda_m)_o;$ $M_{wet,6} = (G_{a,E})_o - \frac{A_o}{2}(\lambda_m)_o;$ $M_{wet,6} = \begin{bmatrix}\beta_1\eta_s A_o(t_g)_o - \frac{A_o}{2}(W_{a,E}+W_{a,L})_o \\ -\frac{\beta_1 A_o(1-\eta_s)}{2}(t_{a,E}+t_{a,L})_o\end{bmatrix}\left(\frac{\partial\lambda_m}{\partial G_{a,E}}\right) + (W_{a,E}-W_{a,L})_o;$ $\xi_{\Delta t_{a,L,wet}} = -\frac{1}{2}\varepsilon_a\rho_a b A_a\frac{\partial\Delta W_{a,E}}{\partial\tau}.$
Eq. (2.31)	$T_g = c_g M_g; Z_{wet,1} = Z_{wet,6} = \frac{A_i}{2}(a_{gw})_o;$ $Z_{wet,2} = Z_{wet,5} = \frac{A_o\eta_s}{2}(a_{ga})_o + \frac{1}{2}\beta_1(1-\eta_s)q_r A_o(\lambda_m)_o;$ $Z_{wet,3} = Z_{wet,9} = \frac{A_o q_r}{2}(\lambda_m)_o;$ $Z_{wet,4} = -A_i(a_{gw})_o - A_o\eta_s(a_{ga})_o - q_r\beta_1\eta_s A_o(\lambda_m)_o;$ $Z_{wet,7} = -A_i\left(t_g - \frac{t_{w,E}+t_{w,L}}{2}\right)_o\left(\frac{\partial a_{gw}}{\partial G_{w,E}}\right)_o;$ $Z_{wet,8} = \eta_s A_o\left(\frac{\partial a_{ga}}{\partial G_{a,E}}\right)_o\left[\frac{1}{2}(t_{a,L}+t_{a,E})_o - (t_g)_o\right]$ $+A_o q_r\left(\frac{\partial\lambda_m}{\partial G_{a,E}}\right)_o\left[\frac{1}{2}\beta_1(1-\eta_s)(t_{a,E}+t_{a,L})_o - \beta_1\eta_s(t_g)_o + \frac{1}{2}(W_{a,E}+W_{a,L})_o\right]$

For the wet-condition process:

$$x_{\text{coil}} = [\Delta t_{\text{w,L}}, \Delta t_{\text{a,L}}, \Delta W_{\text{a,L}}, \Delta t_{\text{g}}]^T; \; y_{\text{coil}} = [\Delta t_{\text{w,L}}, \Delta G_{\text{w,L}}, \Delta t_{\text{a,L}}, \Delta W_{\text{a,L}}, \Delta G_{\text{a,L}}]^T;$$

$$u_{\text{coil}} = [\Delta t_{\text{w,E}}, \Delta G_{\text{w,E}}, \Delta t_{\text{a,E}}, \Delta W_{\text{a,E}}, \Delta G_{\text{a,E}}]^T; \; \xi = \left[\xi_{\Delta t_{\text{w,L,wet}}}, \xi_{\Delta t_{\text{a,L,wet}}}, \xi_{\Delta t_{\text{g,wet}}}, 0\right]^T;$$

$$A_{\text{coil}} = \begin{bmatrix} \frac{X_1}{T_{\text{w}}} & 0 & 0 & \frac{X_2}{T_{\text{w}}} \\ 0 & \frac{Y_{\text{wet},1}}{T_{\text{a}}} & \frac{Y_{\text{wet},2}}{T_{\text{a}}} & \frac{Y_{\text{wet},3}}{T_{\text{a}}} \\ 0 & \frac{M_{\text{wet},1}}{T_{\text{m}}} & \frac{M_{\text{wet},2}}{T_{\text{m}}} & \frac{M_{\text{wet},3}}{T_{\text{m}}} \\ \frac{Z_{\text{wet},1}}{T_{\text{g}}} & \frac{Z_{\text{wet},2}}{T_{\text{g}}} & \frac{Z_{\text{wet},3}}{T_{\text{g}}} & \frac{Z_{\text{wet},4}}{T_{\text{g}}} \end{bmatrix}; \; B_{\text{coil}} = \begin{bmatrix} \frac{X_3}{T_{\text{w}}} & \frac{X_4}{T_{\text{w}}} & 0 & 0 & 0 \\ 0 & 0 & \frac{Y_{\text{wet},4}}{T_{\text{a}}} & \frac{Y_{\text{wet},6}}{T_{\text{a}}} & \frac{Y_{\text{wet},5}}{T_{\text{a}}} \\ 0 & 0 & \frac{M_{\text{wet},4}}{T_{\text{m}}} & \frac{M_{\text{wet},6}}{T_{\text{m}}} & \frac{M_{\text{wet},5}}{T_{\text{m}}} \\ \frac{Z_{\text{wet},6}}{T_{\text{g}}} & \frac{Z_{\text{wet},7}}{T_{\text{g}}} & \frac{Z_{\text{wet},5}}{T_{\text{g}}} & \frac{Z_{\text{wet},9}}{T_{\text{g}}} & \frac{Z_{\text{wet},8}}{T_{\text{g}}} \end{bmatrix}$$

$$C_{\text{coil}} = \begin{bmatrix} 1 & 0 & 0 & 0 \\ 0 & 0 & 0 & 0 \\ 0 & 1 & 0 & 0 \\ 0 & 0 & 1 & 0 \\ 0 & 0 & 0 & 0 \end{bmatrix}; \; D_{\text{coil}} = \begin{bmatrix} 0 & 0 & 0 & 0 & 0 \\ 0 & 1 & 0 & 0 & 0 \\ 0 & 0 & 0 & 0 & 0 \\ 0 & 0 & 0 & 0 & 0 \\ 0 & 0 & 0 & 0 & 1 \end{bmatrix}.$$

In Eq. (2.34), the variables, $\Delta t_{\text{w,E}}, \Delta G_{\text{w,E}}, \Delta t_{\text{a,E}}, \Delta W_{\text{a,E}}$ and $\Delta G_{\text{a,E}}$, are input perturbations, and the variables, $\Delta t_{\text{w,L}}, \Delta G_{\text{w,L}}, \Delta t_{\text{a,L}}, \Delta W_{\text{a,L}}$ and $\Delta G_{\text{a,L}}$, are the response ones. The state variables are $\Delta t_{\text{w,L}}, \Delta t_{\text{a,L}}, \Delta t_{\text{g}}$ for the dry-condition process and $\Delta t_{\text{w,L}}, \Delta t_{\text{a,L}}, \Delta W_{\text{a,L}}, \Delta t_{\text{g}}$ for the wet-condition process.

2.2.1.2 Model Validation

An experimental setup has been built for validating the state-space model of water-to-air heat exchanger. The schematic diagram of the experimental setup is shown in Fig. 2.3. A grid board is placed in the upstream of the test section to reduce the influence of turbulence of inlet air on the test. A louvered mixer is placed after the heat exchanger to guarantee a uniform temperature profile of outlet air. The test instruments and relevant equipments in the experimental system are listed in Table 2.2, and the detailed information of structure about the water-to-air heat exchanger is given in Fig. 2.4 and Table 2.3.

Three experimental cases with the validation time of 1200 s are investigated to verify the heat exchanger's transient behaviors including the transient responses of the exit air temperature and humidity as well as the exit water temperature subjected to different sudden perturbations. These cases are as follows: ① Start up the chiller at the beginning of the system's running; ② the water flow rate has a sudden increase of 0.058 kg/s during the system's running; and ③ Stop the chiller at the end of the system's running. The initial conditions corresponding to these cases (see Table 2.4), which are used for the model simulation, are obtained under the steady state before the perturbations began to apply to the heat exchanger. During the

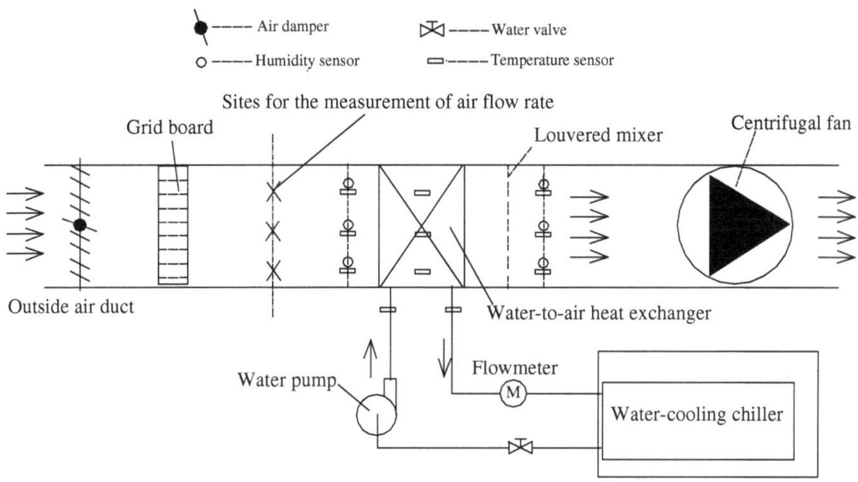

Fig. 2.3 Schematic diagram of experimental setup for water-to-air heat exchanger model validation

Table 2.2 Specific details about the apparatus in the experimental system

Items	Properties or features
Temperature sensor	Copper-constantan thermocouples: ±0.2 °C
Temperature and humidity sensor	Type: HHC2-S; ±0.008 for air humidity ratio and ±0.1 °C for air temperature
Digital anemometer	Type: T425; precision: ±(0.03 + 5 %) m/s
Electromagnetic flowmeter	Type: ETFM-20; precision: ±0.5 %
Water pump	Type: 50T5WA-3; nominal water flow: 4.2 m^3/h; power: 1.7 kW
Ventilator	Type: KT40-2.5; nominal airflow rate: 2000 m^3/h; power: 0.75 kW
Chiller	Nominal cooling capacity: 6.0 kW; power: 2.0 kW

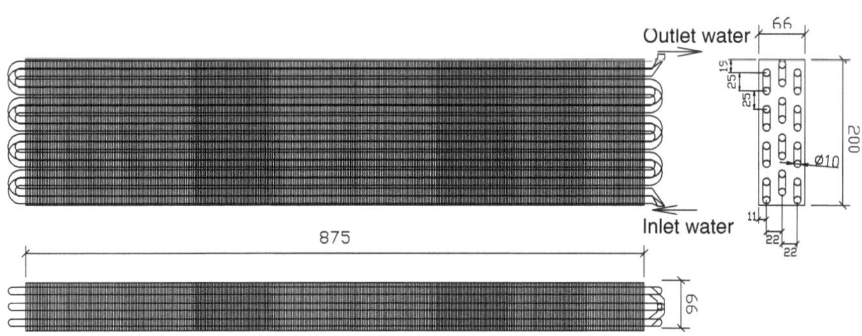

Fig. 2.4 Structure of the water-to-air heat exchanger

Table 2.3 Structural information of the water-to-air heat exchanger

Length of the coil: l (m)	21.0	Area of windward side A_a (m^2)	0.175
Inner diameter of the coil: r_i (m)	0.004	Length along airflow: b (m)	0.66
Inner surface area of the coil: A_i (m^2)	0.5287	Thickness of fin: δ_c (m)	0.0002
External surface area of the coil: A_o (m^2)	8.8065	Total mass: M_g (kg)	17.52
Mean specific heat: c_g [J/(kg °C)]	625	Fin spacing e (m)	0.0024

Table 2.4 Initial conditions for model validation and simulation

Cases / Initial conditions	Case I	Case II	Case III
$(t_{a,E})_o$ (°C)	29.8 ± 0.1	28.3 ± 0.1	28.2 ± 0.1
$(t_{a,L})_o$ (°C)	28.8 ± 0.1	21.3 ± 0.1	20.7 ± 0.1
$(W_{a,E})_o$ (g/kg dry air)	20.1 ± 0.2	18.7 ± 0.2	16.1 ± 0.2
$(W_{a,L})_o$ (g/kg dryair)	20.1 ± 0.2	17.0 ± 0.2	14.3 ± 0.2
$(G_a)_o$ (kg/s)	0.2000 ± 0.01	0.2000 ± 0.01	0.2000 ± 0.01
$(t_{w,E})_o$ (°C)	28.2 ± 0.1	16.7 ± 0.1	15.7 ± 0.1
$(t_{w,L})_o$ (°C)	28.4 ± 0.1	18.9 ± 0.1	18.6 ± 0.1
$(G_w)_o$ (kg/s)	0.2587 ± 0.005	0.2387 ± 0.005	0.1964 ± 0.005
$(t_g)_o$ (°C)	29.3 ± 0.1	19.5 ± 0.1	18.9 ± 0.1

Note: Case I: ***Start-up the chiller***; Case II: G_w ↑ ***0.058 kg/s***; Case III: ***Stop the chiller***

simulation, the dew-point temperature of inlet air is compared with the heat exchanger's wall surface temperature to judge whether the wet-cooling model or the dry-cooling model is used for the calculation.

The average error (AE) is used to evaluate the goodness of the calculated results by the model compared with the experimental data during the transient response process, which is defined by Eq. (2.39).

$$\text{AE(Average error)} = \frac{1}{N}\sum_{i=1}^{N}\left(\frac{|\Delta Y_{\text{m,i}} - \Delta Y_{\text{exp},i}|}{|\Delta Y_{\text{exp},i}|} \times 100\,\%\right) \quad (2.39)$$

where ΔY stands for the variation of the response parameters comparing with the initial value; the subscripts 'm' and 'exp' stand, respectively, for the model result and the experimental ones; i denotes the ith calculated or experimental result at the same time point; and N stands for the total number of the calculated or experimental result during the transient response process.

(1) Experimental case I

For the experimental case I, only changes were made on inlet air temperature and humidity as well as inlet water temperature of the water-to-air heat exchanger. As

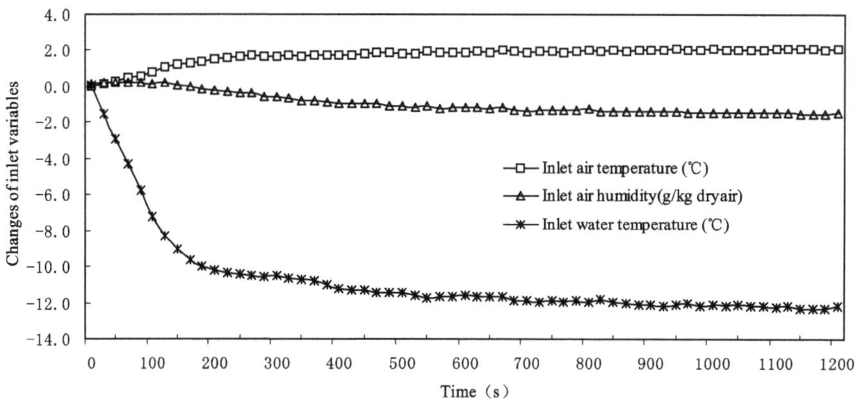

Fig. 2.5 Changes of inlet variables in the case I (measured data)

shown in Fig. 2.5, the inlet air temperature increases gradually till to the steady value; the inlet air humidity increases a little in the beginning (last for about 120 s) and decreases afterward; and the inlet water temperature drops quickly in the first 180 s and then tends to be steady. The ultimate change of the inlet air temperature, the inlet air humidity, and the inlet water temperature is about +2.1 °C, −1.6 g/(kg dry air), and −12.3 °C, respectively. The flow rates of air and water passing through the water-to-air heat exchanger were kept stable as much as possible during the experiment, and they are considered as zero in the model simulation.

Figure 2.6 shows the comparisons of the model results with the experimental data on transient responses of the exit air temperature, the exit air humidity, and the exit water temperature for the case I, and the corresponding average error [AE, calculated by Eq. (2.39)] is estimated as 8.7, 8.8, and 10.3 %, respectively. It is reasonable to see that the exit air temperature and humidity will decrease when subjected to a big decrease in the inlet water temperature while the other inlet variables have small change. The exit water temperature decreases with the decrease in the inlet one, and the ultimate change of the exit water temperature is lower than that of the inlet one. This is because decreasing the inlet water temperature will increase heat transfer quantity of the water-to-air heat exchanger.

(2) Experimental case II

For the experimental case II, the water flow rate had a sudden increase of 0.058 kg/s while the other inlet variables were kept unchanged. The comparisons between model results and experimental data on transient responses of the exit variables are given in Fig. 2.7. The model results are shown to have a good agreement with the experimental data, and the average errors of the simulated results for the case II are all less than 12 %. It is obvious that a larger water flow rate will bring about a larger cooling and dehumidification capacity of the water-to-air heat exchanger and cause the exit air temperature and humidity to

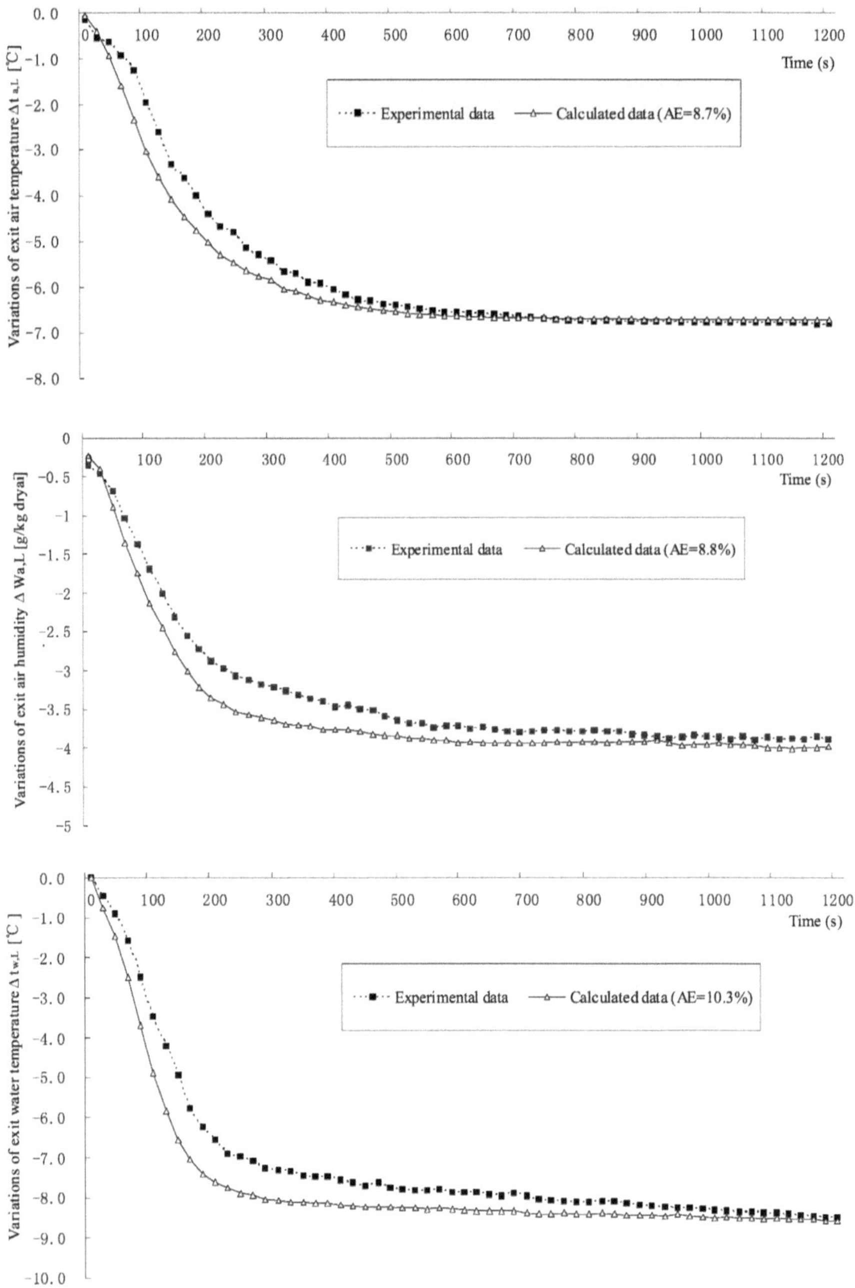

Fig. 2.6 Responses of exit variables to the changes of inlet ones (case I)

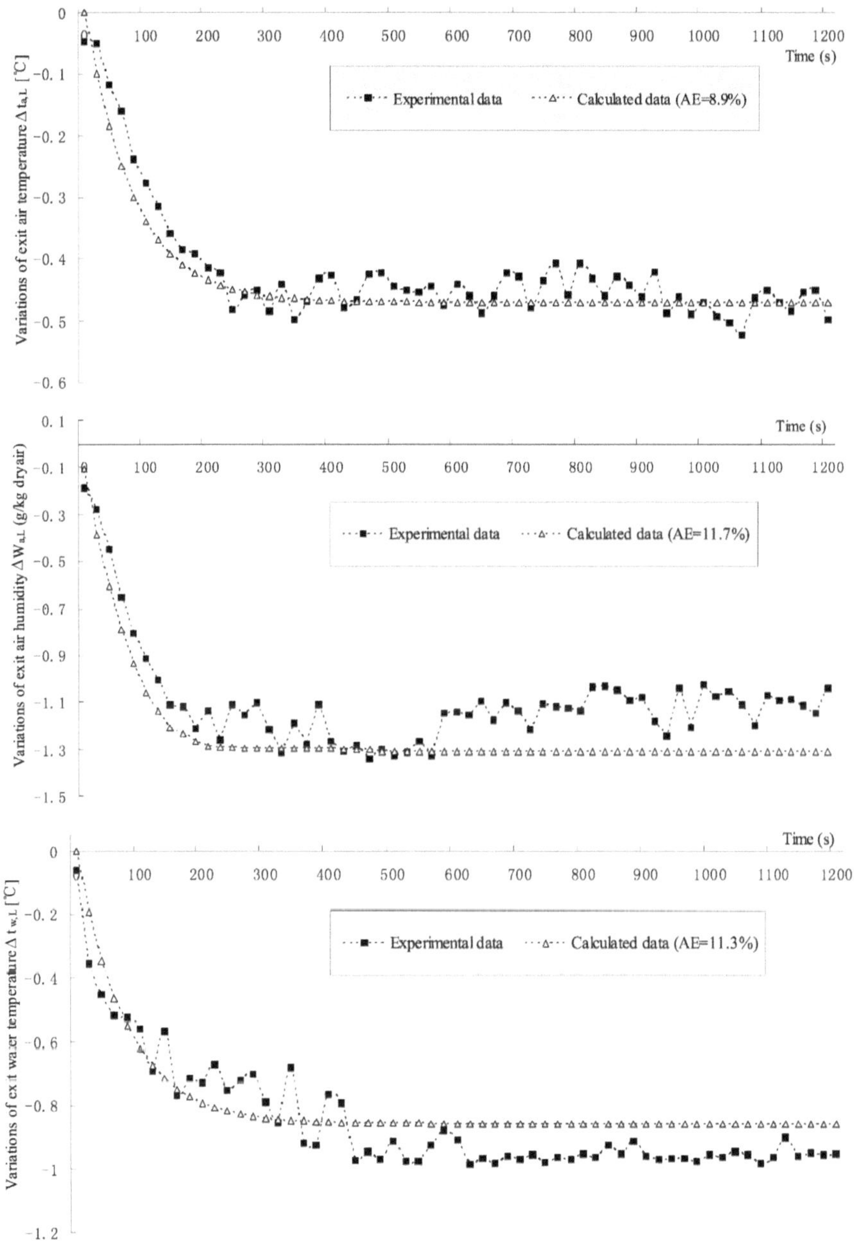

Fig. 2.7 Responses of exit variables to a sudden increase in water flow rate by 0.058 kg/s (case II)

decrease. Meanwhile, the water flow rate affects as well the exit water temperature of water-to-air heat exchanger. In this case, the exit water temperature gradually decreased after the water flow rate had a sudden increase.

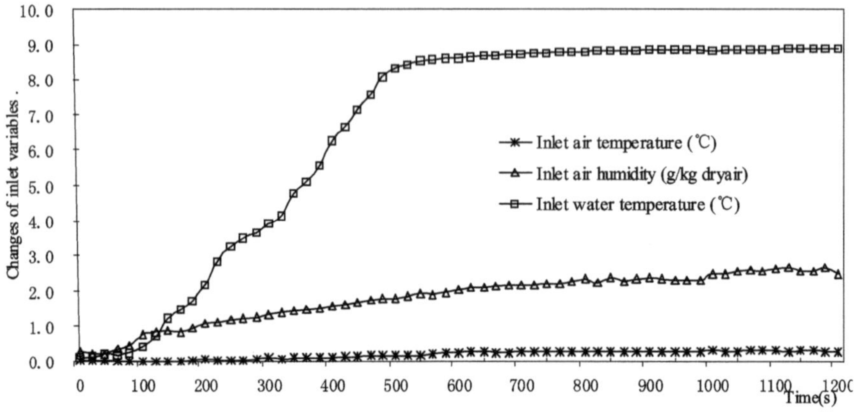

Fig. 2.8 Changes of inlet variables in the case III (measured data)

(3) Experimental case III

In the case III, the inlet air temperature and humidity and the inlet water temperature all gradually increase (see Fig. 2.8), while the water and the airflow rates are kept unchanged. Figure 2.9 shows the calculated and experimental results on the transient responses of exit variables to the changes of inlet variables under initial conditions in the case III. Comparing with the experimental data, the average error of the model results on the transient responses of the exit air temperature and humidity as well as the exit water temperature is estimated as 14.2, 8.9, and 8.2 %, respectively.

As seen from the three study cases (case I, case II, and case III), the simulation errors compared with the experimental data are all less than 15 %, which indicates that the dynamic model developed in this study is capable of predicting well-transient performances of water-to-air heat exchanger under wet conditions.

2.2.2 *Chiller*

Vapor-compression packaged liquid chillers are often employed to provide chilled media for space thermal environment control purposes in commercial and industrial applications. As well known, the space-cooling load often changes and the refrigeration system needs to operate in an unsteady manner and adjust its capacity to guarantee an anticipated space thermal conditions. Usually, a refrigeration system adjusts its capacity mainly through modulating compressor speed with input frequency control. To obtain better control effect of the space thermal conditions, we need to have the knowledge of the dynamic performance of the refrigeration system under different conditions, and a dynamic model for refrigeration system is very necessary for that objective.

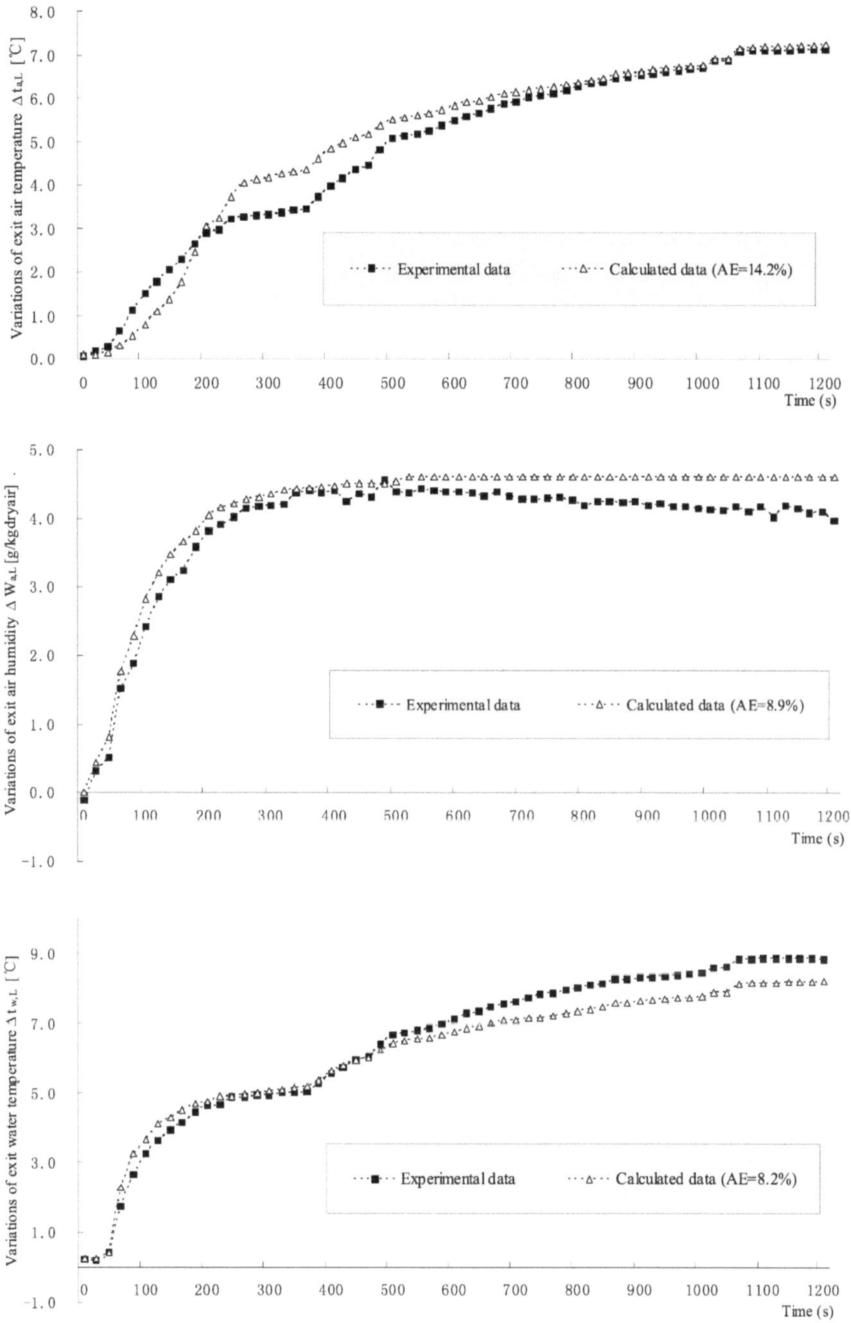

Fig. 2.9 Responses of exit variables to the changes of inlet ones (case III)

2.2.2.1 Model Development [9]

Basically, a conventional refrigeration system mainly consists of four key components (i.e., compressor, condenser, expander, and evaporator) and other accessories. Figure 2.10 shows the theoretical single-stage cycle used as a model for actual vapor-compression chiller. Heat is added to the refrigerant from state '4' to state '1.' The heat capacity of the refrigerant fluid is assumed to be constant in this process. The refrigerant temperature is increased in process from state '1' to state '2'. Process from state '2' to state '3' is a heat rejection process in which the refrigerant temperature decreases linearly with heat transfer. The cycle ends with expansion between states '3' and '4.'

1) Assumptions and fundamental equations

For the convenience of modeling, some assumptions are necessary to be made as below:

(1) The compression (from state '1' to state '2') is assumed to be a isentropic process, and the expansion from state '3' to state '4' is considered to be isenthalpic;
(2) Mass flow rate of refrigerant is uniform throughout the cycle;
(3) Mass of refrigerant in the condenser and evaporator is unchanged during the dynamic response simulation;
(4) Working performance of the compressor is kept invariable during the dynamic response simulation;
(5) Temperature of chilled liquid (passing through the evaporator) and cooling liquid (passing through the condenser) change linearly from the inlet to the outlet of the corresponding heat exchanger. And shell temperature of the heat

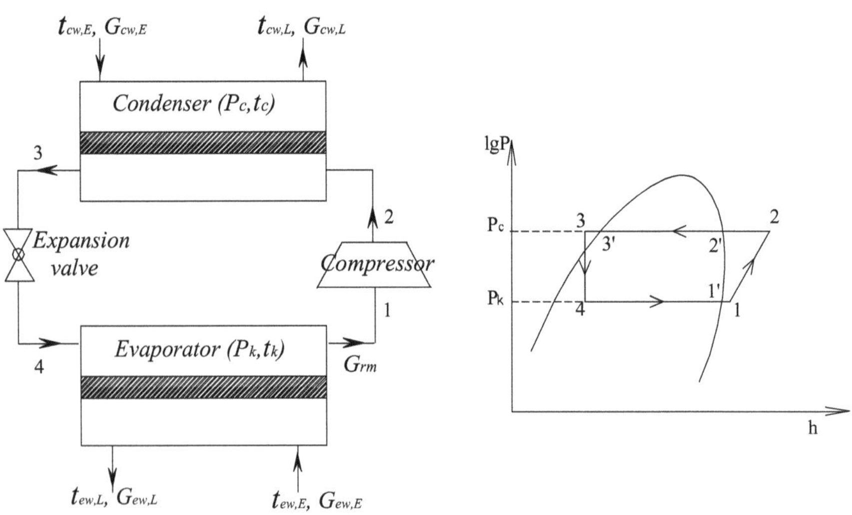

Fig. 2.10 Schematic diagram for the working cycle of vapor-compression chiller

exchangers (i.e., the condenser and the evaporator) is represented by a lumped parameter;

(6) Superheated temperature of gaseous refrigerant after evaporator and subcooled temperature of liquid refrigerant after condenser are assumed to be constant (equal to initial conditions) during the transient response simulation.

Based on the above assumptions, the fundamental equations can be established according to the principle of energy and mass conservation.

Energy equation for refrigerant in the condenser

$$c_{cr}M_{cr}\frac{\mathbf{d}t_c}{\mathbf{d}\tau} = G_{rm}(h_{r,2} - h_{r,3}) + a_{co}A_{co}(t_{cg} - t_c) \tag{2.40}$$

Mass equation for cooling liquid passing through the condenser

$$G_{cw,L} = G_{cw,E} = G_{cw} \tag{2.41}$$

Energy equation for cooling liquid passing through the condenser

$$\frac{1}{2}c_{cw}M_{cw}\frac{\mathbf{d}(t_{cw,L} + t_{cw,E})}{\mathbf{d}\tau} = c_{cw}G_{cw,E}(t_{cw,E} - t_{cw,L}) + a_{cw}A_{cw}\left(t_{cg} - \frac{t_{cw,E} + t_{cw,L}}{2}\right) \tag{2.42}$$

Energy equation for shell of the condenser

$$c_{cg}M_{cg}\frac{\mathbf{d}t_{cg}}{\mathbf{d}\tau} = a_{co}A_{co}(t_c - t_{cg}) + a_{cw}A_{cw}\left(\frac{t_{cw,E} + t_{cw,L}}{2} - t_{cg}\right) \tag{2.43}$$

Energy equation for refrigerant in the evaporator

$$c_{er}M_{er}\frac{\mathbf{d}t_k}{\mathbf{d}\tau} = G_{rm}(h_{r,3} - h_{r,1}) + a_{eo}A_{eo}(t_{eg} - t_k) \tag{2.44}$$

Mass equation for chilled liquid passing through the evaporator

$$G_{ew,L} = G_{ew,E} = G_{ew} \tag{2.45}$$

Energy equation for chilled liquid passing through the evaporator

$$\frac{1}{2}c_{ew}M_{ew}\frac{\mathbf{d}(t_{ew,L} + t_{ew,E})}{\mathbf{d}\tau} = c_{ew}G_{ew,E}(t_{ew,E} - t_{ew,L}) + a_{ew}A_{ew}\left(t_{eg} - \frac{t_{ew,E} + t_{ew,L}}{2}\right) \tag{2.46}$$

Energy equation for shell of the evaporator

$$c_{eg}M_{eg}\frac{\mathbf{d}t_{eg}}{\mathbf{d}\tau} = a_{eo}A_{eo}(t_k - t_{eg}) + a_{ew}A_{ew}\left(\frac{t_{ew,E} + t_{ew,L}}{2} - t_{eg}\right) \tag{2.47}$$

Electric power of compressor

$$N_{\mathrm{com}}=\frac{G_{\mathrm{rm}}}{\eta_{\mathrm{com}}}\frac{R_{\mathrm{r}}T_1}{\kappa_{\mathrm{s}}-1}\left[\left(\frac{p_{\mathrm{c}}}{p_{\mathrm{k}}}\right)^{\frac{\kappa_{\mathrm{s}}-1}{\kappa_{\mathrm{s}}}}-1\right]=\frac{G_{\mathrm{rm}}}{\eta_{\mathrm{com}}}\frac{R_{\mathrm{r}}(t_{\mathrm{k}}+\Delta t_{\mathrm{e,shr}}+273.15)}{\kappa_{\mathrm{s}}-1}\left[\left(\frac{f_{\mathrm{p}}(t_{\mathrm{c}})}{f_{\mathrm{p}}(t_{\mathrm{k}})}\right)^{\frac{\kappa_{\mathrm{s}}-1}{\kappa_{\mathrm{s}}}}-1\right]$$
$$=G_{\mathrm{rm}}f_{N_{\mathrm{com}}}(t_{\mathrm{k}},t_{\mathrm{c}}) \tag{2.48}$$

Energy equation for refrigerant in the expansion process

$$h_{\mathrm{r},3}=h_{\mathrm{r},4} \tag{2.49}$$

Cooling capacity of chiller

$$\begin{aligned}Q_{\mathrm{c}}&=G_{\mathrm{rm}}(h_{\mathrm{r},1}-h_{\mathrm{r},4})=G_{\mathrm{rm}}(f_{\mathrm{h}}(t_{\mathrm{k}})-f_{\mathrm{h}}(t_{\mathrm{c}}))\\&=G_{\mathrm{rm}}f_{Q_{\mathrm{c}}}(t_{\mathrm{k}},t_{\mathrm{c}})\end{aligned} \tag{2.50}$$

Coefficient of performance (COP) of chiller

$$\mathrm{COP}=\frac{Q_{\mathrm{c}}}{N_{\mathrm{com}}}=\frac{f_{Q\mathrm{c}}(t_{\mathrm{k}},t_{\mathrm{c}})}{f_{N_{\mathrm{com}}}(t_{\mathrm{k}},t_{\mathrm{c}})}=f_{\mathrm{COP}}(t_{\mathrm{k}},t_{\mathrm{c}}) \tag{2.51}$$

In Eqs. (2.40) through (2.51), the symbol 'G' stands for flow rate, kg/s; 'c' for mass specific heat, J/ (kg °C); 'A' for area, m^2; 'h' for enthalpy, J/kg; 't_c' for condensing temperature, °C; 't_k' for evaporating temperature, °C; 'M' for mass, kg; 'a' for heat transfer coefficient, W/(m^2 °C); and 'τ' for time, s. The subscript 'cg' for shell wall of condenser; '*eg*' for shell wall of evaporator; 'co' for refrigerant-side surface of condenser; 'cr' for refrigerant in condenser; '*cw*' for cooling liquid in condenser; '*cr*' for refrigerant in condenser; 'eo' for refrigerant-side surface of evaporator; 'er' for refrigerant in evaporator; '*ew*' for coolant liquid in evaporator; '*er*' for refrigerant in evaporator; '*rm*' for circulating refrigerant in refrigeration system; '$\mathrm{N_{com}}$' for power of compressor, W; 'η_{com}' for working efficient of compressor; 'R_r' for gas constant of gaseous refrigerant, J/kg.K; 'κ_s' for adiabatic compression index of compressor; '$\mathrm{T_1}$' for inlet temperature of compressor, K; 'p_c' and 'p_k' are pressure in condenser and evaporator, respectively, Pa; '$\Delta t_{\mathrm{e,shr}}$' for superheat degree of gaseous refrigerant, °C; 'Q_c' for cooling capacity, W; and 'f' denotes function.

2) Key parameters determination

(1) *Heat transfer coefficients*

Four coefficients of heat transfer are concerned in the model equations. They include condensation heat transfer coefficient in condenser (a_{co}), boiling heat transfer coefficient in evaporator (a_{eo}), convective heat transfer coefficient between liquid coolant and condenser (a_{cw}), and that between cold carrier liquid and evaporator (a_{ew}). These coefficients can be usually modeled by Eqs. (2.52) through (2.54), respectively [10] .

$$a_{co} = C_1\left(t_c - t_{cg}\right)^{n_1} \quad (2.52)$$

$$a_{eo} = C_2\left(t_{eg} - t_k\right)^{n_2} \quad (2.53)$$

$$Nu_d = C_3 {Re_d}^{n_3} \quad (2.54)$$

For condenser: $Nu_d = \frac{a_{cw} d_{ci}}{\lambda_w}$, $Re_d = \frac{u_{cw} d_{ci}}{v_w}$; For evaporator: $Nu_d = \frac{a_{ew} d_{ei}}{\lambda_w}$, $Re_d = \frac{u_{ew} d_{ei}}{v_w}$. $C_1, C_2, C_3, n_1, n_2, n_3$ are empirical coefficients determined by field test data.

(2) *Enthalpy calculation for refrigerant at key state in the refrigeration cycle*

According to the diagram of refrigeration cycle (Fig. 2.19), the refrigerant often exists in the form of superheated vapor at the states '1' and '2' and subcooled liquid at states '3' and '4'. For a real refrigerant, the state equations are often expressed by semiempirical ones. The famous Martin-Hou equations are often used to calculate the thermophysical properties of refrigerants [11, 12], based on which the enthalpy of refrigerant at different states can be obtained.

The enthalpy of refrigerant in the gaseous state (h_{vapor}, kJ/kg) can be calculated by:

$$h_{vapor} = h_0 + (p_{r,g} v_{r,g} - R_r T_{r,g}) + \sum_{n=1}^{4} \frac{c_n T_{r,g}^n}{n} + c_5 \ln T_{r,g} - \frac{c_6}{T_{r,g}} + \sum_{n=2}^{5} \frac{A_n}{(n-1)(v_{r,g} - b)^{n-1}} + e^{-K_1 T_{r,g}/T_{cc}}(1 + K_1 T_{r,g}/T_{cc}) \times \sum_{n=2}^{5} \frac{C_n}{(n-1)(v_{r,g} - b)^{n-1}} \quad (2.55)$$

The state equation for the gaseous refrigerant can be expressed as follows:

$$p_{r,g} = \frac{R_r T_{r,g}}{v_{r,g} - b} + \sum_{n=2}^{5} \frac{A_n + B_n T_{r,g} + C_n e^{-K_1 T_{r,g}/T_{cc}}}{(v_{r,g} - b)^n} \quad (2.56)$$

The saturated vapor pressure of refrigerant ($p_{r,bg}$, kPa) can be computed by

$$\ln p_{r,bg} = A + \frac{B}{T_{r,bg}} + \mathrm{C}T_{r,bg} + \mathrm{D}T_{r,bg}^2 + \frac{E(F - T_{r,bg})}{T_{r,bg}} \ln(F - T_{r,bg}) + G \ln T_{r,bg} \quad (2.57)$$

The enthalpy of subcooled liquid refrigerant ($h_{r,l\Delta t}$, kJ/kg) can be calculated by Eq. (2.58).

$$h_{r,l\Delta t} = h_{r,bl} - \frac{p_{r,bl} - p_{atm}}{q_r} \tag{2.58}$$

In Eq. (2.58), the heat of vaporization (q_r, kJ/kg) can be obtained by the Clausius-Clapeyron equation as follows:

$$q_r = T_{r,q}(v_{r,bl} - v_{r,bg})p_{r,q} \left[-\frac{B}{T_{r,q}^2} + C + 2\mathrm{DT}_{r,q} - \frac{\mathrm{EF}}{T_{r,q}^2} \times \ln(F - T_{r,q}) - \frac{\mathrm{EF}}{T_{r,q}} \times \frac{1}{F - T_{r,q}} + E \times \frac{1}{F - T_{r,q}} + \frac{G}{T_{r,q}} \right] \tag{2.59}$$

The enthalpy of saturated liquid refrigerant ($h_{r,bl}$, kJ/kg) is written as follows:

$$h_{r,bl} = h_{r,bg} - q_r \tag{2.60}$$

The density of saturated liquid refrigerant can be counted by Eq. (2.61).

$$\rho_{r,bl} = \rho_{cc} + \sum_{n=1}^{6} D_n (1 - T_{r,bl}/T_{cc})^{\frac{n}{3}} \tag{2.61}$$

In Eqs. (2.55)–(2.61), h stands for enthalpy, kJ/kg; p for pressure, kPa; T for temperature, K; ρ for density, kg/m^3; v for specific volume, m^3/kg; h_o for reference enthalpy, kJ/kg; R_o gas constant of gaseous refrigerant, kJ/kg.K; q_r for vaporization heat of refrigerant, kJ/kg; the subscript 'r, bg' for saturated gaseous refrigerant; the subscript 'r, bl' for saturated liquid refrigerant; the subscript 'r, q' for gaseous refrigerant; the subscript '$r, l\Delta t$' for subcooled liquid refrigerant; the subscript '*atm*' for standard atmospheric pressure; and the subscript '*cc*' for critical state; the rest parameters including *A, B, C, D, E, F, G*, K_1, *b*, $c_1 \sim c_6$, $A_2 \sim A_5$, $B_2 \sim B_5$, $C_2 \sim C_5$, and $D_1 \sim D_6$ are undetermined coefficients.

The enthalpy of refrigerant at states '1' (h_1) and '2' (h_2) can be calculated by using Eqs. (2.55) and (2.56) with respect to temperature and pressure. The pressure at states '1' (p_1) and '2' (p_2) corresponds, respectively, to saturated evaporating pressure and saturated condensation pressure. Known from Eq. (2.57), the saturated evaporating pressure is dependent on the evaporating temperature (T_k), and the saturated condensation one is dependent on the condensing temperature (T_c). Assuming the influence of small perturbations on the superheated temperature of vapor refrigerant is negligible, the variations of h_1 and h_2 are only dependent upon T_k and T_c, respectively. The enthalpy of refrigerant at states '3' (h_3) and '4' (h_4) can be computed by Eqs. (2.56), (2.58), (2.59), (2.60), and (2.61). Likewise, h_3 and h_4 are both dependent upon the condensing temperature (T_c).

(3) *Adiabatic compression process index (* κ_s*)*

The adiabatic compression process index for the real gaseous refrigerant can be expressed by Eq. (2.63) [13].

$$\kappa_{s}=\frac{Z}{Z_{p}-\frac{R_{r}Z_{T}^{2}}{c_{r,p}}} \tag{2.62}$$

where$Z_{p}=Z-p\left(\frac{\partial Z}{\partial p}\right)_{T}$; $Z_{T}=Z+T\left(\frac{\partial Z}{\partial T}\right)_{p}$. Z is the compressibility factor; the subscripts 'p' and 'T' stand, respectively, for the isobaric process and isothermal process. During the dynamic simulation, the adiabatic compression process index (κ_{s}) can be assumed to be constant.

3) State-space model formulation

Through parameter linearization, Eqs. (2.40) through (2.51) can be transformed as follows:

$$T_{cr}\frac{\mathbf{d}\Delta t_{c}}{\mathbf{d}\tau}=X_{cr,1}\Delta t_{c}+X_{cr,2}\Delta t_{cg}+X_{cr,3}\Delta G_{rm} \tag{2.63}$$

$$T_{cw}\frac{\mathbf{d}\Delta t_{cw,L}}{\mathbf{d}\tau}=X_{cw,1}\Delta t_{cw,L}+X_{cw,2}\Delta t_{cg}+X_{cw,3}\Delta t_{cw,E}+X_{cw,4}\Delta G_{cw,E}+\xi_{\Delta t_{cw,L}} \tag{2.64}$$

$$T_{cg}\frac{\mathbf{d}\Delta t_{cg}}{\mathbf{d}\tau}=X_{cg,1}\Delta t_{c}+X_{cg,2}\Delta t_{cw,L}+X_{cg,3}\Delta t_{cg}+X_{cg,4}\Delta t_{cw,E}+X_{cg,5}\Delta G_{cw,E} \tag{2.65}$$

$$T_{er}\frac{\mathbf{d}\Delta t_{k}}{\mathbf{d}\tau}=X_{er,1}\Delta t_{k}+X_{er,2}\Delta t_{eg}+X_{er,3}\Delta G_{rm} \tag{2.66}$$

$$T_{ew}\frac{\mathbf{d}\Delta t_{ew,L}}{\mathbf{d}\tau}=X_{ew,1}\Delta t_{ew,L}+X_{ew,2}\Delta t_{eg}+X_{ew,3}\Delta t_{ew,E}+X_{ew,4}\Delta G_{ew,E}+\xi_{\Delta t_{ew,L}} \tag{2.67}$$

$$T_{eg}\frac{\mathbf{d}\Delta t_{eg}}{\mathbf{d}\tau}=X_{e.g.,1}\Delta t_{k}+X_{e.g.,2}\Delta t_{ew,L}+X_{e.g.,3}\Delta t_{eg}+X_{e.g.,4}\Delta t_{ew,E}+X_{e.g.,5}\Delta G_{ew,E} \tag{2.68}$$

$$\Delta G_{cw,L}=\Delta G_{cw,E}=\Delta G_{cw} \tag{2.69}$$

$$\Delta G_{ew,L}=\Delta G_{ew,E}=\Delta G_{ew} \tag{2.70}$$

$$\Delta N_{com}=X_{Ew,1}\Delta t_{c}+X_{Ew,2}\Delta t_{k}+X_{Ew,3}\Delta G_{rm} \tag{2.71}$$

$$\Delta Q_{c}=X_{Q,1}\Delta t_{c}+X_{Q,2}\Delta t_{k}+X_{Q,3}\Delta G_{rm} \tag{2.72}$$

$$\Delta \mathrm{COP}=X_{cop,1}\Delta t_{c}+X_{cop,2}\Delta t_{k} \tag{2.73}$$

Table 2.5 Coefficients in Eqs. (2.63) through (2.73)

Equation No.	Coefficient expression
Eq. (2.63)	$T_{cr} = (c_{cr}M_{cr})_o$; $X_{cr,1} = \left[\left(\frac{\partial h_{r,2}}{\partial t_c}\right)_o - \left(\frac{\partial h_{r,3}}{\partial t_c}\right)_o\right](G_{rm})_o + A_{co}(t_{cg} - t_c)_o\left(\frac{\partial a_{co}}{\partial t_c}\right)_o - (a_{co}A_{co})_o$; $X_{cr,2} = A_{co}(t_{cg} - t_c)_o\left(\frac{\partial a_{co}}{\partial t_{cg}}\right)_o + (a_{co}A_{co})_o$; $X_{cr,3} = (h_{r,2} - h_{r,3})_o$
Eq. (2.64)	$T_{cw} = c_w M_{cw}/2$; $X_{cw,1} = -c_w(G_{cw,E})_o - A_{cw}(a_{cw})_o/2$; $X_{cw,2} = A_{cw}(a_{cw})_o$; $X_{cw,3} = c_w(G_{cw,E})_o - A_{cw}(a_{cw})_o/2$; $X_{cw,4} = c_w(t_{cw,E} - t_{cw,L})_o + A_{cw}(t_{cg} - \frac{t_{cw,E} + t_{cw,L}}{2})_o\left(\frac{\partial a_{cw}}{\partial G_{cw,E}}\right)_o$; $\xi_{\Delta t_{cw,L}} = -\frac{1}{2}c_w M_{cw}\frac{\partial \Delta t_{cw,E}}{\partial \tau}$
Eq. (2.65)	$T_{cg} = c_{cg}M_{cg}$; $X_{cg,1} = A_{co}(t_c - t_{cg})_o\left(\frac{\partial a_{co}}{\partial t_c}\right)_o + (a_{co}A_{co})_o$; $X_{cg,2} = X_{cg,4} = A_{cw}(a_{cw})/2_o$; $X_{cg,3} = A_{co}(t_c - t_{cg})_o\left(\frac{\partial a_{co}}{\partial t_{cg}}\right)_o - (a_{cw}A_{cw} + a_{co}A_{co})_o$; $X_{cg,5} = -A_{cw}(t_{cg} - \frac{t_{cw,E} + t_{cw,L}}{2})_o\left(\frac{\partial a_{cw}}{\partial G_{cw,E}}\right)_o$
Eq. (2.66)	$T_{er} = (c_{er}M_{er})_o$; $X_{er,1} = -\left(\frac{\partial h_{r,1}}{\partial t_k}\right)_o(G_{rm})_o + A_{eo}(t_{eg} - t_k)_o\left(\frac{\partial a_{eo}}{\partial t_k}\right)_o - (a_{eo}A_{eo})_o$; $X_{er,2} = A_{eo}(t_{eg} - t_k)_o\left(\frac{\partial a_{eo}}{\partial t_{eg}}\right)_o + (a_{eo}A_{eo})_o$; $X_{er,3} = (h_{r,4} - h_{r,1})_o$;
Eq. (2.67)	$T_{ew} = c_w M_{ew}/2$; $X_{ew,1} = -c_w(G_{ew,E})_o - \frac{A_{ew}}{2}(a_{ew})_o$; $X_{ew,2} = A_{ew}(a_{ew})_o$; $X_{ew,3} = c_w\rho_w(G_{ew,E})_o - \frac{A_{ew}}{2}(a_{ew})_o$; $X_{ew,4} = c_w(t_{ew,E} - t_{ew,L})_o + A_{ew}(t_{eg} - \frac{t_{ew,E} + t_{ew,L}}{2})_o\left(\frac{\partial a_{ew}}{\partial G_{ew,E}}\right)_o$; $\xi_{\Delta t_{ew,L}} = -\frac{1}{2}c_w M_{ew}\frac{\partial \Delta t_{ew,E}}{\partial \tau}$
Eq. (2.68)	$T_{eg} = c_{eg}M_{eg}$; $X_{e.g.,1} = A_{eo}(t_k - t_{eg})_o\left(\frac{\partial a_{eo}}{\partial t_k}\right)_o + (a_{eo}A_{eo})_o$; $X_{e.g.,2} = X_{e.g.,4} = \frac{A_{ew}}{2}(a_{ew})_o$; $X_{e.g.,3} = A_{eo}(t_k - t_{eg})_o\left(\frac{\partial a_{eo}}{\partial t_{eg}}\right)_o - (a_{ew}A_{ew} + a_{eo}A_{eo})_o$; $X_{e.g.,5} = -A_{ew}(t_{eg} - \frac{t_{ew,E} + t_{ew,L}}{2})_o\left(\frac{\partial a_{ew}}{\partial G_{ew,E}}\right)_o$
Eq. (2.71)	$X_{Ew,1} = (G_{rm})_o\left(\frac{\partial f_{N_{wom}}(t_k, t_c)}{\partial t_c}\right)_o$; $X_{Ew,2} = (G_{rm})_o\left(\frac{\partial f_{N_{com}}(t_k, t_c)}{\partial t_k}\right)_o$; $X_{Ew,3} = (f_{N_{com}}(t_k, t_c))_o$
Eq. (2.72)	$X_{Q,1} = (G_{rm})_o\left(\frac{\partial f_{Q_c}(t_k, t_c)}{\partial t_c}\right)_o$; $X_{Q,2} = (G_{rm})_o\left(\frac{\partial f_{Q_c}(t_k, t_c)}{\partial t_k}\right)_o$; $X_{Q,3} = (f_{Q_c}(t_k, t_c))_o$
Eq. (2.73)	$X_{cop,1} = \left[\frac{\partial f_{COP}(t_k, t_c)}{\partial t_c}\right]_o$; $X_{cop,2} = \left[\frac{\partial f_{COP}(t_k, t_c)}{\partial t_k}\right]_o$

All coefficients in the above equations are listed in Table 2.5. Combining Eqs. (2.63) through (2.73), the state-space equation for the vapor-compression packaged liquid chiller can be expressed as follows:

$$\Delta\dot{x}_{chiller} = A_{chiller} \cdot \Delta x_{chiller} + B_{chiller} \cdot \Delta u_{chiller} + \xi \tag{2.74}$$

$$\Delta y_{\text{chiller}} = C_{\text{chiller}} \cdot \Delta x_{\text{chiller}} + D_{\text{chiller}} \cdot \Delta u_{\text{chiller}} \tag{2.75}$$

Let $\Delta X_{\text{chiller}} = \Delta x_{\text{chiller}} + A_{\text{chiller}}^{-1} \xi_{\text{chiller}}$, then Eqs. (2.74) and (2.75) can be converted into a standard state-space form as follows:

$$\Delta \dot{X}_{\text{chiller}} = A_{\text{chiller}} \Delta X_{\text{chiller}} + B_{\text{chiller}} \Delta u_{\text{chiller}} \tag{2.76}$$

$$\Delta y_{\text{chiller}} = C_{\text{chiller}} \Delta X_{\text{chiller}} + D_{\text{chiller}} \Delta u_{\text{chiller}} - C_{\text{chiller}} A_{\text{chiller}}^{-1} \xi_{\text{chiller}} \tag{2.77}$$

where

$$\begin{aligned}
\Delta x_{\text{chiller}} &= [\Delta t_{\text{c}}, \Delta t_{\text{cw,L}}, \Delta t_{\text{cg}}, \Delta t_{\text{k}}, \Delta t_{\text{ew,L}}, \Delta t_{\text{eg}}]^T; \\
\Delta y_{\text{chiller}} &= [\Delta t_{\text{cw,L}}, \Delta G_{\text{cw,L}}, \Delta t_{\text{ew,L}}, \Delta G_{\text{ew,L}}, \Delta N_{\text{com}}, \Delta Q_{\text{c}}, \Delta \text{COP}]^T; \\
\Delta u_{\text{chiller}} &= [\Delta t_{\text{cw,E}}, \Delta G_{\text{cw,E}}, \Delta t_{\text{ew,E}}, \Delta G_{\text{ew,E}}, \Delta G_{\text{rm}}]^T; \\
\xi &= [0, \xi_{\Delta t_{\text{cw,L}}}, 0, 0, \xi_{\Delta t_{\text{ew,L}}}, 0]^T;
\end{aligned}$$

$$A_{\text{chiller}} = \begin{bmatrix} \frac{X_{\text{cr,1}}}{T_{\text{cr}}} & 0 & \frac{X_{\text{cr,2}}}{T_{\text{cr}}} & 0 & 0 & 0 \\ 0 & \frac{X_{\text{cw,1}}}{T_{\text{cw}}} & \frac{X_{\text{cw,2}}}{T_{\text{cw}}} & 0 & 0 & 0 \\ \frac{X_{\text{cg,1}}}{T_{\text{cg}}} & \frac{X_{\text{cg,2}}}{T_{\text{cg}}} & \frac{X_{\text{cg,3}}}{T_{\text{cg}}} & 0 & 0 & 0 \\ \frac{X_{\text{er,4}}}{T_{\text{er}}} & 0 & 0 & \frac{X_{\text{er,1}}}{T_{\text{er}}} & 0 & \frac{X_{\text{er,2}}}{T_{\text{er}}} \\ 0 & 0 & 0 & 0 & \frac{X_{\text{ew,1}}}{T_{\text{ew}}} & \frac{X_{\text{ew,2}}}{T_{\text{ew}}} \\ 0 & 0 & 0 & \frac{X_{\text{e.g.,1}}}{T_{\text{eg}}} & \frac{X_{\text{e.g.,2}}}{T_{\text{eg}}} & \frac{X_{\text{e.g.,3}}}{T_{\text{eg}}} \end{bmatrix}; B_{\text{chiller}} = \begin{bmatrix} 0 & 0 & 0 & 0 & \frac{X_{\text{cr,3}}}{T_{\text{cr}}} \\ \frac{X_{\text{cw,3}}}{T_{\text{cw}}} & \frac{X_{\text{cw,4}}}{T_{\text{cw}}} & 0 & 0 & 0 \\ \frac{X_{\text{cg,4}}}{T_{\text{cg}}} & \frac{X_{\text{cg,5}}}{T_{\text{cg}}} & 0 & 0 & 0 \\ 0 & 0 & 0 & 0 & \frac{X_{\text{er,3}}}{T_{\text{er}}} \\ 0 & 0 & \frac{X_{\text{ew,3}}}{T_{\text{ew}}} & \frac{X_{\text{ew,4}}}{T_{\text{ew}}} & 0 \\ 0 & 0 & \frac{X_{\text{e.g.,4}}}{T_{\text{eg}}} & \frac{X_{\text{e.g.,5}}}{T_{\text{eg}}} & 0 \end{bmatrix};$$

$$C_{\text{chiller}} = \begin{bmatrix} 0 & 1 & 0 & 0 & 0 & 0 \\ 0 & 0 & 0 & 0 & 0 & 0 \\ 0 & 0 & 0 & 0 & 1 & 0 \\ 0 & 0 & 0 & 0 & 0 & 0 \\ X_{\text{Ew,1}} & 0 & 0 & X_{\text{Ew,2}} & 0 & 0 \\ X_{\text{Q,1}} & 0 & 0 & X_{\text{Q,2}} & 0 & 0 \\ X_{\text{cop,1}} & 0 & 0 & X_{\text{cop,2}} & 0 & 0 \end{bmatrix}; D_{\text{chiller}} = \begin{bmatrix} 0 & 0 & 0 & 0 & 0 \\ 0 & 1 & 0 & 0 & 0 \\ 0 & 0 & 0 & 0 & 0 \\ 0 & 0 & 0 & 1 & 0 \\ 0 & 0 & 0 & 0 & X_{\text{Ew,3}} \\ 0 & 0 & 0 & 0 & X_{\text{Q,3}} \\ 0 & 0 & 0 & 0 & 0 \end{bmatrix}.$$

By solving the state-space model as in Eqs. (2.76) and (2.77), the response characteristics of evaporating and condensing temperature (Δt_{k} and Δt_{c}), exit coolant temperature of evaporator and condenser ($\Delta t_{\text{cw,L}}$ and $\Delta t_{\text{cw,L}}$), shell wall temperature of evaporator and condenser (Δt_{eg} and Δt_{cg}), input electric power (ΔN_{com}), and cooling output (ΔQ_{c}) as well as COP of chiller can be obtained under different initial conditions and perturbations.

2.2.2.2 Model Validation

(1) Experimental setup

An experimental refrigeration system has been built to validate the state-space dynamic model of chiller, whose schematic diagram is shown in Fig. 2.11. The

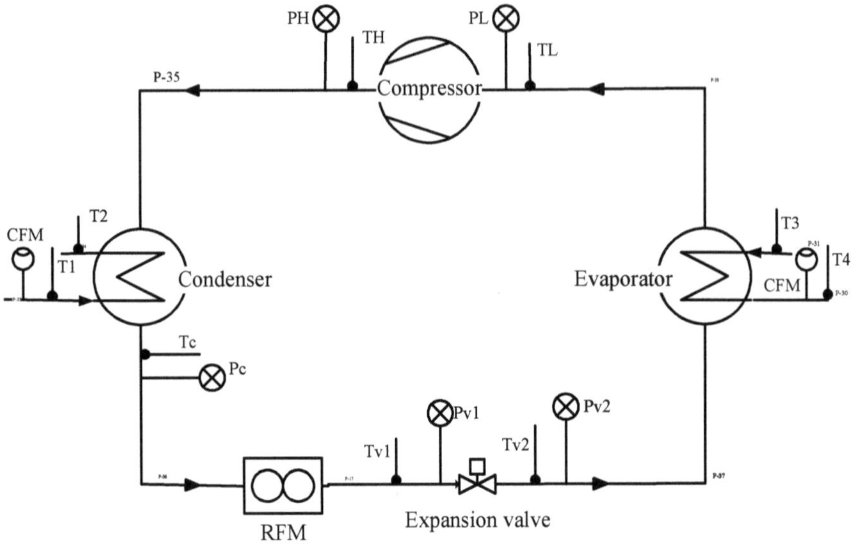

Fig. 2.11 Schematic diagram of experimental setup for the model validation of refrigeration system. *PH* Compressor exhaust pressure sensor; *PL* Compressor inlet pressure sensor; *Pv*1 Expansion valve inlet pressure sensor; *Pv*2 Expansion valve outlet pressure sensor; *Pc* Condenser outlet pressure sensor; *TH* Compressor exhaust temperature sensor; *TL* Compressor inlet temperature sensor; *Tv*1 Expansion valve inlet temperature sensor; *Tv*2 Expansion valve outlet temperature sensor; *Tc* Condenser outlet temperature sensor; *T*1 Condenser inlet coolant temperature sensor; *T*2 Condenser outlet coolant temperature sensor; *T*3 Evaporator inlet cold carrier temperature sensor; *T*4 Evaporator outlet cold carrier temperature sensor; *RFM* Refrigerant flowmeter; *CFM* Coolant or cold carrier flowmeter

condenser and evaporator of the system are of plate-type heat exchanger, and the main structural parameters are listed in Table 2.6. The system uses R-134a (which is admitted to be environmentally friendly) as the working refrigerant. The model coefficients for calculating the thermal properties of the refrigerant R-134a are determined based on the REFPROP 7.01 (a popular calculation program for the refrigerants), which are listed in Table 2.7 [14]. A kind of heat conduction oil is employed as the coolant of the condenser and the cold carrier of the evaporator. The thermal conductivity and the specific heat of the oil versus its temperature ranging from −40 to +60 °C are shown in Fig. 2.12. The density and kinematics viscosity of the heat conduction oil change very small within the temperature range of −40 to +60 °C, which is assumed to be constant (1084 kg/m^3 and 6.5×10^{-6} m^2/s, respectively) in the dynamic response simulation.

The measurement parameters are mainly the refrigerant temperature and pressure at key states of the refrigerant cycle, the inlet and exit coolant (or cold carrier) temperature of the condenser (or the evaporator), the shell wall temperatures of the condenser, and evaporator as well as the corresponding mass flow rate (see Fig. 2.11 for detail). The fluid temperatures are measured by Pt1000 sensors (measurement precision: ±0.1 °C), and the shell wall temperatures are measured by

Table 2.6 Basic information about structural parameters of condenser and evaporator in the experimental chiller system

Structural parameters for the condenser			
Equivalent diameter of channel for coolant (m)	0.0008	Heat transfer area on the coolant side (m^2)	18.33
Heat transfer area on the refrigerant side (m^2)	12.15	Total cross-sectional area of channel for coolant (m^2)	0.006
Coolant volume in the condenser (m^3)	0.0085	Liquid refrigerant volume in the condenser (m^3)	0.0071
Mass of the condenser (kg)	10.86	Specific heat of the condenser (J/(kg °C))	880
Structural parameters for the evaporator			
Equivalent diameter of channel for cold carrier (m)	0.0008	Heat transfer area on the cold carrier side (m^2)	8.87
Heat transfer area on the refrigerant side (m^2)	6.24	Total cross-sectional area of channel for cold carrier (m^2)	0.004
Cold carrier volume in the evaporator (m^3)	0.0038	Liquid refrigerant volume in the evaporator (m^3)	0.0025
Mass of the evaporator (kg)	5.42	Specific heat of the evaporator (J/(kg °C))	880

Table 2.7 Model coefficients for the calculation of thermal properties of the refrigerant R-134a

A	24.8033908	c_3	−0.00000202
A_2	−0.1195051	c_4	0
A_3	0.00014478	c_5	15.8217
A_4	−0.000000105	D	0.0000225
A_5	−6.95E−12	D_1	819.6183
B	−3980.408	D_2	1023.582
B_2	0.000113759	D_3	−1156.757
B_3	−0.0000000894	D_4	789.7191
B_4	0	D_5	0
B_5	1.27E−13	D_6	0
b	0.000345547	E	0.1995548
C	−0.02405332	F	374.8473
C_2	−3.531592	G	0
C_3	0.006469248	K_1	5.475
C_4	0	h_0 (kJ/kg)	209.092133
C_5	−2.05E−9	p_{atm} (kPa)	101.325
c_1	−0.005257455	R (kJ/kg)	0.081488163
c_2	0.00329657	T_c (K)	374.25

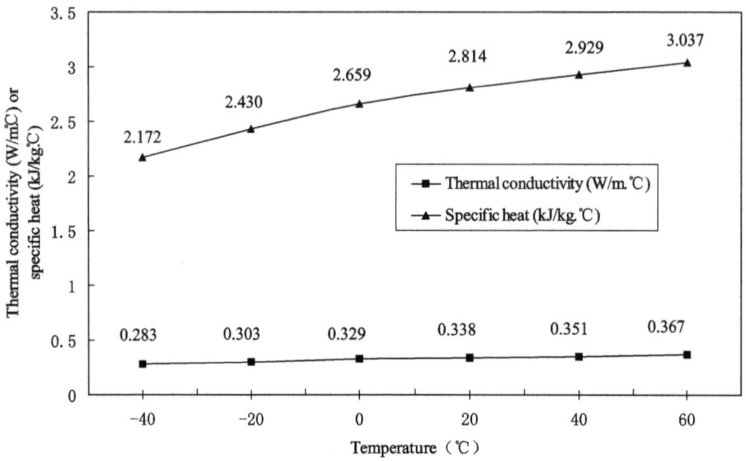

Fig. 2.12 Thermal properties versus temperature of the cooling and cold carrier fluid

copper-constantan thermocouples (measurement precision: ±0.2 °C) which have a direct contact with the exterior surface of the heat exchangers. The pressure sensors are of AKS32 type (measurement precision: (20 ± 0.3 % full scale) Pa). The meter for the measurement of refrigerant mass flow rate is of LSLW type (measurement precision: 0.5 % of full scale) and that for the coolant and cold carrier mass flow rate is of FVC one (measurement precision: 0.5 % of full scale). All the measured parameters were collected automatically by the Keithley 2700 data acquisition system connected with a computer, and the sampling interval was set as 2 s.

(2) Methods for experimental validation

In real situations, the exit cold carrier temperature of evaporator is the key control parameter we care the most. Therefore, the state-space model of chiller is experimentally validated mainly in terms of the transient responses of the exit cold carrier temperature of evaporator under certain initial conditions and perturbations that are chosen arbitrarily.

To begin with, the empirical constants in Eqs. (2.52)–(2.54) need to be determined through experiments by using the following equations:

$$G_{\mathrm{rm}}(h_2 - h_3) = a_{\mathrm{co}}A_{\mathrm{co}}(t_{\mathrm{c}} - t_{\mathrm{cg}}) = C_1A_{\mathrm{co}}(t_{\mathrm{c}} - t_{\mathrm{cg}})^{n_1+1} \tag{2.78}$$

$$G_{\mathrm{rm}}(h_4 - h_1) = a_{\mathrm{eo}}A_{\mathrm{eo}}(t_{\mathrm{eg}} - t_{\mathrm{k}}) = C_2A_{\mathrm{eo}}(t_{\mathrm{eg}} - t_{\mathrm{k}})^{n_2+1} \tag{2.79}$$

$$\begin{aligned} c_{\mathrm{cw}}G_{\mathrm{cw}}(t_{\mathrm{cw,L}} - t_{\mathrm{cw,E}}) &= a_{\mathrm{cw}}A_{\mathrm{cw}}\left[t_{\mathrm{cg}} - (t_{\mathrm{cw,L}} + t_{\mathrm{cw,E}})/2\right] \\ &= A_{\mathrm{cw}}\frac{C_3\lambda_{\mathrm{w}}}{d_{\mathrm{ci}}}\left[t_{\mathrm{cg}} - (t_{\mathrm{cw,L}} + t_{\mathrm{cw,E}})/2\right]Re_{\mathrm{d}}^{\,n_3} \end{aligned} \tag{2.80}$$

As shown in Eqs. (2.78)–(2.80), the measured parameters include the temperature and pressure of refrigerant at the key states (states '1,' '2,' '3,' and '4'), the inlet and exit coolant temperature of condenser ($t_{cw,E}, t_{cw,L}$), the inlet and exit cold carrier temperature of evaporator ($t_{ew,E}, t_{ew,L}$), and the shell wall temperature of condenser and evaporator (t_{eg}, t_{cg}) as well as the mass flow rate of refrigerant, coolant, and cold carrier (G_{rm}, G_{cw}, G_{ew}). All these parameters are measured under different steady conditions in which G_{rm} and G_{cw} ranges, respectively, from 0.05 to 0.30 kg/s and from 1.0 to 2.5 kg/s.

To obtain the empirical constants, $C_1, C_2, C_3, n_1, n_2\, and\, n_3$, Eqs. (2.78)–(2.80) can be written, respectively, as follows:

$$Ln[G_{rm}(h_2 - h_3)] = Ln(C_1 A_{co}) + (n_1 + 1)Ln(t_c - t_{cg}) \tag{2.81}$$

$$Ln[G_{rm}(h_4 - h_1)] = Ln(C_2 A_{eo}) + (n_2 + 1)Ln(t_{eg} - t_k) \tag{2.82}$$

$$Ln\left[c_{cw} G_{cw}(t_{cw,L} - t_{cw,E})\right] = Ln\left[A_{cw}\frac{C_3 \lambda_w}{d_{ci}}\left[t_{cg} - \left(t_{cw,L} + t_{cw,E}\right)/2\right]\right] + n_3 Ln Re_d \tag{2.83}$$

The constants, $C_1, C_2, C_3, n_1, n_2\, and\, n_3$, can be determined, respectively, according to the slope and intercept of the straight line of $Ln[G_{rm}(h_2 - h_3)]$ versus $Ln(t_c - t_{cg})$, $Ln[G_{rm}(h_4 - h_1)]$ versus $Ln(t_{eg} - t_k)$ and $Ln\left[c_{cw} G_{cw}(t_{cw,L} - t_{cw,E})\right]$ versus $Ln Re_d$ fitting to the experimental data. The results are listed as follows: $C_1 = 2558$; $C_2 = 3056$; $n_1 = -0.25$; $n_2 = 0.25$; $C_3 = 0.05$; $n_3 = 0.8$.

Similarly, the average error (AE, Eq. (2.39)) is employed as well to evaluate the validity of the model results compared with the experimental data during the transient response process.

(3) Experimental results

Three experimental cases have been performed for the model validation. The initial conditions of the three cases are listed in Table 2.8, which were measured under the steady states before the input perturbations began. The perturbations corresponding to the three experimental cases are shown in Fig. 2.13, respectively. The perturbations on the inlet cold carrier temperature of evaporator were achieved through adjusting the heat load simulated by an electric heating box, and that on the inlet coolant temperature of condenser were done through changing the cooling capacity of the cooling tower. The refrigerant mass flow rate is closely related to the compressor rotation speed which can be adjusted by a compatible frequency converter.

Figure 2.14 shows the comparisons of the theoretical simulations with the experimental data on the transient responses of the exit cold carrier temperature of evaporator under the perturbations of case *I* (see Fig. 2.13). It is easily understood that the exit cold carrier temperature of evaporator will increase with the inlet cold

Table 2.8 Initial conditions for model validation and simulation

Initial conditions	Experimental cases		
	Case I	Case II	Case III
Evaporating temperature $(t_k)_o$ (°C)	14.5	15.9	14.5
Inlet cold carrier temperature of evaporator $(t_{ew,E})_o$ (°C)	24.8	24.8	25.4
Outlet cold carrier temperature of evaporator $(t_{ew,L})_o$ (°C)	19.5	19.9	19.7
Shell wall temperature of evaporator $(t_{eg})_o$ (°C)	22.5	22.9	22.6
Cold carrier flow rate of evaporator $(G_{ew})_o$ (kg/s)	1.45	1.45	1.45
Condensing temperature $(t_c)_o$ (°C)	36.3	50.1	48.2
Inlet coolant temperature of condenser $(t_{cw,E})_o$ (°C)	29.5	42.2	39.9
Outlet cold carrier temperature of condenser $(t_{cw,L})_o$ (°C)	37.2	49.7	49.4
Coolant flow rate of condenser $(G_{cw})_o$ (kg/s)	1.16	1.05	1.51
Shell wall temperature of condenser $(t_{cg})_o$ (°C)	33.9	42.5	40.9
Compressor inlet temperature $(t_{com,E})_o$ (°C)	24.1	20.2	23.4
Compressor exhaust temperature $(t_{com,L})_o$ (°C)	58.1	51.2	81.5
Refrigerant flow rate $(G_{rm})_o$ (kg/s)	0.116	0.17	0.26

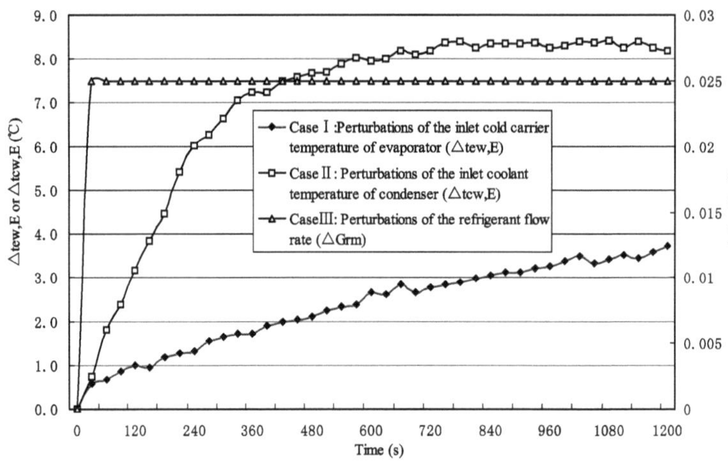

Fig. 2.13 Perturbations for the experimental cases (measured data)

carrier temperature rising. The experimental results manifest that the model can predict well the transient responses of the exit cold carrier temperature of evaporator (the average error (AE) is estimated about 6.4 %) under perturbations of the inlet cold carrier temperature.

Figure 2.15 shows the calculated and experimental data on the transient responses of the exit cold carrier temperature of evaporator under the perturbations of case *II* (see Fig. 2.13). Clearly, the increase in the inlet coolant temperature of

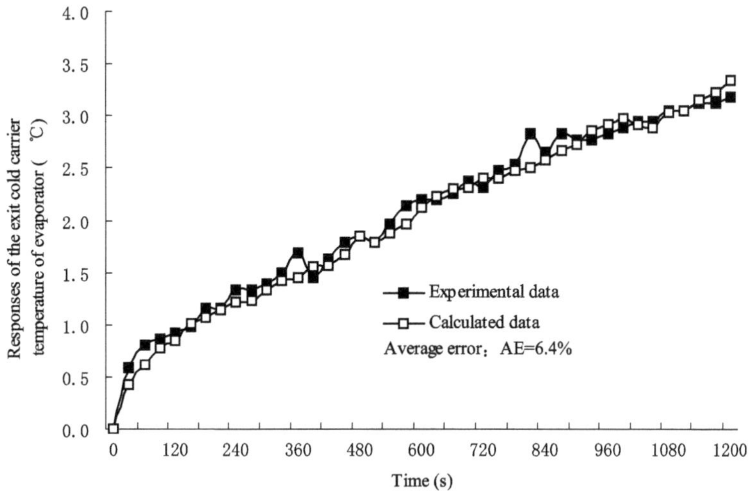

Fig. 2.14 Responses of the exit cold carrier temperature of evaporator ($\Delta t_{ew,L}$) under the perturbations of case I (calculated results vs. experimental data)

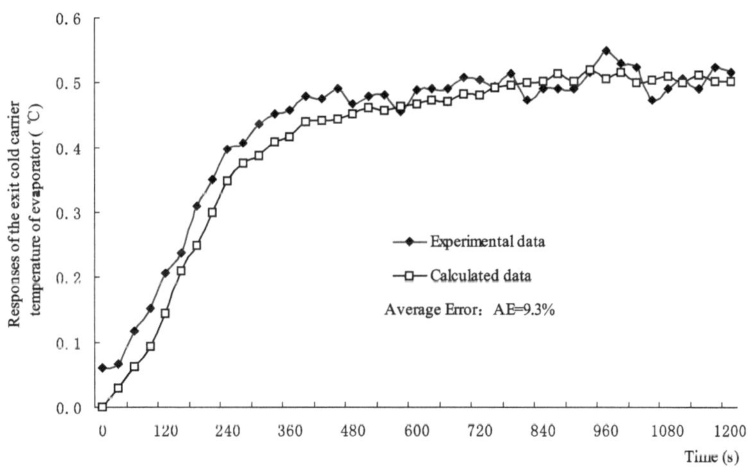

Fig. 2.15 Responses of the exit cold carrier temperature of evaporator ($\Delta t_{ew,L}$) under the perturbations of case II (calculated results vs. experimental data)

condenser will cause the condensing temperature to rise and decrease the chiller's cooling capacity. As a result, the exit cold carrier temperature of evaporator will increase. The average error (AE) of the model results compared with the experimental data for the experimental case *II* is about 9.3 %, which also verifies the

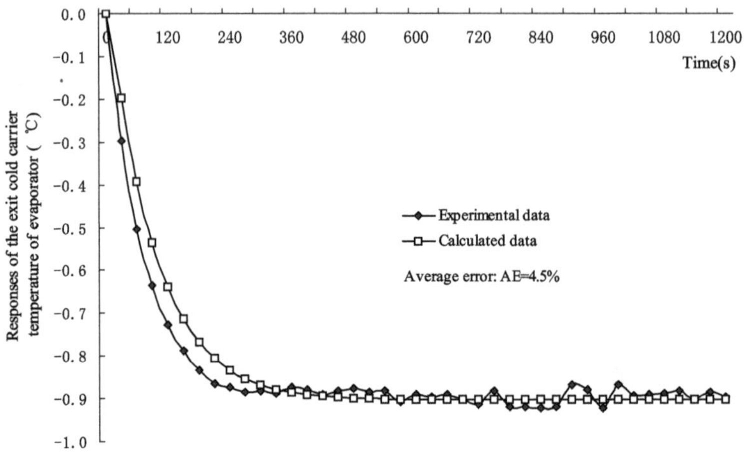

Fig. 2.16 Responses of the exit cold carrier temperature of evaporator ($\Delta t_{ew,L}$) under the perturbations of case III (calculated results vs. experimental data)

effectiveness of the model in predicting the transient responses of the exit cold carrier temperature of evaporator under perturbations of the inlet coolant temperature of condenser.

Figure 2.16 shows the calculated and experimental data on the transient responses of the exit cold carrier temperature of evaporator under the perturbations of case *III* (see Fig. 2.13). It is reasonable to see that the exit cold carrier temperature of evaporator will decrease with the increase in the refrigerant flow rate since the chiller's output cooling capacity increases under such circumstance. The average error (AE) of the model simulation in the experimental case *III* is about 4.5 %, which validates as well the goodness of the proposed model for predicting the transient responses of the exit cold carrier temperature of evaporator under perturbations of the refrigerant flow rate.

As seen from the experimental study cases illustrated above, the errors of model simulation compared with the experimental data are all less than 10 %. Therefore, we can safely conclude that the proposed state-space model is capable of studying the transient response characteristics and dynamic performance of the vapor-compression liquid refrigeration system under different ambient perturbations.

2.2.3 Cooling Tower

A HVAC (heating, ventilating, and air-conditioning) cooling tower is used to dispose of ('reject') unwanted heat from a chiller. Water-cooled chillers are normally more energy efficient than air-cooled chillers due to heat rejection to tower water at or near wet-bulb temperatures. With respect to the heat transfer mechanism employed, the main types are as follows: ① dry-cooling towers operate by heat transfer through a surface that separates the working fluid from ambient air, such as in a tube to air heat exchanger, utilizing convective heat transfer. They do not use evaporation; ② fluid coolers (or closed-circuit cooling towers) are hybrids that pass the working fluid through a tube bundle, upon which clean water is sprayed and a fan-induced draft applied;③ wet-cooling towers (or open-circuit cooling towers) operate on the principle of evaporative cooling. Since the wet-cooling towers have the highest heat transfer performance, and the warm water can be cooled to a temperature lower than the ambient air dry-bulb temperature, the wet-cooling towers are used in a HVAC system more frequently. To achieve better performance (more cooling), a medium called 'fill' is used to increase the surface area and the time of contact between the air and water flows to improve heat transfer.

There are two types of configurations for the wet-cooling tower, one is cross-flow and the other is counterflow. Cross-flow is a design in which the airflow is directed perpendicular to the water flow. Airflow enters one or more vertical faces of the cooling tower to meet the fill material. Water flows (perpendicular to the air) through the fill by gravity. The air continues through the fill and thus past the water flow into an open plenum volume. Lastly, a fan forces the air out into the atmosphere. In a counterflow design, the airflow is directly opposite to the water flow. Airflow first enters an open area beneath the fill media and is then drawn up vertically. The water is sprayed through pressurized nozzles near the top of the tower and then flows downward through the fill, opposite to the airflow. Advantages of the counterflow design include the following: ① Spray water distribution makes the tower more freeze-resistant and ② breakup of water in spray makes heat transfer more efficient. In the following section, the counterflow wet-cooling tower is taken as an example for the state-space modeling.

2.2.3.1 Model Development

1) Assumptions and fundamental equations

As shown in Fig. 2.17, a counterflow wet-cooling tower normally consists of two sections: The upper is fill and the lower is sink. The main assumptions for modeling are as follows:

(1) Moisture air is treated as a mixture of ideal gases;
(2) The exterior wall of the cooling tower is thermally adiabatic;
(3) The single-film theory is applied to the heat and mass transfer between the air and the water film on the fill;
(4) Air temperature and humidity as well as water temperature change linearly from inlet to outlet of the cooling tower, and they are, respectively, mean value of inlet and outlet.

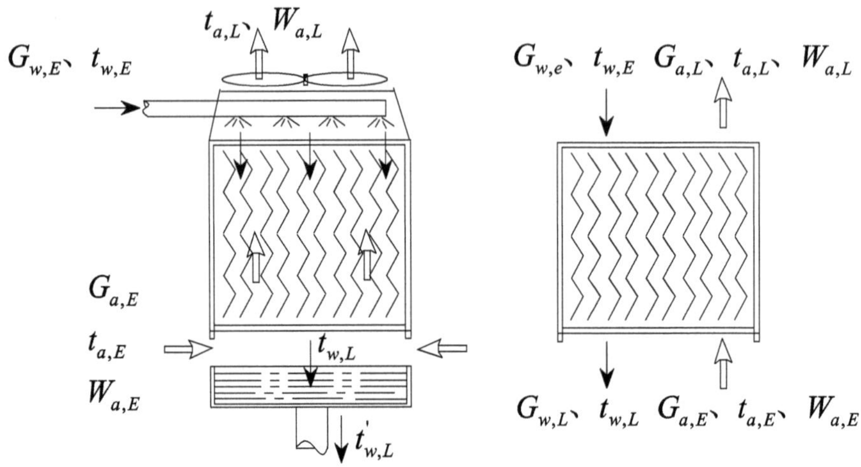

Fig. 2.17 Schematic diagram for counterflow wet-cooling tower

By using the first law of energy and mass conservation, the following equations can be obtained for the counterflow wet-cooling tower.

(1) *Mass and energy equation for air passing through fill of cooling tower*

$$G_{\mathrm{a,E}} = G_{\mathrm{a,L}} = G_{\mathrm{a}} \tag{2.84}$$

$$\begin{aligned}\frac{1}{2}\rho_{\mathrm{a}}\varepsilon_{\mathrm{a}}A_{\mathrm{ta}}H_{\mathrm{t}}\frac{\mathbf{d}(h_{\mathrm{a,E}}+h_{\mathrm{a,L}})}{\mathbf{d}\tau} &= G_{\mathrm{a,E}}(h_{\mathrm{a,E}}-h_{\mathrm{a,L}}) + a_{\mathrm{wa}}A_{\mathrm{t}}\left(\frac{t_{\mathrm{w,E}}+t_{\mathrm{w,L}}}{2}-\frac{t_{\mathrm{a,E}}+t_{\mathrm{a,L}}}{2}\right)\\ &\quad + q_{\mathrm{r}}\lambda_{\mathrm{m}}A_{\mathrm{t}}\left(W_{\mathrm{tb}}-\frac{W_{\mathrm{a,E}}+W_{\mathrm{a,L}}}{2}\right)\end{aligned} \tag{2.85}$$

(2) *Mass equation for air moisture*

$$\frac{1}{2}\rho_{\mathrm{a}}\varepsilon_{\mathrm{a}}A_{\mathrm{ta}}H_{\mathrm{t}}\frac{\mathbf{d}(W_{\mathrm{a,L}}+W_{\mathrm{a,E}})}{\mathbf{d}\tau} = G_{\mathrm{a,E}}(W_{\mathrm{a,E}}-W_{\mathrm{a,L}}) + \lambda_{\mathrm{m}}A_{\mathrm{t}}\left(W_{\mathrm{tb}}-\frac{W_{\mathrm{a,E}}+W_{\mathrm{a,L}}}{2}\right) \tag{2.86}$$

(3) *Mass and energy equation for water passing through fill of cooling tower*

$$G_{\mathrm{w,E}} = G_{\mathrm{w,L}} = G_{\mathrm{w}} \tag{2.87}$$

$$\frac{1}{2}c_{\mathrm{w}}G_{\mathrm{w,E}}\frac{H_{\mathrm{t}}}{u_{\mathrm{w}}}\frac{\mathbf{d}(t_{\mathrm{w,L}}+t_{\mathrm{w,E}})}{\mathbf{d}\tau}=c_{\mathrm{w}}G_{\mathrm{w,E}}(t_{\mathrm{w,E}}-t_{\mathrm{w,L}})+a_{\mathrm{wa}}A_{\mathrm{t}}\left(\frac{t_{\mathrm{a,E}}+t_{\mathrm{a,L}}}{2}-\frac{t_{\mathrm{w,E}}+t_{\mathrm{w,L}}}{2}\right)+q_{\mathrm{r}}\lambda_{\mathrm{m}}A_{\mathrm{t}}\left(\frac{W_{\mathrm{a,E}}+W_{\mathrm{a,L}}}{2}-W_{\mathrm{tb}}\right) \tag{2.88}$$

(4) *Mass equation for water in sink of cooling tower*

$$M_{\mathrm{w,pt}}\frac{\mathbf{d}t'_{\mathrm{w,L}}}{\mathbf{d}\tau}=G_{\mathrm{w,E}}(t_{\mathrm{w,L}}-t'_{\mathrm{w,L}}) \tag{2.89}$$

In Eqs. (2.84) through (2.89), the symbol 'G' stands for flow rate, kg/s; 'ε_{a}'for air volume ratio in cooling tower; 'A_{ta}' for cross-sectional area of cooling tower, m^2; 'A_{t}' for surface area between air and water film in cooling tower, m^2; 'H_{t}' for height of cooling tower, m; 'u_{w}' for falling velocity of water, m/s; 'λ_{m}' for mass transfer coefficient, kg/ ($\mathrm{m}^{2/}$s); and 'a_{wa}' for heat transfer coefficient between air and water film in cooling tower, W/(m^2 °C);

2) Key parameters

According to the single-film theory, the saturated humidity of air on the water film surface of fill (W_{tb}) can be expressed by Eq. (2.90), a function dependent on the average temperature of the water film.

$$W_{\mathrm{tb}}\approx\beta_1\frac{t_{\mathrm{w,E}}+t_{\mathrm{w,L}}}{2} \tag{2.90}$$

The convective heat transfer coefficient (a_{wa}) and convective mass transfer coefficient (λ_{m}) can refer to the correlation equation of heat and mass transfer for plate flow as follows [15]:

$$Nu_{\mathrm{a}}=\frac{a_{\mathrm{wa}}H_{\mathrm{t}}}{\lambda_{\mathrm{a}}}=0.648Re_{\mathrm{a}}^{0.5}\mathrm{Pr}_{\mathrm{a}}^{1/3} \tag{2.91}$$

$$Sh_{\mathrm{m}}=\frac{\lambda_{\mathrm{m}}H_{\mathrm{t}}}{\rho_{\mathrm{a}}D_{\mathrm{w,a}}}=0.648Re_{\mathrm{a}}^{0.5}Sc_{\mathrm{j}}^{1/3} \tag{2.92}$$

where $Re_{\mathrm{a}}=\frac{H_{\mathrm{t}}(u_{\mathrm{a}}+u_{\mathrm{w}})}{v_{\mathrm{a}}}$; H_{t} is height of film in cooling tower, *m*; u_{w} is mean gliding speed of water film on the fill, which can be calculated by Eq. (2.93)

$$u_{\mathrm{w}}=\frac{1}{H_{\mathrm{t}}}\int_0^{H_{\mathrm{t}}}\sqrt{u_{\mathrm{o}}^2+2gH_{\mathrm{t}}dy} \tag{2.93}$$

where u_o is spraying velocity at the spray nozzle, *m/s*. Please note that drag forces of water film is neglected in Eq. (2.93).

3) State-space representation

Through linearization, Eqs. (2.84) through (2.89) can be written as follows:

$$T_{t,ta}\frac{d\Delta t_{a,L}}{d\tau} = X_{t,1}\Delta t_{a,L} + X_{t,2}\Delta W_{a,L} + X_{t,3}\Delta t_{w,L} + X_{t,4}\Delta t_{a,E} + X_{t,5}\Delta t_{w,E} + X_{t,6}W_{a,E} + X_{t,7}\Delta G_{a,E} + \xi_{t_{a,L}} \tag{2.94}$$

$$T_{t,wa}\frac{d\Delta W_{a,L}}{d\tau} = Y_{t,1}\Delta W_{a,L} + Y_{t,2}\Delta t_{w,L} + Y_{t,3}\Delta t_{w,E} + Y_{t,4}\Delta W_{a,E} + Y_{t,5}\Delta G_{a,E} + \xi_{W_{a,L}} \tag{2.95}$$

$$T_{t,tw}\frac{d\Delta t_{w,L}}{d\tau} = Z_{t,1}\Delta t_{a,L} + Z_{t,2}\Delta W_{a,L} + Z_{t,3}\Delta t_{w,L} + Z_{t,4}\Delta t_{a,E} + Z_{t,5}\Delta t_{w,E} + Z_{t,6}W_{a,E} + Z_{t,7}\Delta G_{a,E} + Z_{t,8}\Delta G_{w,E} + \xi_{t_{w,L}} \tag{2.96}$$

$$M_{w,pt}\frac{d\Delta t'_{w,L}}{d\tau} = M_{t,1}\Delta t_{w,L} + M_{t,2}\Delta t'_{w,L} + M_{t,3}\Delta G_{w,E} \tag{2.97}$$

$$\Delta G_{a,L} = \Delta G_{a,E} \tag{2.98}$$

$$\Delta G_{w,L} = \Delta G_{w,E} \tag{2.99}$$

The coefficients in Eqs. (2.94)–(2.97) are described in Table 2.9.

Choosing the variables, $\Delta t_{a,L}, \Delta W_{a,L}, \Delta t_{w,L}, \Delta t'_{w,L}$, in the differential term as state ones, the variables, $\Delta t_{a,E}, \Delta W_{a,E}, \Delta G_{a,E}, \Delta t_{w,E}, \Delta G_{w,E}$, as input ones, and the variables, $\Delta t_{a,L}, \Delta W_{a,L}, \Delta G_{a,L}, \Delta t'_{w,L}, \Delta G_{w,L}$, as output ones, Eqs. (2.94)–(2.99) can be expressed in the form of state-space representation as follows:

$$\dot{x}_{tower} = A_{tower} \cdot x_{tower} + B_{tower} \cdot u_{tower} + \xi_{tower} \tag{2.100}$$

$$y_{tower} = C_{tower} \cdot x_{tower} + D_{tower} \cdot u_{tower} \tag{2.101}$$

Similarly, the standard state-space form can be deduced from Eqs. (2.100) and (2.101) as follows:

$$\dot{X}_{tower} = A_{tower}X_{tower} + B_{tower}u_{tower} \tag{2.102}$$

$$y_{tower} = C_{tower}X_{tower} + D_{tower}u_{tower} - C_{tower}A^{-1}_{tower}\xi_{tower} \tag{2.103}$$

Table 2.9 Coefficients in Eqs. (2.95)–(2.98)

Equation No.	Coefficient expression
Eq. (2.94)	$T_{t,ta}=\frac{1}{2}\varepsilon_a c_a \rho_a A_{t,a} H_t$; $X_{t,1}=-c_a(G_{a,E})_o-A_t(a_{wa})_o/2$; $X_{t,2}=X_{t,6}=-\frac{A_t(q_r-\beta_2)}{2}(\lambda_m)_o$; $X_{t,3}=X_{t,5}=\frac{A_t}{2}\left[(a_{wa})_o+\beta_1(q_r-\beta_2)(\lambda_m)_o\right]$; $X_{t,4}=c_a(G_{a,E})_o-A_t(a_{wa})_o/2$; $X_{t,7}=c_a(t_{a,E}-t_{a,L})_o+\frac{A_t}{2}\left(\frac{\partial a_{wa}}{\partial G_{a,E}}\right)_o(t_{w,L}+t_{w,E}-t_{a,L}-t_{a,E})_o \quad \frac{A_t(q_r-\beta_2)}{2}\left(\frac{\partial \lambda_m}{\partial G_{a,E}}\right)_o[\beta_1(t_{w,E}+t_{w,L})_o-(W_{a,E}+W_{a,L})_o]$; $\xi_{t_{a,L}}=-T_{t,ta}\frac{d\Delta t_{a,E}}{d\tau}$;
Eq. (2.95)	$T_{t,wa}=\frac{1}{2}\varepsilon_a \rho_a A_{t,a} H_t$; $Y_{t,1}=-(G_{a,E})_o-\frac{A_t}{2}(\lambda_m)_o$; $Y_{t,2}=Y_{t,3}=\frac{A_t\beta_1}{2}(\lambda_m)_o$; $Y_{t,4}=(G_{a,E})_o-\frac{A_t}{2}(\lambda_m)_o$; $Y_{t,5}=(W_{a,E}-W_{a,L})_o+\frac{A_t}{2}\left(\frac{\partial \lambda_m}{\partial G_{a,E}}\right)_o[\beta_1(t_{w,L}+t_{w,E})_o-(W_{a,E}+W_{a,L})_o]$; $\xi_{W_{a,L}}=-T_{t,wa}\frac{d\Delta W_{a,E}}{d\tau}$
Eq. (2.96)	$T_{t,tw}=\frac{1}{2}c_w\frac{H_t}{u_w}$; $Z_{t,1}=Z_{t,4}=\frac{A_t}{2(G_{w,E})_o}(a_{wa})_o$; $Z_{t,2}=Z_{t,6}=\frac{A_t q_r}{2(G_{w,E})_o}(\lambda_m)_o$; $Z_{t,3}=-\frac{1}{(G_{w,E})_o}\left(\frac{\beta_1 A_t q_r}{2}(\lambda_m)_o+\frac{A_t}{2}(a_{wa})_o\right)-c_w$; $Z_{t,5}=-\frac{1}{(G_{w,E})_o}\left(\frac{\beta_1 A_t q_r}{2}(\lambda_m)_o+\frac{A_t}{2}(a_{wa})_o\right)+c_w$; $Z_{t,7}=\frac{1}{(G_{w,E})_o}\left[\frac{A_t}{2}\left(\frac{\partial a_{wa}}{\partial G_{a,E}}\right)_o(t_{a,E}+t_{a,L}-t_{w,E}-t_{w,L})_o+\frac{q_r A_t}{2}\left(\frac{\partial \lambda_m}{\partial G_{a,E}}\right)_o[(W_{a,E}+W_{a,L})_o-\beta_1(t_{w,E}+t_{w,L})_o]\right]$; $Z_{t,8}=-\frac{1}{2G_{w,E}^2}\left[(a_{wa})_o A_t\left((t_{a,E}+t_{a,L})_o-(t_{w,E}+t_{w,L})_o\right)+q_r(\lambda_m)_o A_t\left((W_{a,E}+W_{a,L})_o-\beta_1(t_{w,E}+t_{w,L})_o\right)\right]$; $\xi_{t_{w,L}}=-T_{t,tw}\frac{d\Delta t_{w,E}}{d\tau}$;
Eq. (2.97)	$M_{t,1}=-M_{t,2}=(G_{w,E})_o$; $M_{t,3}=(t_{w,L}-t'_{w,L})_o$;

where

$$X_{tower}=x_{tower}+A_{tower}^{-1}\xi_{tower};$$

$$x_{tower}=\left[\Delta t_{a,L},\Delta W_{a,L},\Delta t_{w,L},\Delta t'_{w,L}\right]^T;$$

$$y_{tower}=\left[\Delta t_{a,L},\Delta W_{a,L},\Delta G_{a,L},\Delta t'_{w,L},\Delta G_{w,L}\right]^T;$$

$$u_{tower}=\left[\Delta t_{a,E},\Delta W_{a,E},\Delta G_{a,E},\Delta t_{w,E},\Delta G_{w,E}\right]^T;$$

$$\xi_{tower}=\left[\xi_{t_{a,L}},\xi_{W_{a,L}},\xi_{t_{w,L}},0\right]^T;$$

$$A_{tower}=\begin{bmatrix}\frac{X_{t,1}}{T_{t,ta}} & \frac{X_{t,2}}{T_{t,ta}} & \frac{X_{t,3}}{T_{t,ta}} & 0\\ 0 & \frac{Y_{t,1}}{T_{t,wa}} & \frac{Y_{t,2}}{T_{t,wa}} & 0\\ \frac{Z_{t,1}}{T_{t,tw}} & \frac{Z_{t,2}}{T_{t,tw}} & \frac{Z_{t,3}}{T_{t,tw}} & 0\\ 0 & 0 & \frac{M_{t,1}}{M_{w,pt}} & \frac{M_{t,2}}{M_{w,pt}}\end{bmatrix};B_{tower,a}=\begin{bmatrix}\frac{X_{t,4}}{T_{t,ta}} & \frac{X_{t,6}}{T_{t,ta}} & \frac{X_{t,7}}{T_{t,ta}} & \frac{X_{t,5}}{T_{t,ta}} & \frac{X_{t,8}}{T_{t,ta}}\\ 0 & \frac{Y_{t,4}}{T_{t,wa}} & \frac{Y_{t,5}}{T_{t,wa}} & \frac{Y_{t,3}}{T_{t,wa}} & \frac{Y_{t,6}}{T_{t,wa}}\\ \frac{Z_{t,4}}{T_{t,tw}} & \frac{Z_{t,6}}{T_{t,tw}} & \frac{Z_{t,7}}{T_{t,tw}} & \frac{Z_{t,5}}{T_{t,tw}} & \frac{Z_{t,8}}{T_{t,tw}}\\ 0 & 0 & 0 & 0 & \frac{M_{t,3}}{M_{w,pt}}\end{bmatrix};$$

$$C_{\text{tower}} = \begin{bmatrix} 1 & 0 & 0 & 0 \\ 0 & 1 & 0 & 0 \\ 0 & 0 & 0 & 0 \\ 0 & 0 & 0 & 1 \\ 0 & 0 & 0 & 0 \end{bmatrix}; D_{\text{tower}} = \begin{bmatrix} 0 & 0 & 0 & 0 & 0 \\ 0 & 0 & 0 & 0 & 0 \\ 0 & 0 & 1 & 0 & 0 \\ 0 & 0 & 0 & 0 & 0 \\ 0 & 0 & 0 & 0 & 1 \end{bmatrix}.$$

2.2.3.2 Model Validation

(1) Experimental conditions

A counterflow wet-cooling tower was used for the model validation. The tower is of 1.5 m in diameter and 2.0 m in height. The total surface area of its fill is identified as 285.4 m^2. The test parameters include the following: ①temperature (±0.1 °C in absolute error) and humidity (±0.8 % in absolute error) of air inlet and outlet of cooling tower; ② cross-sectional velocity of air passing through the tower (±2 % of reading data); ③temperature (±0.1 °Cin absolute error) of water inlet and outlet of cooling tower; and ④ water flow rate (±0.5 % of reading data)

Experiments were made on the transient responses of exit air temperature and humidity of cooling tower and exit water temperature of fill and cooling tower when subjected to different disturbances under various initial conditions as shown in Table 2.10. Then, the experimental data were used for the model validation.

(2) Experimental results

For the experimental case I, the water flow rate had a sudden increase of 1.37 kg/s, while the other inlet variables were kept unchanged. The system matrixes

Table 2.10 Initial conditions for cooling tower model validation

Initial conditions	Cases		
	Case I	Case II	Case III
Inlet air temperature $(t_{a,E})_o$ (°C)	32.4	33.1	34.0
Exit air temperature $(t_{a,L})_o$ (°C)	34.3	34.1	34.8
Inlet air humidity $(W_{a,E})_o$ (g/kg dryair)	18.5	19.4	22.4
Exit air humidity $(W_{a,L})_o$ (g/kg dryair)	20.4	21.4	23.8
Airflow rate $(G_a)_o$ (kg/s)	12.62	12.62	12.62
Inlet water temperature $(t_{w,E})_o$ (°C)	40.8	38.8	38.2
Exit water temperature of fill $(t_{w,L})_o$ (°C)	34.5	33.2	32.7
Water flow rate $(G_w)_o$ (kg/s)	3.31	3.31	3.31
Exit water temperature of tower $(t'_{w,L})_o$ (°C)	34.1	32.8	32.3
Mass of water in the sink of tower M_{sump} (kg)	88.5	88.5	88.5

Perturbation for case 1: G_w increases 1.37 kg/s; perturbation for case II: G_a increases 5.60 kg/s; perturbations for case III are shown in Fig. 2.38

of cooling tower ($A_{\text{tower}}, B_{\text{tower}}$) under the initial conditions of experimental case I for the model calculation are counted as follows:

$$A_{\text{tower}} = \begin{bmatrix} -12.0343 & 0.0251 & 2.4203 & 0 \\ 0.0000 & -9.5964 & 0.0002 & 0.0000 \\ 0.5702 & 0.1313 & -3.1097 & 0.0000 \\ 0.0000 & 0.0000 & 0.0374 & -0.0374 \end{bmatrix};$$

$$B_{\text{tower}} = \begin{bmatrix} 7.1580 & 0.0251 & -0.6777 & 2.4203 & 0.0000 \\ 0.0000 & 9.5959 & -0.0013 & 0.0002 & 0.0000 \\ 0.5702 & 0.1313 & -0.2615 & 1.7810 & 3.6122 \\ 0.0000 & 0.0000 & 0.0000 & 0.0000 & 0.0011 \end{bmatrix}.$$

Figure 2.18 compares experimental dada and model results on the transient responses of exit air and water states under the initial conditions and perturbations of case I. The results manifest that the exit air and water temperature of cooling tower will rise as the water flow rate increases, while the impact of water flow rate on the exit air humidity is very small. The response time of exit water temperature of fill is much shorter than that of exit water temperature of tower (or tower's sink), and this is because the quantity of water in the sink is much larger than that in the fill of tower (as well known, a larger mass results in a larger thermal inertia). For the experimental case I, the average error (AE) of model results on the transient

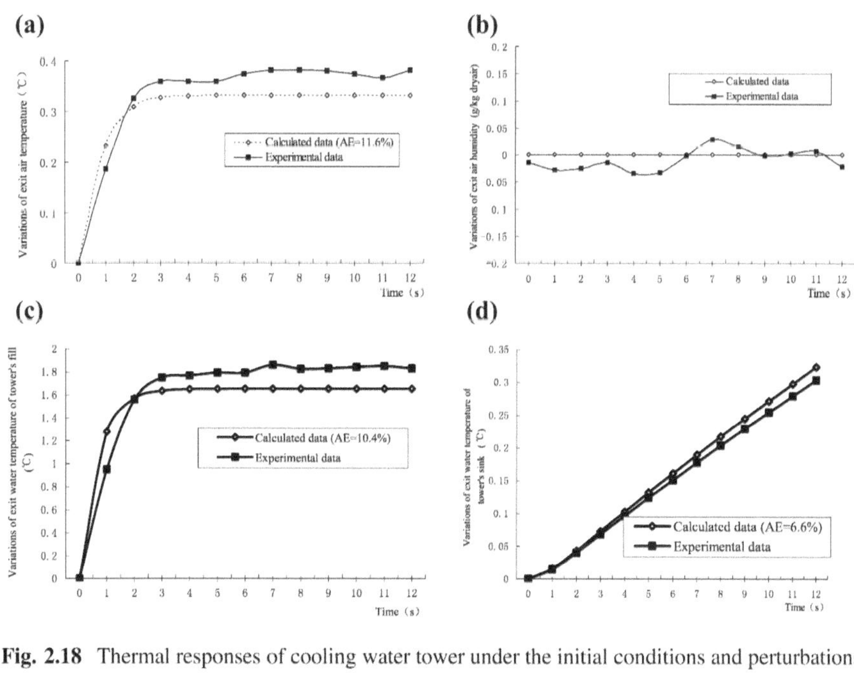

Fig. 2.18 Thermal responses of cooling water tower under the initial conditions and perturbations of case I (calculated results vs. experimental data). **a** Transient response of exit air temperature. **b** Transient response of exit air humidity. **c** Transient response of exit water temperature of tower's fill. **d** Transient response of exit water temperature of tower's sink

response of exit air temperature, exit air humidity, and exit water temperature of fill as well as that of cooling tower's sink is estimated as about 11.6, 10.4, and 6.7 %, respectively.

For the experimental case II, the airflow rate had a sudden increase of 5.60 kg/s while the other inlet variables were kept unchanged. The corresponding system matrixes of cooling tower ($A_{\text{tower}}, B_{\text{tower}}$) under the initial conditions of experimental case II are calculated as follows:

$$A_{\text{tower}} = \begin{bmatrix} -18.0200 & 0.3440 & 8.1797 & 0.0000 \\ 0.0000 & -9.5993 & 0.0023 & 0.0000 \\ 1.9699 & 1.8013 & -5.6933 & 0.0000 \\ 0.0000 & 0.0000 & 0.0374 & -0.0374 \end{bmatrix};$$

$$B_{\text{tower}} = \begin{bmatrix} 1.1722 & 0.3440 & 0.8738 & 8.1797 & 0.0000 \\ 0.0000 & 8.1797 & 0.0013 & 0.0023 & 0.0000 \\ 1.9699 & 1.8013 & -1.4503 & -0.8044 & 30.6488 \\ 0.0000 & 0.0000 & 0.0000 & 0.0000 & 0.0045 \end{bmatrix}.$$

The experimental and simulated results on the variations of response parameters of the cooling tower under the initial conditions and perturbations of case II are shown in Fig. 2.19. As shown in Fig. 2.19, the exit air temperature and the exit

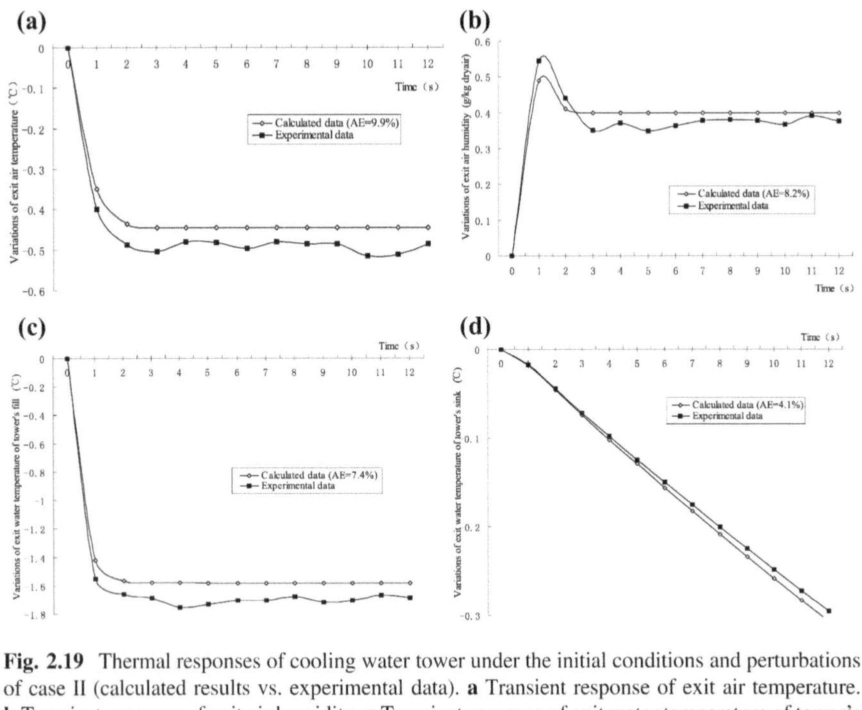

Fig. 2.19 Thermal responses of cooling water tower under the initial conditions and perturbations of case II (calculated results vs. experimental data). **a** Transient response of exit air temperature. **b** Transient response of exit air humidity. **c** Transient response of exit water temperature of tower's fill. **d** Transient response of exit water temperature of tower's sink

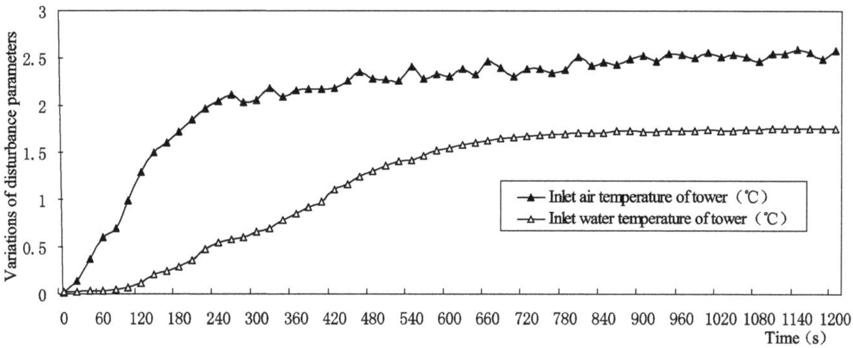

Fig. 2.20 Variations of disturbance parameters for experimental case III

water temperature of the cooling tower's fill and sink tend to decrease as the airflow rate has a step increase, while it is opposite for the exit air humidity. For the experimental case II, the average error (AE) of model results on the transient response of exit air temperature and exit water temperature of fill as well as that of cooling tower's sink is estimated as about 9.9, 8.2, 7.4, and 4.1 %, respectively.

To further validate the cooling tower's model, the perturbation parameters increase gradually under the initial conditions of experimental case III. The variations of the perturbation parameters are shown in Fig. 2.20, and the system matrixes of the cooling tower corresponding to the initial conditions of experimental case III are as follows:

$$A_{\text{tower}} = \begin{bmatrix} -16.4952 & 0.0079 & 4.1736 & 0.0000 \\ 0.0000 & -10.9671 & 0.0001 & 0.0000 \\ 0.6516 & 0.0488 & -7.1269 & 0.0000 \\ 0.0000 & 0.0000 & 0.0299 & -0.0299 \end{bmatrix};$$

$$B_{\text{tower}} = \begin{bmatrix} 9.2031 & 0.0057 & -0.8149 & 2.7824 & 0.0000 \\ 0.0000 & 13.7087 & -0.0015 & 0.0038 & 0.0000 \\ 0.6516 & 0.0300 & -0.2504 & 1,9525 & 1.2299 \\ 0.0000 & 0.0000 & 0.0000 & 0.0000 & 0.0036 \end{bmatrix}.$$

Figure 2.21 compares the simulated results with the experimental data on the transient responses of the concerned parameters under the initial conditions and perturbations of experimental case III. The model results are shown to be of fairly good consistency with the experimental data. The average error (AE) of model results on the transient response of exit air temperature, exit air humidity, and exit water temperature of fill as well as that of cooling tower's sink is about 9.2, 17.2, 10.2, and 11.6 %, respectively.

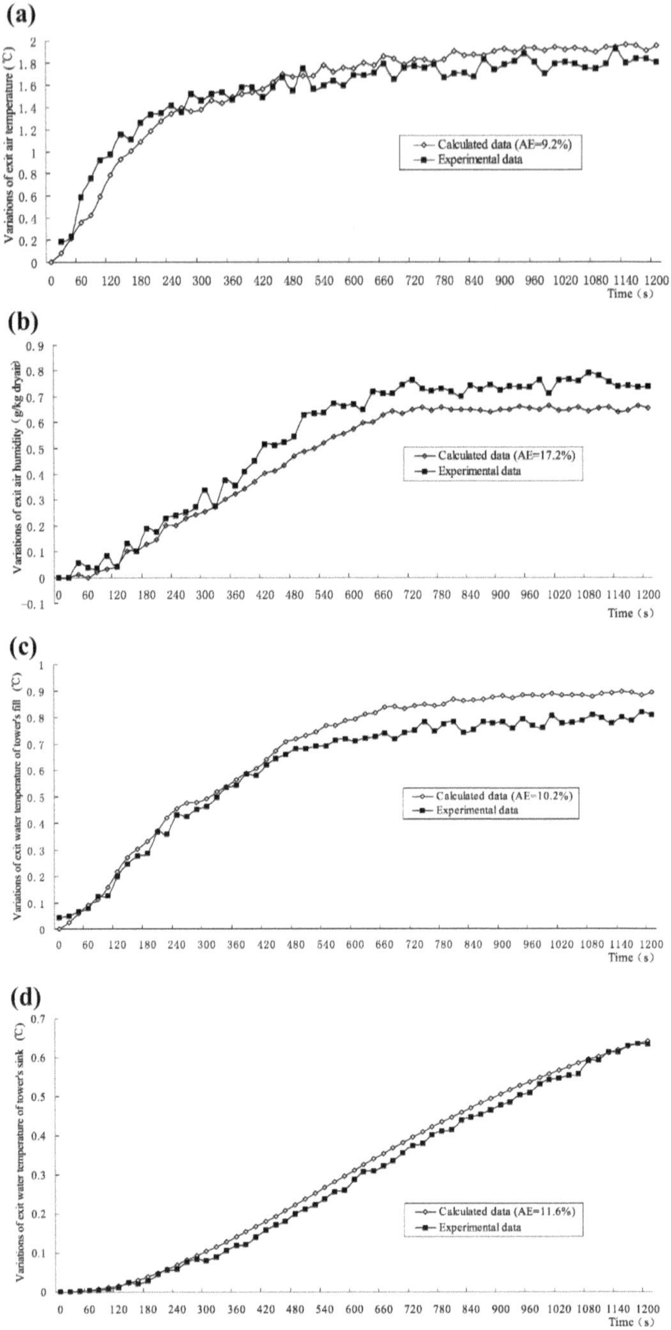

Fig. 2.21 Thermal responses of cooling water tower under the initial conditions and perturbations of case III (calculated results vs. experimental data). **a** Transient response of exit air temperature. **b** Transient response of exit air humidity. **c** Transient response of exit water temperature of tower's fill. **d** Transient response of exit water temperature of tower's sink

2.2.4 *Duct (Pipe) and Fan (Pump)*

Air and water systems transmit and distribute cooling or heating energy through duct and pipe work systems. They are important components of an air-conditioning system, typically including air ducts, water pipes, fans, and pumps. For central air-conditioning systems, the pipes and ducts are normally long, and the heat loss due to the pipes and ducts cannot be neglected during the thermal simulations of the whole air-conditioning system. Meanwhile, the transmission and distribution components connect the air-conditioning equipments, e.g., water-to-air heat exchanger, chiller, and cooling tower, which have been modeled in the state-space form. For the convenience of system modeling, the transmission and distribution components are required to be modeled in the state-space form.

2.2.4.1 Straight Air Duct Modeling

1) Assumptions and fundamental equations

Figure 2.22 shows the schematic diagram for straight air duct, and the main assumptions for modeling are as follows:

- Moisture air is treated as a mixture of ideal gases;
- Thermophysical parameters of air on the transversal surface of duct adopt lumped ones;
- Air temperature and humidity change linearly from inlet to outlet of the duct;
- External surface of duct's thermal insulation is non-condensing as the cooling air passes through the duct.

According to the first law of energy and mass conservation, the following equations can be obtained for the straight air duct.

(1) *Mass equation for moist air passing through straight air duct*

$$G_{\mathrm{da,L}} - G_{\mathrm{da,E}} - G_{\mathrm{da}} \tag{2.104}$$

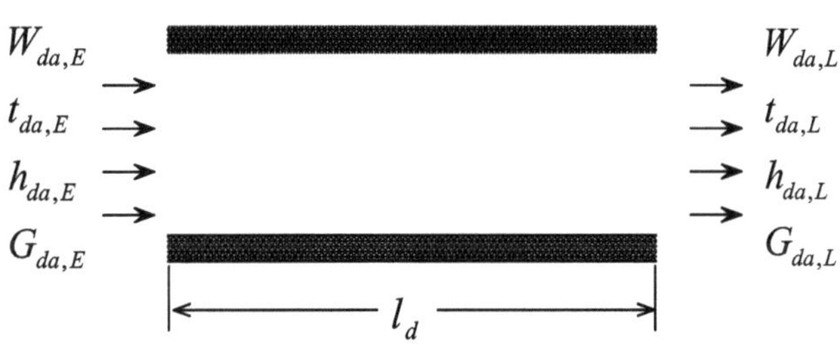

Fig. 2.22 Schematic diagram for straight air duct

(2) *Energy equation for moist air passing through straight air duct*

- In the case of condensation on the interior surface of duct,

$$\frac{1}{2}\rho_{\mathrm{a}}A_{\mathrm{d}}l_{\mathrm{d}}\frac{\mathbf{d}(h_{\mathrm{da,L}}+h_{\mathrm{da,E}})}{\mathbf{d}\tau}=G_{\mathrm{da,E}}(h_{\mathrm{da,E}}-h_{\mathrm{da,L}})+a_{\mathrm{da}}A_{\mathrm{da}}\left(t_{\mathrm{dg}}-\frac{t_{\mathrm{da,E}}+t_{\mathrm{da,L}}}{2}\right)+q_{\mathrm{r}}\lambda_{\mathrm{m}}A_{\mathrm{da}}\left(W_{\mathrm{db}}-\frac{W_{\mathrm{da,E}}+W_{\mathrm{da,L}}}{2}\right) \tag{2.105}$$

- In the case of non-condensation on the interior surface of duct,

$$\frac{1}{2}\rho_{\mathrm{a}}A_{\mathrm{d}}l_{\mathrm{d}}\frac{\mathbf{d}(h_{\mathrm{da,L}}+h_{\mathrm{da,E}})}{\mathbf{d}\tau}=G_{\mathrm{da,E}}(h_{\mathrm{da,E}}-h_{\mathrm{da,L}})+a_{\mathrm{da}}A_{\mathrm{da}}\left(t_{\mathrm{dg}}-\frac{t_{\mathrm{da,E}}+t_{\mathrm{da,L}}}{2}\right) \tag{2.106}$$

(3) *Mass equation for air humidity passing through straight air duct*

- In the case of condensation on the interior surface of duct,

$$\frac{1}{2}\rho_{\mathrm{a}}A_{\mathrm{d}}l_{\mathrm{d}}\frac{\mathbf{d}(W_{\mathrm{da,L}}+W_{\mathrm{da,E}})}{\mathbf{d}\tau}=G_{\mathrm{da,E}}(W_{\mathrm{da,E}}-W_{\mathrm{da,L}})+\lambda_{\mathrm{m}}A_{\mathrm{da}}\left(W_{\mathrm{db}}-\frac{W_{\mathrm{da,E}}+W_{\mathrm{da,L}}}{2}\right) \tag{2.107}$$

- In the case of non-condensation on the interior surface of duct,

$$W_{\mathrm{da,L}}=W_{\mathrm{da,E}} \tag{2.108}$$

(4) *Energy equation for duct wall*

- In the case of condensation on the interior surface of duct,

$$c_{\mathrm{dg}}M_{\mathrm{dg}}\frac{\mathbf{d}t_{\mathrm{dg}}}{\mathbf{d}\tau}=\frac{l_{\mathrm{d}}}{R_{\mathrm{dg}}}(t_{\mathrm{env}}-t_{\mathrm{dg}})+a_{\mathrm{da}}A_{\mathrm{di}}\left(\frac{t_{\mathrm{da,E}}+t_{\mathrm{da,L}}}{2}-t_{\mathrm{dg}}\right)+q_{\mathrm{r}}\lambda_{\mathrm{m}}A_{\mathrm{di}}\left(\frac{W_{\mathrm{da,E}}+W_{\mathrm{da,L}}}{2}-W_{\mathrm{db}}\right) \tag{2.109}$$

- In the case of non-condensation on the interior surface of duct,

$$c_{\mathrm{dg}}M_{\mathrm{dg}}\frac{\mathbf{d}t_{\mathrm{dg}}}{\mathbf{d}\tau}=\frac{l_{\mathrm{d}}}{R_{\mathrm{dg}}}(t_{\mathrm{env}}-t_{\mathrm{dg}})+a_{\mathrm{da}}A_{\mathrm{di}}\left(\frac{t_{\mathrm{da,E}}+t_{\mathrm{da,L}}}{2}-t_{\mathrm{dg}}\right) \tag{2.110}$$

2) State-space representation

As illustrated in previous text, the saturated air humidity (W_{db}) near the wet wall surface of duct in the case of condensation can be approximately expressed by:

$$W_{db} \approx \beta_1 t_{dg} \tag{2.111}$$

where β_1 is a constant coefficient.

Through linearization, Eqs. (2.104) through (2.110) can be converted as follows:

- In the case of condensation on the interior surface of duct,

$$\begin{aligned} T_{d,ta}\frac{\mathbf{d}\Delta t_{da,L}}{\mathbf{d}\tau} = & X_{dwet,1}\Delta t_{da,L} + X_{dwet,2}\Delta W_{da,L} + X_{dwet,3}\Delta t_{dg} + X_{dwet,4}\Delta t_{da,E} \\ & + X_{dwet,5}W_{da,E} + X_{dwet,6}\Delta G_{da,E} + \xi_{t,dweta} \end{aligned} \tag{2.112}$$

$$T_{d,wa}\frac{\mathbf{d}\Delta W_{da,L}}{\mathbf{d}\tau} = Y_{dwet,1}\Delta W_{da,L} + Y_{dwet,2}\Delta t_{dg} + Y_{dwet,3}\Delta W_{da,E} + Y_{dwet,4}\Delta G_{da,E} + \xi_{w,dweta} \tag{2.113}$$

$$\begin{aligned} T_{d,dg}\frac{\mathbf{d}\Delta t_{dg}}{\mathbf{d}\tau} = & Z_{dwet,1}\Delta t_{da,L} + Z_{dwet,2}\Delta W_{da,L} + Z_{dwet,3}\Delta t_{dg} + Z_{dwet,4}\Delta t_{da,E} \\ & + Z_{dwet,5}\Delta W_{da,E} + Z_{dwet,6}\Delta G_{da,E} \end{aligned} \tag{2.114}$$

$$\Delta G_{da,L} = \Delta G_{da,E} \tag{2.115}$$

- In the case of non-condensation on the interior surface of duct,

$$T_{d,ta}\frac{\mathbf{d}\Delta t_{da,L}}{\mathbf{d}\tau} = X_{ddry,1}\Delta t_{da,L} + X_{ddry,2}\Delta t_{dg} + X_{ddry,3}\Delta t_{da,E} + X_{ddry,4}\Delta G_{da,E} + \xi_{t,ddrya} \tag{2.116}$$

$$T_{d,dg}\frac{\mathbf{d}\Delta t_{dg}}{\mathbf{d}\tau} = Z_{ddry,1}\Delta t_{da,L} + Z_{ddry,2}\Delta t_{dg} + Z_{ddry,3}\Delta t_{da,E} + Z_{ddry,4}\Delta G_{da,E} \tag{2.117}$$

$$\Delta W_{da,L} = \Delta W_{da,E} \tag{2.118}$$

$$\Delta G_{da,L} = \Delta G_{da,E} \tag{2.119}$$

Coefficients in Eqs. (2.112) through (2.117) are listed in Table 2.11

Equations (2.112) through (2.119) can be expressed by the state-space representation as follows:

$$\dot{x}_{zduct} = A_{zduct} \cdot x_{zduct} + B_{zduct} \cdot u_{zduct} + \xi_{zduct} \tag{2.120}$$

$$y_{zduct} = C_{zduct} \cdot x_{zduct} + D_{zduct} \cdot u_{zduct} \tag{2.121}$$

Table 2.11 Coefficients in Eqs. (2.112) through (2.117)

Equation No.	Coefficient expression
Eq. (2.112)	$T_{d,ta}=\frac{1}{2}\rho_a c_a A_d l_d$; $X_{dwet,1}=-c_a(G_{da,E})_o-\frac{A_{da}}{2}(a_{da})_o$; $X_{dwet,2}=X_{dwet,5}=-\frac{q_r-\beta_2}{2}A_{da}(\lambda_m)_o$; $X_{dwet,3}=A_{da}(a_{da})_o+\beta_1(q_r-\beta_2)A_{da}(\lambda_m)_o$; $X_{dwet,4}=c_a(G_{da,E})_o-\frac{A_{da}}{2}(a_{da})_o$; $X_{dwet,6}=c_a(t_{da,E}-t_{da,L})_o+A_{da}\left[t_{dg}-\frac{t_{da,E}+t_{da,L}}{2}\right]_o\left(\frac{\partial a_{da}}{\partial G_{da,E}}\right)_o+A_{da}(q_r-\beta_2)\left[\beta_1 t_{dg}-\frac{W_{da,E}+W_{da,L}}{2}\right]_o\left(\frac{\partial \lambda_m}{\partial G_{da,E}}\right)_o$ $\xi_{t,dweta}=-T_{d,ta}\frac{d\Delta t_{da,E}}{d\tau}$
Eq. (2.113)	$T_{d,wa}=\frac{1}{2}\rho_a A_d l_d$; $Y_{dwet,1}=-(G_{da,E})_o-\frac{A_{da}}{2}(\lambda_m)_o$; $Y_{dwet,2}=\beta_1 A_{da}(\lambda_m)_o$; $Y_{dwet,3}=(G_{da,E})_o-\frac{A_{da}}{2}(\lambda_m)_o$; $Y_{dwet,4}=(W_{da,E}-W_{da,L})_o+A_{da}\left[\beta_1 t_{dg}-\frac{W_{da,E}+W_{da,L}}{2}\right]_o\left(\frac{\partial \lambda_m}{\partial G_{da,E}}\right)_o$; $\xi_{w,dweta}=-T_{d,wa}\frac{d\Delta W_{da,E}}{d\tau}$
Eq. (2.114)	$T_{d,dg}=c_{dg}M_{dg}$; $Z_{dwet,1}=Z_{dwet,4}=\frac{A_{da}}{2}(a_{da})_o$; $Z_{dwet,2}=Z_{dwet,5}=\frac{A_{da}}{2}q_r(\lambda_m)_o$; $Z_{dwet,3}=-A_{di}(a_{da})_o-\frac{l_d}{R_{dg}}-q_r\beta_1 A_{da}(\lambda_m)_o$; $Z_{dwet,6}=A_{da}q_r\left[\frac{W_{da,E}+W_{da,L}}{2}-\beta_1 t_{dg}\right]_o\left(\frac{\partial \lambda_m}{\partial G_{da,E}}\right)_o+A_{da}\left[\frac{t_{da,E}+t_{da,L}}{2}-t_{dg}\right]_o\left(\frac{\partial a_{da}}{\partial G_{da,E}}\right)_o$
Eq. (2.116)	$T_{d,ta}=\frac{1}{2}\rho_a c_a A_d l_d$; $X_{ddry,1}=-c_a(G_{da,E})_o-\frac{A_{da}}{2}(a_{da})_o$; $X_{ddry,2}=A_{da}(a_{da})_o$; $X_{ddry,3}=c_a(G_{da,E})_o-\frac{A_{da}}{2}(a_{da})_o$; $X_{ddry,4}=c_a(t_{da,E}-t_{da,L})_o+A_{di}\left[t_{dg}-\frac{t_{da,E}+t_{da,L}}{2}\right]_o\left(\frac{\partial a_{da}}{\partial G_{da,E}}\right)_o$; $\xi_{t,ddrya}=-T_{d,ta}\frac{d\Delta t_{da,E}}{d\tau}$
Eq. (2.117)	$T_{d,dg}=c_{dg}M_{dg}$; $Z_{ddry,1}=Z_{ddry,3}=\frac{A_{da}}{2}(a_{da})_o$; $Z_{ddry,2}=-A_{da}(a_{da})_o-\frac{l_d}{R_{dg}}$; $X_{tar,1}=c_a(G_{a,i}-G_{a,leal})_o$

The standard form of state-space equation corresponding to Eqs. (2.120) and (2.121) is written as follows:

$$\overset{dot}{X_{zduct}}=A_{zduct}X_{zduct}+B_{zduct}u_{zduct} \tag{2.122}$$

$$y_{zduct}=C_{zduct}X_{zduct}+D_{zduct}u_{zduct}-C_{zduct}A_{zduct}^{-1}\xi_{zduct} \tag{2.123}$$

where $X_{zduct}=x_{zduct}+A_{zduct}^{-1}\xi_{zduct}$;

For the case of condensation on the interior surface of duct,

$$
\begin{aligned}
x_{\text{zduct}} &= \left[\Delta t_{\text{da,L}}, \Delta W_{\text{da,L}}, \Delta t_{\text{dg}}\right]^T; y_{\text{zduct}} = \left[\Delta t_{\text{da,L}}, \Delta W_{\text{da,L}}, \Delta G_{\text{da,L}}\right]^T; \\
u_{\text{zduct}} &= \left[\Delta t_{\text{da,E}}, \Delta W_{\text{da,E}}, \Delta G_{\text{da,E}}\right]^T; \xi_{\text{zduct}} = \left[\xi_{\text{t,dweta}}, \xi_{\text{w,dweta}}, 0\right]^T; \\
A_{\text{zduct}} &= \begin{bmatrix} \frac{X_{\text{dwet,1}}}{T_{\text{d,ta}}} & \frac{X_{\text{dwet,2}}}{T_{\text{d,ta}}} & \frac{X_{\text{dwet,3}}}{T_{\text{d,ta}}} \\ 0 & \frac{Y_{\text{dwet,1}}}{T_{\text{d,wa}}} & \frac{Y_{\text{dwet,2}}}{T_{\text{d,wa}}} \\ \frac{Z_{\text{dwet,1}}}{T_{\text{d,dg}}} & \frac{Z_{\text{dwet,2}}}{T_{\text{d,dg}}} & \frac{Z_{\text{dwet,3}}}{T_{\text{d,dg}}} \end{bmatrix}; B_{\text{zduct}} = \begin{bmatrix} \frac{X_{\text{dwet,4}}}{T_{\text{d,ta}}} & \frac{X_{\text{dwet,5}}}{T_{\text{d,ta}}} & \frac{X_{\text{dwet,6}}}{T_{\text{d,ta}}} \\ 0 & \frac{Y_{\text{dwet,3}}}{T_{\text{d,wa}}} & \frac{Y_{\text{dwet,4}}}{T_{\text{d,wa}}} \\ \frac{Z_{\text{dwet,4}}}{T_{\text{d,dg}}} & \frac{Z_{\text{dwet,5}}}{T_{\text{d,dg}}} & \frac{Z_{\text{dwet,6}}}{T_{\text{d,dg}}} \end{bmatrix}; \\
C_{\text{zduct}} &= \begin{bmatrix} 1 & 0 & 0 \\ 0 & 1 & 0 \\ 0 & 0 & 0 \end{bmatrix}; D_{\text{zduct}} = \begin{bmatrix} 0 & 0 & 0 \\ 0 & 0 & 0 \\ 0 & 0 & 1 \end{bmatrix}.
\end{aligned}
$$

For the case of non-condensation on the interior surface of duct,

$$
\begin{aligned}
x_{\text{zduct}} &= \left[\Delta t_{\text{da,L}}, \Delta t_{\text{dg}}\right]^T; y_{\text{zduct}} = \left[\Delta t_{\text{da,L}}, \Delta W_{\text{da,L}}, \Delta G_{\text{da,L}}\right]^T; \\
u_{\text{zduct}} &= \left[\Delta t_{\text{da,E}}, \Delta W_{\text{da,E}}, \Delta G_{\text{da,E}}\right]^T; \xi_{\text{zduct}} = \left[\xi_{\text{t,ddrya}} 0\right]^T; \\
A_{\text{zduct}} &= \begin{bmatrix} \frac{X_{\text{ddry,1}}}{T_{\text{d,ta}}} & \frac{X_{\text{ddry,2}}}{T_{\text{d,ta}}} \\ \frac{Z_{\text{ddry,1}}}{T_{\text{d,dg}}} & \frac{Z_{\text{ddry,2}}}{T_{\text{d,dg}}} \end{bmatrix}; B_{\text{zduct}} = \begin{bmatrix} \frac{X_{\text{ddry,3}}}{T_{\text{d,ta}}} & 0 & \frac{X_{\text{ddry,4}}}{T_{\text{d,ta}}} \\ \frac{Z_{\text{ddry,3}}}{T_{\text{d,dg}}} & 0 & \frac{Z_{\text{ddry,4}}}{T_{\text{d,dg}}} \end{bmatrix}; C_{\text{zduct}} = \begin{bmatrix} 1 & 0 \\ 0 & 0 \\ 0 & 0 \end{bmatrix}; D_{\text{zduct}} = \begin{bmatrix} 0 & 0 & 0 \\ 0 & 1 & 0 \\ 0 & 0 & 1 \end{bmatrix}.
\end{aligned}
$$

3) Model validation

A straight air duct is used to validate the state-space model. The main physical parameters of the duct are shown in Table 2.12. The test parameters include the following: ① inlet and outlet air temperature and humidity (precision: ±0.1 °C in temperature; ±0.8 % in relative humidity); ② duct wall temperature (precision: ±0.2 °C); ③ airflow rate (precision: ±2 % of reading data).

In most cases, water condensation is not allowed in the air duct. So, the experiments were performed under the non-condensation situation. Initial conditions for the model validation are as follows: The inlet and outlet air temperature are 29.3 and 29.5 °C, respectively; the inlet and outlet air humidity are both 20.3 g/(kg dry air); the duct wall temperature is 30.0 °C; the airflow rate is 0.20 kg/s.

Figure 2.24 compares model results with the experimental data on the transient responses of exit air temperature to perturbations as shown in Fig. 2.23. The average error (AE) of straight air duct model in this case is estimated as about 13.8 %.

Table 2.12 Structural parameters of the experimental duct

Length of duct l_d (m)	3.25	Width of duct W_d (m)	0.87
Height of duct H_d (m)	0.25	Specific heat of duct c_{dg} (J/(kg°C))	620
Density of duct material ρ_d (kg/m^3)	7800	Thickness of duct wall δ_d (m)	0.001

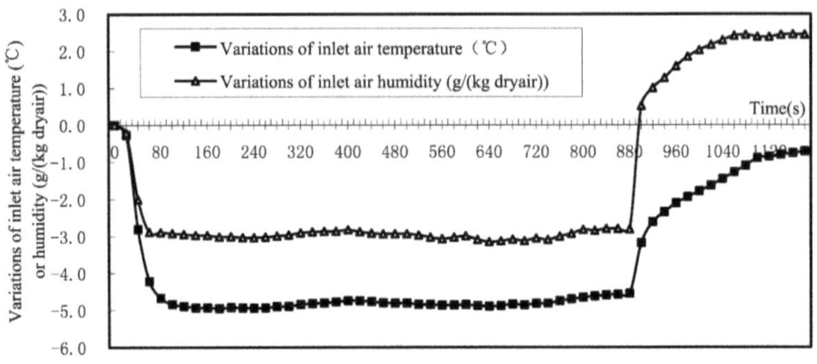

Fig. 2.23 Perturbations of inlet air temperature and humidity (measured data)

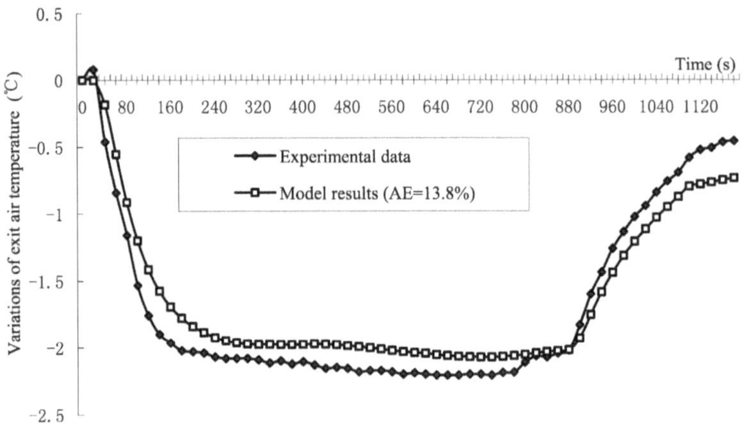

Fig. 2.24 Transient response of exit air temperature to perturbations as shown in Fig. 2.23 (calculated results vs. experimental data)

2.2.4.2 Straight Water Pipe Modeling

1) Fundamental equations

The assumptions for the straight water pipe modeling are basically the same as that for the straight air duct modeling. The schematic diagram for straight water pipe is given in Fig. 2.25, and corresponding equations are obtained as follows:

(1) *Mass equation for water passing through straight water pipe*

$$G_{\text{zpw,L}} = G_{\text{zpw,E}} = G_{\text{zpw}} \tag{2.124}$$

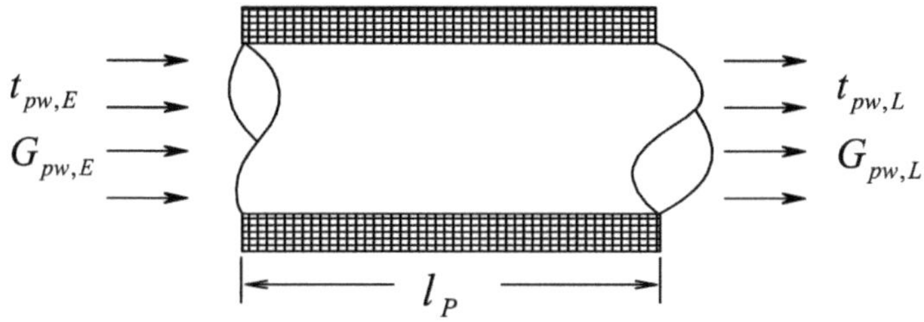

Fig. 2.25 Schematic diagram for straight water pipe

(2) *Energy equation for water passing through straight water pipe*

$$\frac{1}{2}\rho_w c_w A_p l_p \frac{\mathbf{d}(t_{pw,L} + t_{pw,E})}{\mathbf{d}\tau} = G_{pw,E}(t_{pw,E} - t_{pw,L}) + a_{pw}A_{pi}\left(t_{pg} - \frac{t_{pw,E} + t_{pw,L}}{2}\right) \quad (2.125)$$

(3) *Energy equation for pipe wall*

$$c_{pg}M_{pg}\frac{\mathbf{d}t_{pg}}{\mathbf{d}\tau} = \frac{l_p}{R_{pg}}(t_{env} - t_{pg}) + a_{pw}A_{pi}\left(\frac{t_{pw,E} + t_{pw,L}}{2} - t_{pg}\right) \quad (2.126)$$

2) State-space representation

Through linearization, Eqs. (2.124) through (2.126) can be written as follows:

$$T_{p,tw}\frac{\mathbf{d}\Delta t_{pw,L}}{\mathbf{d}\tau} = X_{p,1}\Delta t_{pw,L} + X_{p,2}\Delta t_{pg} + X_{p,3}\Delta t_{pw,E} + X_{p,4}\Delta G_{pw,E} + \xi_{tpw,L} \quad (2.127)$$

$$T_{p,pg}\frac{\mathbf{d}\Delta t_{pg}}{\mathbf{d}\tau} - Y_{p,1}\Delta t_{pw,L} + Y_{p,2}\Delta t_{pg} + Y_{p,3}\Delta t_{pw,E} + Y_{p,4}\Delta G_{pw,E} \quad (2.128)$$

$$\Delta G_{pw,L} = \Delta G_{pw,E} \quad (2.129)$$

Coefficients in Eqs. (2.127) and (2.128) are listed in Table 2.13.

Thus, the state-space model for the straight-through pipe can be expressed by Eqs. (2.130) and (2.131).

$$\dot{X}_{zpipe} = A_{zpipe}X_{zpipe} + B_{zpipe}u_{zpipe} \quad (2.130)$$

$$y_{zpipe} = C_{zpipe}X_{zpipe} + D_{zpipe}u_{zpipe} - C_{zpipe}A_{zpipe}^{-1}\xi_{zpipe} \quad (2.131)$$

Table 2.13 Coefficients in Eqs. (2.127) and (2.128)

Equation No.	Coefficient expression
Eq. (2.127)	$T_{p,tw} = \frac{1}{2}\rho_w c_w A_p l_p$; $X_{p,1} = -(G_{pw,E})_o - \frac{A_{pi}}{2}(a_{pw})_o$; $X_{p,2} = A_{pi}(a_{pw})_o$; $X_{p,3} = (G_{pw,E})_o - \frac{A_{pi}}{2}(a_{pw})_o$; $X_{p,4} = (t_{pw,E} - t_{pw,L})_o + A_{pi}\left(t_{pg} - \frac{t_{pw,E}+t_{pw,L}}{2}\right)_o \left(\frac{\partial a_{pw}}{\partial G_{pw,E}}\right)_o$; $\xi_{tpw,L} = -T_{p,tw}\frac{d_{tpw,E}}{d\tau}$
Eq. (2.128)	$T_{p,pg} = c_{pg}M_{pg}$; $Y_{p,1} = Y_{p,3} = \frac{A_{pi}}{2}(a_{pw})_o$; $Y_{p,2} = -A_{pi}(a_{pw})_o - \frac{l_p}{R_{pg}}$; $Y_{p,4} = A_{pi}\left(\frac{t_{pw,E}+t_{pw,L}}{2} - t_{pg}\right)_o \left(\frac{\partial a_{pw}}{\partial G_{pw,E}}\right)_o$

where

$$X_{zpipe} = x_{zpipe} + A_{zpipe}^{-1}\xi_{zpipe}; x_{zpipe} = [\Delta t_{pw,L}, \Delta t_{pg}]^T;$$

$$y_{zpipe} = [\Delta t_{pw,L}, \Delta G_{pw,L}]^T; u_{zpipe} = [\Delta t_{pw,E}, \Delta G_{pw,E}]^T; \xi_{zpipe} = [\xi_{tpw,L}, 0]^T;$$

$$A_{zpipe} = \begin{bmatrix} \frac{X_{p,1}}{T_{p,tw}} & \frac{X_{p,2}}{T_{p,tw}} \\ \frac{Y_{p,1}}{T_{p,pg}} & \frac{Y_{p,2}}{T_{p,pg}} \end{bmatrix}; B_{zpipe} = \begin{bmatrix} \frac{X_{p,3}}{T_{p,tw}} & \frac{X_{p,4}}{T_{p,tw}} \\ \frac{Y_{p,3}}{T_{p,pg}} & \frac{Y_{p,4}}{T_{p,pg}} \end{bmatrix}; C_{zpipe} = \begin{bmatrix} 1 & 0 \\ 0 & 0 \end{bmatrix}; D_{zpipe} = \begin{bmatrix} 0 & 0 \\ 0 & 1 \end{bmatrix}.$$

3) Model validation

A straight water pipe is used for the model validation. The key physical information on the experimental water pipe is given in Table 2.14.

The test parameters include the following: ① inlet and outlet water temperature (precision: ±0.2 °C in temperature); ② pipe wall temperature (precision: ±0.2 °C); and ③ water flow rate (precision: 0.2 level). Initial conditions for model validation are as follows: the inlet and outlet water temperature are 26.2 and 27.4 °C, respectively; the pipe wall temperature is 29.4 °C; and the water flow rate is 0.205 kg/s.

Figure 2.27 shows the experimental data and model results on the transient response of exit water temperature to the perturbations of inlet water temperature as shown in Fig. 2.26. The average error (AE) of model results compared with the experimental data is estimated as 7.4 % in this case.

Table 2.14 Key physical information on the experimental water pipe

Length of water pipe l_p (m)	7.0	Inner diameter of water pipe D_p (m)	0.02
Specific heat of water pipe c_{pg} (J/(kg °C))	1569	Density of pipe material ρ_{pg} (kg/m^3)	1083
Thickness of water pipe δ_p (m)	0.005		

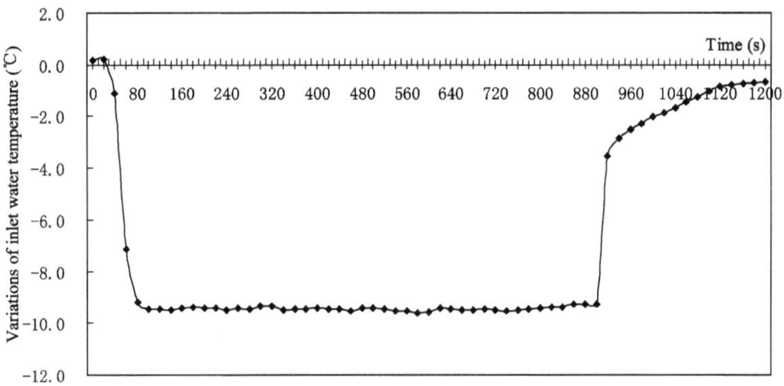

Fig. 2.26 Perturbations of inlet water temperature (measured data)

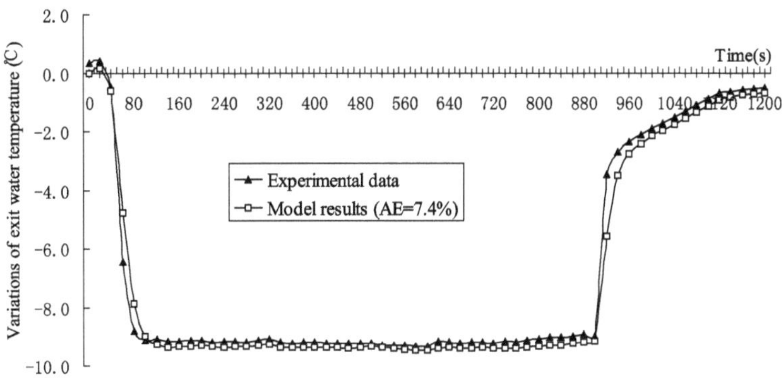

Fig. 2.27 Transient response of exit water temperature to perturbations as shown in Fig. 2.26 (calculated results vs. experimental data)

2.2.4.3 Three-Way Duct/Pipe Modeling

Three-way ducts or pipes are important links in the air duct or water pipe system in an air-conditioning system. Normally, the three-way duct or pipe has two types: One is confluent flow and the other is split flow (see Figs. 2.28 and 2.29). Since it takes very short time for fluid passing through the three-way duct or pipe, thermal loss of fluid caused by the three-way duct or pipe can be neglected.

(1) Three-way air duct

For the confluent flow, the following equations come into existence:

$$G_{\mathrm{da3,L}} = G_{\mathrm{da1,E}} + G_{\mathrm{da2,E}} \tag{2.132}$$

$$G_{\mathrm{da3,L}} W_{\mathrm{da3,L}} = G_{\mathrm{da1,E}} W_{\mathrm{da1,E}} + G_{\mathrm{da2,E}} W_{\mathrm{da2,E}} \tag{2.133}$$

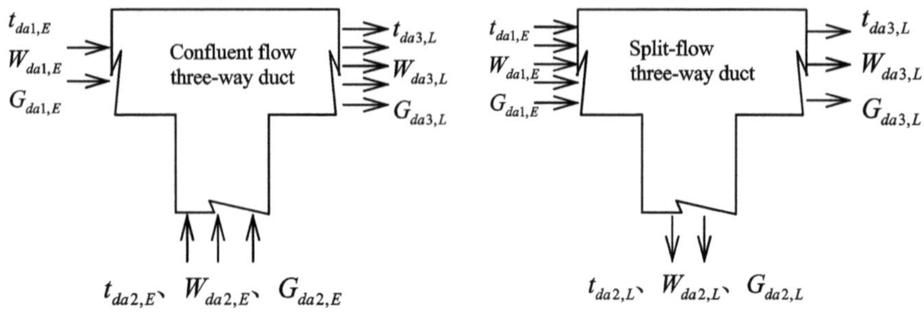

Fig. 2.28 Schematic diagram for three-way air duct

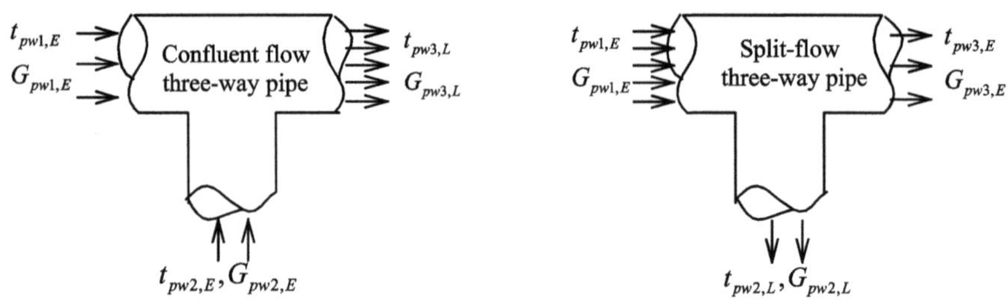

Fig. 2.29 Schematic diagram for three-way water pipe

$$G_{\mathrm{da3,L}} t_{\mathrm{da3,L}} = G_{\mathrm{da1,E}} t_{\mathrm{da1,E}} + G_{\mathrm{da2,E}} t_{\mathrm{da2,E}} \tag{2.134}$$

For the split flow, we have the following equations:

$$t_{\mathrm{da2,L}} = t_{\mathrm{da1,E}} \tag{2.135}$$

$$W_{\mathrm{da2,L}} = W_{\mathrm{da1,E}} \tag{2.136}$$

$$G_{\mathrm{da2,L}} = \kappa_{\mathrm{a}} G_{\mathrm{da1,E}} \tag{2.137}$$

$$t_{\mathrm{da3,L}} = t_{\mathrm{da1,E}} \tag{2.138}$$

$$W_{\mathrm{da3,L}} = W_{\mathrm{da1,E}} \tag{2.139}$$

$$G_{\mathrm{da3,L}} = (1 - \kappa_{\mathrm{a}}) G_{\mathrm{da1,E}} \tag{2.140}$$

Then, the dynamic relationships between the inlet variables and the outlet ones of the three-way air duct can be expressed by Eq. (2.141).

$$y_{\mathrm{sduct}} = D_{\mathrm{sduct}} \cdot u_{\mathrm{sduct}} \tag{2.141}$$

where for the confluent flow,

$$\begin{aligned} y_{\text{sduct}} &= \left[\Delta t_{\text{da3,L}}, \Delta W_{\text{da3,L}}, \Delta G_{\text{da3,L}}\right]^T; u_{\text{sduct}} = \left[u_{\text{sduct,1}}, u_{\text{sduct,2}}\right]^T; \\ u_{\text{sduct,1}} &= \left[\Delta t_{\text{da1,E}}, \Delta W_{\text{da1,E}}, \Delta G_{\text{da1,E}}\right]^T; u_{\text{sduct,2}} = \left[\Delta t_{\text{da2,E}}, \Delta W_{\text{da2,E}}, \Delta G_{\text{da2,E}}\right]^T; \\ D_{\text{sduct}} &= \left[D_{\text{da1,E}}, D_{\text{da2,E}}\right]; \end{aligned}$$

$$D_{\text{da1,E}} = \begin{bmatrix} \frac{(G_{\text{da1,E}})_o}{(G_{\text{da3,L}})_o} & 0 & \frac{[(t_{\text{da1,E}})_o - (t_{\text{da3,L}})_o]}{(G_{\text{da3,L}})_o} \\ 0 & \frac{(G_{\text{da1,E}})_o}{(G_{\text{da3,L}})_o} & \frac{[(W_{\text{da1,E}})_o - (W_{\text{da3,L}})_o]}{(G_{\text{da3,L}})_o} \\ 0 & 0 & 1 \end{bmatrix};$$

$$D_{\text{da2,E}} = \begin{bmatrix} \frac{(G_{\text{da2,E}})_o}{(G_{\text{da3,L}})_o} & 0 & \frac{[(t_{\text{da2,E}})_o - (t_{\text{da3,L}})_o]}{(G_{\text{da3,L}})_o} \\ 0 & \frac{(G_{\text{da2,E}})_o}{(G_{\text{da3,L}})_o} & \frac{[(W_{\text{da2,E}})_o - (W_{\text{da3,L}})_o]}{(G_{\text{da3,L}})_o} \\ 0 & 0 & 1 \end{bmatrix};$$

and for the split flow,

$$\begin{aligned} y_{\text{sduct}} &= \left[y_{\text{sduct,2}}, y_{\text{sduct,3}}\right]^T; u_{\text{sduct}} = \left[\Delta t_{\text{da1,E}}, \Delta W_{\text{da1,E}}, \Delta G_{\text{da1,E}}\right]^T; \\ y_{\text{sduct,2}} &= \left[\Delta t_{\text{da2,L}}, \Delta W_{\text{da2,L}}, \Delta G_{\text{da2,L}}\right]^T; y_{\text{sduct,3}} = \left[\Delta t_{\text{da3,L}}, \Delta W_{\text{da3,L}}, \Delta G_{\text{da3,L}}\right]^T; \\ D_{\text{sduct}} &= \begin{bmatrix} D_{\text{fda12,E}} \\ D_{\text{fda13,E}} \end{bmatrix}; D_{\text{fda12,E}} = \begin{bmatrix} 1 & 0 & 0 \\ 0 & 1 & 0 \\ 0 & 0 & \kappa_a \end{bmatrix}; D_{\text{fda13,E}} = \begin{bmatrix} 1 & 0 & 0 \\ 0 & 1 & 0 \\ 0 & 0 & 1 - \kappa_a \end{bmatrix}. \end{aligned}$$

(2) Three-way water pipe

For the confluent flow, the following equations can be obtained:

$$G_{\text{pw3,L}} = G_{\text{pw1,E}} + G_{\text{pw2,E}} \tag{2.142}$$

$$G_{\text{pw3,L}} t_{\text{pw3,L}} - G_{\text{pw1,E}} t_{\text{pw1,E}} + G_{\text{pw2,E}} t_{\text{pw2,E}} \tag{2.143}$$

For the split flow, we have the following equations:

$$t_{\text{pw2,L}} = t_{\text{pw1,E}} \tag{2.144}$$

$$G_{\text{pw2,L}} = \kappa_w G_{\text{pw1,E}} \tag{2.145}$$

$$t_{\text{pw3,L}} = t_{\text{pw1,E}} \tag{2.146}$$

$$G_{\text{pw3,L}} = (1 - \kappa_w) G_{\text{pw1,E}} \tag{2.147}$$

Likewise, the dynamic relationships between the inlet variables and the outlet ones of the three-way air duct can be expressed by:

$$y_{\text{spipe}} = D_{\text{spipe}} \cdot u_{\text{spipe}} \tag{2.148}$$

where
for the confluent flow,

$$y_{\text{spipe}} = \left[\Delta t_{\text{pw3,L}}, \Delta G_{\text{pw3,L}}\right]^T; u_{\text{spipe}} = \left[\Delta t_{\text{pw1,E}}, \Delta G_{\text{pw1,E}}, \Delta t_{\text{pw2,E}}, \Delta G_{\text{pw2,E}}\right]^T;$$

$$D_{\text{spipe}} = \begin{bmatrix} \frac{(G_{\text{pw1,E}})_o}{(G_{\text{pw3,L}})_o} & \frac{[(t_{\text{pw1,E}})_o-(t_{\text{pw3,L}})_o]}{(G_{\text{pw3,L}})_o} & \frac{(G_{\text{pw2,E}})_o}{(G_{\text{pw3,L}})_o} & \frac{[(t_{\text{pw2,E}})_o-(t_{\text{pw3,L}})_o]}{(G_{\text{pw3,L}})_o} \\ 0 & 1 & 0 & 1 \end{bmatrix};$$

and for the split flow,

$$y_{\text{spipe}} = \left[\Delta t_{\text{pw2,L}}, \Delta G_{\text{pw2,L}}, \Delta t_{\text{pw3,L}}, \Delta G_{\text{pw3,L}}\right]^T; u_{\text{sduct}} = \left[\Delta t_{\text{pw1,E}}, \Delta G_{\text{pw1,E}}\right]^T;$$

$$D_{\text{spipe}} = \begin{bmatrix} 1 & 0 \\ 0 & \kappa_{\text{w}} \\ 1 & 0 \\ 0 & 1-\kappa_{\text{w}} \end{bmatrix}.$$

2.2.4.4 Fan and Pump

Fan and pump are key power equipments for fluid transportation in an air-conditioning system. They are also the main control objects in a variable air volume (VAV) or variable water volume (VWV) air-conditioning system. The fan or pump is normally driven by an alternating current motor, as shown in Fig. 2.30.

Assuming that input electric voltage of motor be U_{m}, and electric resistance and inductance and rotary inertia of motor be R_{a}, L_{a} and J_{m}, respectively, we have the following equations [16]:

(1) *Angular speed equation*

$$\frac{\mathbf{d}\theta_{\text{m}}}{\mathbf{d}\tau} = \omega_{\text{m}} \tag{2.149}$$

where θ_{m} is angular displacement, rad; ω_{m} is angular velocity, rad/s.

(2) *Voltage equation*

$$L_{\text{a}}\frac{\mathbf{d}i_{\text{a}}}{\mathbf{d}\tau} = -R_{\text{a}}i_{\text{a}} - K_{\text{v}}\omega_{\text{m}} + U_{\text{m}} \tag{2.150}$$

where i_{a} is electric current, *A*; K_{v} is coefficient of counterpotential, V s/rad;

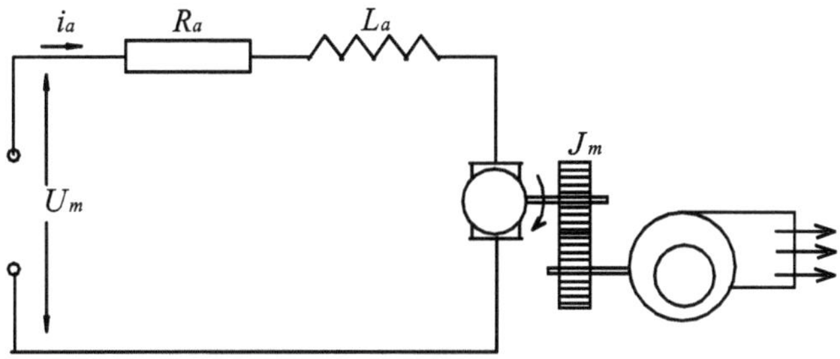

Fig. 2.30 Schematic diagram for fan/pump's motor

(3) *Moment equation*

$$J_{\mathrm{m}}\frac{\mathbf{d}\omega_{\mathrm{m}}}{\mathbf{d}\tau}=K_{\mathrm{m}}i_{\mathrm{a}}-B_{\mathrm{m}}\omega_{\mathrm{m}}-F_{\mathrm{l}} \tag{2.151}$$

where J_{m} is rotary inertia of motor, kg m^2; K_{m} is coefficient of torsion torque, N m rad/A; B_{m} is equivalent friction coefficient, kg m^2/s^2; and F_{l} is load of motor, kg m^2 rad/s^2.

The load of motor, F_{l}, is the function dependent on the outlet pressure of fan or pump ($P_{\mathrm{f/p}}$) and fluid flow rate ($G_{\mathrm{f/p}}$) as follows:

$$F_{\mathrm{l}}\omega_{\mathrm{m}}\eta_{\mathrm{f/p}}=\frac{G_{\mathrm{f/p}}P_{\mathrm{f/p}}}{\rho_{\mathrm{a/w}}} \tag{2.152}$$

$$G_{\mathrm{f/p}}=\chi_{\mathrm{f/p,G}}\chi_{\mathrm{n}}\rho_{\mathrm{a/w}}d_{\mathrm{f/p}}^{2}b_{\mathrm{f/p}}n_{\mathrm{m}} \tag{2.153}$$

$$P_{\mathrm{f/p}}=\chi_{\mathrm{f/p,P}}\chi_{\mathrm{n}}^{2}\rho_{\mathrm{a/w}}d_{\mathrm{f/p}}^{2}n_{\mathrm{m}}^{2} \tag{2.154}$$

Thus, Eq. (2.151) can be written as follows:

$$J_{\mathrm{m}}\frac{\mathbf{d}\omega_{\mathrm{m}}}{\mathbf{d}\tau}=K_{\mathrm{m}}i_{\mathrm{a}}-B_{\mathrm{m}}\omega_{\mathrm{m}}-\frac{\varepsilon_{\mathrm{G}}\varepsilon_{\mathrm{P}}\omega_{\mathrm{m}}^{2}}{(2\pi)^{3}\eta_{\mathrm{f/p}}} \tag{2.155}$$

where $\varepsilon_{\mathrm{G}}=\chi_{\mathrm{f/p,G}}\chi_{\mathrm{n}}\rho_{\mathrm{a/w}}d_{\mathrm{f/p}}^{2}b_{\mathrm{f/p}}$, $\varepsilon_{\mathrm{P}}=\chi_{\mathrm{f/p,P}}\chi_{\mathrm{n}}^{2}\rho_{\mathrm{a/w}}d_{\mathrm{f/p}}^{2}$, $\omega_{\mathrm{m}}=2\pi n_{\mathrm{m}}$

(4) *Power equation*

$$N_{\mathrm{m}}=\alpha_{\mathrm{m}}U_{\mathrm{m}}i_{\mathrm{a}} \tag{2.156}$$

where N_{m} is power of motor, W; $\alpha_{\mathrm{m}}=\sqrt{3}\phi_{\mathrm{m}}$, φ_{m} is power factor.

Through linearization, Eqs. (2.149), (2.150), (2.151), (2.153), and (2.156) can be changed into incremental form as follows:

$$\frac{\mathbf{d}\Delta\theta_{\mathrm{m}}}{\mathbf{d}\tau} = \Delta\omega_{\mathrm{m}} \tag{2.157}$$

$$L_{\mathrm{a}}\frac{\mathbf{d}\Delta i_{\mathrm{a}}}{\mathbf{d}\tau} = -R_{\mathrm{a}}\Delta i_{\mathrm{a}} - K_{\mathrm{v}}\Delta\omega_{\mathrm{m}} + \Delta U_{\mathrm{m}} \tag{2.158}$$

$$J_{\mathrm{m}}\frac{\mathbf{d}\Delta\omega_{\mathrm{m}}}{\mathbf{d}\tau} = K_{\mathrm{m}}\Delta i_{\mathrm{a}} - \left[B_{\mathrm{m}} + \frac{2\varepsilon_{\mathrm{G}}\varepsilon_{\mathrm{P}}\omega_{\mathrm{m}}}{(2\pi)^3\eta_{\mathrm{f/p}}}\right]_{\mathrm{o}}\Delta\omega_{\mathrm{m}} \tag{2.159}$$

$$\Delta G_{\mathrm{f/p}} = \frac{(\varepsilon_{\mathrm{G}})_{\mathrm{o}}}{2\pi}\Delta\omega_{\mathrm{m}} \tag{2.160}$$

$$\Delta N_{\mathrm{m}} = \alpha_{\mathrm{m}}(U_{\mathrm{m}})_{\mathrm{o}}\Delta i_{\mathrm{a}} + \alpha_{\mathrm{m}}(i_{\mathrm{a}})_{\mathrm{o}}\Delta U_{\mathrm{m}} \tag{2.161}$$

Assume that the inlet air conditions of fan equal to the outlet ones, i.e.,

$$\Delta t_{\mathrm{fan,L}} = \Delta t_{\mathrm{fan,E}} \tag{2.162}$$

$$\Delta W_{\mathrm{fan,L}} = \Delta W_{\mathrm{fan,E}} \tag{2.163}$$

$$\Delta G_{\mathrm{fan,L}} = \Delta G_{\mathrm{fan,E}} = \Delta G_{\mathrm{f/p}} = \frac{(\varepsilon_{\mathrm{G}})_{\mathrm{o}}}{2\pi}\Delta\omega_{\mathrm{m}} \tag{2.164}$$

Choosing $\Delta\theta_{\mathrm{m}}, \Delta\omega_{\mathrm{m}}$ and Δi_{a} as state-space variables, $\Delta t_{\mathrm{fan,E}}, \Delta W_{\mathrm{fan,E}}, \Delta G_{\mathrm{fan,E}}$ and ΔU_{m} as input ones, and $\Delta t_{\mathrm{fan,L}}, \Delta W_{\mathrm{fan,L}}, \Delta G_{\mathrm{fan,L}}$ and ΔN_{m} as output ones, the state-space model for fan is expressed by Eqs. (2.165) and (2.166)

$$\dot{x}_{\mathrm{fan}} = A_{\mathrm{fan}} \cdot x_{\mathrm{fan}} + B_{\mathrm{fan}} \cdot u_{\mathrm{fan}} \tag{2.165}$$

$$y_{\mathrm{fan}} = C_{\mathrm{fan}} \cdot x_{\mathrm{fan}} + D_{\mathrm{fan}} \cdot u_{\mathrm{fan}} \tag{2.166}$$

where

$$\begin{aligned}
x_{\mathrm{fan}} &= [\Delta\theta_{\mathrm{m}}, \Delta i_{\mathrm{a}}, \Delta\omega_{\mathrm{m}}]^T; y_{\mathrm{fan}} = [\Delta t_{\mathrm{fan,L}}, \Delta W_{\mathrm{fan,L}}, \Delta G_{\mathrm{fan,L}}, \Delta N]^T;\\
u_{\mathrm{fan}} &= [\Delta t_{\mathrm{fan,E}}, \Delta W_{\mathrm{fan,E}}, \Delta G_{\mathrm{fan,E}}, \Delta U_{\mathrm{m}}]^T;\\
A_{\mathrm{fan}} &= \begin{bmatrix} 0 & 0 & 1 \\ 0 & -\frac{R_{\mathrm{a}}}{L_{\mathrm{a}}} & -\frac{K_{\mathrm{v}}}{L_{\mathrm{a}}} \\ 0 & \frac{K_{\mathrm{m}}}{J_{\mathrm{m}}} & -\left(\frac{B_{\mathrm{m}}}{J_{\mathrm{m}}} + \frac{2\varepsilon_{\mathrm{G}}\varepsilon_{\mathrm{P}}\omega_{\mathrm{m}}}{(2\pi)^3 J_{\mathrm{m}}\eta_{\mathrm{f/p}}}\right)_{\mathrm{o}} \end{bmatrix};\\
B_{\mathrm{fan}} &= [B_{\mathrm{fan,a}}, B_{\mathrm{fan,U}}]; C_{\mathrm{fan}} = \begin{bmatrix} C_{\mathrm{fan,a}} \\ C_{\mathrm{fan,U}} \end{bmatrix}; D_{\mathrm{fan}} = \begin{bmatrix} D_{\mathrm{fan,a}} \\ D_{\mathrm{fan,U}} \end{bmatrix};
\end{aligned}$$

$$B_{\mathrm{fan,a}} = \begin{bmatrix} 0 & 0 & 0 \\ 0 & 0 & 0 \\ 0 & 0 & 0 \end{bmatrix}; B_{\mathrm{fan,U}} = \begin{bmatrix} 0 \\ 1/L_{\mathrm{a}} \\ 0 \end{bmatrix}; C_{\mathrm{fan,a}} = \begin{bmatrix} 0 & 0 & 0 \\ 0 & 0 & 0 \\ 0 & 0 & (\varepsilon_{\mathrm{G}})_{\mathrm{o}}/2\pi \end{bmatrix};$$

$$C_{\mathrm{fan,U}} = \begin{bmatrix} 0 & a_{\mathrm{m}}(U_{\mathrm{m}})_{\mathrm{o}} & 0 \end{bmatrix}; D_{\mathrm{fan,a}} = \begin{bmatrix} 1 & 0 & 0 & 0 \\ 0 & 1 & 0 & 0 \\ 0 & 0 & 0 & 0 \end{bmatrix}; D_{\mathrm{fan,U}} = \begin{bmatrix} 0 & 0 & 0 & a_{\mathrm{m}}(i_{\mathrm{a}})_{\mathrm{o}} \end{bmatrix}.$$

Likewise, the state-space model for pump is written as follows:

$$\dot{x}_{\mathrm{pump}} = A_{\mathrm{pump}} \cdot x_{\mathrm{pump}} + B_{\mathrm{pump}} \cdot u_{\mathrm{pump}} \tag{2.167}$$

$$y_{\mathrm{pump}} = C_{\mathrm{pump}} \cdot x_{\mathrm{pump}} + D_{\mathrm{pump}} \cdot u_{\mathrm{pump}} \tag{2.168}$$

where

$$x_{\mathrm{pump}} = [\Delta\theta_{\mathrm{m}}, \Delta i_{\mathrm{a}}, \Delta\omega_{\mathrm{m}}]^T; y_{\mathrm{pump}} = [\Delta t_{\mathrm{pump,L}}, \Delta G_{\mathrm{pump,L}}, \Delta N]^T;$$

$$u_{\mathrm{pump}} = [\Delta t_{\mathrm{pump,E}}, \Delta G_{\mathrm{pump,E}}, \Delta U_{\mathrm{m}}]^T;$$

$$A_{\mathrm{pump}} = \begin{bmatrix} 0 & 0 & 1 \\ 0 & -\frac{R_{\mathrm{a}}}{L_{\mathrm{a}}} & -\frac{K_{\mathrm{v}}}{L_{\mathrm{a}}} \\ 0 & \frac{K_{\mathrm{m}}}{J_{\mathrm{m}}} & -\left(\frac{B_{\mathrm{m}}}{J_{\mathrm{m}}} + \frac{2\varepsilon_{\mathrm{G}}\varepsilon_{\mathrm{P}}\omega_{\mathrm{m}}}{(2\pi)^3 J_{\mathrm{m}}\eta_{\mathrm{f/p}}}\right)_{\mathrm{o}} \end{bmatrix};$$

$$B_{\mathrm{pump}} = [B_{\mathrm{pump,w}}, B_{\mathrm{pump,U}}]; C_{\mathrm{pump}} = \begin{bmatrix} C_{\mathrm{pump,w}} \\ C_{\mathrm{pump,U}} \end{bmatrix}; D_{\mathrm{pump}} = \begin{bmatrix} D_{\mathrm{pump,w}} \\ D_{\mathrm{pump,U}} \end{bmatrix};$$

$$B_{\mathrm{pump,w}} = \begin{bmatrix} 0 & 0 \\ 0 & 0 \\ 0 & 0 \end{bmatrix}; B_{\mathrm{pump,U}} = \begin{bmatrix} 0 \\ 1/L_{\mathrm{a}} \\ 0 \end{bmatrix}; C_{\mathrm{pump,w}} = \begin{bmatrix} 0 & 0 & 0 \\ 0 & 0 & (\varepsilon_{\mathrm{G}})_{\mathrm{o}}/2\pi \end{bmatrix};$$

$$C_{\mathrm{pump,U}} = \begin{bmatrix} 0 & a_{\mathrm{m}}(U_{\mathrm{m}})_{\mathrm{o}} & 0 \end{bmatrix}; D_{\mathrm{pump,w}} = \begin{bmatrix} 1 & 0 & 0 \\ 0 & 0 & 0 \end{bmatrix}; D_{\mathrm{pump,U}} = \begin{bmatrix} 0 & 0 & a_{\mathrm{m}}(i_{\mathrm{a}})_{\mathrm{o}} \end{bmatrix}.$$

2.2.5 Air-Conditioned Room Modeling

Although air-conditioned room itself is not part of an air-conditioning system, its model is indispensable for the simulation or analysis of the air-conditioning system because the ultimate objective of an air-conditioning system is to keep an anticipated thermal environment indoors. In this section, a state-space model for air-conditioned room is established. The room model can be used to analyze thermal response characteristics of air in a room.

2.2.5.1 Model Development [17]

1) Assumptions

The following assumptions are made for the room model development:

① Room is separated into three typical zones, i.e., the air-supply, the work, and the air-return zone, as shown in Fig. 2.31. Each air zone is fully mixed and is described by one state.
② The surface temperature of each inner and external wall is all described with a lumped value. The heat transfer from the external walls only affects the air in the work zone, and there is no heat and mass transfer between the air-supply and the air-return zones.
③ In the air-supply zone, light bulbs are the only heat sources defined by a constant surface temperature during the transient response simulation. No moisture sources exist in the air-supply zone.
④ In the work zone, the heat sources are mainly electrical appliances and people indoors, and the moisture sources are evaporative water from the skin and from the respiration of human indoors. The surface temperature of all the heat sources is assumed to be constant during the dynamic response simulation.
⑤ There are no heat and moisture sources in the air-return zone.
⑥ The radiant heat transfer between walls and objects in the room is negligible. The convective heat transfer coefficients between the walls or the heat sources and the adjacent air are considered to be constant during the dynamic response simulation.

2) Basic equations

Based on the above assumptions, the following equations for each indoor air zone can be obtained according to the principle of energy and mass conservation.

(1) *For the air-supply zone*

$$\rho_{\mathrm{a}} V_{\mathrm{a,s}} \frac{\mathrm{d}h_{\mathrm{a,s}}}{\mathrm{d}\tau} = G_{\mathrm{a}}(h_{\mathrm{a,i}} - h_{\mathrm{a,s}}) + \Delta Q_{\mathrm{a,s}} \tag{2.169}$$

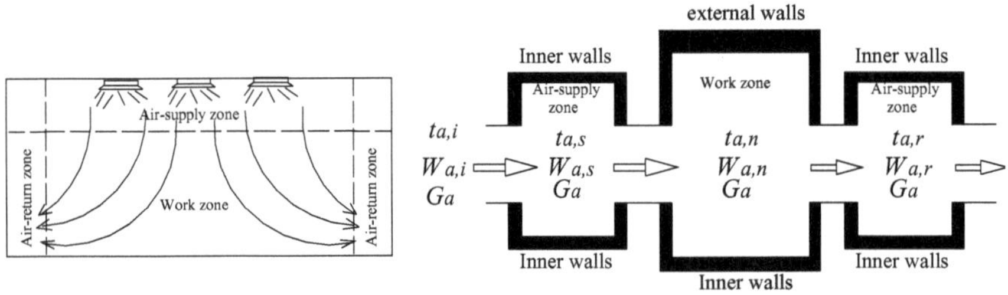

Fig. 2.31 Schematic diagram for air-conditioned room

$$\rho_a V_{a,s} \frac{\mathbf{d}W_{a,s}}{\mathbf{d}\tau} = G_a(W_{a,i} - W_{a,s}) \tag{2.170}$$

$$c_{riw,s}\rho_{riw} V_{riw,s} \frac{\mathbf{d}t_{riw,s}}{\mathbf{d}\tau} = a_{riw,s}A_{riw,s}(t_{a,s} - t_{riw,s}) \tag{2.171}$$

(2) *For the work zone*

$$\rho_a V_{a,n} \frac{\mathbf{d}h_{a,n}}{\mathbf{d}\tau} = G_a(h_{a,s} - h_{a,n}) + \Delta Q_{a,n} \tag{2.172}$$

$$\rho_a V_{a,n} \frac{\mathbf{d}W_{a,n}}{\mathbf{d}\tau} = G_a(W_{a,s} - W_{a,n}) + \Delta M_{Wa,n} \tag{2.173}$$

$$c_{riw,n}\rho_{riw} V_{riw,n} \frac{\mathbf{d}t_{riw,n}}{\mathbf{d}\tau} = a_{riw,n}A_{riw,n}(t_{a,n} - t_{riw,n}) \tag{2.174}$$

$$c_{rew,n}\rho_{rew} V_{rew,n} \frac{\mathbf{d}t_{rew,n}}{\mathbf{d}\tau} = a_{rew,n}A_{rew,n}(t_{a,n} - t_{rew,n}) + \frac{\lambda_{rew}A_{rew,n}}{\delta_{rew}}(t_{rew,o} - t_{rew,n}) \tag{2.175}$$

(3) *For the air-return zone*

$$\rho_a V_{a,r} \frac{\mathbf{d}h_{a,r}}{\mathbf{d}\tau} = G_a(h_{a,n} - h_{a,r}) + \Delta Q_{a,r} \tag{2.176}$$

$$\rho_a V_{a,r} \frac{\mathbf{d}W_{a,r}}{\mathbf{d}\tau} = G_a(W_{a,n} - W_{a,r}) \tag{2.177}$$

$$c_{riw,r}\rho_{riw} V_{riw,r} \frac{\mathbf{d}t_{riw,r}}{\mathbf{d}\tau} = a_{riw,r}A_{riw,r}(t_{a,r} - t_{riw,r}) \tag{2.178}$$

3) Key parameters

The enthalpy of air in different zones can be calculated by Eq. (2.179).

$$h_a = c_a t_a + \beta_2 W_a \tag{2.179}$$

where β_2 equals to 2.5×10^6.

Energy gain rate of the air in the air-supply zone, $\Delta Q_{a,s}$, consists of the following three parts: (1) heat gain rate from the internal walls, $\Delta Q_{a,s,1}$; (2) heat gain rate from the adjacent air zone (the work zone), $\Delta Q_{a,s,2}$; and (3) heat gain rate from the indoor heat sources, $\Delta Q_{a,s,3}$.

$$\Delta Q_{a,s,1} = a_{riw,s} A_{riw,s} (t_{riw,s} - t_{a,s}) \tag{2.180}$$

$$\Delta Q_{a,s,2} = a_{as-an} A_{as-an} (t_{a,n} - t_{a,s}) \tag{2.181}$$

$$\Delta Q_{a,s,3} = a_{rq,s} A_{rq,s} (t_{rq,s} - t_{a,s}) \tag{2.182}$$

Energy gain rate of the air in the work zone, $\Delta Q_{a,n}$, consists of the following four parts: ① heat gain rate from the internal and external walls, $\Delta Q_{a,n,1}$; ② heat gain rate from the adjacent air zones, $\Delta Q_{a,n,2}$; ③heat gain rate from the indoor heat sources, $\Delta Q_{a,n,3}$; and ④ heat gain rate due to the exhaled air from the occupants, $\Delta Q_{a,n,4}$.

$$\Delta Q_{a,n,1} = a_{riw,n} A_{riw,n} (t_{riw,n} - t_{a,n}) + a_{rew,n} A_{rew,n} (t_{rew,n} - t_{a,n}) \tag{2.183}$$

$$\Delta Q_{a,n,2} = a_{as-an} A_{as-an} (t_{a,s} - t_{a,n}) + a_{an-ar} A_{an-ar} (t_{a,r} - t_{a,n}) \tag{2.184}$$

$$\Delta Q_{a,n,3} = \sum_{i=1}^{k} \left[a_{rq,n} A_{rq,n}^{(i)} \left(t_{rq,n}^{(i)} - t_{a,n} \right) \right] \tag{2.185}$$

$$\Delta Q_{a,n,4} = G_{res} (h_{exhale} - h_{a,n}) \tag{2.186}$$

In the work zone, there are different kinds of indoor heat sources with different surface temperatures. The superscript 'i' in Eq. (2.185) stands for the ith indoor heat source. The enthalpy of the exhaled air, h_{exhale}, is related to the temperature (t_{exhale}) and humidity ratio (W_{exhale}) of the exhaled air. For typical indoor environments ($t_{a,n} = 25\,°C$), the exhaled temperature and humidity ratio are given in terms of ambient conditions [8]:

$$t_{exhale} = 32.6 + 0.066 t_{a,n} \tag{2.187}$$

$$W_{exhale} = 0.02933 + 0.2 W_{a,n} \tag{2.188}$$

Please note that the unit of $W_{a,n}$ in Eq. (2.188) is kg/(kg dry air). The pulmonary ventilation rate, G_{res}, is primarily a function of metabolic rate as follows [8]:

$$G_{res} = C_{res} E_{body} A_{rq,n}^{(1)} \tag{2.189}$$

where C_{res} is a proportionality constant (1.43×10^{-6}kg/J); E_{body} is metabolic rate, W/m_2; and $A_{rq,n}^{(1)}$ is body surface area of occupant indoors (m^2), which is calculated by Eq. (2.190) [8].

$$A_{rq,n}^{(1)} = 0.202 M_{body}^{0.425} H_{body}^{0.725} \tag{2.190}$$

where M_{body} and H_{body} are the weight and the height of the occupant.

Moisture gain rate of the air in the work zone, $\Delta M_{\mathrm{Wa,n}}$, is calculated by:

$$\Delta M_{\mathrm{Wa,n}} = G_{\mathrm{res}}(W_{\mathrm{exhale}} - W_{\mathrm{a,n}}) = C_{\mathrm{res}} E_{\mathrm{body}} A^{(1)}_{\mathrm{rq,n}}(0.02933 - 0.8W_{\mathrm{a,n}}) \quad (2.191)$$

Energy gain rate of the air in the air-return zone, $\Delta Q_{\mathrm{a,r}}$, consists of two parts: (1) heat gain rate from the internal walls, $\Delta Q_{\mathrm{a,r,1}}$; and (2) heat gain rate from the adjacent air zone (the work zone), $\Delta Q_{\mathrm{a,r,2}}$.

$$\Delta Q_{\mathrm{a,r,1}} = a_{\mathrm{riw,r}} A_{\mathrm{riw,r}}(t_{\mathrm{riw,r}} - t_{\mathrm{a,r}}) \quad (2.192)$$

$$\Delta Q_{\mathrm{a,r,2}} = a_{\mathrm{an-ar}} A_{\mathrm{an-ar}}(t_{\mathrm{a,n}} - t_{\mathrm{a,r}}) \quad (2.193)$$

The external surface temperature of external walls, $t_{\mathrm{rew,o}}$, is affected by the ambient air temperature and solar radiation intensity on the walls, which can be calculated by:

$$t_{\mathrm{rew,o}} = t_{\mathrm{a,out}} + \alpha_{\rho,rew} I_{\mathrm{sol}} / a_{\mathrm{rew,o}} \quad (2.194)$$

The division of the room air volume into air zones ($V_{\mathrm{a,s}}, V_{\mathrm{a,n}}, V_{\mathrm{a,r}}$) can be determined based on the airflow pattern and the steady-state temperature field calculated with the CFD method [18, 19]. Meanwhile, the actual situations should be fully considered in the division of the indoor air zones. Normally, the work zone is less than 2.0 m in height.

The other critical parameters of the zonal model include the heat transfer coefficients between the air and the internal surface of the room walls ($a_{\mathrm{riw,s}}$, $a_{\mathrm{riw,n}}$, $a_{\mathrm{riw,r}}$),the indoor heat sources ($a_{\mathrm{rq,s}}$, $a_{\mathrm{rq,n}}$), and the heat transfer coefficients between two adjacent air zones ($a_{\mathrm{an-ar}}$ and $a_{\mathrm{as-an}}$). Basically, these heat transfer coefficients are originally obtained from the relevant literature [20, 21] and need to be adjusted according to the comparisons of the calculated results and the experimental data.

4) State-space representation

Through linearization, Eqs. (2.169)–(2.178) can be written as follows:

$$T_{\mathrm{tas}} \frac{\mathbf{d}\Delta t_{\mathrm{a,s}}}{\mathbf{d}\tau} = X_{\mathrm{tas,1}} \Delta t_{\mathrm{a,s}} + X_{\mathrm{tas,2}} \Delta t_{\mathrm{a,i}} + X_{\mathrm{tas,3}} \Delta t_{\mathrm{riw,s}} + X_{\mathrm{tas,4}} \Delta t_{\mathrm{a,n}} + X_{\mathrm{tas,5}} \Delta G_{\mathrm{a,i}} \quad (2.195)$$

$$T_{\mathrm{was}} \frac{\mathbf{d}\Delta W_{\mathrm{a,s}}}{\mathbf{d}\tau} = X_{\mathrm{was,1}} \Delta W_{\mathrm{a,s}} + X_{\mathrm{was,2}} \Delta W_{\mathrm{a,i}} + X_{\mathrm{was,3}} \Delta G_{\mathrm{a,i}} \quad (2.196)$$

$$T_{triws} \frac{\mathbf{d}\Delta t_{\mathrm{riw,s}}}{\mathbf{d}\tau} = X_{triws,1} \Delta t_{\mathrm{a,s}} + X_{triws,2} \Delta t_{\mathrm{riw,s}} \quad (2.197)$$

$$T_{\mathrm{tan}}\frac{\mathbf{d}\Delta t_{\mathrm{a,n}}}{\mathbf{d}\tau} = X_{\mathrm{tan,1}}\Delta t_{\mathrm{a,s}} + X_{\mathrm{tan,2}}\Delta t_{\mathrm{a,n}} + X_{\mathrm{tan,3}}\Delta t_{\mathrm{riw,n}} + X_{\mathrm{tan,4}}\Delta t_{\mathrm{rew,n}} + X_{\mathrm{tan,5}}\Delta t_{\mathrm{a,r}} + X_{\mathrm{tan,6}}\Delta G_{\mathrm{a,i}} + \sum_{i=1}^{k}\left[X_{\mathrm{tan,7}}^{(i)}\Delta A_{\mathrm{rq,n}}^{(i)}\right] \tag{2.198}$$

$$T_{\mathrm{wan}}\frac{\mathbf{d}\Delta W_{\mathrm{a,n}}}{\mathbf{d}\tau} = X_{\mathrm{wan,1}}\Delta W_{\mathrm{a,s}} + X_{\mathrm{wan,2}}\Delta W_{\mathrm{a,n}} + X_{\mathrm{wan,3}}\Delta G_{\mathrm{a,i}} + X_{\mathrm{wan,4}}\Delta A_{\mathrm{rq,n}}^{(1)} \tag{2.199}$$

$$T_{\mathrm{triwn}}\frac{\mathbf{d}\Delta t_{\mathrm{riw,n}}}{\mathbf{d}\tau} = X_{\mathrm{triwn,1}}\Delta t_{\mathrm{a,n}} + X_{\mathrm{triwn,2}}\Delta t_{\mathrm{riw,n}} \tag{2.200}$$

$$T_{\mathrm{trewn}}\frac{\mathbf{d}\Delta t_{\mathrm{rew,n}}}{\mathbf{d}\tau} = X_{\mathrm{trewn,1}}\Delta t_{\mathrm{a,n}} + X_{\mathrm{trewn,2}}\Delta t_{\mathrm{rew,n}} + X_{\mathrm{trewn,3}}\Delta t_{\mathrm{a,out}} + X_{\mathrm{trewn,4}}\Delta I_{\mathrm{sol}} \tag{2.201}$$

$$T_{\mathrm{tar}}\frac{\mathbf{d}\Delta t_{\mathrm{a,r}}}{\mathbf{d}\tau} = X_{\mathrm{tar,1}}\Delta t_{\mathrm{a,n}} + X_{\mathrm{tar,2}}\Delta t_{\mathrm{a,r}} + X_{\mathrm{tar,3}}\Delta t_{\mathrm{riw,r}} + X_{\mathrm{tar,4}}\Delta G_{\mathrm{a,i}} \tag{2.202}$$

$$T_{\mathrm{war}}\frac{\mathbf{d}\Delta W_{\mathrm{a,r}}}{\mathbf{d}\tau} = X_{\mathrm{war,1}}\Delta W_{\mathrm{a,n}} + X_{\mathrm{war,2}}\Delta W_{\mathrm{a,r}} + X_{\mathrm{war,3}}\Delta G_{\mathrm{a,i}} \tag{2.203}$$

$$T_{\mathrm{triwr}}\frac{\mathbf{d}\Delta t_{\mathrm{riw,r}}}{\mathbf{d}\tau} = X_{\mathrm{triwr,1}}\Delta t_{\mathrm{a,r}} + X_{\mathrm{triwr,2}}\Delta t_{\mathrm{riw,r}} \tag{2.204}$$

The coefficients in Eqs. (2.195)–(2.204) are listed in Table 2.15.

Thus, the state-space model for air-conditioned room can be written as follows:

$$\dot{x}_{\mathrm{room}} = A_{\mathrm{room}} \cdot x_{\mathrm{room}} + B_{\mathrm{room}} \cdot u_{\mathrm{room}} \tag{2.205}$$

$$y_{\mathrm{room}} = C_{\mathrm{room}} x_{\mathrm{room}} \tag{2.206}$$

where

$$x_{room} = [\Delta t_{a,s}\ \ \Delta W_{a,s}\ \ \Delta t_{riw,s}\ \ \Delta t_{a,n}\ \ \Delta W_{a,n}\ \ \Delta t_{riw,n}\ \ \Delta t_{rew,n}\ \ \Delta t_{a,r}\ \ \Delta W_{a,r}\ \ \Delta t_{riw,r}]^T;$$

$$y_{room} = [\Delta t_{a,s}、\Delta W_{a,s}、\Delta t_{a,n}、\Delta W_{a,n}、\Delta t_{a,r}、\Delta W_{a,r}]^T;$$

$$u_{room} = [\Delta t_{a,i}\ \ \Delta W_{a,i}\ \ \Delta G_{a,i}\ \ \Delta A^{(1)}{}_{rq,n}\ \ \Delta A^{(2)}{}_{rq,n}\ \ \Delta A^{(i)}{}_{rq,n}\ \ \Delta t_{a,out}\ \ \Delta I_{sol}]^T;$$

$$A_{room} = \begin{bmatrix}
X_{tas,1}/T_{tas} & 0 & X_{tas,3}/T_{tas} & X_{tas,4}/T_{tas} & 0 & 0 & 0 & 0 & 0 & 0 \\
0 & X_{was,1}/T_{was} & 0 & 0 & 0 & 0 & 0 & 0 & 0 & 0 \\
X_{triws,1}/T_{triws} & 0 & X_{triws,2}/T_{triws} & 0 & 0 & 0 & 0 & 0 & 0 & 0 \\
X_{tan,1}/T_{tan} & 0 & 0 & X_{tan,2}/T_{tan} & 0 & X_{tan,3}/T_{tan} & X_{tan,4}/T_{tan} & X_{tan,5}/T_{tan} & 0 & 0 \\
0 & X_{wan,1}/T_{wan} & 0 & 0 & X_{wan,2}/T_{wan} & 0 & 0 & 0 & 0 & 0 \\
0 & 0 & 0 & X_{triwn,1}/T_{triwn} & 0 & X_{triwn,2}/T_{triwn} & 0 & 0 & 0 & 0 \\
0 & 0 & 0 & X_{trewn,1}/T_{trewn} & 0 & 0 & X_{trewn,2}/T_{trewn} & 0 & 0 & 0 \\
0 & 0 & 0 & X_{tar,1}/T_{tar} & 0 & 0 & 0 & X_{tar,2}/T_{tar} & 0 & X_{tar,3}/T_{tar} \\
0 & 0 & 0 & 0 & X_{war,1}/T_{war} & 0 & 0 & 0 & X_{war,2}/T_{war} & 0 \\
0 & 0 & 0 & 0 & 0 & 0 & 0 & X_{triwr,1}/T_{triwr} & 0 & X_{triwr,2}/T_{triwr}
\end{bmatrix}$$

Table 2.15 Coefficients in Eqs. (2.195) through (2.204)

Equation No.	Coefficient expression
Eq. (2.195)	$T_{\mathrm{tas}} = c_{\mathrm{a}}\rho_{\mathrm{a}}V_{\mathrm{a,s}};\ X_{\mathrm{tas,1}} = -(c_{\mathrm{a}}G_{\mathrm{a}} + a_{\mathrm{riw,s}}A_{\mathrm{riw,s}} + a_{\mathrm{as-an}}A_{\mathrm{as-an}} + a_{\mathrm{rq,s}}A_{\mathrm{rq,s}})_{\mathrm{o}};$ $X_{\mathrm{tas,2}} = c_{\mathrm{a}}(G_{\mathrm{a}})_{\mathrm{o}};\ X_{\mathrm{tas,3}} = a_{\mathrm{riw,s}}A_{\mathrm{riw,s}};\ X_{\mathrm{tas,4}} = a_{\mathrm{as-an}}A_{\mathrm{as-an}};$ $X_{\mathrm{tas,5}} = c_{\mathrm{a}}(t_{\mathrm{a,i}} - t_{\mathrm{a,s}})_{\mathrm{o}}$
Eq. (2.196)	$T_{\mathrm{was}} = \rho_{\mathrm{a}}V_{\mathrm{a,s}};\ X_{\mathrm{was,1}} = -(G_{\mathrm{a}})_{\mathrm{o}};\ X_{\mathrm{was,2}} = (G_{\mathrm{a}})_{\mathrm{o}};$ $X_{\mathrm{was,3}} = (W_{\mathrm{a,i}} - W_{\mathrm{a,s}})_{\mathrm{o}}$
Eq. (2.197)	$T_{triws} = c_{\mathrm{riw,s}}\rho_{\mathrm{riw}}V_{\mathrm{riw,s}};\ X_{\mathrm{triws,1}} = a_{\mathrm{riw,s}}A_{\mathrm{riw,s}};\ X_{\mathrm{triws,2}} = -a_{\mathrm{riw,s}}A_{\mathrm{riw,s}}$
Eq. (2.198)	$T_{\mathrm{tan}} = c_{\mathrm{a}}\rho_{\mathrm{a}}V_{\mathrm{a,n}};\ X_{\mathrm{tan,1}} = c_{\mathrm{a}}(G_{\mathrm{a}})_{\mathrm{o}} + a_{\mathrm{as-an}}A_{\mathrm{as-an}};$ $X_{\mathrm{tan,2}} = -\left[\begin{array}{l} c_{\mathrm{a}}(G_{\mathrm{a}})_{\mathrm{o}} + 0.934c_{\mathrm{a}}C_{\mathrm{res}}E_{\mathrm{body}}A^{(1)}_{\mathrm{rq,n}} + a_{\mathrm{riw,n}}A_{\mathrm{riw,n}} \\ \quad + a_{\mathrm{rew,n}}A_{\mathrm{rew,n}} + a_{\mathrm{as-an}}A_{\mathrm{as-an}} + a_{\mathrm{an-ar}}A_{\mathrm{an-ar}} + \sum_{i=1}^{k}\left[a_{\mathrm{rq,n}}(A^{(i)}_{\mathrm{rq,n}})_{\mathrm{o}}\right] \end{array}\right];$ $X_{\mathrm{tan,3}} = a_{\mathrm{riw,n}}A_{\mathrm{riw,n}};\ X_{\mathrm{tan,4}} = a_{\mathrm{rew,n}}A_{\mathrm{rew,n}};\ X_{\mathrm{tan,5}} = a_{\mathrm{an-ar}}A_{\mathrm{an-ar}};$ $X_{\mathrm{tan,6}} = c_{\mathrm{a}}(t_{\mathrm{a,s}} - t_{\mathrm{a,n}})_{\mathrm{o}}; X^{(i)}_{\mathrm{tan,7}} = a_{\mathrm{rq,n}}\left[t^{(i)}_{\mathrm{rq,n}} - t_{\mathrm{a,n}}\right]_{\mathrm{o}}$
Eq. (2.199)	$T_{\mathrm{wan}} = \rho_{\mathrm{a}}V_{\mathrm{a,n}}; X_{\mathrm{wan,1}} = (G_{\mathrm{a}})_{\mathrm{o}};\ X_{\mathrm{wan,2}} = \left[-0.8C_{\mathrm{res}}E_{\mathrm{body}}A^{(1)}_{\mathrm{rq,n}} - G_{\mathrm{a}}\right]_{\mathrm{o}};$ $X_{\mathrm{wan,3}} = (W_{\mathrm{a,s}} - W_{\mathrm{a,n}})_{\mathrm{o}}; X_{\mathrm{wan,4}} = C_{\mathrm{res}}E_{\mathrm{body}}(0.02933 - 0.8W_{\mathrm{a,n}})_{\mathrm{o}}$
Eq. (2.200)	$T_{\mathrm{triwn}} = c_{\mathrm{riw,n}}\rho_{\mathrm{riw}}V_{\mathrm{riw,n}};\ X_{\mathrm{triwn,1}} = a_{\mathrm{riw,n}}A_{\mathrm{riw,n}};\ X_{\mathrm{triwn,2}} = -a_{\mathrm{riw,n}}A_{\mathrm{riw,n}}$
Eq. (2.201)	$T_{\mathrm{trewn}} = c_{\mathrm{rew,n}}\rho_{\mathrm{rew}}V_{\mathrm{rew,n}}; X_{\mathrm{trewn,1}} = a_{\mathrm{rew,n}}A_{\mathrm{rew,n}};$ $X_{\mathrm{trewn,2}} = -a_{\mathrm{rew,n}}A_{\mathrm{rew,n}} - \lambda_{\mathrm{rew}}A_{\mathrm{rew,n}}/\delta_{\mathrm{rew}}; X_{\mathrm{trewn,3}} = \lambda_{\mathrm{rew}}A_{\mathrm{rew,n}}/\delta_{\mathrm{rew}};$ $X_{\mathrm{trewn,4}} = \alpha_{\rho,rew}\lambda_{\mathrm{rew}}A_{\mathrm{rew,n}}/(a_{\mathrm{rew,o}}\delta_{\mathrm{rew}})$
Eq. (2.202)	$T_{\mathrm{tar}} = c_{\mathrm{a}}\rho_{\mathrm{a}}V_{\mathrm{a,r}};\ X_{\mathrm{tar,1}} = c_{\mathrm{a}}(G_{\mathrm{a}})_{\mathrm{o}} + a_{\mathrm{an-ar}}A_{\mathrm{an-ar}};$ $X_{\mathrm{tar,2}} = -[a_{\mathrm{riw,r}}A_{\mathrm{riw,r}} + c_{\mathrm{a}}(G_{\mathrm{a}})_{\mathrm{o}} + a_{\mathrm{an-ar}}A_{\mathrm{an-ar}}];$ $X_{\mathrm{tar,3}} = a_{\mathrm{riw,r}}A_{\mathrm{riw,r}}; X_{\mathrm{tar,4}} = c_{\mathrm{a}}(t_{\mathrm{a,n}} - t_{\mathrm{a,r}})_{\mathrm{o}}$
Eq. (2.203)	$T_{\mathrm{war}} = \rho_{\mathrm{a}}V_{\mathrm{a,r}}; X_{\mathrm{tar,1}} = c_{\mathrm{a}}(G_{\mathrm{a}})_{\mathrm{o}} + a_{\mathrm{an-ar}}A_{\mathrm{an-ar}};$ $X_{\mathrm{war,1}} = (G_{\mathrm{a}})_{\mathrm{o}};\ X_{\mathrm{war,2}} = -(G_{\mathrm{a}})_{\mathrm{o}};\ X_{\mathrm{war,3}} = (W_{\mathrm{a,n}} - W_{\mathrm{a,r}})_{\mathrm{o}}$
Eq. (2.204)	$T_{\mathrm{triwr}} = c_{\mathrm{riw,r}}\rho_{\mathrm{riw}}V_{\mathrm{riw,r}};\ X_{\mathrm{triwr,1}} = a_{\mathrm{riw,r}}A_{\mathrm{riw,r}}; X_{\mathrm{triwr,2}} = -a_{\mathrm{riw,r}}A_{\mathrm{riw,r}}$

$$
B_{room} = \begin{bmatrix}
X_{tas,2}/T_{tas} & 0 & X_{tas,5}/T_{tas} & 0 & 0 & 0 & 0 & 0 & 0 \\
0 & X_{was,2}/T_{was} & X_{was,3}/T_{was} & 0 & 0 & 0 & 0 & 0 & 0 \\
0 & 0 & 0 & 0 & 0 & 0 & 0 & 0 & 0 \\
0 & 0 & X_{tan,6}/T_{tan} & X^{(1)}{}_{tan,7}/T_{tan} & X^{(2)}{}_{tan,7}/T_{tan} & \cdots & X^{(i)}{}_{tan,7}/T_{tan} & 0 & 0 \\
0 & 0 & X_{wan,3}/T_{wan} & X_{wan,4}/T_{wan} & 0 & 0 & 0 & 0 & 0 \\
0 & 0 & 0 & 0 & 0 & 0 & 0 & 0 & 0 \\
0 & 0 & 0 & 0 & 0 & 0 & 0 & X_{trewn,3}/T_{trewn} & X_{trewn,4}/T_{trewn} \\
0 & 0 & X_{tar,4}/T_{tar} & 0 & 0 & 0 & 0 & 0 & 0 \\
0 & 0 & X_{war,3}/T_{war} & 0 & 0 & 0 & 0 & 0 & 0 \\
0 & 0 & 0 & 0 & 0 & 0 & 0 & 0 & 0
\end{bmatrix}
$$

$$
C_{room} = \begin{bmatrix}
1 & 0 & 0 & 0 & 0 & 0 & 0 & 0 & 0 & 0 \\
0 & 1 & 0 & 0 & 0 & 0 & 0 & 0 & 0 & 0 \\
0 & 0 & 0 & 1 & 0 & 0 & 0 & 0 & 0 & 0 \\
0 & 0 & 0 & 0 & 1 & 0 & 0 & 0 & 0 & 0 \\
0 & 0 & 0 & 0 & 0 & 0 & 0 & 1 & 0 & 0 \\
0 & 0 & 0 & 0 & 0 & 0 & 0 & 0 & 1 & 0
\end{bmatrix}
$$

The matrix A_{room} in the state-space model contains different heat transfer coefficients which are affected by airflow rate and turbulence.

2.2.5.2 Model Validation

(1) Experimental system and conditions

To validate the room model, experiments have been conducted in a full-size air-conditioned room. The internal walls of the room are mainly made from bricks and limestone with a thickness of 16.8 cm. The air-handling system consists of a water-to-air surface heat exchanger, a ventilator, and air-supply/return ducts. Two air-supply diffusers are located in the upper space of the room for a favorable air distribution in the room. The supply air temperature can be controlled through adjusting supply water temperature of the heat exchanger.

The state-space model is validated experimentally mainly in terms of transient responses of air temperature in different zones and that of air humidity in the return air zone to perturbations of supply air temperature and humidity. According to the airflow patterns obtained by the standard $k-\varepsilon$ CFD model, the size of the air-supply, the work, and the air-return zone is identified as $348 \times 270 \times 70$, $348 \times 270 \times 180\,\text{cm}$, and $80 \times 270 \times 180\,\text{cm}$ (length × width × height),

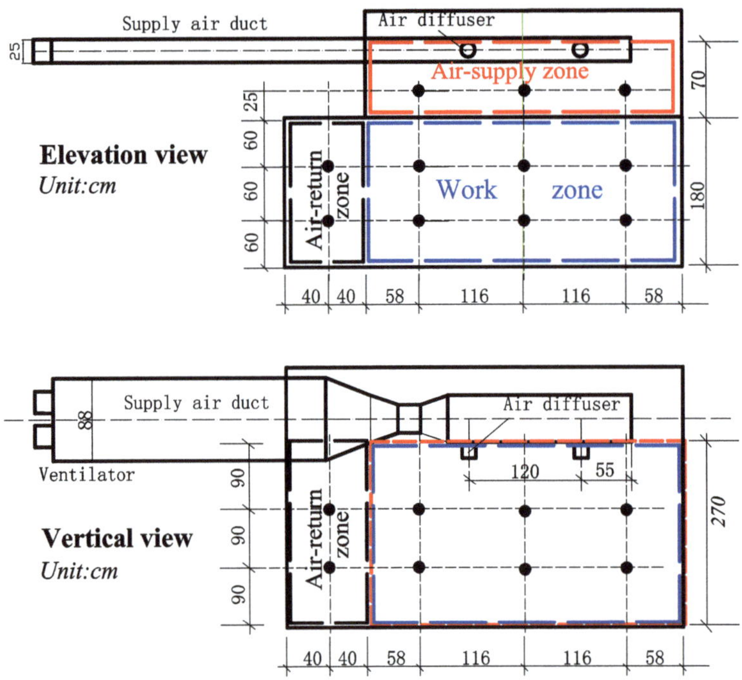

Fig. 2.32 Test room and detailed positions of temperature sensors

respectively, as shown in Fig. 2.23. The internal surface area of walls corresponding to the three zones is 13.72, 20.52, and 3.60 m^2 respectively. The contact area between the work zone and the air-supply zone and the air-return zone is 9.40 m^2 and 4.90 m^2, respectively. Figure 2.32 also shows detailed locations of thermocouples (measurement precision: ±0.1 °C) in the room. The measured air temperature of each zone is obtained through taking average value of all test points in corresponding zone. A humidity sensor (measurement precision: ±0.8 % of the humidity ratio) is placed in the air-return zone to observe the air humidity. The supply airflow rate is measured with a hot-wire anemometer with a measurement precision of ±0.015 m/s. Being one of the initial conditions, the internal surface temperatures of the walls are measured by thermocouples (measurement precision: ±0.1 °C) embedded in the internal surface of the walls. The heat sources in the air-supply zone are fluorescent lamps whose surface area is estimated as 0.3 m^2. In the work zone, there are two adults (total body surface area: 3.65 m^2) and a constant-temperature plate heater (surface area: 0.28 m^2). An infrared thermometer (measurement precision: ±0.2 °C) is used to measure surface temperature of the indoor heat sources. The surface temperature was measured to be 39.6, 35.2, and 43.9 °C, respectively, for the lamps, the human bodies, and the plate heater during the dynamic response experiments.

Two experimental cases have been investigated for a time period of 2400 s. The initial conditions for the model validation (see Table 2.16) were obtained before the

Table 2.16 Initial conditions for the room model validation

Experimental cases Parameters	Case I	Case II
Air-supply temperature $(t_{a,i})_o$ (°C)	31.4	26.4
Air-supply humidity $(W_{a,i})_o$ (g/(kg dry air))	20.3	18.1
Air temperature in the air-supply zone $(t_{a,s})_o$ (°C)	32.1	28.0
Air humidity in the air-supply zone $(W_{a,s})_o$ (g/(kg dry air))	20.3	18.1
Internal surface temperature of walls in the air-supply zone $(t_{riw,s})_o$ (°C)	32.7	30.2
Air temperature in the work zone $(t_{a,n})_o$ (°C)	33.0	29.5
Air humidity in the work zone $(W_{a,n})_o$ (g/(kg dry air))	20.5	18.3
Internal surface temperature of walls in the work zone $(t_{riw,n})_o$ (°C)	33.2	30.6
Air temperature in the air-return zone $(t_{a,r})_o$ (°C)	33.1	29.9
Air humidity in the air-return zone $(W_{a,r})_o$ (g/(kg dry air))	20.5	18.3
Internal surface temperature of walls in the air-return zone $(t_{riw,r})_o$ (°C)	33.6	31.1
Supply airflow rate $(G_a)_o$ (kg/s)	0.15	0.20

Other parameters for the model calculation: $c_a = 1005$ J/(kg °C); $c_{riw} = 1250$ J/(kg °C); $\rho_a = 1.18$ kg/m^3; $\rho_{riw} = 1800$ kg/m^3; $\delta_{riw} = 0.22$ m

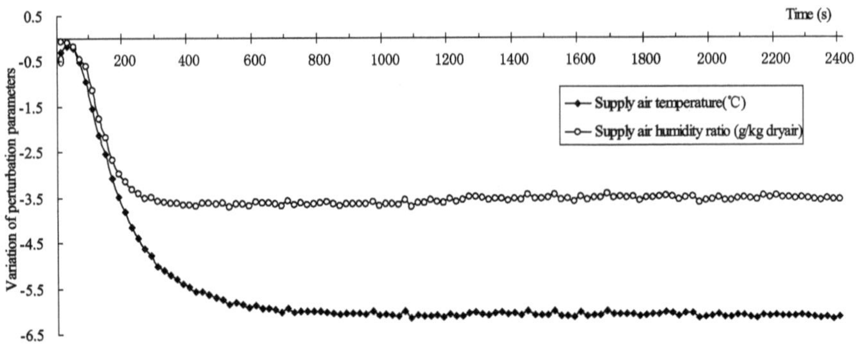

Fig. 2.33 Variation of perturbation parameters in the case I (measured data)

response experiments began. All the values of air temperature and humidity as well as wall surface temperatures were collected with a data acquisition system at a sample interval of two seconds.

(2) Experimental results

Experimental case I

Figure 2.33 shows the variation of perturbation parameters in the experimental case *I*, in which the supply air temperature and humidity ratio decrease till to steady value. The calculated and experimental results on the transient response of indoor air temperature and humidity in different air zones under the initial conditions of case *I* are shown in Figs. 2.34 and 2.35. In the model calculation, the convective heat transfer coefficients are taken as 9.2, 11.6 and 3.2 W/(m^2.K), respectively, for the room walls, the indoor heat sources, and the air layers between the adjacent indoor air zones. As shown in Figs. 2.34 and 2.35, the calculated response curves of the indoor air temperatures and humidity ratios are favorably consistent with the experimental ones, and the average errors of the calculated results compared with the experimental data are all lower than 10 % .

Experimental case II

Figure 2.36 shows the perturbations under the initial conditions of the case *II*, in which the supply air temperature and humidity ratio increase. The corresponding results on the response of the indoor air temperature and humidity in different air zones are shown in Figs. 2.37 and 2.38. In this case, the heat transfer coefficients of internal wall surface and indoor heat sources as well as the air layers between the adjacent indoor air zones are adjusted, respectively, as 10.7, 12.3, and 3.6 W/(m^2 K) for the model calculations, which are a little larger than that in the case *I* due to the higher supply airflow rate in the case *II* (0.20 kg/s). The average errors of the calculated results compared with the experimental data are no more than 12 % for the case *II*.

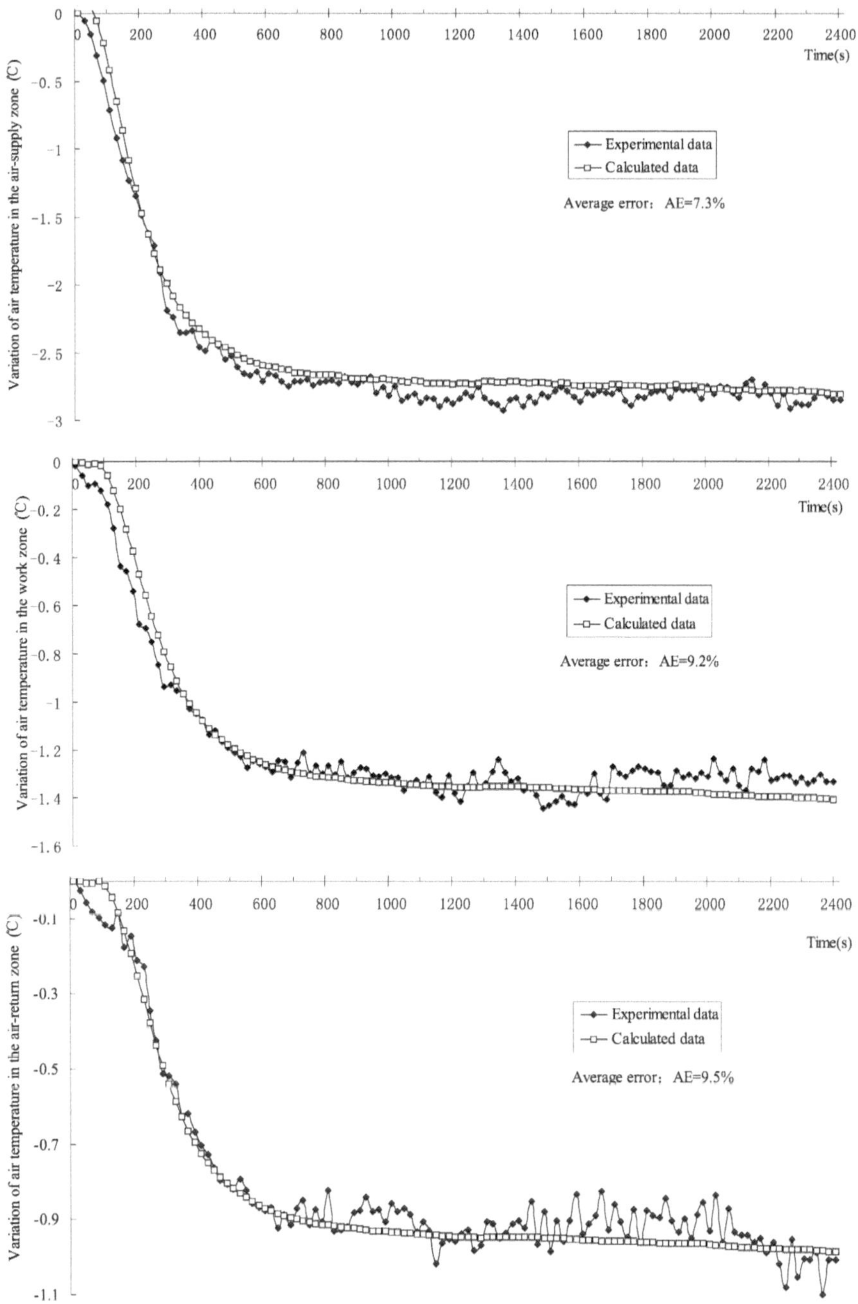

Fig. 2.34 Dynamic responses of air temperature in different zones in the case I (calculation vs. measurement)

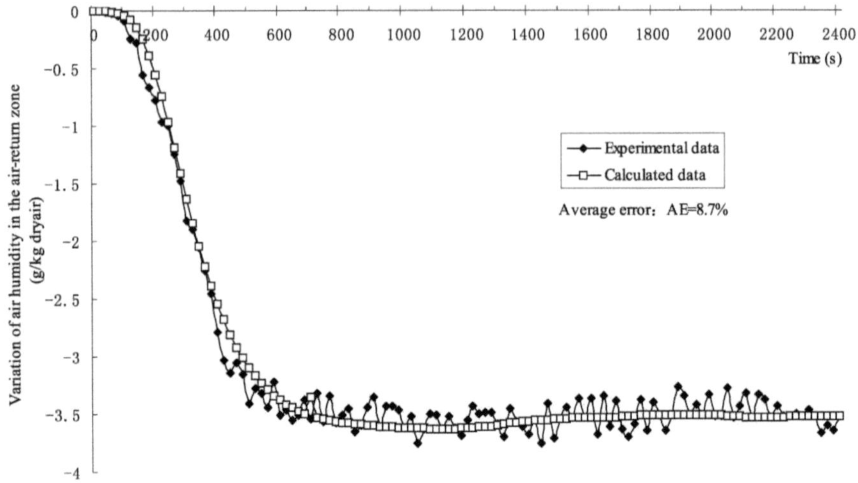

Fig. 2.35 Dynamic responses of air humidity in the air-return zone in the case I (calculation vs. measurement)

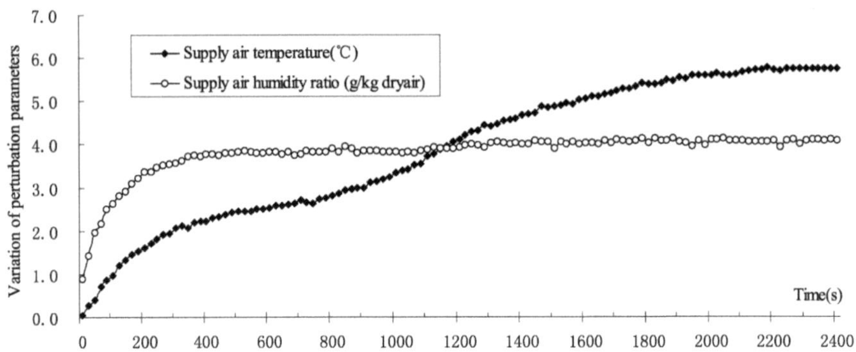

Fig. 2.36 Variation of perturbation parameters in the case II (measured data)

2.3 Modeling for HVAC System

2.3.1 Component Model Connection

HVAC system comprises of numerous components or subsystems. These components or subsystems are generally connected in parallel or series, as shown in Fig. 2.39.

In Fig. 2.39, the components S_1 and S_2 are expressed, respectively, by Eqs. (2.207) and (2.208)

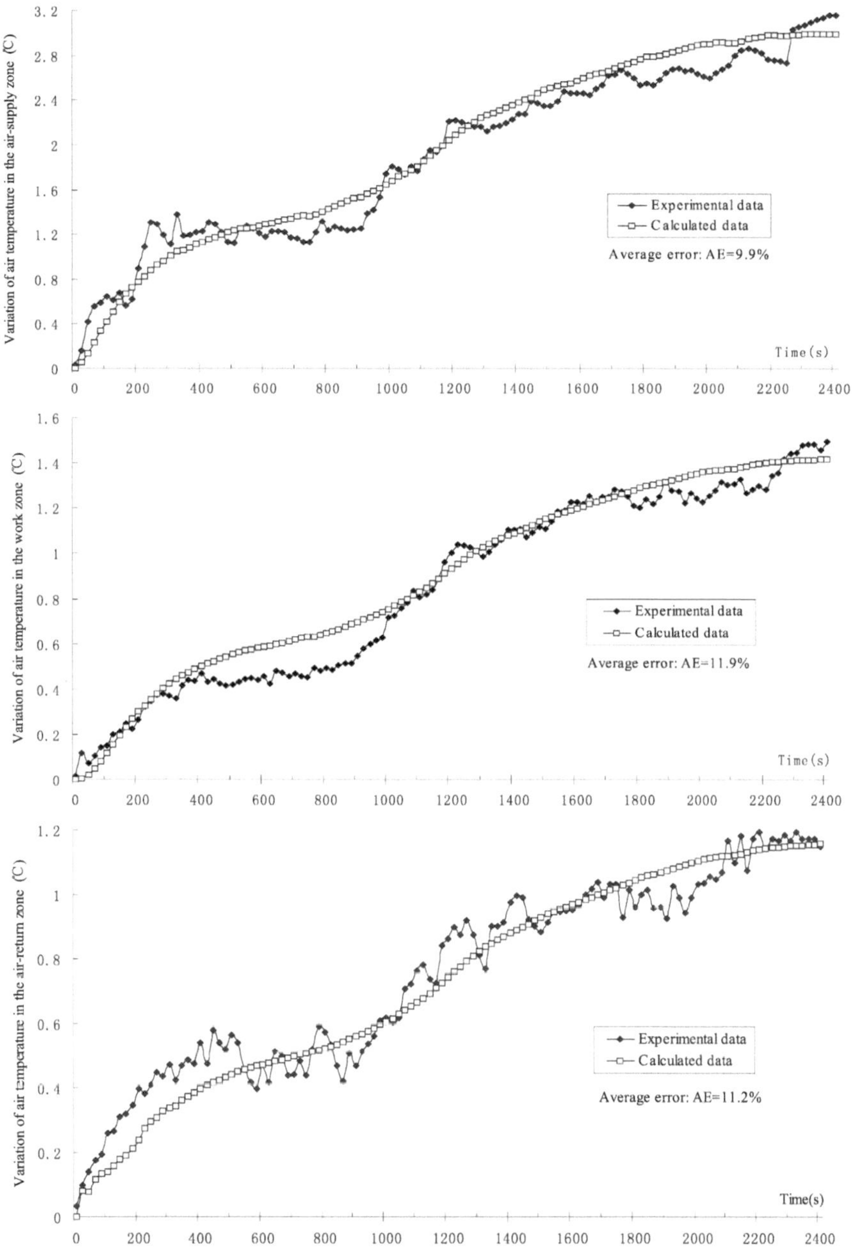

Fig. 2.37 Dynamic responses of air temperature in different zones in the case II (calculation vs. measurement)

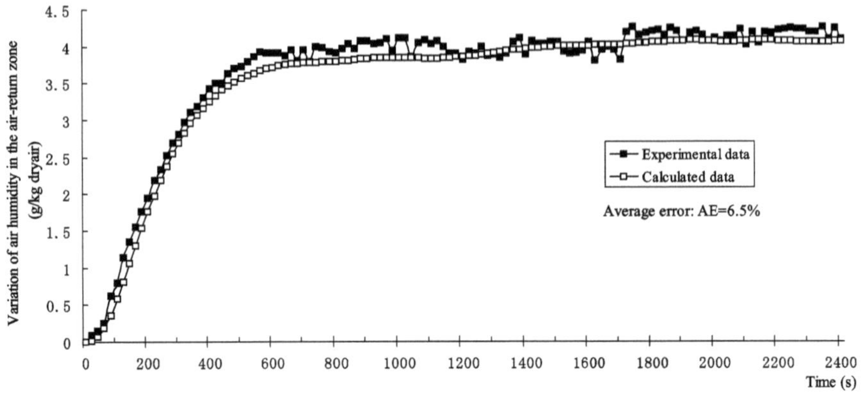

Fig. 2.38 Dynamic responses of air humidity in the air-return zone in the case II (calculation vs. measurement)

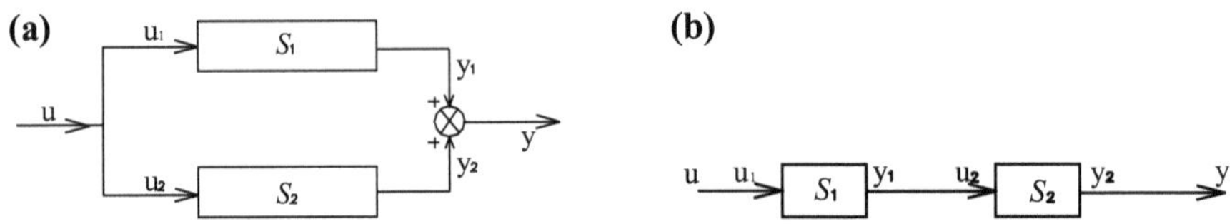

Fig. 2.39 Schematic diagram for the component model connection. **a** Parallel connection. **b** Series connection

$$\text{Component}\,S_1\colon \begin{cases} \dot{x}_1 = A_1x_1 + B_1u \\ y_1 = C_1x_1 + D_1u_1 \end{cases} \tag{2.207}$$

$$\text{Component}\,S_2\colon \begin{cases} \dot{x}_2 = A_2x_2 + B_2u_2 \\ y_2 = C_2x_2 + D_2u_2 \end{cases} \tag{2.208}$$

2.3.1.1 Parallel Connection

When the type of output parameters and dimensions of the two subsystems are simultaneously **homogeneous**, the various parameters in the combined system output vector $\boldsymbol{y}$ is equal to the sum of corresponding parameter values of output vector y_1 for subsystem S_1 and output vector y_2 for subsystem S_2, i.e.,

$$[y] = [y_1] + [y_2] \tag{2.209}$$

In such case, the state-space equation of combined system in parallel connection (Fig. 2.39a) is described as follows:

$$\begin{bmatrix} \dot{x}_1 \\ \dot{x}_2 \end{bmatrix} = \begin{bmatrix} A_1 & 0 \\ 0 & A_2 \end{bmatrix} \cdot \begin{bmatrix} x_1 \\ x_2 \end{bmatrix} + \begin{bmatrix} B_1 & 0 \\ 0 & B_2 \end{bmatrix} \cdot \begin{bmatrix} u_1 \\ u_2 \end{bmatrix} \tag{2.209a}$$

$$y = [C_1 C_2] \cdot \begin{bmatrix} x_1 \\ x_2 \end{bmatrix} + [D_1 D_2] \cdot \begin{bmatrix} u_1 \\ u_2 \end{bmatrix} \tag{2.209b}$$

When the type of output parameters and dimensions of the two subsystems are not simultaneously homogeneous (i.e., not similar), the combined system output vector $\boldsymbol{y}$ is the union of output vectors y_1 and y_2 of subsystems S_1 and S_2, respectively, i.e.,

$$[y] = \begin{bmatrix} y_1 \\ y_2 \end{bmatrix} \tag{2.210}$$

In such case, the state-space model of the parallel system (Fig. 2.39a) can be obtained as follows:

$$\begin{bmatrix} \dot{x}_1 \\ \dot{x}_2 \end{bmatrix} = \begin{bmatrix} A_1 & 0 \\ 0 & A_2 \end{bmatrix} \cdot \begin{bmatrix} x_1 \\ x_2 \end{bmatrix} + \begin{bmatrix} B_1 & 0 \\ 0 & B_2 \end{bmatrix} \cdot \begin{bmatrix} u_1 \\ u_2 \end{bmatrix} \tag{2.211a}$$

$$[y] = \begin{bmatrix} C_1 & 0 \\ 0 & C_2 \end{bmatrix} \cdot \begin{bmatrix} x_1 \\ x_2 \end{bmatrix} + \begin{bmatrix} D_1 & 0 \\ 0 & D_2 \end{bmatrix} \cdot \begin{bmatrix} u_1 \\ u_2 \end{bmatrix}. \tag{2.211b}$$

2.3.1.2 Series Connection

The combined system in series connection (Fig. 2.39b) has input vector $u = u_1$ and output vector $y = y_2$; and the output vector y_1 of subsystem S_1 is equal to the input vector u_2 of subsystem S_2. Thus, we have

$$\dot{x}_1 = A_1 x_1 + B_1 u_1 \tag{2.212}$$

$$y_1 = C_1 x_1 + D_1 u_1 \tag{2.213}$$

$$\dot{x}_2 = A_2 x_2 + B_2 u_2 = A_2 x_2 + B_2 y_1 = A_2 x_2 + B_2 C_1 x_1 + B_2 D_1 u \tag{2.214}$$

$$y_2 = C_2 x_2 + D_2 u_2 = C_2 x_2 + D_2 y_1 = C_2 x_2 + D_2 C_1 x_1 + D_2 D_1 u \tag{2.215}$$

So, the combined system equation of series connection of S_1 and S_2 is as follows:

$$\begin{bmatrix} \dot{x}_1 \\ \dot{x}_2 \end{bmatrix} = \begin{bmatrix} A_1 & 0 \\ B_2C_1 & A_2 \end{bmatrix} \cdot \begin{bmatrix} x_1 \\ x_2 \end{bmatrix} + \begin{bmatrix} B_1 \\ B_2D_1 \end{bmatrix} \cdot u \tag{2.215}$$

$$y = \begin{bmatrix} D_2C_1 & C_2 \end{bmatrix} \cdot \begin{bmatrix} x_1 \\ x_2 \end{bmatrix} + D_2D_1u \tag{2.216}$$

2.3.2 *State-Space Representation for HVAC System*

Taking a general HVAC system (see Fig. 2.40) for example, in order to make the system be of hierarchy, the system can be divided into five subsystems: ① refrigeration system represented by S_{refri}; ② cooling coolant system represented by S_{conw} which consists of cooling coolant pump, cooling tower, and pipes; ③ chilled coolant system represented by S_{chw} which consists of chilled coolant pump and pipes; ④ air-handling system represented by S_{ahu} which consists of surface heat exchanger and supply air fan; ⑤ supply air system represented by $S_{air,s}$; and exhaust-return air system represented by $S_{air,er}$. Based on the above principle of model connection, we can get state-space model of the subsystems in the HVAC system.

2.3.2.1 Mathematical Dynamic Model of Cooling Coolant System (S_{conw})

$$\dot{X}_{conw} = A_{conw}X_{conw} + B_{conw}U_{conw} \tag{2.217}$$

$$Y_{conw} = C_{zpipe,2}x_{zpipe,2} + D_{zpipe,2}C_{tower,w}x_{tower} + R_3u_{tower,w} \tag{2.218}$$

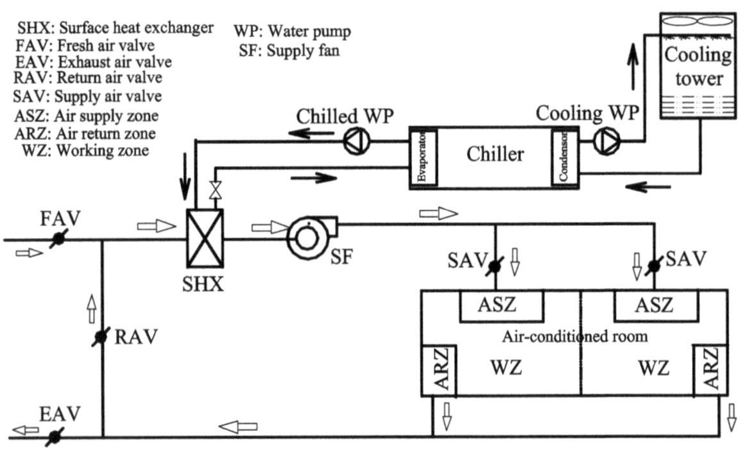

Fig. 2.40 Schematic diagram of a HVAC system

where

$$X_{\rm conw} = [x_{\rm pump,1}, x_{\rm zpipe,1}, x_{\rm tower}, x_{\rm zpipe,2}]^T;\ U_{\rm conw} = [u_{\rm pump,w,1}, \Delta U_{\rm m}, u_{\rm tower,1}, u_{\rm tower,w}]^T;$$
$$Y_{\rm conw} = y_{\rm zpipe,2};$$
$$A_{\rm conw} = \begin{bmatrix} A_{\rm pump,1} & 0 & 0 & 0 \\ B_{\rm zpipe,1} C_{\rm pump,w,1} & A_{\rm zpipe,1} & 0 & 0 \\ B_{\rm tower,w} D_{\rm zpipe,1} C_{\rm pump,w,1} & B_{\rm tower,w} C_{\rm zpipe,1} & A_{\rm tower} & 0 \\ 0 & 0 & B_{\rm zpipe,2} C_{\rm tower,w} & A_{\rm zpipe,2} \end{bmatrix};$$
$$U_{\rm conw} = \begin{bmatrix} B_{\rm pump,w,1} & B_{\rm pump,U,1} & 0 & 0 \\ R_1 & 0 & 0 & 0 \\ 0 & 0 & B_{\rm tower,a} & 0 \\ 0 & 0 & 0 & R_2 \end{bmatrix};$$
$$R_1 = \begin{bmatrix} \left(\frac{X_{\rm p,3}}{T_{\rm p,tw}}\right)_1 & 0 \\ \left(\frac{Y_{\rm p,3}}{T_{\rm p,pg}}\right)_1 & 0 \end{bmatrix}; R_2 = \begin{bmatrix} 0 & \left(\frac{X_{\rm p,4}}{T_{\rm p,tw}}\right)_2 \\ 0 & \left(\frac{Y_{\rm p,4}}{T_{\rm p,pg}}\right)_2 \end{bmatrix}; R_3 = \begin{bmatrix} 0 & 0 \\ 0 & 1 \end{bmatrix}.$$

2.3.2.2 Mathematical Dynamic Model of Chilled Coolant System ($S_{\rm chw}$)

The chilled coolant system consists of water-to-air heat exchanger, pipes, and pump which are in series connection. The state-space model for the chilled coolant system ($S_{\rm chw}$) can be written as follows:

$$\dot{X}_{\rm chw} = A_{\rm chw} X_{\rm chw} + B_{\rm chw} U_{\rm chw} \tag{2.219}$$

$$\begin{aligned} Y_{\rm chw} = {} & C_{\rm zpipe,4} x_{\rm zpipe,4} + D_{\rm zpipe,4} C_{\rm coil,w} x_{\rm coil} + D_{\rm zpipe,4} R_5 C_{\rm zpipe,3} x_{\rm zpipe,3} \\ & + D_{\rm zpipe,4} R_5 D_{\rm zpipe,3} C_{\rm pump,w,2} x_{\rm pump,2} \end{aligned} \tag{2.220}$$

where

$$X_{\rm chw} = [x_{\rm pump,2}, x_{\rm zpipe,3}, x_{\rm coil}, x_{\rm zpipe,4}]^T;\ U_{\rm chw} = [u_{\rm pump,w,2}, \Delta U_{\rm m}]^T;\ Y_{\rm chw} = y_{\rm zpipe,4};$$
$$A_{\rm chw} = \begin{bmatrix} A_{\rm pump,2} & 0 & 0 & 0 \\ B_{\rm zpipe,3} C_{\rm pump,w,2} & A_{\rm zpipe,3} & 0 & 0 \\ B_{\rm coil,w} D_{\rm zpipe,3} C_{\rm pump,w} & B_{\rm coil,w} C_{\rm zpipe,3} & A_{\rm coil} & 0 \\ B_{\rm zpipe,4} R_5 D_{\rm zpipe,3} C_{\rm pump,w,2} & B_{\rm zpipe,4} R_5 C_{\rm zpipe,3} & B_{\rm zpipe,4} C_{\rm coil,w} & A_{\rm zpipe,4} \end{bmatrix};$$

$$B_{\rm chw} = \begin{bmatrix} B_{\rm pump,w,2} & B_{\rm pump,U,2} \\ R_4 & 0 \\ 0 & 0 \\ 0 & 0 \end{bmatrix}; R_4 = \begin{bmatrix} \left(\frac{X_{\rm p,3}}{T_{\rm p,tw}}\right)_3 & 0 \\ \left(\frac{Y_{\rm p,3}}{T_{\rm p,pg}}\right)_3 & 0 \end{bmatrix}; R_5 = \begin{bmatrix} 0 & 0 \\ 0 & 1 \end{bmatrix}.$$

2.3.2.3 Mathematical Dynamic Model of Supply Air System ($S_{\rm air,s}$)

$$\dot{X}_{\rm air,s} = A_{\rm air,s} X_{\rm air,s} + B_{\rm air,s} U_{\rm air,s} \tag{2.221}$$

$$Y_{\rm air,s} = C_{\rm air,s} X_{\rm air,s} + D_{\rm air,s} U_{\rm air,s} \tag{2.222}$$

where

$$X_{\rm air,s} = [x_{\rm zduct,1}, x_{\rm zduct,2}]^T; Y_{\rm air,s} = [(y_{\rm sduct,2})_1, y_{\rm zduct,2}]^T; U_{\rm air,s} = u_{\rm zduct,1};$$
$$A_{\rm air,s} = \begin{bmatrix} A_{\rm zduct,1} & 0 \\ B_{\rm zduct,2}(C_{\rm fda13,E})_1 C_{\rm zduct,1} & A_{\rm zduct,2} \end{bmatrix}; B_{\rm air,s} = \begin{bmatrix} B_{\rm zduct,1} \\ B_{\rm zduct,2}(C_{\rm fda13,E})_1 D_{\rm zduct,1} \end{bmatrix};$$
$$C_{\rm air,s} = \begin{bmatrix} (C_{\rm fda12,E}) C_{\rm zduct,1} & 0 \\ D_{\rm zduct,2}(C_{\rm fda13,E})_1 C_{\rm zduct,1} & C_{\rm zduct,2} \end{bmatrix}; D_{\rm air,s} = \begin{bmatrix} (C_{\rm fda12,E})_1 D_{\rm zduct,1} \\ D_{\rm zduct,2}(C_{\rm fda13,E})_1 D_{\rm zduct,1} \end{bmatrix}.$$

2.3.2.4 Mathematical Dynamic Model of Exhaust-Return Air System ($S_{\rm air,Er}$)

$$\dot{X}_{\rm air,er} = A_{\rm air,er} X_{\rm air,er} + B_{\rm air,er} U_{\rm air,er} \tag{2.223}$$

$$Y_{\rm air,er} = C_{\rm air,er} X_{\rm air,er} + D_{\rm air,er} U_{\rm air,er} \tag{2.224}$$

where

$$X_{\rm air,er} = x_{\rm zduct,3}; Y_{\rm air,er} = [(y_{\rm sduct,2})_3, (y_{\rm sduct,3})_3]^T; U_{\rm air,er} = (u_{\rm sduct})_2;$$
$$A_{\rm air,er} = A_{\rm zduct,3}; B_{\rm air,er} = B_{\rm zduct,3}(C_{\rm sduct})_2;$$
$$C_{\rm air,er} = [(C_{\rm fda12,E})_3 C_{\rm zduct,3}, (C_{\rm fda13,E})_3 C_{\rm zduct,3}]^T;$$
$$C_{\rm air,er} = [(C_{\rm fda12,E})_3 D_{\rm zduct,3}(C_{\rm sduct})_2, (C_{\rm fda13,E})_3 D_{\rm zduct,3}(C_{\rm sduct})_2]^T;$$
$$(u_{\rm sduct})_2 = [(u_{\rm sduct,1})_2, (u_{\rm sduct,2})_2]; (C_{\rm sduct})_2 = [(C_{\rm da1,E})_2, (C_{\rm da2,E})_2].$$

2.3.2.5 Mathematical Dynamic Model of Air-Handling System (S_{ahu})

$$\dot{X}_{\text{ahu}} = A_{\text{ahu}} X_{\text{ahu}} + B_{\text{ahu}} U_{\text{ahu}} \tag{2.225}$$

$$Y_{\text{ahu}} = C_{\text{coil,a}} x_{\text{coil}} + R_7 C_{\text{fan,a}} x_{\text{fan}} + R_7 R_6 (C_{\text{sduct}})_4 (u_{\text{sduct}})_4 \tag{2.226}$$

where

$$X_{\text{ahu}} = [x_{\text{fan}}, x_{\text{coil}}]^T;\ Y_{\text{ahu}} = y_{\text{coil,a}};\ U_{\text{ahu}} = [\Delta U_{\text{m}}, u_{\text{coil,w}}, (u_{\text{sduct}})_4]^T;$$
$$(u_{\text{sduct}})_4 = [(u_{\text{sduct,1}})_4, (u_{\text{sduct,2}})_4];\ (C_{\text{sduct}})_4 = [(C_{\text{da1,E}})_4, (C_{\text{da2,E}})_4];$$

$$A_{\text{ahu}} = \begin{bmatrix} A_{\text{fan}} & 0 \\ B_{\text{coil,a}} C_{\text{fan,a}} & A_{\text{coil}} \end{bmatrix};\ B_{\text{ahu}} = \begin{bmatrix} B_{\text{fan,U}} & 0 & 0 \\ 0 & B_{\text{coil,w}} & B_{\text{coil,a}} R_6 (C_{\text{sduct}})_4 \end{bmatrix};\ R_6 = \begin{bmatrix} 1 & 0 & 0 \\ 0 & 1 & 0 \\ 0 & 0 & 0 \end{bmatrix};$$

Under dry condition: $R_7 = \begin{bmatrix} 0 & 0 & 0 \\ 0 & 1 & 0 \\ 0 & 0 & 1 \end{bmatrix}$; Under wet condition: $R_7 = \begin{bmatrix} 0 & 0 & 0 \\ 0 & 0 & 0 \\ 0 & 0 & 1 \end{bmatrix}$.

2.3.3 *Case Study*

2.3.3.1 System Model Description

On the basis of state-space models of components and connection principle, this section gives an example on the system model integration. The system considered in this study consists of a water-to-air surface heat exchanger, an outlet water pipe, and an outlet air duct. According to the actual situation, the water outlet pipe and air outlet duct are in parallel, and they are connected to the heat exchanger in series, as shown in Fig. 2.41.

According to the state-space representation of component models in the previous chapters, the state-space model of water-to-air surface heat exchanger can be represented by:

$$\dot{X}_{\text{coil}} = A_{\text{coil}} X_{\text{coil}} + B_{\text{coil}} u_{\text{coil}} \tag{2.227}$$

$$Y_{\text{coil}} = C_{\text{coil}} X_{\text{coil}} + D_{\text{coil}} u_{\text{coil}} - C_{\text{coil}} A_{\text{coil}}^{-1} \xi_{\text{coil}} \tag{2.228}$$

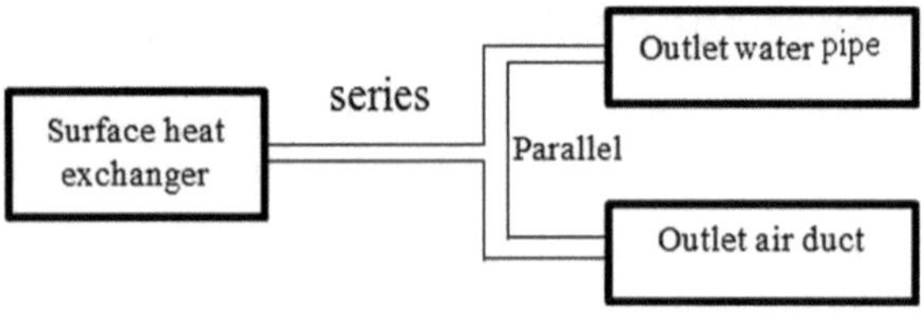

Fig. 2.41 Multi-component system in this case study

The state-space model for water pipe can be represented by:

$$\dot{X}_{zpipe} = A_{zpipe}X_{zpipe} + B_{zpipe}u_{zpipe} \tag{2.229}$$

$$Y_{zpipe} = C_{zpipe}X_{zpipe} + D_{zpipe}u_{zpipe} - C_{zpipe}A_{zpipe}^{-1}\xi_{zpipe} \tag{2.230}$$

The state-space model of air duct can be written as follows:

$$\dot{X}_{zduct} = A_{zduct}X_{zduct} + B_{zduct}u_{zduct} \tag{2.231}$$

$$Y_{zduct} = C_{zduct}X_{zduct} + D_{zduct}u_{zduct} - C_{zduct}A_{zduct}^{-1}\xi_{zduct} \tag{2.232}$$

To begin with, the state-space model of parallel connection of air duct and water pipe are obtained as follows:

$$\dot{X}_{duct\&pipe} = \begin{bmatrix} \dot{X}_{zpipe} \\ \dot{X}_{zduct} \end{bmatrix} = \dot{A}_{duct\&pipe}\begin{bmatrix} X_{zpipe} \\ X_{zduct} \end{bmatrix} + \dot{B}_{duct\&pipe}\begin{bmatrix} u_{zpipe} \\ u_{zduct} \end{bmatrix} \tag{2.233}$$

$$Y_{duct\&pipe} = \begin{bmatrix} Y_{zpipe} \\ Y_{zduct} \end{bmatrix} = \dot{C}_{duct\&pipe}\begin{bmatrix} X_{zpipe} \\ X_{zduct} \end{bmatrix} + \dot{D}_{duct\&pipe}\begin{bmatrix} u_{zpipe} \\ u_{zduct} \end{bmatrix} - \zeta_{duct\&pipe} \tag{2.234}$$

where

$$A_{duct\&pipe} = \begin{bmatrix} A_{zpipe} & 0 \\ 0 & A_{zduct} \end{bmatrix}; B_{duct\&pipe} = \begin{bmatrix} B_{zpipe} & 0 \\ 0 & B_{zduct} \end{bmatrix}; C_{duct\&pipe} = \begin{bmatrix} C_{zpipe} & 0 \\ 0 & C_{zduct} \end{bmatrix}; D_{duct\&pipe} = \begin{bmatrix} D_{zpipe} & 0 \\ 0 & D_{zduct} \end{bmatrix}; \zeta_{duct\&pipe} = \begin{bmatrix} C_{zpipe}A_{zpipe}^{-1}\xi_{zpipe} \\ C_{zduct}A_{zduct}^{-1}\xi_{zduct} \end{bmatrix}.$$

Then, the above state-space model in parallel is integrated with the water-to-air heat exchanger in series, and we have:

$$\dot{X}_{sys} = A_{sys}X_{sys} + B_{sys}\,\dot{u}_{coil} - \xi_{Xsys} \tag{2.235}$$

$$Y_{sys} = Y_{duct\&pipe} = C_{sys}\dot{X}_{sys} + D_{sys}\dot{u}_{coil} - \xi_{Ysys} \tag{2.236}$$

where

$$X_{sys} = \begin{bmatrix} X_{coil} \\ X_{duct\&pipe} \end{bmatrix}; A_{sys} = \begin{bmatrix} A_{coil} & 0 \\ B_{duct\&pipe}C_{coil} & A_{duct\&pipe} \end{bmatrix}; B_{sys} = \begin{bmatrix} B_{coil} \\ B_{duct\&pipe}D_{coil} \end{bmatrix};$$

$$C_{sys} = \begin{bmatrix} D_{duct\&pipe}C_{coil} & C_{duct\&pipe} \end{bmatrix}; D_{sys} = D_{duct\&pipe}D_{coil};$$

$$\xi_{Xsys} = \begin{bmatrix} 0 \\ B_{duct\&pipe}C_{coil}A_{coil}^{-1}\xi_{coil} \end{bmatrix}; \xi_{Ysys} = D_{duct\&pipe}C_{coil}A_{coil}^{-1}\xi_{coil} + \xi_{duct\&pipe};$$

For the convenience of calculation, the system model, i.e., Eqs. (2.235) and (2.236), needs to be transformed into a standard form as follows:

$$\dot{\overline{X_{sys}}} = A_{sys}\overline{X_{sys}} + B_{sys}u_{coil} \tag{2.237}$$

$$Y_{sys} = C_{sys}\overline{X_{sys}} + D_{sys}u_{sys} + \xi_{sys} \tag{2.238}$$

where $\overline{X_{sys}} = X_{sys} - A_{sys}^{-1}\xi_{Xsys}$; $\xi_{sys} = C_{sys}A_{sys}^{-1}\zeta_{Xsys} - \zeta_{Ysys}$.

2.3.3.2 System Model Validation

An experimental air-conditioning system is built for the system model validation. As shown in Fig. 2.42, the system mainly consists of a water-to-air surface heat exchanger, a heat pump (for chilled or hot water supply), a circulating water pump, fans, air ducts, and water pipes. The test instruments mainly include the temperature

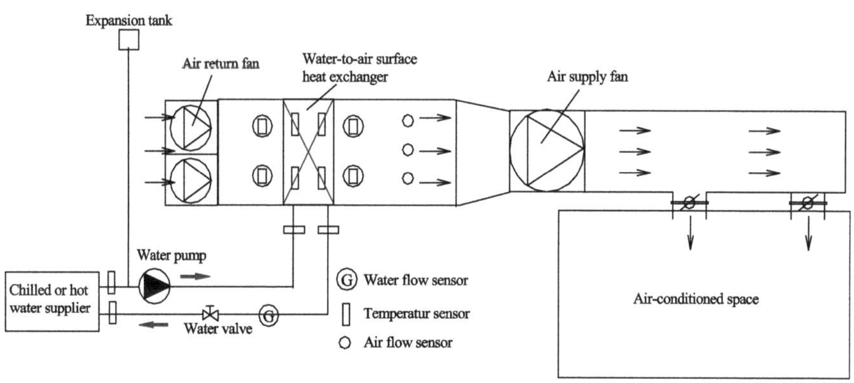

Fig. 2.42 Schematic diagram for the experimental system

Table 2.17 Initial conditions for system model validation

Initial conditions of parameters	Initial value
Inlet air temperature of surface heat exchanger, $(t_{a,E})_o$ (°C)	28.8
Inlet air humidity of surface heat exchanger, $(W_{a,E})_o$ (g/kg)	20.0
Airflow rate, $(G_a)_o$ (kg/s)	0.2002
Inlet water temperature of surface heat exchanger, $(t_{w,E})_o$ (°C)	29.1
Water flow rate, $(G_w)_o$ (kg/s)	0.930
External wall temperature of plate and fin pipe, $(t_g)_o$ (°C)	28.7
Wall temperature of air duct, $(t_{dg})_o$ (°C)	28.5
Exit air temperature of air duct, $(t_{da,L})_o$ (°C)	28.9
Exit air humidity of air duct, $(W_{da,L})_o$ (g/kg)	20.0
Wall temperature of water pipe, $(t_{pg})_o$ (°C)	29.4
Exit water temperature of water pipe, $(t_{pw,L})_o$ (°C)	27.5

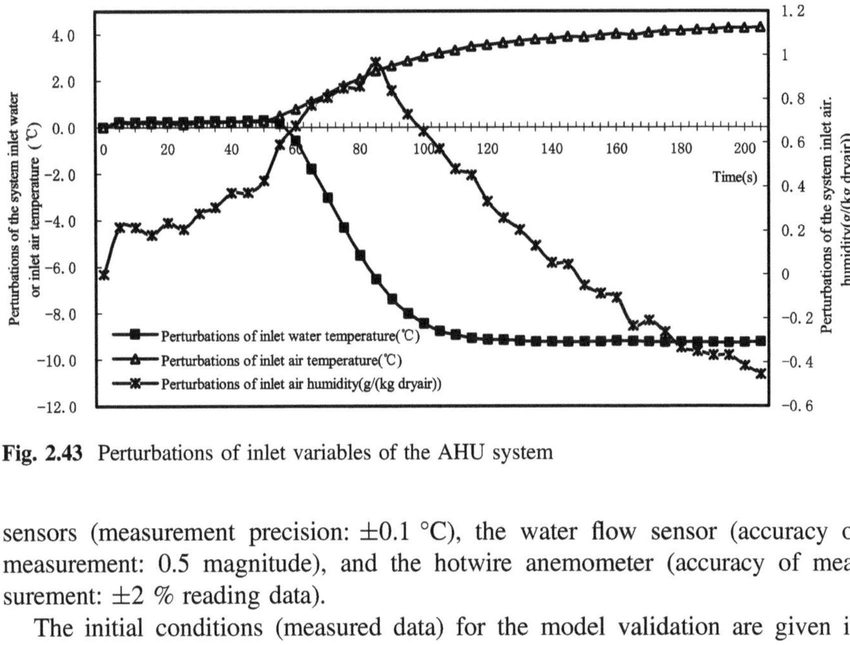

Fig. 2.43 Perturbations of inlet variables of the AHU system

sensors (measurement precision: ±0.1 °C), the water flow sensor (accuracy of measurement: 0.5 magnitude), and the hotwire anemometer (accuracy of measurement: ±2 % reading data).

The initial conditions (measured data) for the model validation are given in Table 2.17. The perturbations of inlet variables of the system are shown in Fig. 2.43. The experimental and calculated results of the change responses of the system exit variables are plotted in Fig. 2.44. As shown in Fig. 2.44, the calculated responses of the system exit variables under the perturbations of inlet variables have a favorable agreement with the experimental ones. The average error of the model calculation is estimated as 12.6, 22.2, and 9.5 %, respectively, for the response of the exit air temperature, the exit air humidity, and the exit water temperature of the system.

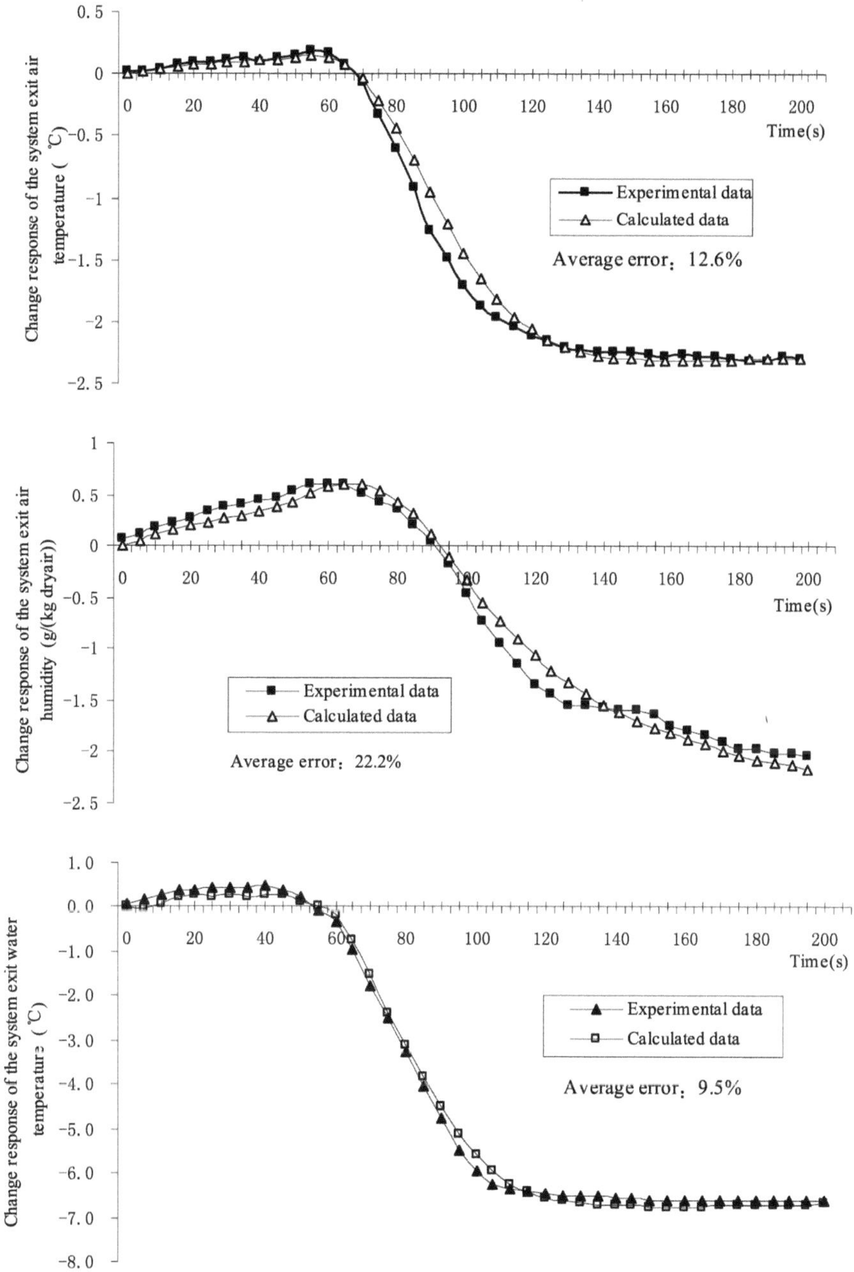

Fig. 2.44 Responses of exit variables of the AHU system (experimental vs. calculated)

References

1. Hangos, K.M., Lakner, R., Gerzson, M.: Intelligent Control Systems: An Introduction with Examples. Springer, Berlin (2001)
2. Hangos, K.M., Bokor, J., Szederkényi, G.: Analysis and Control of Nonlinear Process Systems. Springer, Berlin (2004)
3. Brogan, W.L.: Modern Control Theory (1st edn). Quantum Publishers Inc. (1974)
4. http://en.wikipedia.org/wiki/State-space_representation
5. ASHRAE. ASHRAE Handbook: Heating, Ventilating, and Air-conditioning Systems and Equipment (SI Edition). Atlanta (USA), GA (2004)
6. Yao, Y., Huang, M., Mo, J., Dai, S.: State-space model for transient behavior of water-to-air surface heat exchanger. Int. J. Heat Mass Transf. **66**(9), 173–192 (2013)
7. Vaisi, A., Talebi, S., Esmaeilpour, M.: Transient behavior simulation of fin-and-tube heat exchangers for the variation of the inlet temperatures of both fluids. Int. Commun. Heat Mass Transfer **38**(5), 951–957 (2011)
8. ASHRAE. ASHRAE Handbook—Fundamentals, Chap. 6. Atlanta (USA), GA (2005)
9. Yao, Y., Huang, M., Chen, J.: State-space model for dynamic behavior of vapor compression liquid chiller. Int. J. Refrig. **36**(8), 2128–2147 (2013)
10. Browne, M.W., Bansal, P.K.: An elemental NTU-ε model for vapour-compression liquid chillers. Int. J. Refrig. **24**(5), 612–627 (2001)
11. Chan, C.Y., Haselden, G.G.: Computer-based refrigerant thermodynamic properties, part 1: basic equations. Int. J. Refrig. **4**(1), 7–12 (1981)
12. Chan, C.Y., Haselden, G.G.: Computer-based refrigerant thermodynamic properties, part 2: program listings. Int. J. Refrig. **4**(1), 52–60 (1981)
13. Su, C.: Advanced Engineering Thermodynamics. Higher Education Press (HEP), Beijing (1987)
14. Lemmon, E.W., McLinden, M.O., Huber, M.L.: NIST Reference Fluid Thermodynamic and Transport Properties—REFPRO Users' Guide (version 7.0). National Institute of Standards and Technology (2002)
15. Zhang, X.M., Ren, Z.F.: Heat Transfer. China Architecture & Building Press, Beijing (2006)
16. Badescu, V.: Dynamic model of a complex system including PV cells, electric battery, electrical motor and water pump. Energy **28**, 1165–1181 (2003)
17. Yao, Y., Yang, K., Huang, M.: A state-space model for dynamic response of indoor air temperature and humidity. Build. Environ. **64**(6), 26–37 (2013)
18. Inard, C., Bouia, H., Dalicieux, P.: Prediction of air temperature distribution in buildings with a zonal model. Energy Build. **24**(2), 125–132 (1996)
19. Peng, X., van Paassen, A.H.C.: A state space model for predicting and controlling the temperature responses of indoor air zones. Energy Build. **28**(1), 197–203 (1998)
20. Liu, W., Lian, Z., Yao, Y.: Optimization on indoor air diffusion of floor-standing type room air-conditioners. Energy Build. **40**(2), 59–70 (2008)
21. Gao, J., Zhang, X., Zhao, J.N., Gao, F.S.: A heat transfer parameter at air interfaces in the BLOCK model for building thermal environment. Int. J. Therm. Sci. **49**(2), 463–470 (2010)

Chapter 3
Dynamic Simulations with State-Space Models

In the chapter, characteristics of dynamic thermal response of HVAC components are simulated with corresponding state-space model. The dynamic response characteristics are usually described by two parameters: One is amplification coefficient (i.e., proportionality coefficient of response parameters to the perturbation ones); the other is time constant (i.e., the time required for the response parameters to arrive at a newly steady state after the perturbation occurs). We define the dimensionless response parameter as:

$$Y^* = \frac{(Y - Y_o)}{(Y_s - Y_o)} \tag{3.1}$$

where Y_o and Y_s are initial and newly steady values of the response variables, respectively. Y is transient value of the response variables. In the following simulation analysis, the steady state is considered to be achieved for $Y^* = 0.99$.

3.1 On Water-to-Air Surface Heat Exchanger

3.1.1 Subjected to Different Perturbations

The experimental heat exchanger, whose detailed physical structure is shown in Table 2.3, is used for the dynamic response study. The system matrixes for the surface heat exchanger (A_{coil}, B_{coil}) corresponding to the initial conditions of the case II in Table 2.4 are computed as below:

Y. Yao and Y. Yu, *Modeling and Control in Air-conditioning Systems*,
Energy and Environment Research in China, DOI 10.1007/978-3-662-53313-0_3

$$A_{\mathrm{coil}} = \begin{bmatrix} -1.0513 & 0 & 0 & 0.4686 \\ 0 & -7.1687 & 0.5052 & 3.2979 \\ 0 & -0.0011 & -4.7779 & 0.0069 \\ 0.0748 & 0.0460 & 0.2083 & -0.3520 \end{bmatrix};$$

$$B_{\mathrm{coil}} = \begin{bmatrix} 0.3485 & -3.8809 & 0 & 0 & 0 \\ 0 & 0 & 2.9864 & 0.5052 & 44.6301 \\ 0 & 0 & -0.0011 & 4.7606 & 0.0460 \\ 0.0748 & -3.4501 & 0.0460 & 0.2083 & 0.0257 \end{bmatrix}.$$

The transient responses of the state variables ($\Delta t_{\mathrm{a,L}}, \Delta W_{\mathrm{a,L}}, \Delta t_{\mathrm{w,L}}$ and Δt_{g}) subjected to the step perturbation of the inlet ones are simulated, respectively, under above initial conditions.

(1) *Step change of inlet air temperature*
The transient responses of the state variables ($\Delta t_{\mathrm{a,L}}, \Delta W_{\mathrm{a,L}}, \Delta t_{\mathrm{w,L}}$ and Δt_{g}) to a sudden increase of the inlet air temperature by 1.0 °C are presented in Fig. 3.1. According to the simulation results, the proportionality coefficient of the state variables, $\Delta t_{\mathrm{a,L}}, \Delta W_{\mathrm{a,L}}, \Delta t_{\mathrm{w,L}}$ and Δt_{g}, to the step change of the inlet air temperature ($\Delta t_{\mathrm{a,E}}$) is estimated as 0.5172 °C/°C, 0.1000 (g/kg dry air)/°C, 0.0975 °C/°C and 0.2187 °C/°C, respectively. And the corresponding time constant is estimated as 140, 80, 220, and 220 s, respectively. The increase of the inlet air temperature results in more heat transfer, and hence, causes the exit water temperature ($\Delta t_{\mathrm{w,L}}$) and the heat exchanger's wall temperature (Δt_{g}) to rise. Meanwhile, the air moisture ratio near the wall increases with the increase of the wall temperature, which results in the reduction of the dehumidification capacity of the heat exchanger and the increase of the exit air humidity ($\Delta W_{\mathrm{a,L}}$).

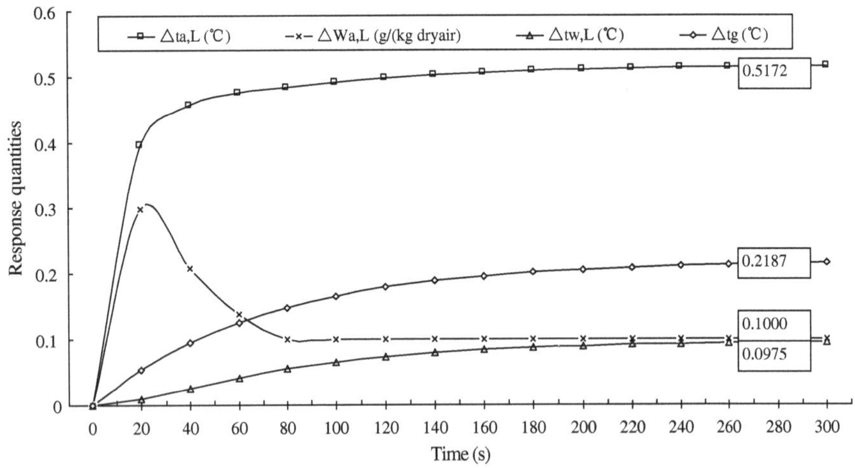

Fig. 3.1 Responses of the state variables to sudden increase of inlet air temperature by 1.0 °C

(2) *Step change of inlet water temperature*
Figure 3.2 presents the transient responses of the state variables ($\Delta t_{a,L}, \Delta W_{a,L}, \Delta t_{w,L}$ and Δt_g) to a sudden increase of the inlet water temperature by 1.0 °C. The results show that all the state variables increase gradually till to the new steady value. The proportionality coefficient of the state variables, $\Delta t_{a,L}, \Delta W_{a,L}, \Delta t_{w,L}$ and Δt_g, to the step change of the inlet water temperature ($\Delta t_{w,E}$) is estimated as 0.1542 °C/°C, 0.4080 (g/kg dry air)/°C, 0.4808 °C/°C, and 0.3352 °C/°C, respectively. And the corresponding time constant is estimated as 220, 160, 220, and 220 s, respectively.

(3) *Step change of water flow rate*
Figure 3.3 shows the transient responses of the state variables ($\Delta t_{a,L}, \Delta W_{a,L}, \Delta t_{w,L}$ and Δt_g) to the step change of the water flow rate ($\Delta G_w = 0.1$ kg/s). With respect to the simulation results, the proportionality coefficient of the state variables, $\Delta t_{a,L}, \Delta W_{a,L}, \Delta t_{w,L}$ and Δt_g, to the step change of the water flow rate (ΔG_w) is estimated as −5.767 °C/(kg/s), −17.000 (g/kg dry air)/(kg/s), −9.277 °C/(kg/s), and −12.533 °C/(kg/s), respectively. And the corresponding time constant is identically as 260 s. The increase of the chilled water flow rate heat improves the water-side transfer coefficient and increases the cooling and dehumidification capacity of the heat exchanger. Hence, the exit air temperature and humidity as well as the wall temperature will decrease. The change of the exit water temperature depends on the increased water flow rate and the increased heat gain from the air. In this case, the exit water temperature decreases with the step increase of the water flow rate by 0.1 kg/s.

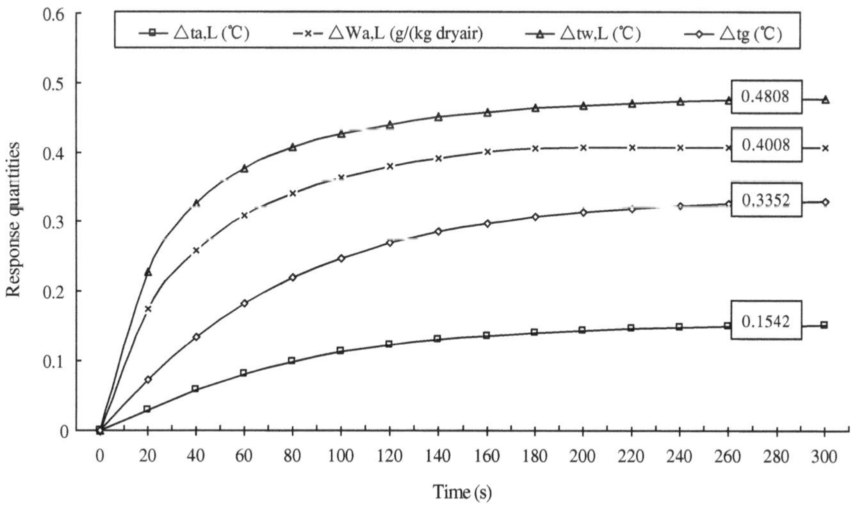

Fig. 3.2 Responses of the state variables to sudden increase of inlet water temperature by 1.0 °C

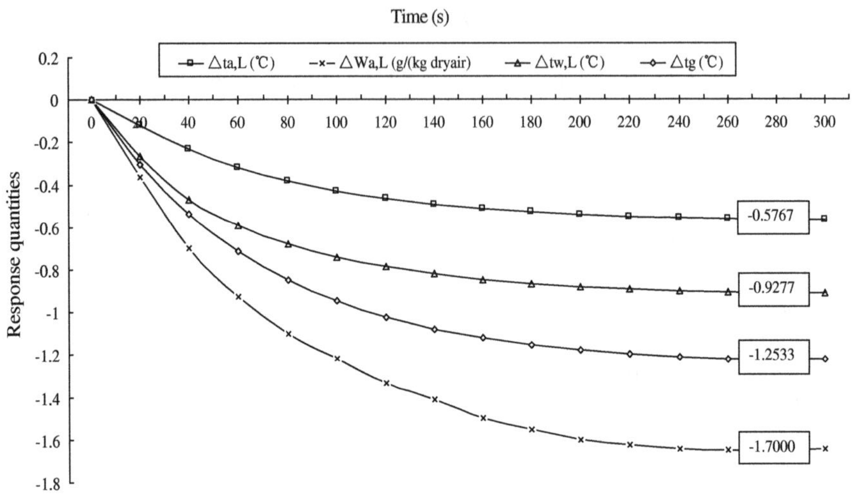

Fig. 3.3 Responses of the state variables to sudden increase of water flow rate by 0.1 kg/s

(4) *Step change of airflow rate*

Figure 3.4 shows the transient responses of the state variables ($\Delta t_{a,L}, \Delta W_{a,L}, \Delta t_{w,L}$ and Δt_g) to the step change of the airflow rate ($\Delta G_a = 0.1$ kg/s). According to the results in Fig. 3.4, the proportionality coefficient of the state variables, $\Delta t_{a,L}, \Delta W_{a,L}, \Delta t_{w,L}$ and Δt_g, to the step change of the airflow rate (ΔG_a) is estimated

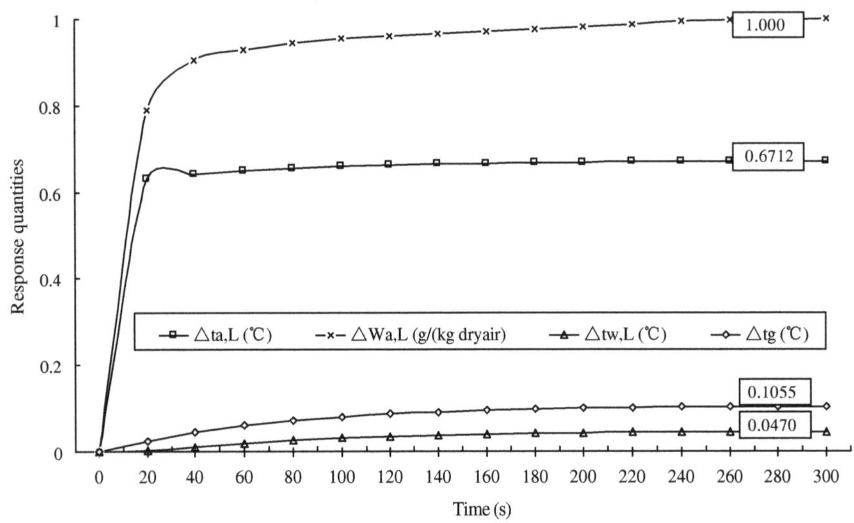

Fig. 3.4 Responses of the state variables to sudden increase of airflow rate by 0.1 kg/s

as 6.712 °C/(kg/s), 10.000 (g/kg dry air)/(kg/s), 0.470 °C/(kg/s), and 1.055 °C/(kg/s), respectively. And the corresponding time constant is estimated as 60, 180, 220, and 220 s, respectively.

3.1.2 For Different Initial Conditions

The influence of initial conditions on the transient thermal response characteristic of the heat exchanger is investigated as well by using the state-space model.

(1) *Influence of initial inlet air temperature* $(t_{a,E})_o$

Three initial values of inlet air temperature (i.e., 30.0, 32.0, and 34.0 °C) combined with the other initial conditions ($(t_{w,E})_o$ = 15.0 °C, $(W_{a,E})_o$ = 20.0 g/(kg dry air), $(G_a)_o$ = 0.20 kg/s, $(G_w)_o$ = 0.20 kg/s) are given for the dynamic thermal response simulation of water-to-air heat exchanger, and the corresponding system matrixes for the heat exchanger (A_{coil}, B_{coil}) under these initial conditions are calculated as below:

$$A_{coil,\,t_{a,E}=30\,^\circ C} = A_{coil,\,t_{a,E}=32\,^\circ C} = A_{coil,\,t_{a,E}=34\,^\circ C} = \begin{bmatrix} -0.5327 & 0 & 0 & 0.3072 \\ 0 & -4.4031 & 0.4860 & 1.9125 \\ 0 & -0.0011 & -2.9432 & 0.0066 \\ 0.0491 & 0.0380 & 0.2004 & -0.2803 \end{bmatrix};$$

$$B_{coil,\,t_{a,E}=30\,^\circ C} = \begin{bmatrix} 0.0720 & -8.4627 & 0 & 0 & 0 \\ 0 & 0 & 1.8499 & 0.4860 & 29.3115 \\ 0 & 0 & -0.0011 & 2.9266 & 0.1325 \\ 0.0491 & -7.3084 & 0.0380 & 0.2004 & -0.0863 \end{bmatrix};$$

$$B_{coil,\,t_{a,E}=32\,^\circ C} = \begin{bmatrix} 0.0720 & -8.7574 & 0 & 0 & 0 \\ 0 & 0 & 1.8499 & 0.4860 & 37.6962 \\ 0 & 0 & -0.0011 & 2.9266 & 0.1309 \\ 0.0491 & -8.1784 & 0.0380 & 0.2004 & -0.0236 \end{bmatrix};$$

$$B_{coil,\,t_{a,E}=34\,^\circ C} = \begin{bmatrix} 0.0720 & -8.8831 & 0 & 0 & 0 \\ 0 & 0 & 1.8499 & 0.4860 & 42.9023 \\ 0 & 0 & -0.0011 & 2.9266 & 0.1293 \\ 0.0491 & -9.225 & 0.0380 & 0.2004 & -0.1267 \end{bmatrix}.$$

Figure 3.5 shows the dynamic responses of exit air temperature and humidity of the water-to-air heat exchanger under different initial inlet air temperatures. As shown in Fig. 3.5, the dynamic response characteristics of exit air temperature and humidity to inlet air temperature and humidity as well as inlet water temperature will not affected by initial values of inlet air temperature. However, when subjected to the perturbations of water and airflow rate, the dynamic response characteristics of exit air temperature and humidity will be impacted by initial values of inlet air temperature. The results in Fig. 3.5 manifest that the amplification coefficient of response of exit air temperature and humidity to the perturbations of water and

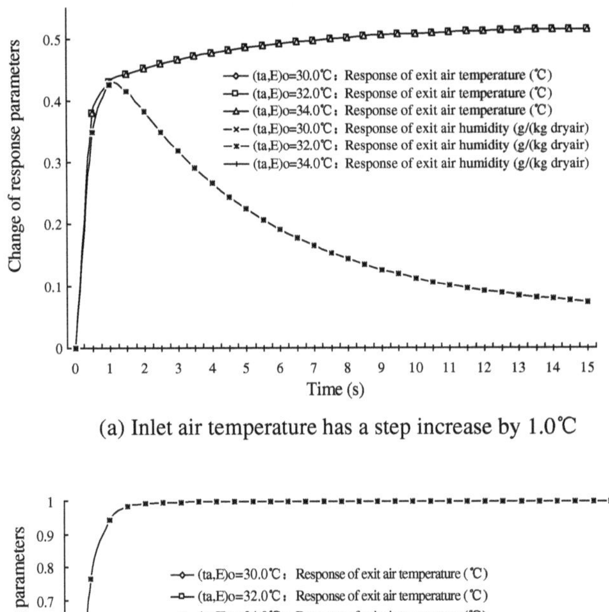

(a) Inlet air temperature has a step increase by 1.0℃

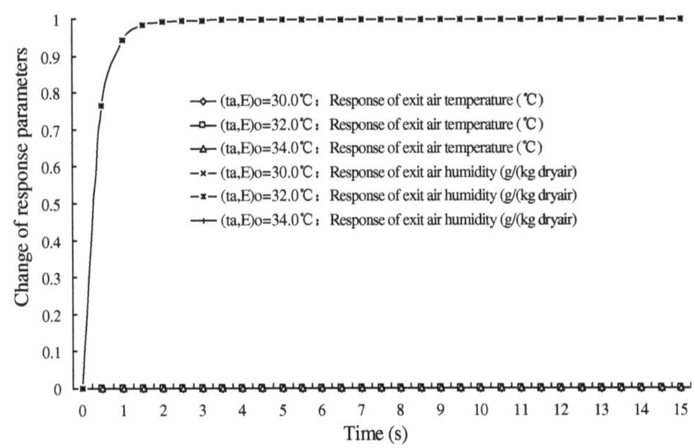

(b) Inlet air humidity has a step increase by 1.0 g/(kg dryair)

Fig. 3.5 Dynamic responses of exit air temperature and humidity to perturbations under different initial values of inlet air temperature **a** Inlet air temperature has a step increase by 1.0 °C. **b** Inlet air humidity has a step increase by 1.0 g/(kg dry air). **c** Airflow rate has a step increase by 0.1 kg/s. **d** Inlet water temperature has a step increase by 1.0 °C. **e** Water flow rate has a step increase by 0.1 kg/s

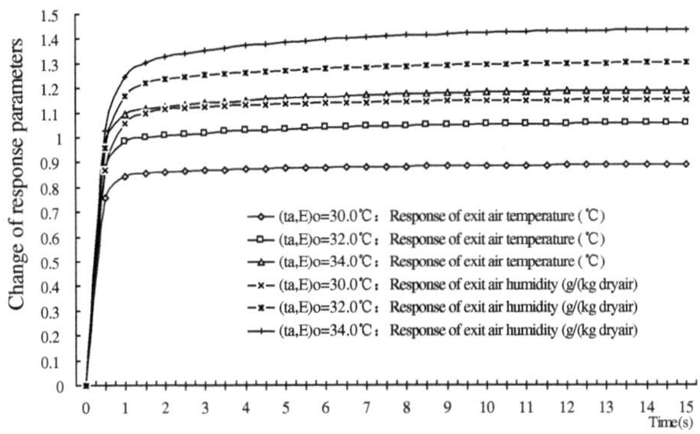

(c) Air flow rate has a step increase by 0.1 kg/s

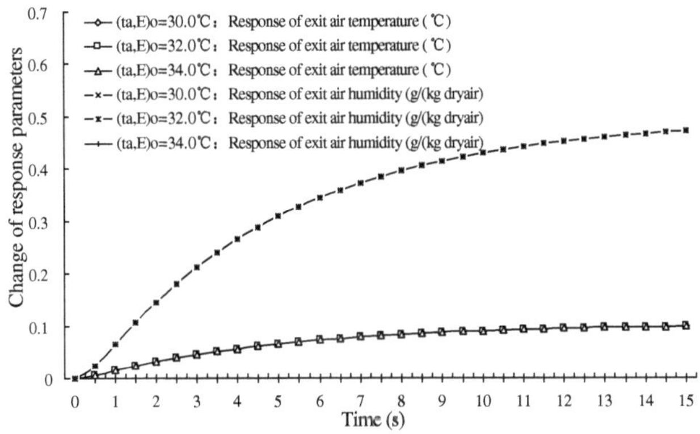

(d) Inlet water temperature has a step increase by 1.0 ℃

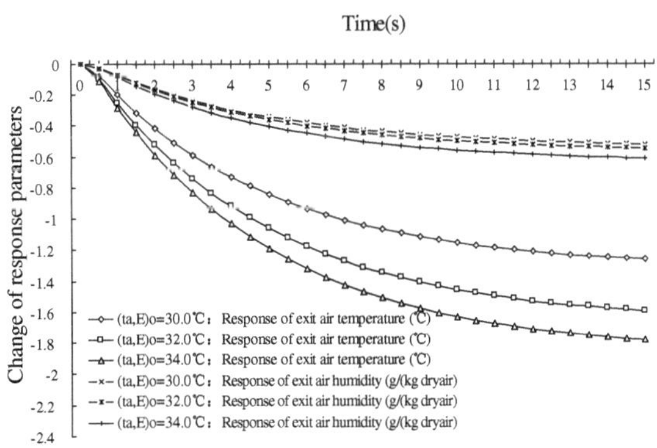

(e) Water flow rate has a step increase by 0.1 kg/s

Fig. 3.5 (continued)

airflow rate will be larger under the higher initial value of inlet air temperature, and this indicates that the impact of water and airflow rate of heat exchanger on its exit air temperature and humidity will be more significant under the higher inlet air temperature.

(2) *Influence of initial inlet air humidity* $(W_{a,\,E})_o$

Three initial values of inlet air humidity (i.e., 18.0, 20.0, and 22.0 g/(kg dry air)) combined with the other initial conditions (i.e., $(t_{a,\,E})_o = 30.0$ °C, $(t_{w,\,E})_o = 15.0$ °C, $(G_a)_o = 0.20$ kg/s,$(G_w)_o = 0.2$ kg/s) are given for the thermal dynamic response simulation of water-to-air heat exchanger, and the corresponding system matrixes for the heat exchanger (A_{coil}, B_{coil}) under these initial conditions are computed as below:

$$A_{coil,\,W_{a,E}=18.0(kg/kg)} = A_{coil,\,W_{a,E}=20.0(kg/kg)} = A_{coil,\,W_{a,E}=22.0(kg/kg)} = \begin{bmatrix} -0.5327 & 0 & 0 & 0.3072 \\ 0 & -4.4031 & 0.4860 & 1.9125 \\ 0 & -0.0011 & -2.9432 & 0.0066 \\ 0.0491 & 0.0380 & 0.2004 & -0.2803 \end{bmatrix};$$

$$B_{coil,\,W_{a,E}=18.0(kg/kg)} = \begin{bmatrix} 0.0720 & -8.3370 & 0 & 0 & 0 \\ 0 & 0 & 1.8499 & 0.4860 & 32.9884 \\ 0 & 0 & -0.0011 & 2.9266 & 0.1241 \\ 0.0491 & -6.2643 & 0.0380 & 0.2004 & -0.0393 \end{bmatrix};$$

$$B_{coil,\,W_{a,E}=20.0(kg/kg)} = \begin{bmatrix} 0.0720 & -8.4627 & 0 & 0 & 0 \\ 0 & 0 & 1.8499 & 0.4860 & 29.3115 \\ 0 & 0 & -0.0011 & 2.9266 & 0.1325 \\ 0.0491 & -7.3084 & 0.0380 & 0.2004 & -0.0863 \end{bmatrix};$$

$$B_{coil,\,W_{a,E}=22.0(kg/kg)} = \begin{bmatrix} 0.0720 & -9.4096 & 0 & 0 & 0 \\ 0 & 0 & 1.8499 & 0.4860 & 17.9014 \\ 0 & 0 & -0.0011 & 2.9266 & 0.1550 \\ 0.0491 & -8.8745 & 0.0380 & 0.2004 & -0.1673 \end{bmatrix}$$

Figure 3.6 shows the dynamic responses of exit air temperature and humidity of the water-to-air heat exchanger under different initial values of inlet air humidity. It is shown in Fig. 3.6 that the influence of initial inlet air humidity on the dynamic response characteristics of exit air temperature and humidity of water-to-air heat exchanger depends on the kind of disturbance parameters. The dynamic response characteristics of exit air temperature and humidity to the perturbations of inlet air temperature and humidity as well as inlet water temperature will not be affected by initial value of inlet air humidity. However, the initial value of inlet air humidity will make influence on the dynamic response characteristics of exit air temperature

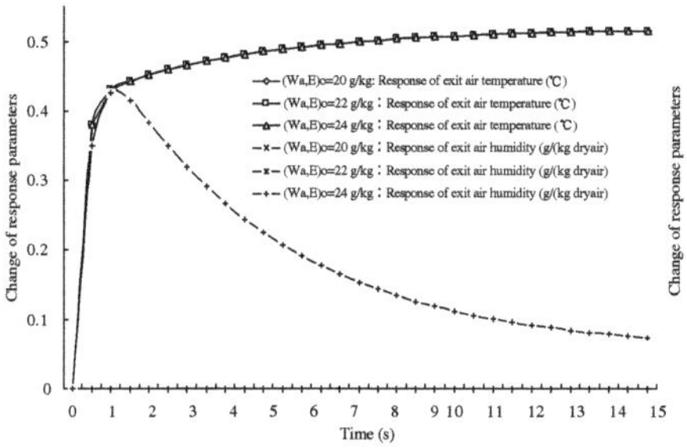

(a) Inlet air temperature has a step increase by 1.0 ℃

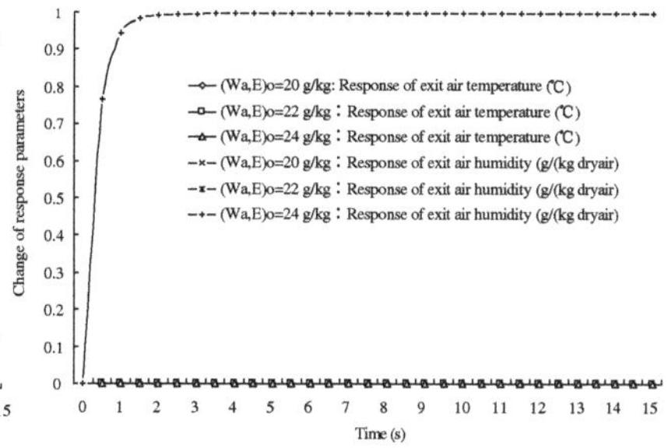

(b) Inlet air humidity has a step increase by 1.0 g/(kg dryair)

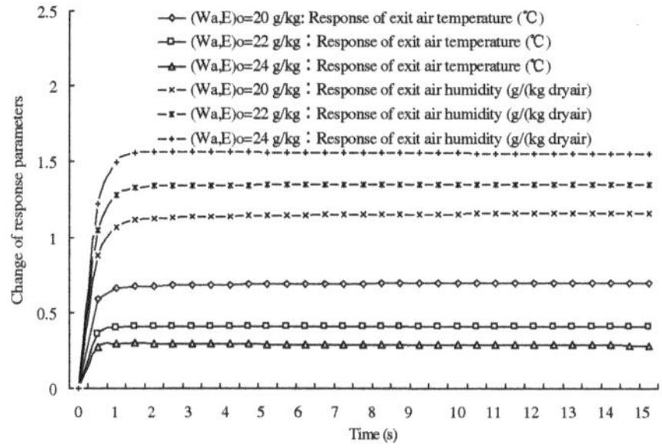

(c) Air flow rate has a step increase by 0.1 kg/s

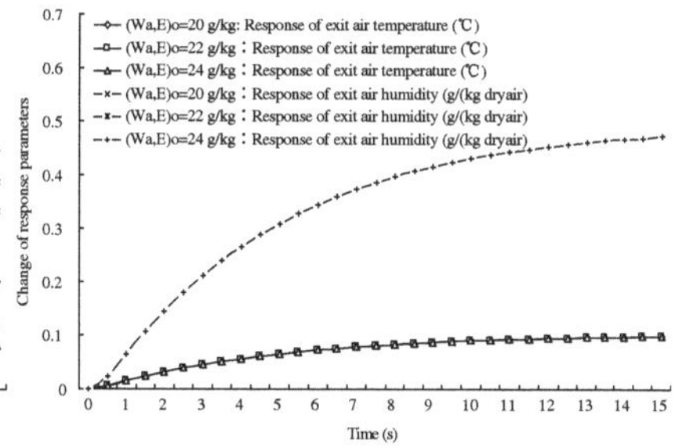

(d) Inlet water temperature has a step increase by 1.0℃

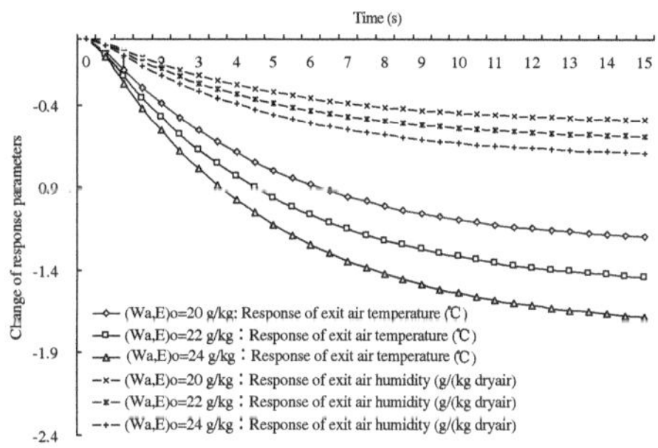

(e) Water flow rate has a step increase by 0.1 kg/s

Fig. 3.6 Dynamic responses of exit air temperature and humidity to perturbations under different initial values of inlet air humidity. **a** Inlet air temperature has a step increase by 1.0 °C. **b** Inlet air humidity has a step increase by 1.0 g/(kg dry air). **c** Airflow rate has a step increase by 0.1 kg/s. **d** Inlet water temperature has a step increase by 1.0°C (**c**) Airflow rate has a step increase by 0.1 kg/s . **e** Water flow rate has a step increase by 0.1 kg/s

and humidity when subjected to the perturbations of water and airflow rate. As shown in Fig. 3.6c, e, the amplification coefficient of response of exit air temperature and humidity to the perturbations of water and airflow rate will increase with the initial inlet air humidity increasing.

(3) *Influence of initial airflow rate* $(G_a)_o$

Three initial values of airflow rate (i.e., 0.20, 0.22 and 0.24 kg/s) combined with the other initial conditions (i.e., $(t_{a,\ E})_o = 30.0$ °C, $(t_{w,\ E})_o = 15.0$ °C, $(W_{a,\ E})_o = 18.0$ g/(kg dry air), $(G_w)_o = 0.20$ kg/s) are used to investigate the influence of initial airflow rate on dynamic thermal response of water-to-air heat exchanger. The corresponding system matrixes for the heat exchanger (A_{coil} and B_{coil}) under these initial conditions are computed as below:

$$A_{\mathrm{coil,\,G_a=0.20\,kg/s}} = \begin{bmatrix} -0.5327 & 0 & 0 & 0.3072 \\ 0 & -4.4031 & 0.4860 & 1.9125 \\ 0 & -0.0011 & -2.9432 & 0.0066 \\ 0.0491 & 0.0380 & 0.2004 & -0.2803 \end{bmatrix};$$

$$B_{\mathrm{coil,\,G_a=0.20\,kg/s}} = \begin{bmatrix} 0.0720 & -8.3370 & 0 & 0 & 0 \\ 0 & 0 & 1.8499 & 0.4860 & 32.9884 \\ 0 & 0 & -0.0011 & 2.9266 & 0.1241 \\ 0.0491 & -6.2643 & 0.0380 & 0.2004 & -0.0393 \end{bmatrix};$$

$$A_{\mathrm{coil,\,G_a=0.22\,kg/s}} = \begin{bmatrix} -0.5327 & 0 & 0 & 0.3072 \\ 0 & -4.8462 & 0.4897 & 2.1346 \\ 0 & -0.0011 & -3.2368 & 0.0067 \\ 0.0491 & 0.0393 & 0.2019 & -0.2837 \end{bmatrix};$$

$$B_{\mathrm{coil,\,G_a=0.22\,kg/s}} = \begin{bmatrix} 0.0720 & -9.4744 & 0 & 0 & 0 \\ 0 & 0 & 2.0313 & 0.4897 & 30.5393 \\ 0 & 0 & -0.0011 & 3.2000 & 0.1265 \\ 0.0491 & -6.2643 & 0.0393 & 0.2019 & -0.0076 \end{bmatrix};$$

$$A_{\mathrm{coil,\,G_a=0.24\,kg/s}} = \begin{bmatrix} -0.5327 & 0 & 0 & 0.3072 \\ 0 & -5.2890 & 0.4931 & 2.3566 \\ 0 & -0.0011 & -3.5303 & 0.0067 \\ 0.0491 & 0.0406 & 0.2033 & -0.2871 \end{bmatrix};$$

$$B_{\mathrm{coil,\,G_a=0.24\,kg/s}} = \begin{bmatrix} 0.0720 & -10.8436 & 0 & 0 & 0 \\ 0 & 0 & 2.2129 & 0.4931 & 28.0517 \\ 0 & 0 & -0.0011 & 3.5135 & 0.1306 \\ 0.0491 & -6.0903 & 0.0406 & 0.2033 & -0.0162 \end{bmatrix}.$$

Figure 3.7 gives the dynamic responses of exit air temperature and humidity to perturbations under different initial airflow rates. The results indicate that the dynamic response characteristics of exit air temperature to perturbations of inlet air temperature and humidity as well as inlet water temperature will not be affected by

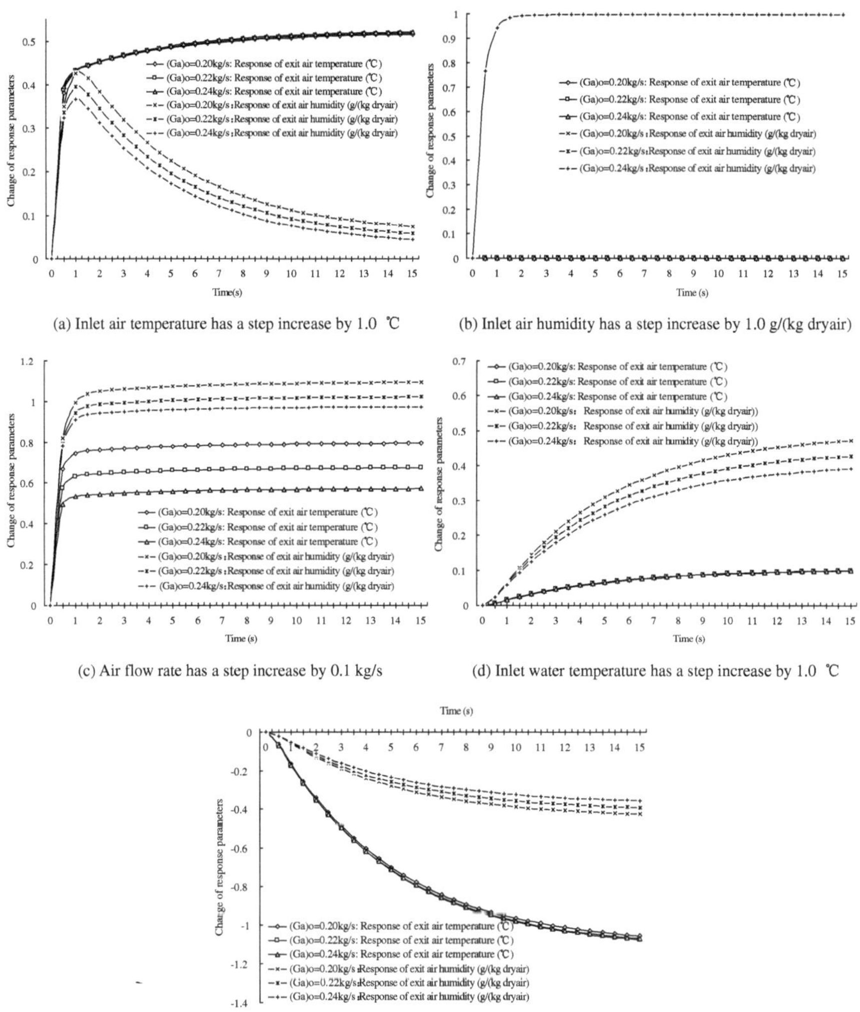

(a) Inlet air temperature has a step increase by 1.0 ℃

(b) Inlet air humidity has a step increase by 1.0 g/(kg dryair)

(c) Air flow rate has a step increase by 0.1 kg/s

(d) Inlet water temperature has a step increase by 1.0 ℃

(e) Water flow rate has a step increase by 0.1 kg/s

Fig. 3.7 Dynamic responses of exit air temperature and humidity to perturbations under different initial values of airflow rate. **a** Inlet air temperature has a step increase by 1.0 °C. **b** Inlet air humidity has a step increase by 1.0 g/(kg dry air). **c** Airflow rate has a step increase by 0.1 kg/s. **d** Inlet water temperature has a step increase by 1.0 °C. **e** Water flow rate has a step increase by 0.1 kg/s

initial value of airflow rate, and that to the other kind of disturbance parameters (such as air and water flow rate) will be affected by initial value of airflow rate. The amplification coefficient of response of exit air temperature to the perturbations of airflow rate decreases with the initial airflow rate increasing, and so does the amplification coefficient of response of exit air humidity to the perturbations of inlet air (or water) temperature and flow rate.

(4) *Influence of initial inlet water temperature* $(t_{w,\ E})_o$

Three initial values of inlet water temperature (i.e., 11.0, 13.0, and 15.0 °C) combined with the other initial conditions including $(t_{a,\ E})_o = 30.0$ °C, $(W_{a,\ E})_o = 18.0$ g/(kg dry air), $(G_a)_o = 0.20$ kg/s,$(G_w)_o = 0.20$ kg/s are used for the dynamic response simulations, and the corresponding system matrixes for the heat exchanger (A_{coil}, B_{coil}) are obtained below:

$$A_{coil,\, t_{w,E}=11\,°C} = A_{coil,\, t_{w,E}=13\,°C} = A_{coil,\, t_{w,E}=15\,°C} = \begin{bmatrix} -0.5327 & 0 & 0 & 0.3072 \\ 0 & -4.4031 & 0.4860 & 1.9125 \\ 0 & -0.0011 & -2.9432 & 0.0066 \\ 0.0491 & 0.0380 & 0.2004 & -0.2803 \end{bmatrix};$$

$$B_{coil,\, t_{w,E}=11\,°C} = \begin{bmatrix} 0.0720 & -11.4094 & 0 & 0 & 0 \\ 0 & 0 & 1.8499 & 0.4860 & 86.2240 \\ 0 & 0 & -0.0011 & 2.9266 & 0.1577 \\ 0.0491 & -10.7885 & 0.0380 & 0.2004 & -0.0202 \end{bmatrix};$$

$$B_{coil,\, t_{w,E}=13\,°C} = \begin{bmatrix} 0.0720 & -10.1050 & 0 & 0 & 0 \\ 0 & 0 & 1.8499 & 0.4860 & 66.7030 \\ 0 & 0 & -0.0011 & 2.9266 & 0.1418 \\ 0.0491 & -8.3524 & 0.0380 & 0.2004 & -0.0335 \end{bmatrix};$$

$$B_{coil,\, t_{w,E}=15\,°C} = \begin{bmatrix} 0.0720 & -8.3370 & 0 & 0 & 0 \\ 0 & 0 & 1.8499 & 0.4860 & 32.9884 \\ 0 & 0 & -0.0011 & 2.9266 & 0.1241 \\ 0.0491 & -6.2643 & 0.0380 & 0.2004 & -0.0393 \end{bmatrix}.$$

The dynamic responses of exit air temperature and humidity of heat exchanger to perturbations under different initial inlet water temperatures are shown in Fig. 3.8. The results indicate that the dynamic response characteristics of exit air temperature and humidity to the perturbations of inlet air temperature and humidity as well as inlet water temperature will not affected by initial value of inlet water temperature. However, the initial inlet water temperature will impact the dynamic response characteristics of exit air temperature and humidity when subjected to the perturbations of water and airflow rate. As shown in Fig. 3.8c, e, the amplification coefficient of response of exit air temperature to the perturbations of water and airflow rate decreases with the increase of initial water temperature.

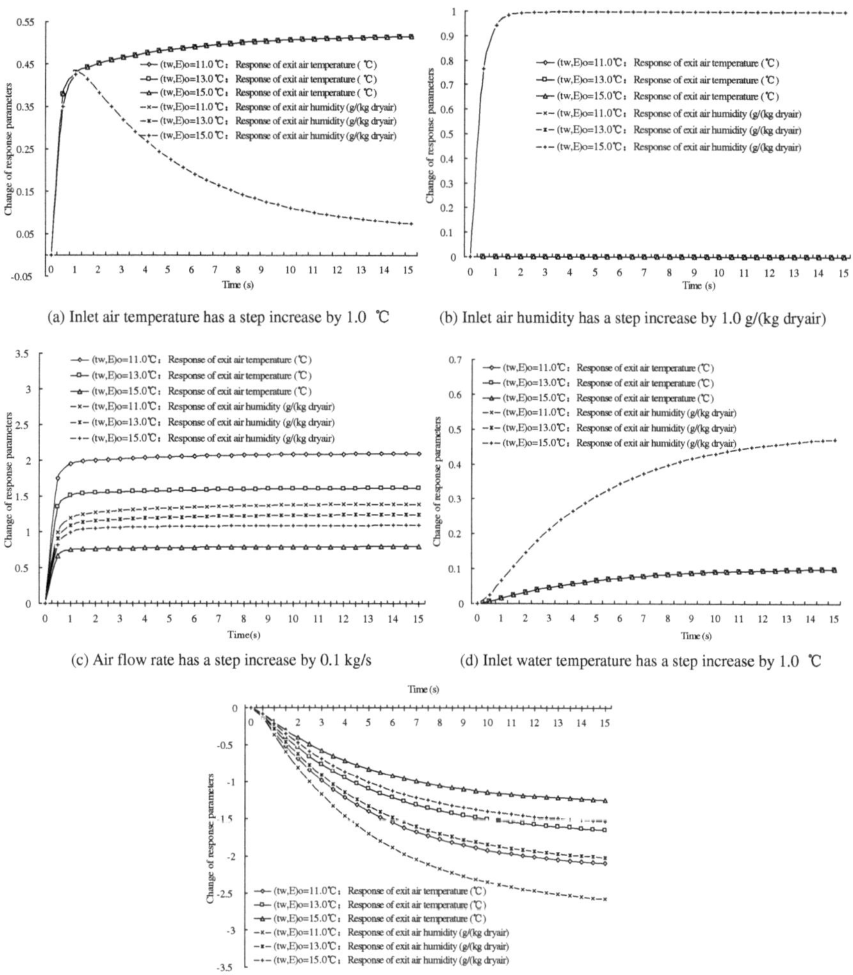

(a) Inlet air temperature has a step increase by 1.0 ℃

(b) Inlet air humidity has a step increase by 1.0 g/(kg dryair)

(c) Air flow rate has a step increase by 0.1 kg/s

(d) Inlet water temperature has a step increase by 1.0 ℃

(e) Water flow rate has a step increase by 0.1 kg/s

Fig. 3.8 Dynamic responses of exit air temperature and humidity to perturbations under different initial values of inlet water temperature. **a** Inlet air temperature has a step increase by 1.0 °C. **b** Inlet air humidity has a step increase by 1.0 g/(kg dry air). **c** Airflow rate has a step increase by 0.1 kg/s. **d** Inlet water temperature has a step increase by 1.0 °C. **e** Water flow rate has a step increase by 0.1 kg/s

(5) *Influence of initial water flow rate* $(G_w)_o$

Likewise, the influences of initial water flow rate on the dynamic response characteristics of exit air temperature and humidity of water-to-air heat exchanger are investigated under different initial water flow rates (i.e., 0.20,0.22 and

0.24 kg/s) combined with the other initial conditions including $(t_{a,\,E})_o = 30.0$ °C, $(t_{w,\,E})_o = 15.0$ °C, $(W_{a,\,E})_o = 18.0$ g/(kg dry air), and $(G_a)_o = 0.20$ kg/s. The corresponding system matrixes for the heat exchanger (A_{coil}, B_{coil}) are computed as follows:

$$A_{\mathrm{coil,\,G_w=0.02\,kg/s}} = \begin{bmatrix} -0.5789 & 0 & 0 & 0.3237 \\ 0 & -4.4031 & 0.4860 & 1.9125 \\ 0 & -0.0011 & -2.9432 & 0.0066 \\ 0.0517 & 0.0380 & 0.2004 & -0.2803 \end{bmatrix};$$

$$B_{\mathrm{coil,G_w=0.02\,kg/s}} = \begin{bmatrix} 0.0720 & -8.3370 & 0 & 0 & 0 \\ 0 & 0 & 1.8499 & 0.4860 & 32.9884 \\ 0 & 0 & -0.0011 & 2.9266 & 0.1241 \\ 0.0491 & -6.2643 & 0.0380 & 0.2004 & -0.0393 \end{bmatrix};$$

$$A_{\mathrm{coil,\,G_w=0.02\,kg/s}} = \begin{bmatrix} -0.5327 & 0 & 0 & 0.3072 \\ 0 & -4.4031 & 0.4860 & 1.9125 \\ 0 & -0.0011 & -2.9432 & 0.0066 \\ 0.0491 & 0.0380 & 0.2004 & -0.2803 \end{bmatrix};$$

$$B_{\mathrm{coil,\,G_w=0.022\,kg/s}} = \begin{bmatrix} 0.0934 & -8.4136 & 0 & 0 & 0 \\ 0 & 0 & 1.8499 & 0.4860 & 45.8849 \\ 0 & 0 & -0.0011 & 2.9266 & 0.1295 \\ 0.0517 & -5.1678 & 0.0380 & 0.2004 & -0.0379 \end{bmatrix};$$

$$A_{\mathrm{coil,\,G_w=0.02\,kg/s}} = \begin{bmatrix} -0.6247 & 0 & 0 & 0.3072 \\ 0 & -4.4031 & 0.4860 & 1.9125 \\ 0 & -0.0011 & -2.9432 & 0.0066 \\ 0.0542 & 0.0380 & 0.2004 & -0.2906 \end{bmatrix};$$

$$B_{\mathrm{coil,\,G_w=0.022\,kg/s}} = \begin{bmatrix} 0.1154 & -8.3161 & 0 & 0 & 0 \\ 0 & 0 & 1.8499 & 0.4860 & 47.5348 \\ 0 & 0 & -0.0011 & 2.9266 & 0.1343 \\ 0.0542 & -4.0076 & 0.0380 & 0.2004 & 0.0163 \end{bmatrix}.$$

Figure 3.9 presents the results of dynamic responses of exit air temperature and humidity to different perturbations under different initial water flow rates. From Fig. 3.9, we can conclude that the initial water flow rate makes little influence on the dynamic response characteristics of exit air temperature and humidity when subjected to the disturbances of inlet air temperature and humidity, but the situation is different when subjected to the disturbances of water and airflow rate as well as inlet water temperature. The coefficient of response of exit air temperature to the

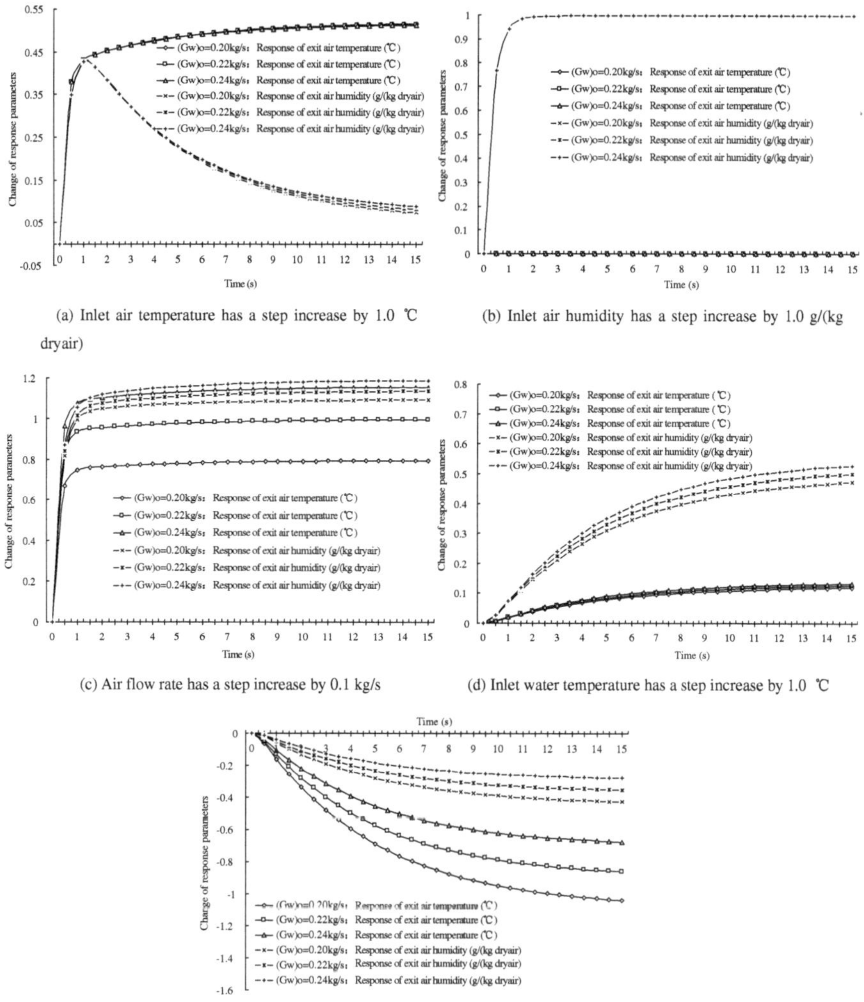

(a) Inlet air temperature has a step increase by 1.0 ℃ dryair)

(b) Inlet air humidity has a step increase by 1.0 g/(kg

(c) Air flow rate has a step increase by 0.1 kg/s

(d) Inlet water temperature has a step increase by 1.0 ℃

(e) Water flow rate has a step increase by 0.1 kg/s

Fig. 3.9 Dynamic responses of exit air temperature and humidity to perturbations under different initial values of water flow rate. **a** Inlet air temperature has a step increase by 1.0 °C. **b** Inlet air humidity has a step increase by 1.0 g/(kg dry air). **c** Airflow rate has a step increase by 0.1 kg/s. **d** Inlet water temperature has a step increase by 1.0 °C. **e** Water flow rate has a step increase by 0.1 kg/s

perturbations of airflow rate and inlet water temperature increases with the increase of initial water flow rate, and it is opposite for that to the perturbations of water flow rate itself.

3.2 On Chiller

3.2.1 Subjected to Different Perturbations

Taking the experimental chiller (see Fig. 2.11) and the initial conditions of case *I* in Table 2.8 for example, the transient responses of the exit cold carrier temperature and COP of the chiller under different perturbations are investigated. The system model matrixes $(A_{\text{chiller}}, B_{\text{chiller}}, C_{\text{chiller}}, D_{\text{chiller}})$ corresponding to the case *I* are computed as below:

$$A_{\text{chiller}} = \begin{bmatrix} -5.1837 & 0 & 5.1790 & 0 & 0 & 0 \\ 0 & -1.4962 & 2.0831 & 0 & 0 & 0 \\ 1.0117 & 0.2026 & -1.4169 & 0 & 0 & 0 \\ 0.0540 & 0 & 0 & -30.6731 & 0 & 30.6338 \\ 0 & 0 & 0 & 0 & -2.1864 & 1.8934 \\ 0 & 0 & 0 & 4.4032 & 0.1520 & -4.7073 \end{bmatrix};$$

$$B_{\text{chiller}} = \begin{bmatrix} 0 & 0 & 0 & 0 & 27.1594 \\ -0.5870 & -0.0142 & 0 & 0 & 0 \\ 0.2026 & -0.9294 & 0 & 0 & 0 \\ 0 & 0 & 0 & 0 & -56.8937 \\ 0 & 0 & 0.0454 & 5.9930 & 0 \\ 0 & 0 & 0.1508 & -0.2387 & 0 \end{bmatrix};$$

$$C_{\text{chiller}} = \begin{bmatrix} 0 & 1.0000 & 0 & 0 & 0 & 0 \\ 0 & 0 & 0 & 0 & 0 & 0 \\ 0 & 0 & 0 & 0 & 1.0000 & 0 \\ 0 & 0 & 0 & 0 & 0 & 0 \\ 136.1257 & 0 & 0 & -125.3410 & 0 & 0 \\ -172.1875 & 0 & 0 & 125.3410 & 0 & 0 \\ -0.2992 & 0 & 0 & 0.2662 & 0 & 0 \end{bmatrix};$$

$$D_{\text{chiller}} = \begin{bmatrix} 0 & 0 & 0 & 0 & 0 \\ 0 & 1 & 0 & 0 & 0 \\ 0 & 0 & 0 & 0 & 0 \\ 0 & 0 & 0 & 1 & 0 \\ 0 & 0 & 0 & 0 & 0 \\ 0 & 0 & 0 & 0 & 27502 \\ 0 & 0 & 0 & 0 & 181333 \end{bmatrix}.$$

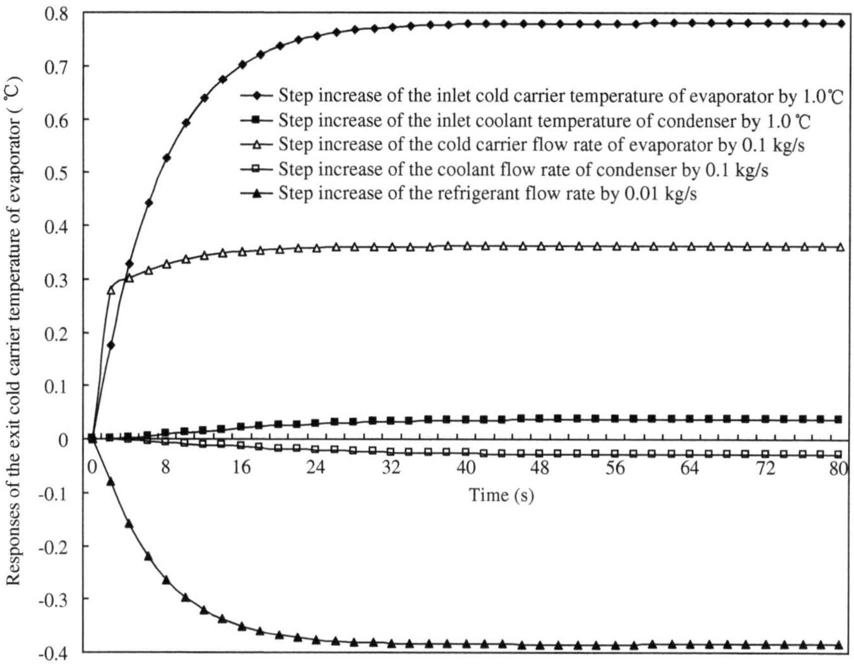

Fig. 3.10 Responses of the exit cold carrier temperature of evaporator subjected to different perturbations under the same initial conditions of case I

The exit cold carrier temperature of evaporator subjected to different perturbations under the same initial conditions of case *I* is presented in Fig. 3.10. As clearly shown in Fig. 3.10, the evaporator-side perturbations affect the exit cold carrier temperature of evaporator much more greatly than the condenser-side ones do. According to the results, the proportionality coefficient of the exit cold carrier temperature of evaporator ($\Delta t_{ew,L}$) to the step change of the inlet cold carrier temperature of evaporator ($\Delta t_{ew,E}$) and the inlet coolant temperature of condenser ($\Delta t_{cw,E}$) is estimated as about 0.77 °C/°C and 0.038 °C/°C, respectively. And that of $\Delta t_{ew,L}$ to the step change of the cold carrier flow rate of evaporator (ΔG_{ew}) and the inlet coolant flow rate of condenser (ΔG_{cw}) is estimated as about 3.7 °C/(kg/s) and −0.28 °C/(kg/s), respectively. The perturbations of refrigerant flow rate (ΔG_{rm}) will produce great impact on the exit cold carrier temperature of evaporator ($\Delta t_{ew,L}$). The proportionality coefficient of $\Delta t_{ew,L}$ to ΔG_{rm} is estimated as about −38.2 °C/(kg/s). The time constant of response of the exit cold carrier temperature of evaporator ($\Delta t_{ew,L}$) to the evaporator-side and the condenser-side perturbations is identified as about 25 and 45 s, respectively, with respect to the calculated results.

The transient responses of compressor power, cooling capacity, and COP of the refrigeration system subjected to different step disturbances (i.e., $\Delta t_{ew,E}, \Delta G_{ew}, \Delta t_{cw,E}, \Delta G_{cw}, \Delta G_{rm}$) are also examined under the initial conditions of

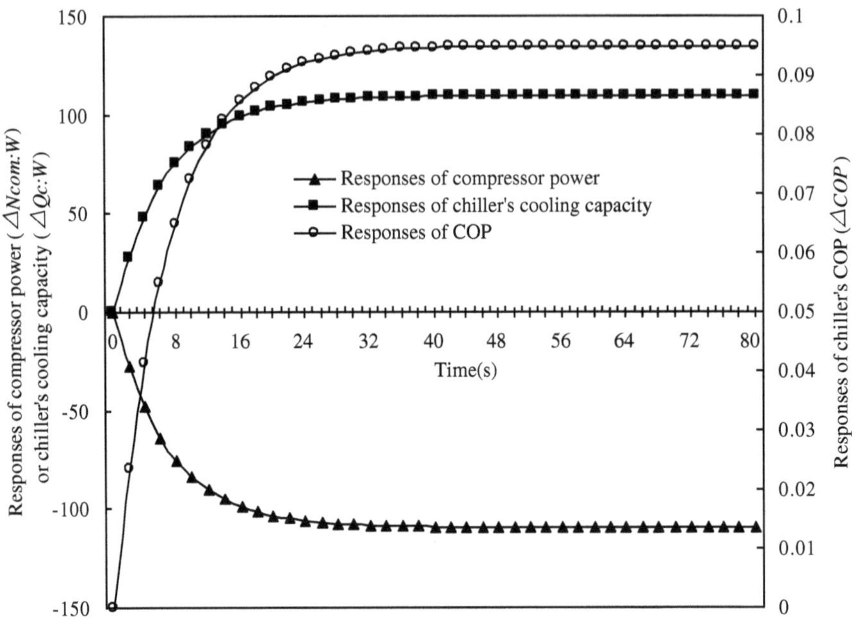

Fig. 3.11 Responses of compressor power, cooling capacity, and COP of the chiller to step increase of inlet cold carrier temperature of evaporator by 1.0 °C

case *I*. As shown in Figs. 3.11 and 3.12, the cooling capacity of the refrigeration system increases gradually when the inlet cold carrier temperature or flow rate of evaporator has a step increase. This is very reasonable since the increased inlet cold carrier temperature or flow rate of evaporator will cause the heat transfer quantity (e.g., the cooling capacity) to increase. Meanwhile, the compressor suction temperature and pressure increase as the inlet cold carrier temperature or flow rate of evaporator rises, which will bring about the decrease of the compressor power. Consequently, the coefficient of performance (COP) of the refrigeration system will increase. According to the calculated results, the COP ultimately increases about 0.09 and 0.027, respectively, under the step increase of the inlet cold carrier temperature by 1.0 °C and the step increase of the cold carrier flow rate by 0.1 kg/s.

Figure 3.13 presents the dynamic responses of the chiller's performance under the step increase of inlet coolant temperature of condenser by 1.0 °C. The increased inlet coolant temperature of condenser, on one hand, makes the compressor outlet pressure rise and results in the increased compressor power, and on the other hand, makes the chiller's cooling capacity decrease. As a consequence, the COP of the chiller finally decreases by about 0.11 when the inlet coolant temperature of condenser has a step increase of 1.0 °C. The situation is the opposite when the coolant flow rate of condenser increases. As shown in Fig. 3.14, the chiller's COP finally

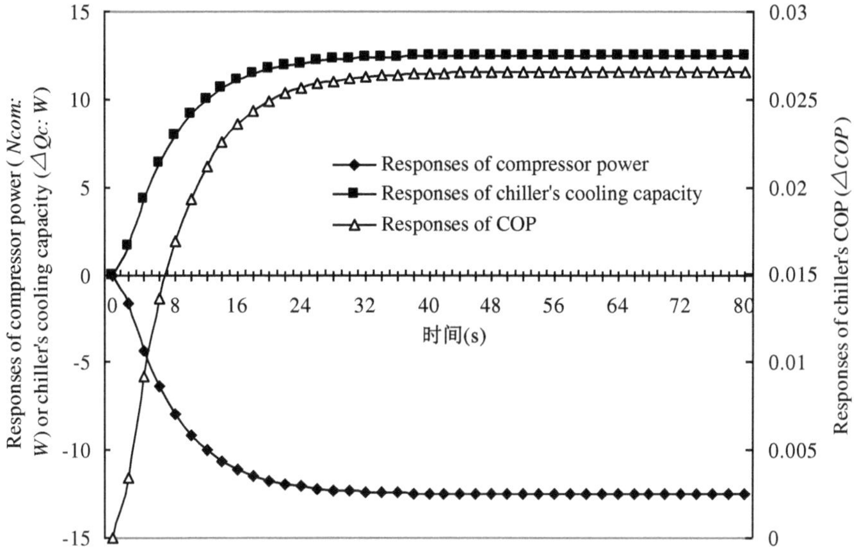

Fig. 3.12 Responses of compressor power, cooling capacity, and COP of the chiller to step increase of cold carrier flow rate of evaporator by 0.1 kg/s

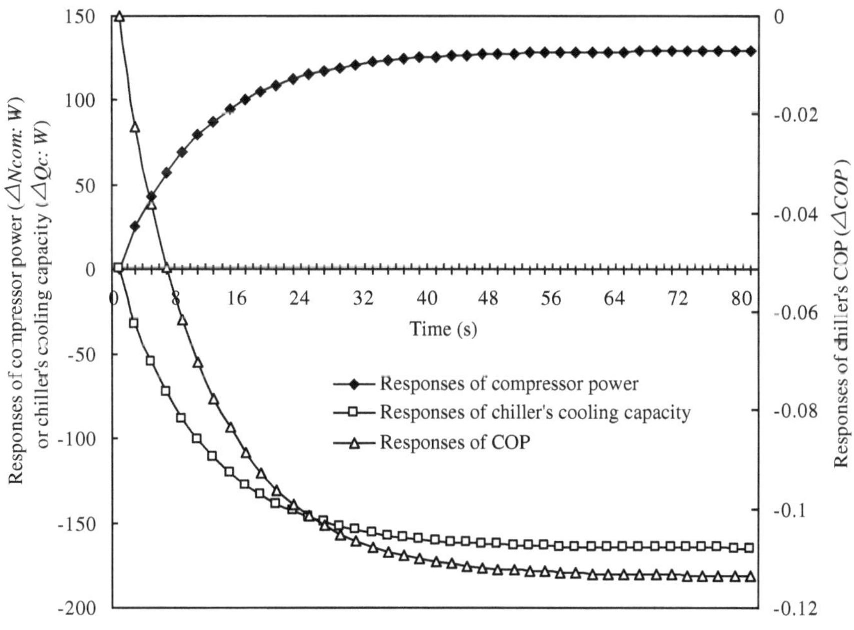

Fig. 3.13 Responses of compressor power, cooling capacity, and COP of the chiller to step increase of inlet coolant temperature of condenser by 1.0 °C

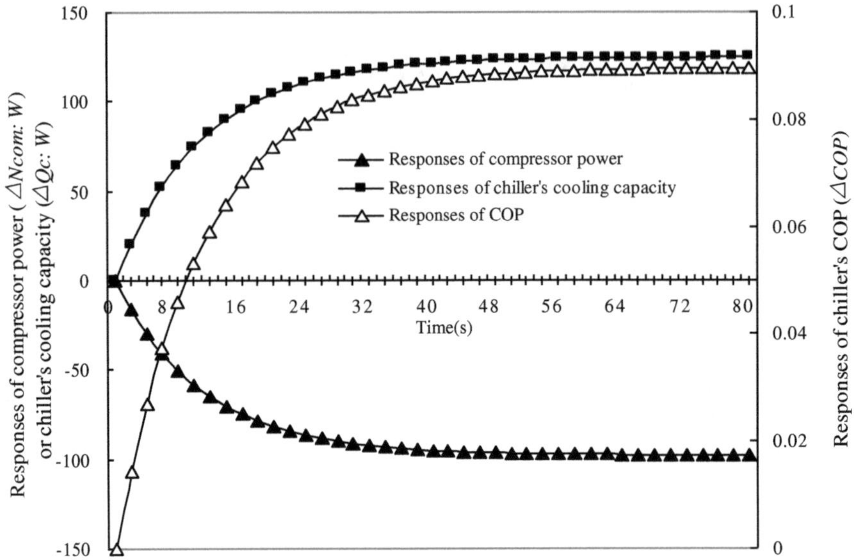

Fig. 3.14 Responses of compressor power, cooling capacity, and COP of the chiller to step increase of coolant flow rate of condenser by 0.1 kg/s

increases by about 0.09 after the coolant flow rate of condenser increased by 0.1 kg/s. Figure 3.15 gives the dynamic performance of the refrigeration system subjected to the step increase of refrigerant flow rate by 0.01 kg/s. As shown in Fig. 3.15, the compressor power increases suddenly by about 280 W at the start and then followed by a slow rise. It coincides with the fact that the increase of the refrigerant flow rate is achieved through increasing the compressor's rotation speed which is accompanied by the sudden increase of the compressor power. There is no doubt that the chiller's cooling capacity increases with the refrigerant flow rate rising. The gradual increase in the cooling capacity is caused by the thermal inertia of the evaporator and the refrigerant inside. In this case, the variation of the chiller's COP has a step decrease by about -0.18 at the start and increase gradually afterward. The final increase in the COP is about 0.018 with respect to the simulated results.

3.2.2 For Different Initial Conditions

The influence of the initial conditions on the transient response characteristic of the refrigeration system is investigated by using the state-space model. Three simulation cases (Please see Table 3.1) are performed for this investigation. The initial conditions listed in Table 3.1 are all measured data based on which the rest initial conditions can be obtained including the shell wall temperature of evaporator

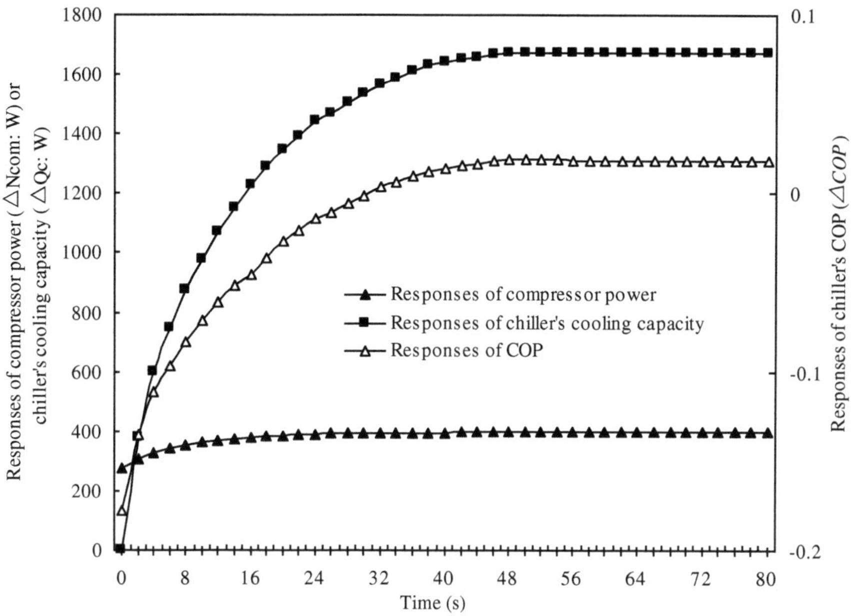

Fig. 3.15 Responses of compressor power, cooling capacity, and COP of the chiller to step increase of refrigerant flow rate by 0.01 kg/s

$((t_{eg})_o)$ and condenser $((t_{cg})_o)$, the exit coolant temperature of condenser $((t_{cw,L})_o)$, and the exit cold carrier temperature of evaporator $((t_{ew,L})_o)$.

(1) *Simulation case (a)*

In the simulation case (a), the response calculations are done with step disturbances under three initial temperatures of inlet cold carrier of evaporator (i.e., $(t_{ew,E})_o$ = 24.6, 29.3, 36.2 °C). The corresponding system matrixes for the refrigeration system ($A_{chiller}$, $B_{chiller}$, $C_{chiller}$, $D_{chiller}$) under the different initial conditions of simulation case (a) are calculated, respectively, as follows:

For the initial condition of $(t_{ew,E})_o - 24.6\,°C$::

$$A_{chiller} = \begin{bmatrix} -5.2500 & 0 & 5.2454 & 0 & 0 & 0 \\ 0 & -1.4962 & 2.0831 & 0 & 0 & 0 \\ 1.0247 & 0.2026 & -1.4299 & 0 & 0 & 0 \\ 0.0541 & 0 & 0 & -31.2970 & 0 & 31.2579 \\ 0 & 0 & 0 & 0 & -2.1785 & 1.8934 \\ 0 & 0 & 0 & 4.4929 & 0.1508 & -4.7945 \end{bmatrix};$$

Table 3.1 Experimental initial conditions for model simulation

Simulation case (a)		
Study purpose: Investigations into the dynamic response characteristics of the exit cold carrier temperature of evaporator under different initial temperatures of inlet cold carrier of evaporator $(t_{ew,E})_o$.		
$(t_{ew,\ E})_o = 24.6$ °C	$(t_{ew,\ E})_o = 29.3$ °C	$(t_{ew,\ E})_o = 36.2$ °C
Compressor inlet temp.: 24.1 (°C) Compressor outlet temp.: 56.1 (°C)	Compressor inlet temp.: 28.2 (°C) Compressor outlet temp.: 61.5 (°C)	Compressor inlet temp.: 34.5 (°C) Compressor outlet temp.: 67.8 (°C)
Other initial conditions shared: Evaporating temp. $(t_k)_o$ = 16.3 (°C); Cold carrier flow rate of evaporator $(G_{ew})_o$ = 1.45 (kg/s); Condensing temp. $(t_k)_o$ = 36.7 (°C); Inlet coolant temp. of condenser $(t_{cw,E})_o$ = 29.3 (°C); Coolant flow rate of condenser $(G_{cw})_o$ = 1.16 (kg/s); Refrigerant flow rate $(G_{rm})_o$ = 0.13 (kg/s).		
Simulation case (b)		
Study purpose: Investigations into the dynamic response characteristics of the exit cold carrier temperature of evaporator under different initial temperatures of inlet coolant of condenser $(t_{cw,E})_o$		
$(t_{cw,\ E})_o = 29.6$ °C	$(t_{cw,\ E})_o = 40.7$ °C	$(t_{cw,\ E})_o = 49.7$ °C
Compressor inlet temp.: 21.9 (°C) Compressor outlet temp.: 51.5(°C) Condensing Temp.: $(t_c)_o$ = 37.1 (°C)	Compressor inlet temp.: 22.2 (°C) Compressor outlet temp.: 57.5 (°C) Condensing Temp.:$(t_c)_o$ = 49.5 (°C)	Compressor inlet temp.: 22.4 (°C) Compressor outlet temp.: 59.7 (°C) Condensing Temp.: $(t_c)_o$ = 55.2 (°C)
Other initial conditions shared: Evaporating temp. $(t_k)_o$ = 18.8 (°C); Cold carrier flow rate of evaporator $(G_{ew})_o$ = 1.45 (kg/s); Inlet cold carrier temp. of evaporator $(t_{ew,E})_o$ = 24.1 (°C); Coolant flow rate of condenser $(G_{cw})_o$ = 1.16 (kg/s); Refrigerant flow rate $(G_{rm})_o$ = 0.13 (kg/s)		
Simulation case (c)		
Study purpose: Investigations into the dynamic response characteristics of the exit cold carrier temperature of evaporator under different initial refrigerant flow rate $(G_{rm})_o$		
$(G_{rm})_o = 0.176$ *kg/s*	$(G_{rm})_o = 0.199$ *kg/s*	$(G_{rm})_o = 0.238$ *kg/s*
Other initial conditions shared: Evaporating temp. $(t_k)_o$ = 15.1 (°C); Inlet cold carrier temp. of evaporator $(t_{ew,E})_o$ = 24.7 (°C); Cold carrier flow rate of evaporator $(G_{ew})_o$ = 2.14 (kg/s); Condensing temp. $(t_c)_o$ = 41.2 (°C); Inlet coolant temp. of condenser $(t_{cw,E})_o$ = 33.4 (°C); Coolant flow rate of condenser $(G_{cw})_o$ = 2.14 (kg/s); Compressor inlet temp.: $(t_{com,E})_o$ = 23.7 (°C); Compressor outlet temp. $(t_{com,L})_o$ = 71.3 (°C)		

$$B_{\text{chiller}} = \begin{bmatrix} 0 & 0 & 0 & 0 & 26.7888 \\ -0.5870 & -1.7765 & 0 & 0 & 0 \\ 0.2026 & -0.2879 & 0 & 0 & 0 \\ 0 & 0 & 0 & 0 & -56.8612 \\ 0 & 0 & 0.0454 & 6.4199 & 0 \\ 0 & 0 & 0.1508 & -0.3203 & 0 \end{bmatrix};$$

$$C_{\text{chiller}} = \begin{bmatrix} 0 & 1.0000 & 0 & 0 & 0 & 0 \\ 0 & 0 & 0 & 0 & 0 & 0 \\ 0 & 0 & 0 & 0 & 1.0000 & 0 \\ 0 & 0 & 0 & 0 & 0 & 0 \\ 136.6685 & 0 & 0 & -124.5601 & 0 & 0 \\ -172.2758 & 0 & 0 & 124.5601 & 0 & 0 \\ -0.3644 & 0 & 0 & 0.3221 & 0 & 0 \end{bmatrix};$$

$$D_{\text{chiller}} = \begin{bmatrix} 0 & 0 & 0 & 0 & 0 \\ 0 & 1 & 0 & 0 & 0 \\ 0 & 0 & 0 & 0 & 0 \\ 0 & 0 & 0 & 1 & 0 \\ 0 & 0 & 0 & 0 & 0 \\ 0 & 0 & 0 & 0 & 24760 \\ 0 & 0 & 0 & 0 & 181230 \end{bmatrix}.$$

For the initial condition of $(t_{\text{ew,E}})_{\text{o}} = 29.3\,^{\circ}\text{C}$:

$$A_{\text{chiller}} = \begin{bmatrix} -4.9140 & 0 & 4.9092 & 0 & 0 & 0 \\ 0 & -1.4962 & 2.0831 & 0 & 0 & 0 \\ 0.9591 & 0.2026 & -1.3642 & 0 & 0 & 0 \\ 0.054 & 0 & 0 & -31.4166 & 0 & 31.3771 \\ 0 & 0 & 0 & 0 & -2.1785 & 1.8934 \\ 0 & 0 & 0 & 4.5100 & 0.1508 & -4.8116 \end{bmatrix};$$

$$B_{\text{chiller}} = \begin{bmatrix} 0 & 0 & 0 & 0 & 27.6178 \\ -0.5870 & 0.7468 & 0 & 0 & 0 \\ 0.2026 & -0.7649 & 0 & 0 & 0 \\ 0 & 0 & 0 & 0 & -57.9874 \\ 0 & 0 & 0.0454 & 5.2185 & 0 \\ 0 & 0 & 0.1508 & -0.0166 & 0 \end{bmatrix};$$

$$C_{\text{chiller}} = \begin{bmatrix} 0 & 1.0000 & 0 & 0 & 0 & 0 \\ 0 & 0 & 0 & 0 & 0 & 0 \\ 0 & 0 & 0 & 0 & 1.0000 & 0 \\ 0 & 0 & 0 & 0 & 0 & 0 \\ 135.8178 & 0 & 0 & -125.9451 & 0 & 0 \\ -172.1875 & 0 & 0 & 125.9451 & 0 & 0 \\ -0.3027 & 0 & 0 & 0.2712 & 0 & 0 \end{bmatrix};$$

$$\mathrm{D}_{\text{chiller}} = \begin{bmatrix} 0 & 0 & 0 & 0 & 0 \\ 0 & 1 & 0 & 0 & 0 \\ 0 & 0 & 0 & 0 & 0 \\ 0 & 0 & 0 & 1 & 0 \\ 0 & 0 & 0 & 0 & 0 \\ 0 & 0 & 0 & 0 & 27540 \\ 0 & 0 & 0 & 0 & 184820 \end{bmatrix}.$$

For the initial condition of $(t_{\text{ew,E}})_{\text{o}} = 36.2\,°\text{C}$:

$$A_{\text{chiller}} = \begin{bmatrix} -5.1311 & 0 & 5.1264 & 0 & 0 & 0 \\ 0 & -1.4962 & 2.0831 & 0 & 0 & 0 \\ 1.0015 & 0.2026 & -1.4066 & 0 & 0 & 0 \\ 0.0541 & 0 & 0 & -33.5511 & 0 & 33.5114 \\ 0 & 0 & 0 & 0 & -2.1785 & 1.8934 \\ 0 & 0 & 0 & 4.8168 & 0.1508 & -5.1184 \end{bmatrix};$$

$$B_{\text{chiller}} = \begin{bmatrix} 0 & 0 & 0 & 0 & 28.2526 \\ -0.5870 & -1.3140 & 0 & 0 & 0 \\ 0.2026 & -0.4129 & 0 & 0 & 0 \\ 0 & 0 & 0 & 0 & -59.3983 \\ 0 & 0 & 0.0454 & 4.9352 & 0 \\ 0 & 0 & 0.1508 & 0.0366 & 0 \end{bmatrix};$$

$$C_{\text{chiller}} = \begin{bmatrix} 0 & 1.0000 & 0 & 0 & 0 & 0 \\ 0 & 0 & 0 & 0 & 0 & 0 \\ 0 & 0 & 0 & 0 & 1.0000 & 0 \\ 0 & 0 & 0 & 0 & 0 & 0 \\ 136.1326 & 0 & 0 & -126.4971 & 0 & 0 \\ -172.4500 & 0 & 0 & 126.4971 & 0 & 0 \\ -0.3017 & 0 & 0 & 0.2711 & 0 & 0 \end{bmatrix};$$

$$D_{\text{chiller}} = \begin{bmatrix} 0 & 0 & 0 & 0 & 0 \\ 0 & 1 & 0 & 0 & 0 \\ 0 & 0 & 0 & 0 & 0 \\ 0 & 0 & 0 & 1 & 0 \\ 0 & 0 & 0 & 0 & 0 \\ 0 & 0 & 0 & 0 & 27920 \\ 0 & 0 & 0 & 0 & 189320 \end{bmatrix}.$$

Figure 3.16 shows the influences of the initial inlet cold carrier temperature on the transient response characteristics of the exit cold carrier temperature subjected to different step disturbances. As seen from the simulation results, the proportionality coefficient of the exit cold carrier temperature to step disturbance of the inlet coolant and inlet cold carrier temperature is not affected by the initial condition of the inlet cold carrier temperature. However, the absolute proportionality coefficient of the exit cold carrier temperature to step disturbance of the coolant, the cold carrier, and the refrigerant flow rate will increase with the increase of the initial inlet cold carrier temperature.

(2) *Simulation case (b)*

In the simulation case (b), the transient response characteristics of the exit cold carrier temperature to step disturbances are investigated under different initial values of the inlet coolant temperature of condenser (i.e., $(t_{\text{cw,E}})_{\text{o}}$ = 29.6, 40.7, 49.7 °C). The corresponding system matrixes for the refrigeration system (A_{chiller}, B_{chiller}, C_{chiller}, D_{chiller}) under the different initial conditions of simulation case (b) are obtained, respectively, as follows:

For the initial condition of $(t_{\text{cw,E}})_{\text{o}} = 29.6\,°\text{C}$:

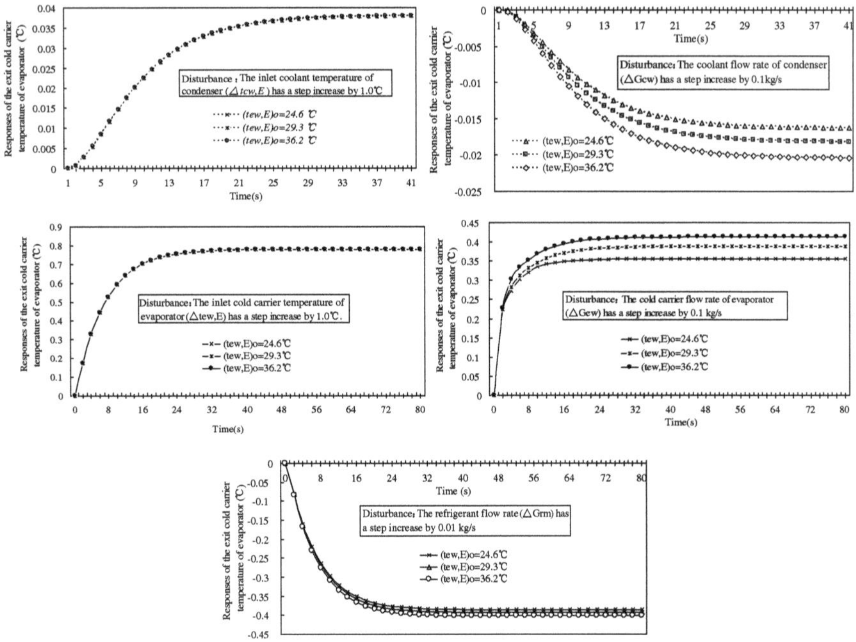

Fig. 3.16 Transient response characteristics of the exit cold carrier temperature of evaporator under different initial temperatures of inlet cold carrier of evaporator (Simulation case (a))

$$A_{\text{chiller}} = \begin{bmatrix} -5.4059 & 0 & 5.4015 & 0 & 0 & 0 \\ 0 & -1.4962 & 2.0831 & 0 & 0 & 0 \\ 1.0552 & 0.2026 & -1.4604 & 0 & 0 & 0 \\ 0.0541 & 0 & 0 & -30.6431 & 0 & 30.6037 \\ 0 & 0 & 0 & 0 & -2.1785 & 1.8934 \\ 0 & 0 & 0 & 4.3989 & 0.1508 & -4.7004 \end{bmatrix};$$

$$B_{\text{chiller}} = \begin{bmatrix} 0 & 0 & 0 & 0 & 25.4475 \\ -0.5870 & -4.0857 & 0 & 0 & 0 \\ 0.2026 & 0.1826 & 0 & 0 & 0 \\ 0 & 0 & 0 & 0 & -54.7588 \\ 0 & 0 & 0.0454 & 5.8103 & 0 \\ 0 & 0 & 0.1508 & -0.1772 & 0 \end{bmatrix};$$

$$C_{\text{chiller}} = \begin{bmatrix} 0 & 1.0000 & 0 & 0 & 0 & 0 \\ 0 & 0 & 0 & 0 & 0 & 0 \\ 0 & 0 & 0 & 0 & 1.0000 & 0 \\ 0 & 0 & 0 & 0 & 0 & 0 \\ 138.7992 & 0 & 0 & -125.3369 & 0 & 0 \\ -172.3632 & 0 & 0 & 125.3369 & 0 & 0 \\ -0.4796 & 0 & 0 & 0.4220 & 0 & 0 \end{bmatrix};$$

$$D_{\text{chiller}} = \begin{bmatrix} 0 & 0 & 0 & 0 & 0 \\ 0 & 1 & 0 & 0 & 0 \\ 0 & 0 & 0 & 0 & 0 \\ 0 & 0 & 0 & 1 & 0 \\ 0 & 0 & 0 & 0 & 0 \\ 0 & 0 & 0 & 0 & 21140 \\ 0 & 0 & 0 & 0 & 174530 \end{bmatrix}.$$

For the initial condition of $(t_{\text{cw,E}})_{\text{o}} = 40.7\,°\text{C}$

$$A_{\text{chiller}} = \begin{bmatrix} -5.7379 & 0 & 5.7351 & 0 & 0 & 0 \\ 0 & -1.4962 & 2.0831 & 0 & 0 & 0 \\ 1.1204 & 0.2026 & -1.5255 & 0 & 0 & 0 \\ 0.0549 & 0 & 0 & -27.8795 & 0 & 27.8393 \\ 0 & 0 & 0 & 0 & -2.1785 & 1.8934 \\ 0 & 0 & 0 & 4.0015 & 0.1508 & -4.3031 \end{bmatrix};$$

$$B_{\text{chiller}} = \begin{bmatrix} 0 & 0 & 0 & 0 & 23.4127 \\ -0.5870 & -6.3773 & 0 & 0 & 0 \\ 0.2026 & 0.7205 & 0 & 0 & 0 \\ 0 & 0 & 0 & 0 & -50.2580 \\ 0 & 0 & 0.0454 & 3.0071 & 0 \\ 0 & 0 & 0.1508 & 0.2246 & 0 \end{bmatrix};$$

$$C_{\text{chiller}} = \begin{bmatrix} 0 & 1.0000 & 0 & 0 & 0 & 0 \\ 0 & 0 & 0 & 0 & 0 & 0 \\ 0 & 0 & 0 & 0 & 1.0000 & 0 \\ 0 & 0 & 0 & 0 & 0 & 0 \\ 153.9727 & 0 & 0 & -128.0791 & 0 & 0 \\ -175.1204 & 0 & 0 & 128.0791 & 0 & 0 \\ -0.5498 & 0 & 0 & 0.4505 & 0 & 0 \end{bmatrix};$$

$$D_{\text{chiller}} = \begin{bmatrix} 0 & 0 & 0 & 0 & 0 \\ 0 & 1 & 0 & 0 & 0 \\ 0 & 0 & 0 & 0 & 0 \\ 0 & 0 & 0 & 1 & 0 \\ 0 & 0 & 0 & 0 & 0 \\ 0 & 0 & 0 & 0 & 19840 \\ 0 & 0 & 0 & 0 & 160180 \end{bmatrix}.$$

For the initial condition of $(t_{\text{cw,E}})_{\text{o}} = 49.7\,°\text{C}$:

$$A_{\text{chiller}} = \begin{bmatrix} -5.1288 & 0 & 5.1268 & 0 & 0 & 0 \\ 0 & -1.4962 & 2.0831 & 0 & 0 & 0 \\ 1.0015 & 0.2026 & -1.4067 & 0 & 0 & 0 \\ 0.0553 & 0 & 0 & -27.9825 & 0 & 27.9421 \\ 0 & 0 & 0 & 0 & -2.1785 & 1.8934 \\ 0 & 0 & 0 & 4.0163 & 0.1508 & -4.3179 \end{bmatrix};$$

$$B_{\text{chiller}} = \begin{bmatrix} 0 & 0 & 0 & 0 & 23.0057 \\ -0.5870 & -8.1105 & 0 & 0 & 0 \\ 0.2026 & 0.9936 & 0 & 0 & 0 \\ 0 & 0 & 0 & 0 & -48.3970 \\ 0 & 0 & 0.0454 & 3.2998 & 0 \\ 0 & 0 & 0.1508 & 0.0873 & 0 \end{bmatrix};$$

$$C_{\text{chiller}} = \begin{bmatrix} 0 & 1.0000 & 0 & 0 & 0 & 0 \\ 0 & 0 & 0 & 0 & 0 & 0 \\ 0 & 0 & 0 & 0 & 1.0000 & 0 \\ 0 & 0 & 0 & 0 & 0 & 0 \\ 160.7841 & 0 & 0 & -128.8777 & 0 & 0 \\ -176.4054 & 0 & 0 & 128.8777 & 0 & 0 \\ -0.4320 & 0 & 0 & 0.3420 & 0 & 0 \end{bmatrix};$$

$$D_{\text{chiller}} = \begin{bmatrix} 0 & 0 & 0 & 0 & 0 \\ 0 & 1 & 0 & 0 & 0 \\ 0 & 0 & 0 & 0 & 0 \\ 0 & 0 & 0 & 1 & 0 \\ 0 & 0 & 0 & 0 & 0 \\ 0 & 0 & 0 & 0 & 22640 \\ 0 & 0 & 0 & 0 & 154250 \end{bmatrix}.$$

Figure 3.17 gives the transient responses of the exit cold carrier temperature to step disturbances under different initial inlet coolant temperatures in the simulation case (b). The results manifest that the initial value of the inlet coolant temperature

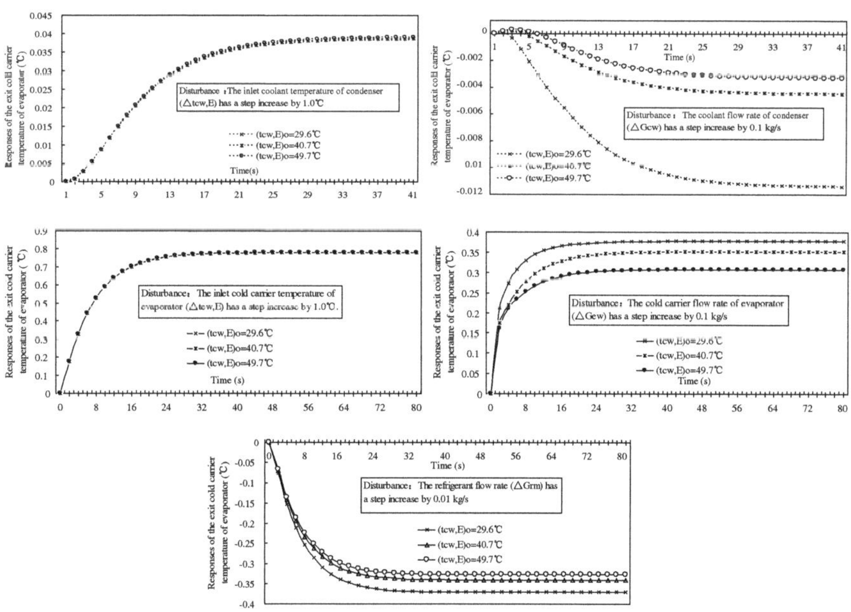

Fig. 3.17 Transient response characteristics of the exit cold carrier temperature of evaporator under different initial temperatures of inlet coolant of condenser (Simulation case (b))

produces little influence on the response characteristics of the exit cold carrier temperature subjected to the perturbations of the inlet coolant and inlet cold carrier temperature, but it may affect that subjected to the perturbations of the coolant, the cold carrier, and the refrigerant flow rate. Basically, the absolute proportionality coefficient of the exit cold carrier temperature to step disturbance of the coolant, the cold carrier, and the refrigerant flow rate will decrease with the increase of the initial value of the inlet coolant temperature.

(3) *Simulation case (c)*

In the simulation case (c), the transient response characteristics of the exit cold carrier temperature to step disturbances are investigated under different initial refrigerant flow rates (i.e., $(G_{rm})_o$ = 0.176, 0.199, 0.237 kg/s). The corresponding system matrixes for the refrigeration system ($A_{chiller}$,$B_{chiller}$,$C_{chiller}$,$D_{chiller}$) under the different initial refrigerant flow rates are listed as below:

For the initial condition of $(G_{rm})_o = 0.176\,\text{kg/s}$:

$$A_{chiller} = \begin{bmatrix} -6.0491 & 0 & 6.0430 & 0 & 0 & 0 \\ 0 & -2.5387 & 3.4000 & 0 & 0 & 0 \\ 1.1805 & 0.3306 & -1.8418 & 0 & 0 & 0 \\ 0.0736 & 0 & 0 & -30.9453 & 0 & 30.8924 \\ 0 & 0 & 0 & 0 & -3.1105 & 2.5852 \\ 0 & 0 & 0 & 4.4404 & 0.2059 & -4.8521 \end{bmatrix};$$

$$B_{chiller} = \begin{bmatrix} 0 & 0 & 0 & 0 & 27.3712 \\ -0.8614 & -7.8559 & 0 & 0 & 0 \\ 0.3306 & 1.0027 & 0 & 0 & 0 \\ 0 & 0 & 0 & 0 & -55.1646 \\ 0 & 0 & 0.0837 & 4.9467 & 0 \\ 0 & 0 & 0.2059 & -0.1208 & 0 \end{bmatrix};$$

$$C_{chiller} = \begin{bmatrix} 0 & 1.0000 & 0 & 0 & 0 & 0 \\ 0 & 0 & 0 & 0 & 0 & 0 \\ 0 & 0 & 0 & 0 & 1.0000 & 0 \\ 0 & 0 & 0 & 0 & 0 & 0 \\ 187.6315 & 0 & 0 & -168.6454 & 0 & 0 \\ -234.5960 & 0 & 0 & 168.6454 & 0 & 0 \\ -0.1947 & 0 & 0 & 0.1681 & 0 & 0 \end{bmatrix};$$

$$D_{\text{chiller}} = \begin{bmatrix} 0 & 0 & 0 & 0 & 0 \\ 0 & 1 & 0 & 0 & 0 \\ 0 & 0 & 0 & 0 & 0 \\ 0 & 0 & 0 & 1 & 0 \\ 0 & 0 & 0 & 0 & 0 \\ 0 & 0 & 0 & 0 & 34640 \\ 0 & 0 & 0 & 0 & 175820 \end{bmatrix};$$

For the initial condition of $(G_{\text{rm}})_{\text{o}} = 0.199\,\text{kg/s}$:

$$A_{\text{chiller}} = \begin{bmatrix} -5.6901 & 0 & 5.6831 & 0 & 0 & 0 \\ 0 & -2.5387 & 3.4000 & 0 & 0 & 0 \\ 1.1204 & 0.3306 & -1.7715 & 0 & 0 & 0 \\ 0.0833 & 0 & 0 & -31.6092 & 0 & 31.5499 \\ 0 & 0 & 0 & 0 & -3.1105 & 2.5852 \\ 0 & 0 & 0 & 4.5349 & 0.2059 & -4.9466 \end{bmatrix};$$

$$B_{\text{chiller}} = \begin{bmatrix} 0 & 0 & 0 & 0 & 27.8736 \\ -0.8614 & -5.2382 & 0 & 0 & 0 \\ 0.3306 & 0.4395 & 0 & 0 & 0 \\ 0 & 0 & 0 & 0 & -55.0432 \\ 0 & 0 & 0.0837 & 5.8096 & 0 \\ 0 & 0 & 0.2059 & -0.1608 & 0 \end{bmatrix};$$

$$C_{\text{chiller}} = \begin{bmatrix} 0 & 1.0000 & 0 & 0 & 0 & 0 \\ 0 & 0 & 0 & 0 & 0 & 0 \\ 0 & 0 & 0 & 0 & 1.0000 & 0 \\ 0 & 0 & 0 & 0 & 0 & 0 \\ 212.1554 & 0 & 0 & -188.9406 & 0 & 0 \\ -265.4966 & 0 & 0 & 188.9406 & 0 & 0 \\ -0.1580 & 0 & 0 & 0.1345 & 0 & 0 \end{bmatrix};$$

$$D_{\text{chiller}} = \begin{bmatrix} 0 & 0 & 0 & 0 & 0 \\ 0 & 1 & 0 & 0 & 0 \\ 0 & 0 & 0 & 0 & 0 \\ 0 & 0 & 0 & 1 & 0 \\ 0 & 0 & 0 & 0 & 0 \\ 0 & 0 & 0 & 0 & 38890 \\ 0 & 0 & 0 & 0 & 175440 \end{bmatrix}.$$

For the initial condition of $(G_{\mathrm{rm}})_{\mathrm{o}} = 0.237\,\mathrm{kg/s}$:

$$A_{\mathrm{chiller}} = \begin{bmatrix} -5.5747 & 0 & 5.5665 & 0 & 0 & 0 \\ 0 & -2.5387 & 3.4000 & 0 & 0 & 0 \\ 1.0874 & 0.3306 & -1.7487 & 0 & 0 & 0 \\ 0.0995 & 0 & 0 & -30.7540 & 0 & 30.6831 \\ 0 & 0 & 0 & 0 & -3.1105 & 2.5852 \\ 0 & 0 & 0 & 4.4103 & 0.2059 & -4.8220 \end{bmatrix};$$

$$B_{\mathrm{chiller}} = \begin{bmatrix} 0 & 0 & 0 & 0 & 27.8087 \\ -0.8614 & -4.9823 & 0 & 0 & 0 \\ 0.3306 & 0.2998 & 0 & 0 & 0 \\ 0 & 0 & 0 & 0 & -54.5852 \\ 0 & 0 & 0.0837 & 5.5628 & 0 \\ 0 & 0 & 0.2059 & -0.0092 & 0 \end{bmatrix};$$

$$C_{\mathrm{chiller}} = \begin{bmatrix} 0 & 1.0000 & 0 & 0 & 0 & 0 \\ 0 & 0 & 0 & 0 & 0 & 0 \\ 0 & 0 & 0 & 0 & 1.0000 & 0 \\ 0 & 0 & 0 & 0 & 0 & 0 \\ 253.9461 & 0 & 0 & -226.1909 & 0 & 0 \\ -317.1475 & 0 & 0 & 226.1909 & 0 & 0 \\ -0.1506 & 0 & 0 & 0.1282 & 0 & 0 \end{bmatrix};$$

$$D_{\mathrm{chiller}} = \begin{bmatrix} 0 & 0 & 0 & 0 & 0 \\ 0 & 1 & 0 & 0 & 0 \\ 0 & 0 & 0 & 0 & 0 \\ 0 & 0 & 0 & 1 & 0 \\ 0 & 0 & 0 & 0 & 0 \\ 0 & 0 & 0 & 0 & 39850 \\ 0 & 0 & 0 & 0 & 173980 \end{bmatrix}.$$

Figure 3.18 presents the transient responses of the exit cold carrier temperature to step disturbances under different initial refrigerant flow rates in the simulation case (c). As known from Fig. 3.18, the absolute proportionality coefficient of the exit cold carrier temperature to the perturbations of the inlet coolant temperature or the cold carrier flow rate increases with the initial value of the refrigerant flow rate, and the change trend is opposite for the perturbations of the coolant flow rate. However, the response characteristics of the exit cold carrier temperature to the perturbations of the inlet cold carrier temperature or the refrigerant flow rate are affected little by the initial value of the refrigerant flow rate.

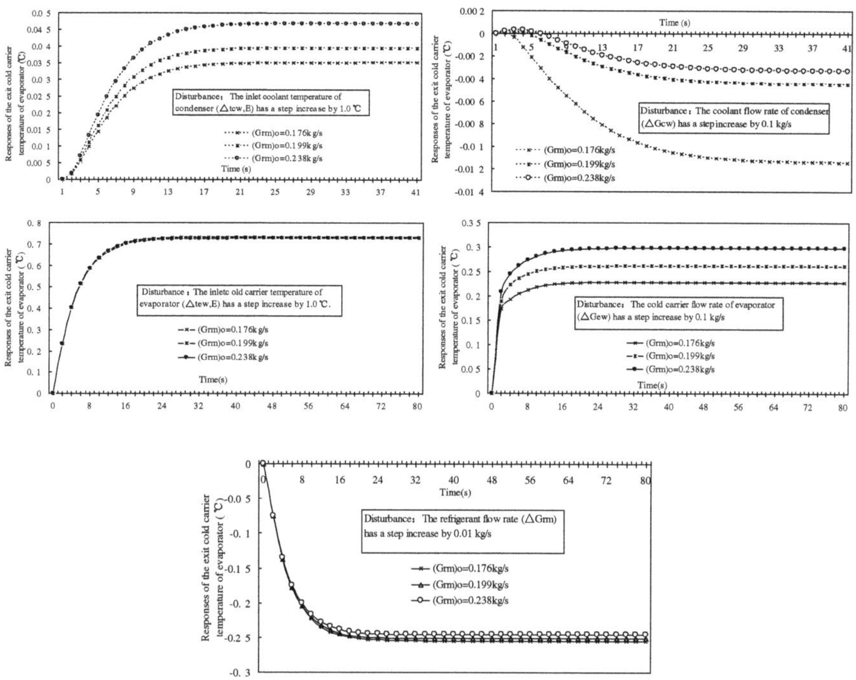

Fig. 3.18 Transient response characteristics of the exit cold carrier temperature of evaporator under different initial refrigerant flow rates (Simulation case (c))

3.3 On Cooling Tower

3.3.1 Subjected to Different Perturbations

Since only exit water temperature of cooling tower will produce influence on the performance of refrigeration system, the transient response characteristics of the tower's exit water temperature are to be investigated under different perturbations. Taking the experimental cooling tower and the initial conditions of case I in Table 2. 10 for example, the system matrixes of cooling tower (A_{tower}, B_{tower}) are counted as,

$$A_{\text{tower}} = \begin{bmatrix} -12.0343 & 0.0251 & 2.4203 & 0 \\ 0.0000 & -9.5964 & 0.0002 & 0.0000 \\ 0.5702 & 0.1313 & -3.1097 & 0.0000 \\ 0.0000 & 0.0000 & 0.0374 & -0.0374 \end{bmatrix};$$

$$B_{\text{tower}} = \begin{bmatrix} 7.1580 & 0.0251 & -0.6777 & 2.4203 & 0.0000 \\ 0.0000 & 9.5959 & -0.0013 & 0.0002 & 0.0000 \\ 0.5702 & 0.1313 & -0.2615 & 1.7810 & 3.6122 \\ 0.0000 & 0.0000 & 0.0000 & 0.0000 & 0.0011 \end{bmatrix}$$

The following perturbations are given for the dynamic response investigation: ① Inlet air temperature has a step increase by 1.0 °C; ② inlet air humidity has a step increase by 1.0 g/(kg dry air); ③ inlet water temperature has a step increase by 1.0 °C; ④ airflow rate has a step increase by 1.0 kg/s; and ⑤ water flow rate has a step increase by 1.0 kg/s.

Figure 3.19 shows the transient responses of exit water temperature of cooling tower under the above perturbations. As shown in Fig. 3.19, the increase in the inlet air temperature and humidity and inlet water temperature as well as water flow rate will cause the exit water temperature of cooling tower to rise, while the increase in the airflow rate will lead to the decrease of the exit water temperature. Increasing inlet air temperature and humidity will weaken the water-cooling effect by air, and this leads to an increase in the exit water temperature. Increasing inlet water temperature and water flow rate means increasing water thermal energy, and this also results in an increase in the exit water temperature if the other heat transfer conditions are unchanged. A higher airflow rate will bring about a larger quantity of heat transfer. So, the exit water temperature of cooling tower will decrease as the airflow rate increases. According to the simulated results under the initial conditions of case I, the proportionality coefficient of the exit water temperature to step disturbance of the inlet air temperature, the inlet air humidity, the inlet water

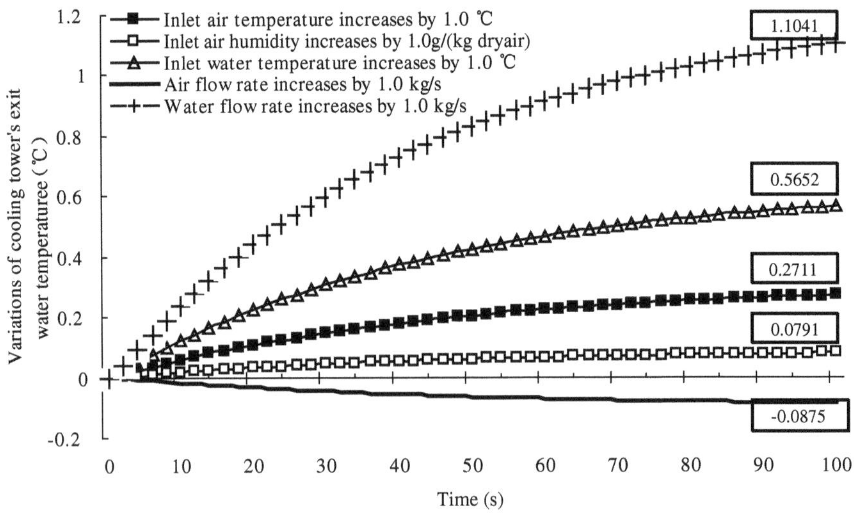

Fig. 3.19 Transient responses of exit water temperature of cooling tower under different perturbations

temperature, and airflow rate as well as water flow rate is calculated as 0.2711 °C/°C, 0.0791°C/(g/kg dry air), 0.5652 °C/°C,−0.0875 °C/(kg/s), and 1.1041 °C/(kg/s), respectively.

3.3.2 *For Different Initial Conditions*

The following initial conditions are used to investigate the influence of initial values of inlet variables on the dynamic response characteristics of the cooling tower's exit water temperature:

Case (a) Three initial inlet air temperatures, i.e., $(t_{a,\ E})_o$ = 30.0, 32.0 and 34.0 °C, combined with the other initial conditions: $(t_{w,\ E})_o$ = 38.2 °C, $(W_{a,\ E})_o$ = 22.4 g/(kg dry air), $(G_a)_o$ = 12.62 kg/s, $(G_w)_o$ = 3.31 kg/s;
Case (b) Three initial inlet air humidity ratios, i.e., $(W_{a,\ E})_o$ = 22.0,24.0, and 26.0 g/(kg dry air), combined with the other initial conditions: $(t_{a,\ E})_o$ = 34.0 °C, $(t_{w,\ E})_o$ = 38.2 °C, $(G_a)_o$ = 12.62 kg/s,$(G_w)_o$ = 3.31 kg/s;
Case (c) Three initial inlet water temperatures, i.e., $(t_{w,\ E})_o$ = 38.0, 40.0 and 42.0 °C, combined with the other initial conditions: $(t_{a,\ E})_o$ = 34.0 °C, $(W_{a,\ E})_o$ = 22.0 g/(kg dry air), $(G_a)_o$ = 12.62 kg/s,$(G_w)_o$ = 3.31 kg/s;

The mass of water in the sink of cooling tower (M_{sump}) is identically given as 88.5 kg for all the simulation cases. The initial values of the exit variables can be obtained with respect to the initial conditions of inlet variables.

(1) *Different initial values of inlet air temperature* $(t_{a,\ E})_o$

The system matrixes of cooling tower under the initial conditions of case (a) are calculated as below:

$$A_{\text{tower},\,t_{a,E}=30\,^{\circ}\text{C}} = A_{\text{tower},\,t_{a,E}=32\,^{\circ}\text{C}} = A_{\text{tower},\,t_{a,E}=34\,^{\circ}\text{C}} = \begin{bmatrix} -12.0343 & 0.0251 & 2.4203 & 0.0000 \\ 0.0000 & -9.5964 & 0.0002 & 0.0000 \\ 0.5702 & 0.1313 & -3.1079 & 0.0000 \\ 0.0000 & 0.0000 & 0.0374 & -0.0374 \end{bmatrix};$$

$$B_{\text{tower},\,t_{a,E}=30\,^{\circ}\text{C}} = \begin{bmatrix} 7.1580 & 0.0251 & -0.1633 & 2.4203 & 0.0000 \\ 0.0000 & 9.5959 & -0.0017 & 0.0002 & 0.0000 \\ 0.5702 & 0.1313 & -0.2509 & 1.7810 & 3.4163 \\ 0.0000 & 0.0000 & 0.0000 & 0.0000 & 0.0045 \end{bmatrix};$$

$$B_{\text{tower},\, t_{a,E}=32\,^\circ\text{C}} = \begin{bmatrix} 7.1580 & 0.0251 & -0.2823 & 2.4203 & 0.0000 \\ 0.0000 & 9.5959 & -0.0015 & 0.0002 & 0.0000 \\ 0.5702 & 0.1313 & -0.1882 & 1.7810 & 2.9005 \\ 0.0000 & 0.0000 & 0.0000 & 0.0000 & 0.0045 \end{bmatrix};$$

$$B_{\text{tower},\, t_{a,E}=34\,^\circ\text{C}} = \begin{bmatrix} 7.1580 & 0.0251 & -0.4100 & 2.4203 & 0.0000 \\ 0.0000 & 9.5959 & -0.0014 & 0.0002 & 0.0000 \\ 0.5702 & 0.1313 & -0.1234 & 1.7810 & 2.3646 \\ 0.0000 & 0.0000 & 0.0000 & 0.0000 & 0.0045 \end{bmatrix}.$$

Figure 3.20 presents transient responses of the cooling tower's exit water temperature subjected to various disturbances under different initial values of inlet air temperature. The results manifest that the transient response characteristics of the cooling tower's exit water temperature to the perturbations of inlet air temperature and humidity as well as inlet water temperature will not affected by initial value of inlet air temperature. However, when subjected to the perturbations of water or airflow rate, the initial value of inlet air temperature will impact on the transient response characteristics of the cooling tower's exit water temperature. As shown in Fig. 3.20d, e, the absolute proportionality coefficient of the exit water temperature to the water or airflow rate decreases with the initial inlet air temperature increasing.

(2) *Different initial values of inlet air humidity ratios*

The system matrixes of cooling tower corresponding to the initial conditions of case (b) are calculated as below:

$$A_{\text{tower},\, W_{a,E}=22\,\text{g/kg}} = A_{\text{tower},\, W_{a,E}=24\,\text{g/kg}} = A_{\text{tower},\, W_{a,E}=26\,\text{g/kg}} = \begin{bmatrix} -12.0343 & 0.0251 & 2.4203 & 0.0000 \\ 0.0000 & -9.5964 & 0.0002 & 0.0000 \\ 0.5702 & 0.1313 & -3.1079 & 0.0000 \\ 0.0000 & 0.0000 & 0.0374 & -0.0374 \end{bmatrix};$$

$$B_{\text{tower},\, W_{a,E}=22\,\text{g/kg}} = \begin{bmatrix} 7.1580 & 0.0251 & -0.4539 & 2.4203 & 0.0000 \\ 0.0000 & 9.5959 & -0.0015 & 0.0002 & 0.0000 \\ 0.5702 & 0.1313 & -0.1126 & 1.7810 & 2.2641 \\ 0.0000 & 0.0000 & 0.0000 & 0.0000 & 0.0045 \end{bmatrix};$$

$$B_{\text{tower},\, W_{a,E}=24} = \begin{bmatrix} 7.1580 & 0.0251 & -0.3749 & 2.4203 & 0.0000 \\ 0.0000 & 9.5959 & -0.0013 & 0.0002 & 0.0000 \\ 0.5702 & 0.1313 & -0.1320 & 1.7810 & 2.4448 \\ 0.0000 & 0.0000 & 0.0000 & 0.0000 & 0.0045 \end{bmatrix};$$

$$B_{\text{tower},\, W_{a,E}=26\,\text{g/kg}} = \begin{bmatrix} 7.1580 & 0.0251 & 0.0701 & 2.4203 & 0.0000 \\ 0.0000 & 9.5959 & -0.0010 & 0.0002 & 0.0000 \\ 0.5702 & 0.1313 & -0.1661 & 1.7810 & 2.7551 \\ 0.0000 & 0.0000 & 0.0000 & 0.0000 & 0.0045 \end{bmatrix}.$$

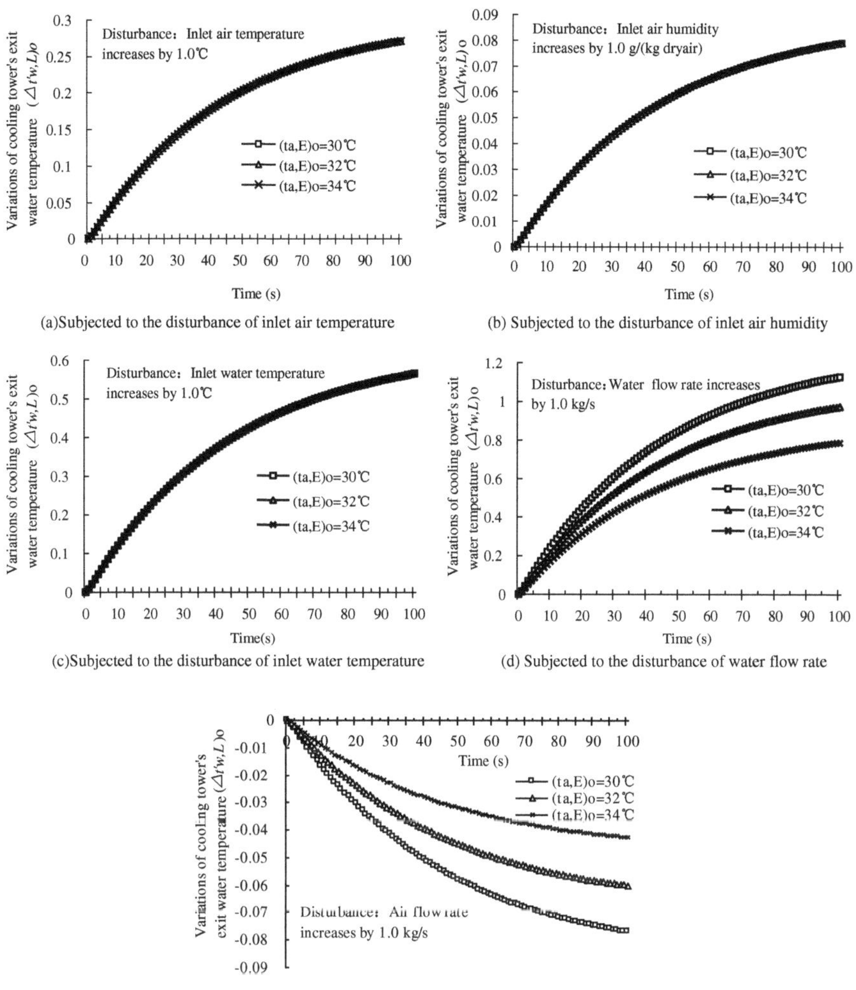

Fig. 3.20 Responses of cooling tower's exit water temperature to various disturbances under different initial values of inlet air temperature. **a** Subjected to the disturbance of inlet air temperature. **b** Subjected to the disturbance of inlet air humidity. **c** Subjected to the disturbance of inlet water temperature. **d** Subjected to the disturbance of water flow rate. **e** Subjected to the disturbance of airflow rate

The influences of initial inlet air humidity on the transient response of cooling tower's exit water temperature to various disturbances are shown in Fig. 3.21. The results manifest that the initial inlet air humidity will not affect the transient response characteristics of the cooling tower's exit water temperature when subjected to the perturbations of inlet air temperature and humidity as well as inlet water temperature. However, the transient response characteristics of the cooling tower's exit water temperature to the perturbations of water or airflow rate will be affected by the initial inlet air humidity. As shown in Fig. 3.21d, e, the absolute proportionality coefficient of the exit water temperature to the water or airflow rate increases with the initial inlet air humidity increasing.

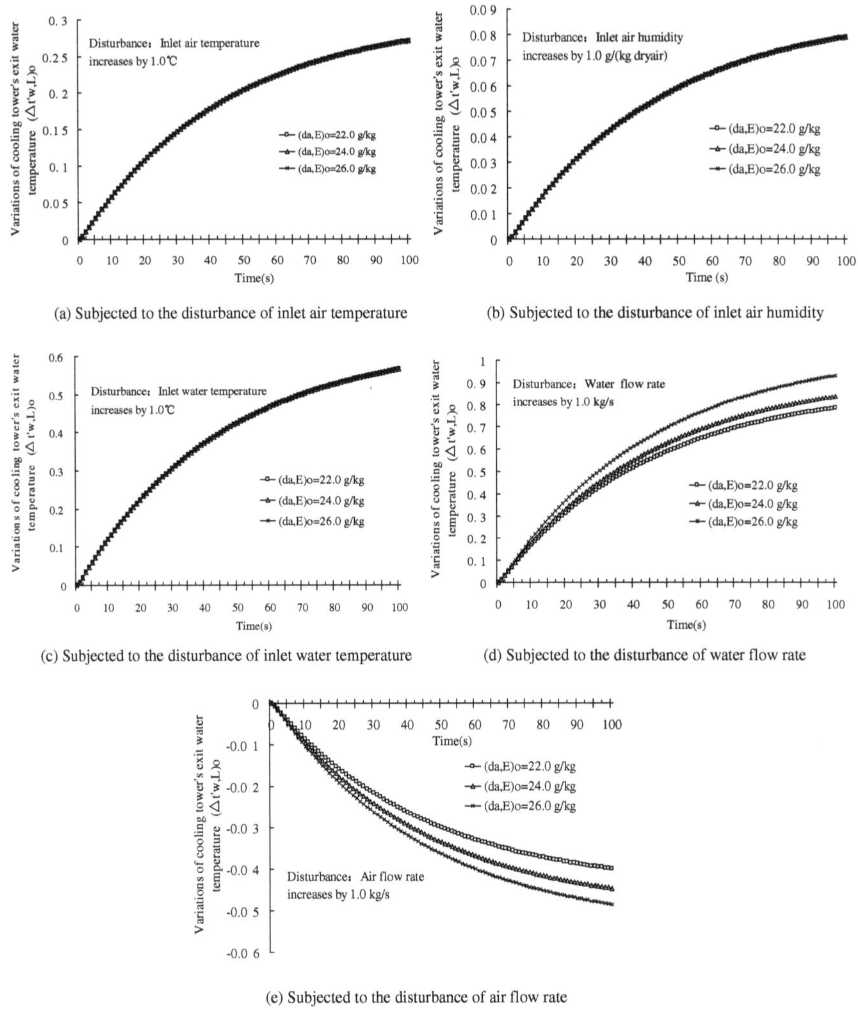

Fig. 3.21 Responses of cooling tower's exit water temperature to various disturbances under different initial values of inlet air humidity. **a** Subjected to the disturbance of inlet air temperature. **b** Subjected to the disturbance of inlet air humidity. **c** Subjected to the disturbance of inlet water temperature. **d** Subjected to the disturbance of water flow rate. **e** Subjected to the disturbance of airflow rate

(3) *Different initial values of inlet water temperatures*

The system matrixes of cooling tower under the initial conditions of case (c) are as follows:

$$A_{\text{tower},\,t_{w,E}=38\,^{\circ}\text{C}} = A_{\text{tower},\,t_{w,E}=40\,^{\circ}\text{C}} = A_{\text{tower},\,t_{w,E}=42\,^{\circ}\text{C}} = \begin{bmatrix} -12.0343 & 0.0251 & 2.4203 & 0.0000 \\ 0.0000 & -9.5964 & 0.0002 & 0.0000 \\ 0.5702 & 0.1313 & -3.1079 & 0.0000 \\ 0.0000 & 0.0000 & 0.0374 & -0.0374 \end{bmatrix}$$

$$B_{\text{tower},\,t_{w,E}=38\,^{\circ}\text{C}} = \begin{bmatrix} 7.1580 & 0.0251 & -0.3866 & 2.4203 & 0.0000 \\ 0.0000 & 9.5959 & -0.0015 & 0.0002 & 0.0000 \\ 0.5702 & 0.1313 & -0.1103 & 1.7810 & 2.2411 \\ 0.0000 & 0.0000 & 0.0000 & 0.0000 & 0.0045 \end{bmatrix}$$

$$B_{\text{tower},\,t_{w,E}=40\,^{\circ}\text{C}} = \begin{bmatrix} 7.1580 & 0.0251 & 0.1080 & 2.4203 & 0.0000 \\ 0.0000 & 9.5959 & -0.0017 & 0.0002 & 0.0000 \\ 0.5702 & 0.1313 & -0.1943 & 1.7810 & 3.0196 \\ 0.0000 & 0.0000 & 0.0000 & 0.0000 & 0.0045 \end{bmatrix}$$

$$B_{\text{tower},\,t_{w,E}=40\,^{\circ}\text{C}} = \begin{bmatrix} 7.1580 & 0.0251 & 0.6700 & 2.4203 & 0.0000 \\ 0.0000 & 9.5959 & -0.0019 & 0.0002 & 0.0000 \\ 0.5702 & 0.1313 & -0.2761 & 1.7810 & 3.7749 \\ 0.0000 & 0.0000 & 0.0000 & 0.0000 & 0.0045 \end{bmatrix}$$

The responses of cooling tower's exit water temperature to various disturbances under different initial values of inlet water temperature are presented in Fig. 3.22. The influence of initial inlet water temperature on the transient response characteristics of the cooling tower's exit water temperature shares the same regular patterns with the influence of initial inlet air humidity on that as shown in Fig. 3.21.

3.4 On Duct and Pipe

3.4.1 On Straight-Through Duct

In this section, the transient responses of straight-through duct with various lengths are investigated. Under the experimental initial conditions (refer to '2.2.4 Duct (pipe) and fan (pump)'), the system matrixes of the straight-through duct with the length of 1.0, 3.0, and 5.0 m are calculated as below:

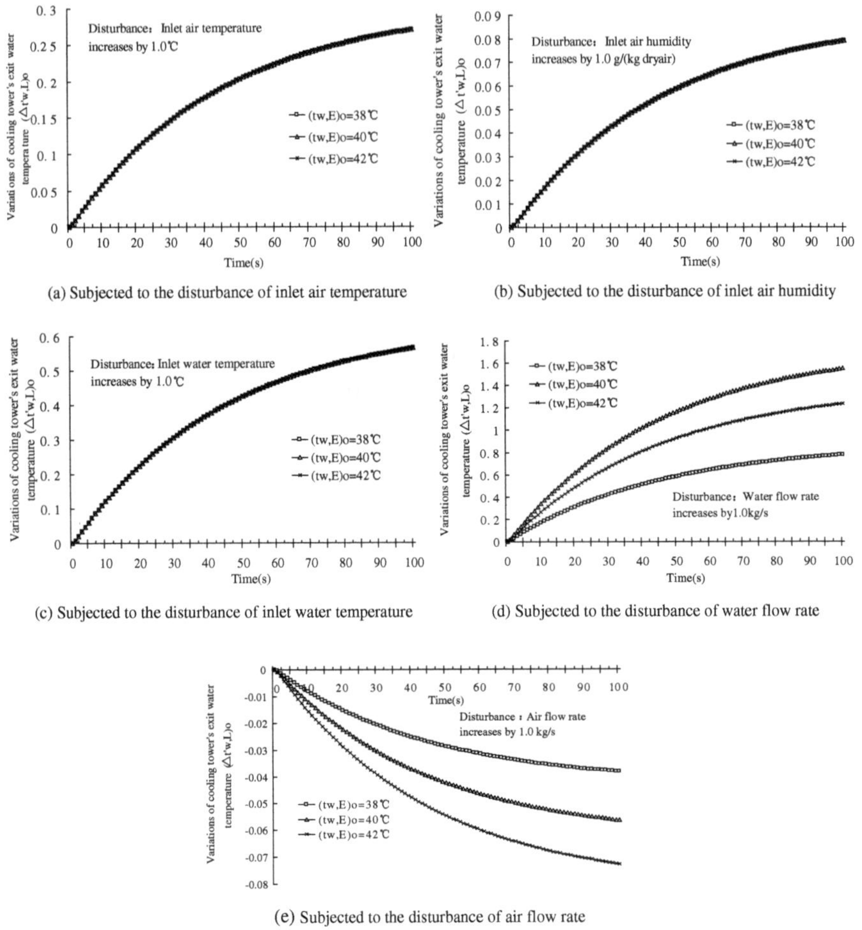

Fig. 3.22 Responses of cooling tower's exit water temperature to various disturbances under different initial values of inlet water temperature. **a** Subjected to the disturbance of inlet air temperature. **b** Subjected to the disturbance of inlet air humidity. **c** Subjected to the disturbance of inlet water temperature. **d** Subjected to the disturbance of water flow rate. **e** Subjected to the disturbance of airflow rate

$$A_{\text{zduct, ld}=1.0\,\text{m}} = \begin{bmatrix} -1.7678 & 0.0000 & 0.4153 \\ 0.0000 & -1.5601 & 0.0000 \\ 0.0025 & 0.0000 & -0.0062 \end{bmatrix},$$

$$B_{\text{zduct, ld}=1.0\,\text{m}} = \begin{bmatrix} -1.3525 & 0.0000 & -0.5276 \\ 0.0000 & 1.5601 & 0.0000 \\ 0.0025 & 0.0000 & -0.0123 \end{bmatrix};$$

$$A_{\text{zduct, ld=3.0 m}} = \begin{bmatrix} -0.7277 & 0.0000 & 0.4153 \\ 0.0000 & -0.5200 & 0.0000 \\ 0.0025 & 0.0000 & -0.0062 \end{bmatrix},$$

$$B_{\text{zduct, ld=3.0 m}} = \begin{bmatrix} 0.3124 & 0.0000 & 0.5115 \\ 0.0000 & 0.5200 & 0.0000 \\ 0.0025 & 0.0000 & -0.0123 \end{bmatrix};$$

$$A_{\text{zduct, ld=5.0 m}} = \begin{bmatrix} -0.519 & 0.0000 & 0.4153 \\ 0.0000 & -0.3120 & 0.0000 \\ 0.0025 & 0.0000 & -0.0062 \end{bmatrix},$$

$$B_{\text{zduct, ld=5.0 m}} = \begin{bmatrix} 0.1044 & 0.0000 & 0.7193 \\ 0.0000 & 0.3120 & 0.0000 \\ 0.0025 & 0.0000 & -0.0123 \end{bmatrix}.$$

Figures 3.23 and 3.24 present transient response of exit air temperature of duct with different lengths to step increase of airflow rate by 0.1 kg/s and to step increase of inlet air temperature by 1.0 °C, respectively. When subjected to the step increase of airflow rate, the exit air of duct with the length of 3.0 and 5.0 m increases rapidly in the beginning and then gradually decreases to a new equilibrium value. However, there is a small difference for the duct with the length of 1.0 m in which the exit air temperature deceases since the start of perturbation of the airflow rate. When subjected to the step increase of inlet air temperature, the exit air temperatures of duct with different lengths all increase with the time, and the increase rate in the beginning is much larger than that in the later time. The variation of exit air temperature depends on the change of inlet air temperature, airflow rate, and heat

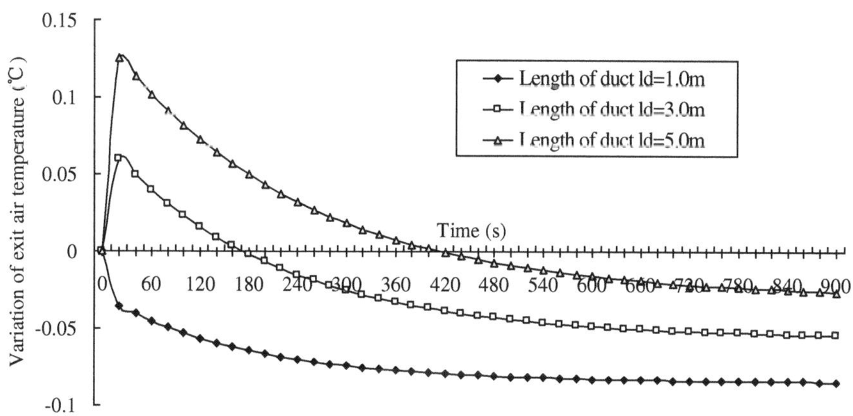

Fig. 3.23 Transient response of exit air temperature of straight-through duct with different lengths to step increase of airflow rate by 0.1 kg/s

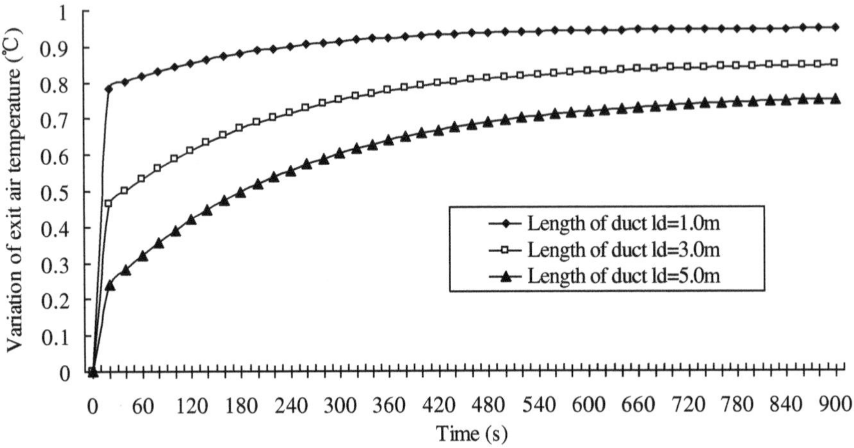

Fig. 3.24 Transient response of exit air temperature of straight-through duct with different lengths to step increase of inlet air temperature by 1.0 °C

exchange quantity between the air and the duct. And the heat exchange quantity is related to contact surface and convective heat transfer coefficient (associated with airflow rate) as well as temperature difference between the air and the duct wall.

From the results in Figs. 3.23 and 3.24, we can conclude that the response time of exit air temperature to inlet air perturbations will increase with the increase of duct length.

3.4.2 On Straight-Through Pipe

The influence of pipe length on the transient response characteristic of pipe's exit water temperature is investigated here. The length of straight-through water pipe is assumed to be 1.0, 3.0, and 5.0 m, respectively, for the transient response simulations, and the corresponding system matrixes of pipe's model ($A_{\text{zpipe}}, B_{\text{zpipe}}$) are as below:

$$A_{\text{zpipe,lp=1.0m}} = \begin{bmatrix} -1.4634 & 0.3205 \\ 0.1971 & -0.4467 \end{bmatrix}; \quad B_{\text{zpipe,lp=1.0m}} = \begin{bmatrix} 1.1429 & -5.4440 \\ 0.1971 & -2.7050 \end{bmatrix};$$
$$A_{\text{zpipe,lp=3.0m}} = \begin{bmatrix} -0.5946 & 0.3205 \\ 0.1971 & -0.4117 \end{bmatrix}; \quad B_{\text{zpipe,lp=3.0m}} = \begin{bmatrix} 0.2741 & -0.3485 \\ 0.1971 & -2.7050 \end{bmatrix};$$
$$A_{\text{zpipe,lp=5.0m}} = \begin{bmatrix} -0.4209 & 0.3205 \\ 0.1971 & -0.4047 \end{bmatrix}; \quad B_{\text{zpipe,lp=5.0m}} = \begin{bmatrix} 0.1004 & 0.6706 \\ 0.1971 & -2.7050 \end{bmatrix}.$$

The transient responses of exit water temperature of straight-through pipe with different lengths to perturbations of water flow rate and inlet water temperature are shown in Figs. 3.25 and 3.26, respectively. Likewise, the response time of exit water temperature to inlet water perturbations will increase with the increase of pipe length. Compared with the air duct, the influence of pipe length on the transient response characteristic of pipe's exit water temperature is much smaller.

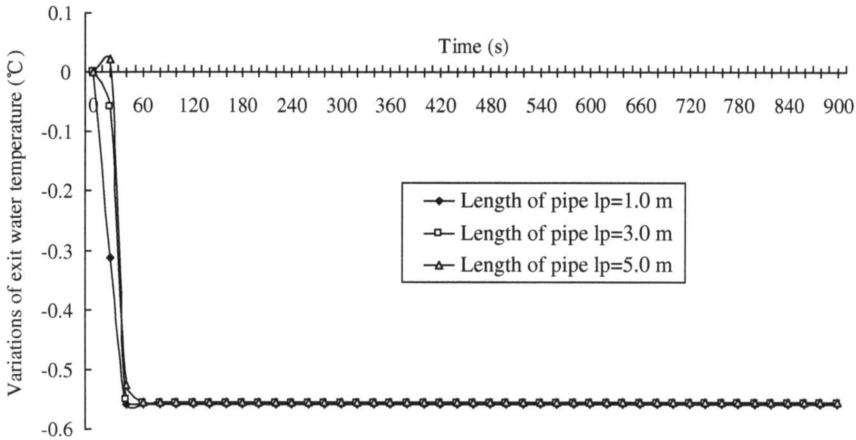

Fig. 3.25 Transient response of exit water temperature of straight-through pipe with different lengths to step increase of water flow rate by 0.1 kg/s

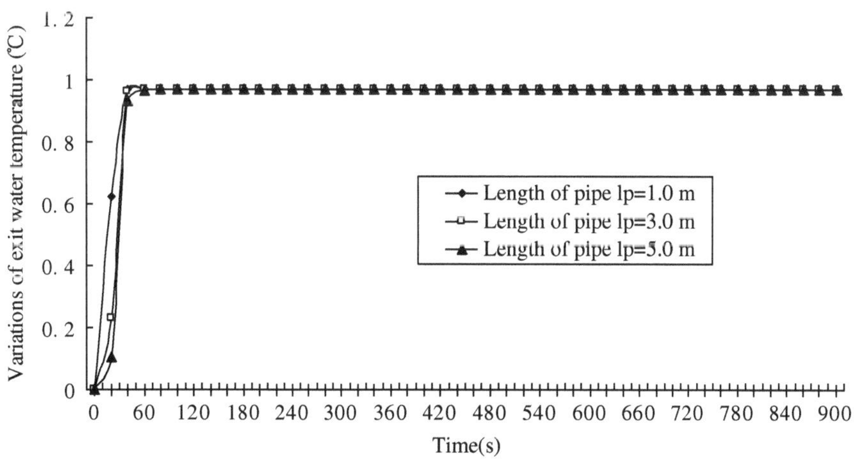

Fig. 3.26 Transient response of exit water temperature of straight-through pipe with different lengths to step increase of inlet water temperature by 1.0 °C

3.5 On Air-Conditioned Room

3.5.1 Basic Conditions

The experimental cases above have proven the validity of the state-space room model. In the following section, the transient responses of the indoor air temperature and humidity subjected to different perturbations will be investigated by using the room model.

The model simulations are made under the initial conditions of the case II (Table 2.16), and the corresponding system matrixes of the room state-space model (A_{room},B_{room}) are computed as below:

$$A_{\text{room}} = \begin{bmatrix} -0.0522 & 0 & 0.0223 & 0.0036 & 0 & 0 & 0 & 0 & 0 & 0 \\ 0 & -0.0258 & 0 & 0 & 0 & 0 & 0 & 0 & 0 & 0 \\ 0.0000 & 0 & 0.0000 & 0 & 0 & 0 & 0 & 0 & 0 & 0 \\ 0.0114 & 0 & 0 & -0.0253 & 0 & 0.0130 & 0 & 0.0007 & 0 & 0 \\ 0 & 0.0100 & 0 & 0 & -0.0100 & 0 & 0 & 0 & 0 & 0 \\ 0 & 0 & 0 & 0.0000 & 0 & -0.0000 & 0 & 0 & 0 & 0 \\ 0 & 0 & 0 & 0 & 0 & 0 & 0 & 0 & 0 & 0 \\ 0 & 0 & 0 & 0.0436 & 0 & 0 & 0 & -0.0536 & 0 & 0.0068 \\ 0 & 0 & 0 & 0 & 0.0436 & 0 & 0 & 0 & -0.0436 & 0 \\ 0 & 0 & 0 & 0 & 0 & 0 & 0 & 0.0000 & 0 & -0.0000 \end{bmatrix}$$

$$B_{\text{room}} = \begin{bmatrix} 0.0258 & 0 & -0.2062 & 0 & 0 \\ 0 & 0.0258 & 0 & 0 & 0 \\ 0 & 0 & 0 & 0 & 0 \\ 0 & 0 & -0.0752 & 0 & 0 \\ 0 & 0 & 0 & 0 & 0 \\ 0 & 0 & 0 & 0 & 0 \\ 0 & 0 & 0 & 0 & 0 \\ 0 & 0 & -0.1526 & 0 & 0 \\ 0 & 0 & 0 & 0 & 0 \\ 0 & 0 & 0 & 0 & 0 \end{bmatrix}$$

3.5.2 Subjected to Different Perturbations

(1) *Step change of supply air temperature*

The transient responses of the indoor air temperatures subjected to the step increase of supply air temperature by 1.0 °C are presented in Fig. 3.27. According to the simulation results, the proportionality coefficient of air temperature in the three indoor zones to the step change of supply air temperature is about 0.5111 °C/°C, 0.2530 °C/°C, and 0.1902 °C/°C, respectively. The corresponding time constant is estimated as 68, 152, and 172 s. The results show that the perturbations of supply air temperature will produce more impact on the air temperature in the air-supply

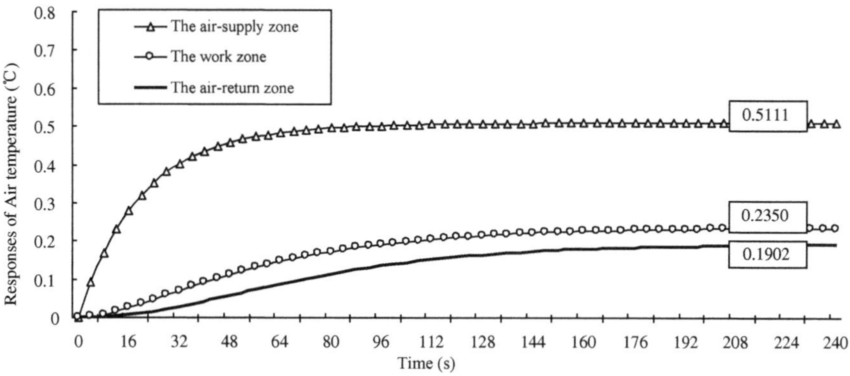

Fig. 3.27 Dynamic responses of air temperature in different air zones when subjected to a step increase of supply air temperature by 1.0 °C

zone than on the work and the air-return zone. Obviously, the perturbations of supply air temperature will not affect the indoor air humidity.

(2) *Step change of supply airflow rate*

Figure 3.28 shows the transient responses of the indoor air temperatures subject to the step increase of supply airflow rate by 0.1 kg/s. Under the conditions of case II, increasing supply airflow rate means increasing the cooling capacity. Therefore, the indoor air temperatures will decrease when the other conditions remain unchanged. The proportionality coefficient of air temperature in the air-supply, the work, and the air-return zone is estimated as −4.373 °C/(kg/s), −5.736 °C/(kg/s), and −7.511 °C/(kg/s), respectively. The corresponding time constant is about 68 s, 140 s, and 140 s. This indicates that the change of supply airflow rate affects more

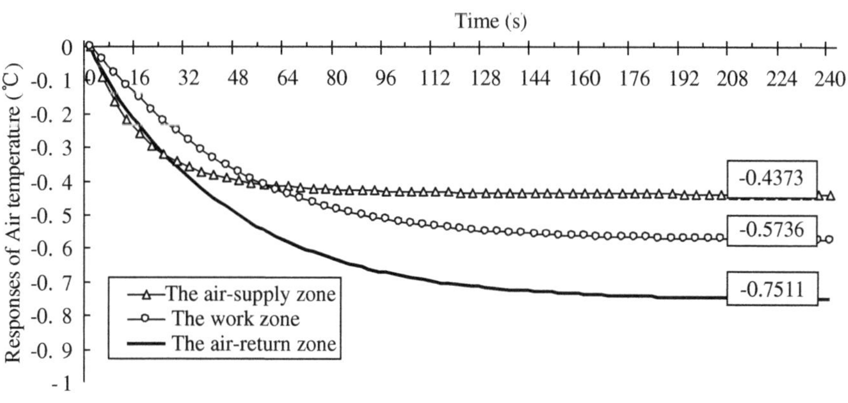

Fig. 3.28 Dynamic responses of air temperature in different zones when subjected to a step increase of supply airflow rate by 0.1 kg/s

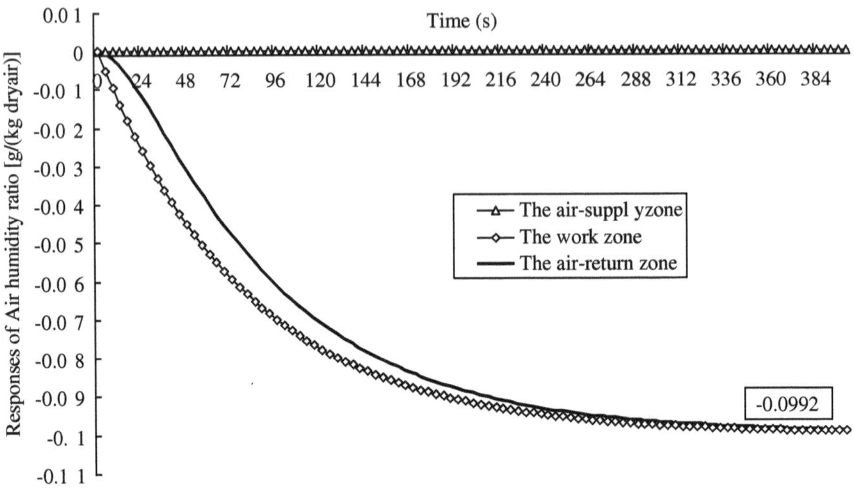

Fig. 3.29 Dynamic responses of air humidity ratio in different zones when subjected to a step increase of supply airflow rate by 0.1 kg/s

on the air temperature in the work and the air-return zone than on that in the air-supply zone.

Figure 3.29 shows the transient responses of the indoor air humidity ratio subjected to a step increase of supply airflow rate by 0.1 kg/s. The results manifest that the air humidity in the work zone and in the air-supply zone shares a similar response characteristic with the air temperature. The proportionality coefficient and time constant of response is identified as about -0.992 (g/kg dry air)/(kg/s) and 320 s, respectively. However, the air humidity in the air-supply zone has no response to the perturbations of supply airflow rate. This is because there is no moisture load in the air-supply zone. Meanwhile, the air humidity ratio in the air-supply zone is not affected by the air in the other zones.

(3) *Step change of indoor occupant number*

The dynamic responses of air temperature and humidity ratio subjected to a step increase of occupant number are also investigated. In the model, the number change of indoor occupants is simulated by changing body's surface of occupants indoors. Figures 3.30 and 3.31 show the calculated responses of indoor air conditions (i.e., temperature and humidity) when the total area of indoor occupants increases by 10 m^2. As indicated by the results, the air conditions in the work zone are affected most by the perturbations of indoor occupants, and then followed by that in the air-return and air-supply zone. Meanwhile, the perturbation of indoor occupant number impacts more on the indoor air temperature than it does on the indoor air humidity.

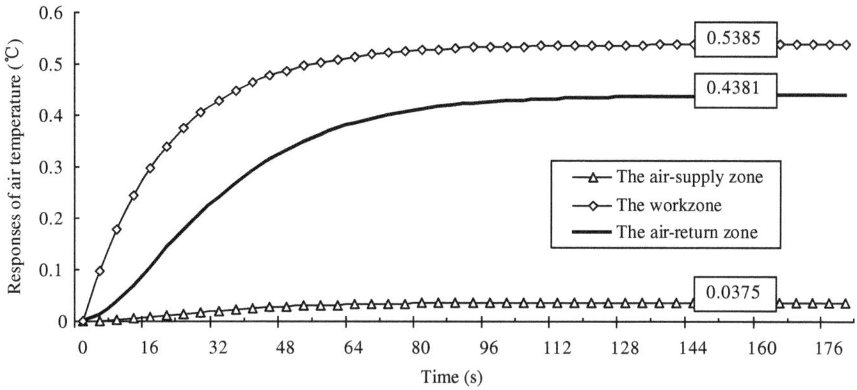

Fig. 3.30 Dynamic responses of air temperature in different zones when subjected to a step increase of occupant number ($A_{\text{rq,n}}^{(1)}$ ↑ × 10 m^2)

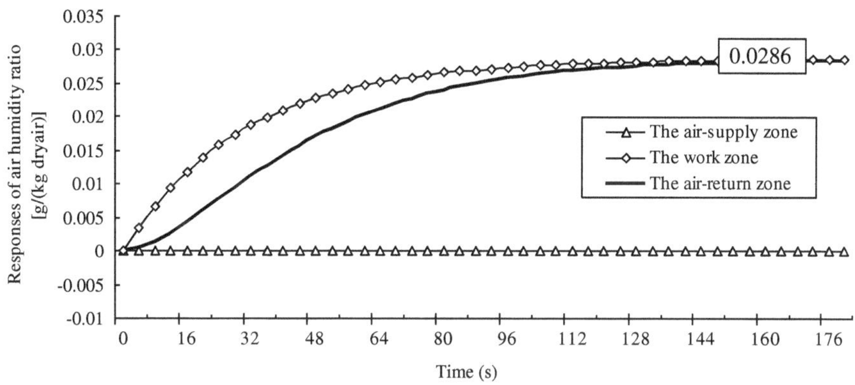

Fig. 3.31 Dynamic responses of air humidity ratio in different zones when subjected to a step increase of occupant number ($A_{\text{rq,n}}^{(1)}$ ↑ × 10 m^2)

(4) *Step change of ambient temperature and solar intensity*

The ambient temperature and the solar intensity affect indoor thermal environment mainly through the external building envelopes. For the convenience of simulation, we assume the surface area and thickness of the room's external wall are 10 m^2 and 0.3 m, respectively. The convective heat transfer coefficient and the solar radiation absorption coefficient of the external surface are assumed to be 13.5 W/(m^2.°C) and 0.65, respectively. The initial ambient temperature is 33.0 °C, and the initial solar radiation intensity on the external wall is 400 W/m^2. The other initial conditions can be referred to the case II (Table 2.16). The influences of external wall's thermal insulation on the thermal response characteristic of indoor

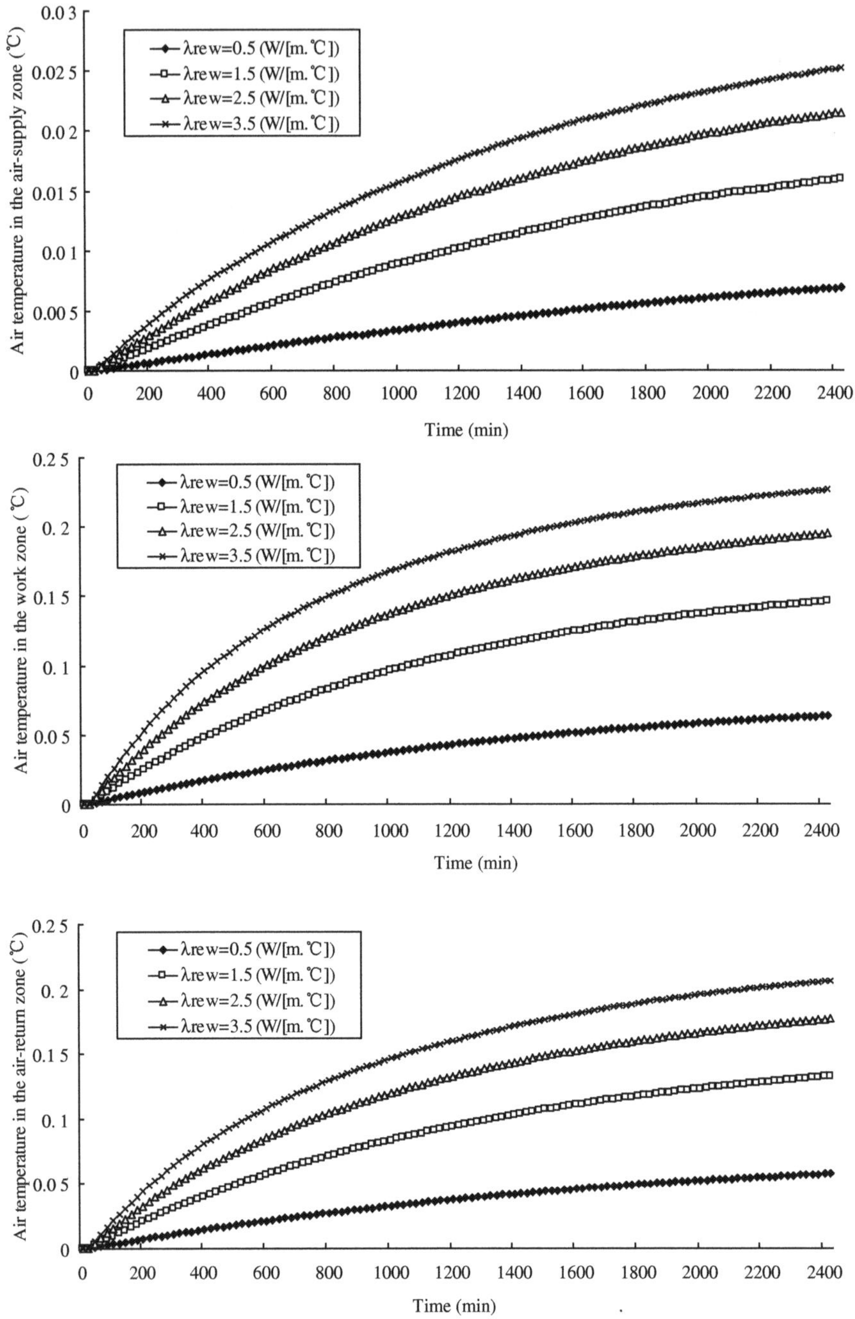

Fig. 3.32 Dynamic responses of air temperature in different air zones when subjected to a step increase of ambient temperature by 1.0 °C

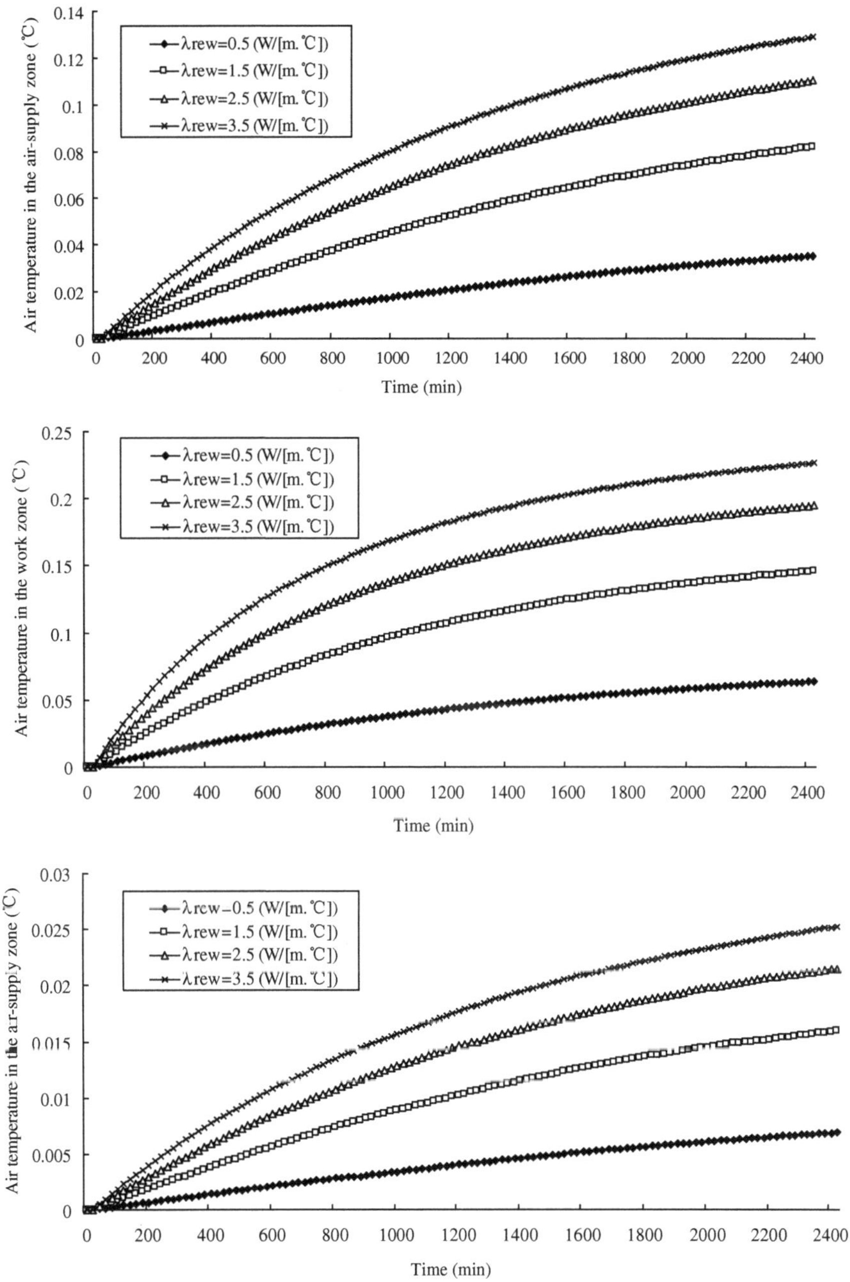

Fig. 3.33 Dynamic responses of air temperature in different air zones when subjected to a step increase of solar radiant intensity on the external wall by 100 W/m^2

air to the perturbations of ambient temperature and solar intensity are investigated with the model simulation.

Figure 3.32 shows the dynamic responses of air temperature in different air zones to a step increase of ambient temperature by 1.0 °C, and Fig. 3.33 shows that to a step increase of solar radiant intensity by 100 W/m^2. The results indicate that a larger thermal resistance of external wall will result in a smaller influence of ambient perturbations on the indoor air temperatures. Comparing with the other perturbations including the supply air conditions, the supply airflow rate, and the indoor heat load, the response time of indoor air temperatures to the ambient perturbations will be much longer.

Chapter 4
Graph-Theory Modeling and Structure-Matrix Analysis

In this chapter, the graph theory is employed for modeling the HVAC components and system. By using the graph-theory models, the direct or indirect relationships among model's input and output variables are clearly described. Meanwhile, the structure-matrix method is used to analyze the dynamic characteristics of various parts in the HVAC system, including reachability, controllability, and observability.

4.1 Graph-Theory Modeling for HVAC Component State-Space Models

4.1.1 Fundamental Rules

Graph theory is a branch of mathematics [1], and its use in system engineering is relatively broad [2, 3]. The graph theory can be used to describe system dynamic model and make it easier to illustrate relationships between input and output signals of each unit component of a complex system as well as relationships among unit components. The focus of this method does not lie in the details of composition of basic element of the modeled system, but in the system structure itself with focusing on the mutual relationship between the system variables. In addition, the zero elements of the state-space model matrix do not appear in the graphical representation; therefore, graph theory is a more concise description compared to the mathematical representation like state-space model.

The basic description of a state-space model is as follows:

$$\dot{x}_i(\tau) = X_i\{x_1(\tau), x_2(\tau), \ldots, x_n(\tau); u_1(\tau), u_2(\tau), \ldots, u_m(\tau); \tau\} \tag{4.1}$$

$$y_j(\tau) = Y_j\{x_1(\tau), x_2(\tau), \ldots, x_n(\tau); u_1(\tau), u_2(\tau), \ldots, u_m(\tau); \tau\} \tag{4.2}$$

Y. Yao and Y. Yu, *Modeling and Control in Air-conditioning Systems*,
Energy and Environment Research in China, DOI 10.1007/978-3-662-53313-0_4

The state-space model can be expressed by a directed graph G^s which consists of a set of points with m input points (i.e.,$u_1, u_2, \ldots, u_m$), n state points (i.e.,$x_1, x_2, \ldots, x_n$), and r output points $y_1, y_2, \ldots, y_r$ as well as a set of edges. The edges are given based on the following rules:

(1) If state variable x_i appears in $X_i(x, u, \tau)$, i.e., $\partial X_i/\partial x_j \neq 0$, then there are edges from state point x_j to state point x_i;
(2) If input variable u_k appears in $X_i(x, u, \tau)$, i.e., $\partial X_i/\partial u_k \neq 0$ there is an edge at output point u_k to state point x_i;
(3) If state variable x_i appears in $Y_j(x, u, \tau)$ i.e., $Y_j/\partial x_i \neq 0$, then there are edges from state point x_i to output point y_j.

4.1.2 *Case Study*

4.1.2.1 Water-to-Air Heat Exchanger

As known from Chap. 2, the state-space model for water-to-air heat exchanger can be represented by the following:

$$\dot{X}_{\text{coil}} = A_{\text{coil}} X_{\text{coil}} + B_{\text{coil}} u_{\text{coil}} \tag{4.3}$$

$$y_{\text{coil}} = C_{\text{coil}} X_{\text{coil}} + D_{\text{coil}} u_{\text{coil}} - C_{\text{coil}} A_{\text{coil}}^{-1} \xi_{\text{coil}} \tag{4.4}$$

For the dry-condition process:

$$x_{\text{coil}} = [\Delta t_{\text{w,L}}, \Delta t_{\text{a,L}}, \Delta t_{\text{g}}]^{\text{T}}; \quad y_{\text{coil}} = [\Delta t_{\text{w,L}}, \Delta G_{\text{w,L}}, \Delta t_{\text{a,L}}, \Delta W_{\text{a,L}}, \Delta G_{\text{a,L}}]^{\text{T}};$$

$$u_{\text{coil}} = [\Delta t_{\text{w,E}}, \Delta G_{\text{w,E}}, \Delta t_{\text{a,E}}, \Delta W_{\text{a,E}}, \Delta G_{\text{a,E}}]^{\text{T}}; \quad \xi = \left[\xi_{\Delta t_{\text{w,L}},\text{dry}}, \xi_{\Delta t_{\text{a,L}},\text{dry}}, 0\right]^{\text{T}};$$

$$A_{\text{coil}} = \begin{bmatrix} \frac{X_1}{T_{\text{w}}} & 0 & \frac{X_2}{T_{\text{w}}} \\ 0 & \frac{Y_{\text{dry},1}}{T_{\text{a}}} & \frac{Y_{\text{dry},2}}{T_{\text{a}}} \\ \frac{Z_{\text{dry},1}}{T_{\text{g}}} & \frac{Z_{\text{dry},2}}{T_{\text{g}}} & \frac{Z_{\text{dry},3}}{T_{\text{g}}} \end{bmatrix}; \quad B_{\text{coil}} = \begin{bmatrix} \frac{X_3}{T_{\text{w}}} & \frac{X_4}{T_{\text{w}}} & 0 & 0 & 0 \\ 0 & 0 & \frac{Y_{\text{dry},3}}{T_{\text{a}}} & 0 & \frac{Y_{\text{dry},4}}{T_{\text{a}}} \\ \frac{Z_{\text{dry},5}}{T_{\text{g}}} & \frac{Z_{\text{dry},6}}{T_{\text{g}}} & \frac{Z_{\text{dry},4}}{T_{\text{g}}} & 0 & \frac{Z_{\text{dry},7}}{T_{\text{g}}} \end{bmatrix};$$

$$C_{\text{coil}} = \begin{bmatrix} 1 & 0 & 0 \\ 0 & 0 & 0 \\ 0 & 1 & 0 \\ 0 & 0 & 0 \\ 0 & 0 & 0 \end{bmatrix}; \quad D_{\text{coil}} = \begin{bmatrix} 0 & 0 & 0 & 0 & 0 \\ 0 & 1 & 0 & 0 & 0 \\ 0 & 0 & 0 & 0 & 0 \\ 0 & 0 & 0 & 1 & 0 \\ 0 & 0 & 0 & 0 & 1 \end{bmatrix}.$$

For the wet-condition process:

$$x_{\text{coil}} = [\Delta t_{\text{w,L}}, \Delta t_{\text{a,L}}, \Delta W_{\text{a,L}}, \Delta t_{\text{g}}]^{\text{T}}; \quad y_{\text{coil}} = [\Delta t_{\text{w,L}}, \Delta G_{\text{w,L}}, \Delta t_{\text{a,L}}, \Delta W_{\text{a,L}}, \Delta G_{\text{a,L}}]^{\text{T}};$$

$$u_{\text{coil}} = [\Delta t_{\text{w,E}}, \Delta G_{\text{w,E}}, \Delta t_{\text{a,E}}, \Delta W_{\text{a,E}}, \Delta G_{\text{a,E}}]^{\text{T}}; \quad \xi = \left[\xi_{\Delta t_{\text{w,L}},\text{wet}}, \xi_{\Delta t_{\text{a,L}},\text{wet}}, \xi_{\Delta t_{\text{g}},\text{wet}}, 0\right]^{\text{T}};$$

$$A_{\text{coil}} = \begin{bmatrix} \frac{X_1}{T_{\text{w}}} & 0 & 0 & \frac{X_2}{T_{\text{w}}} \\ 0 & \frac{Y_{\text{wet},1}}{T_{\text{a}}} & \frac{Y_{\text{wet},2}}{T_{\text{a}}} & \frac{Y_{\text{wet},3}}{T_{\text{a}}} \\ 0 & \frac{M_{\text{wet},1}}{T_{\text{m}}} & \frac{M_{\text{wet},2}}{T_{\text{m}}} & \frac{M_{\text{wet},3}}{T_{\text{m}}} \\ \frac{Z_{\text{wet},1}}{T_{\text{g}}} & \frac{Z_{\text{wet},2}}{T_{\text{g}}} & \frac{Z_{\text{wet},3}}{T_{\text{g}}} & \frac{Z_{\text{wet},4}}{T_{\text{g}}} \end{bmatrix}; \quad B_{coil} = \begin{bmatrix} \frac{X_3}{T_{\text{w}}} & \frac{X_4}{T_{\text{w}}} & 0 & 0 & 0 \\ 0 & 0 & \frac{Y_{\text{wet},4}}{T_{\text{a}}} & \frac{Y_{\text{wet},6}}{T_{\text{a}}} & \frac{Y_{\text{wet},5}}{T_{\text{a}}} \\ 0 & 0 & \frac{M_{\text{wet},4}}{T_{\text{m}}} & \frac{M_{\text{wet},6}}{T_{\text{m}}} & \frac{M_{\text{wet},5}}{T_{\text{m}}} \\ \frac{Z_{\text{wet},6}}{T_{\text{g}}} & \frac{Z_{\text{wet},7}}{T_{\text{g}}} & \frac{Z_{\text{wet},5}}{T_{\text{g}}} & \frac{Z_{\text{wet},9}}{T_{\text{g}}} & \frac{Z_{\text{wet},8}}{T_{\text{g}}} \end{bmatrix}$$

$$C_{\text{coil}} = \begin{bmatrix} 1 & 0 & 0 & 0 \\ 0 & 0 & 0 & 0 \\ 0 & 1 & 0 & 0 \\ 0 & 0 & 1 & 0 \\ 0 & 0 & 0 & 0 \end{bmatrix}; \quad D_{\text{coil}} = \begin{bmatrix} 0 & 0 & 0 & 0 & 0 \\ 0 & 1 & 0 & 0 & 0 \\ 0 & 0 & 0 & 0 & 0 \\ 0 & 0 & 0 & 0 & 0 \\ 0 & 0 & 0 & 0 & 1 \end{bmatrix}.$$

According to the rules of graph-theory modeling, the state-space model of water-to-air heat exchanger can be described by a directed graph, as shown in Fig. 4.1. From Fig. 4.1, we can easily see that the perturbation of inlet air temperature $\Delta t_{\text{a,E}}$ and inlet air flow rate $\Delta G_{\text{a,E}}$ produces direct influence on the exit air temperature $\Delta t_{\text{a,L}}$, while that of inlet water temperature $\Delta t_{\text{w,E}}$ and flow rate $\Delta G_{\text{w,E}}$ affects $\Delta t_{\text{a,L}}$ in an indirect way through coli's shell wall temperature Δt_{g}. Similarly, the influence of inlet water conditions on the exit water temperature $\Delta t_{\text{w,L}}$ is direct, while that of inlet air conditions on the exit water temperature $\Delta t_{\text{w,L}}$ is indirect. Under dry working conditions, the exit air humidity $\Delta W_{\text{a,L}}$ is only directly influenced by inlet air humidity $\Delta W_{\text{a,E}}$, while under wet conditions, it is not only directly influenced by the inlet air humidity $\Delta W_{\text{a,E}}$ and inlet air flow rate $\Delta G_{\text{a,E}}$, but also indirectly influenced by other input parameters.

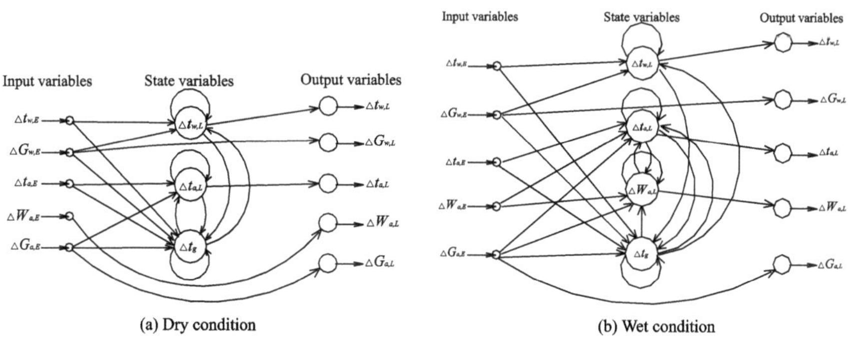

Fig. 4.1 Directed graph for water-to-air heat exchanger model. **a** Dry condition, **b** Wet condition

4.1.2.2 Chiller

The state-space model for vapor compression packaged liquid chiller can be expressed by the following:

$$\Delta\dot{X}_{\text{chiller}} = A_{\text{chiller}}\Delta X_{\text{chiller}} + B_{\text{chiller}}\Delta u_{\text{chiller}} \tag{4.5}$$

$$\Delta y_{\text{chiller}} = C_{\text{chiller}}\Delta X_{\text{chiller}} + D_{\text{chiller}}\Delta u_{\text{chiller}} - C_{\text{chiller}}A^{-1}_{\text{chiller}}\xi_{\text{chiller}} \tag{4.6}$$

where

$$\begin{aligned}
\Delta x_{\text{chiller}} &= [\Delta t_{\text{c}}, \Delta t_{\text{cw,L}}, \Delta t_{\text{cg}}, \Delta t_{\text{k}}, \Delta t_{\text{ew,L}}, \Delta t_{\text{eg}}]^{\text{T}};\\
\Delta y_{\text{chiller}} &= [\Delta t_{\text{cw,L}}, \Delta G_{\text{cw,L}}, \Delta t_{\text{ew,L}}, \Delta G_{\text{ew,L}}, \Delta N_{\text{com}}, \Delta Q_{\text{c}}, \Delta \text{COP}]^{\text{T}};\\
\Delta u_{\text{chiller}} &= [\Delta t_{\text{cw,E}}, \Delta G_{\text{cw,E}}, \Delta t_{\text{ew,E}}, \Delta G_{\text{ew,E}}, \Delta G_{\text{rm}}]^{\text{T}};\\
\xi &= [0, \xi_{\Delta t_{\text{cw,L}}}, 0, 0, \xi_{\Delta t_{\text{ew,L}}}, 0]^{\text{T}};
\end{aligned}$$

$$A_{\text{chiller}} = \begin{bmatrix}
\frac{X_{\text{cr},1}}{T_{\text{cr}}} & 0 & \frac{X_{\text{cr},2}}{T_{\text{cr}}} & 0 & 0 & 0\\
0 & \frac{X_{\text{cw},1}}{T_{\text{cw}}} & \frac{X_{\text{cw},2}}{T_{\text{cw}}} & 0 & 0 & 0\\
\frac{X_{\text{cg},1}}{T_{\text{cg}}} & \frac{X_{\text{cg},2}}{T_{\text{cg}}} & \frac{X_{\text{cg},3}}{T_{\text{cg}}} & 0 & 0 & 0\\
\frac{X_{\text{er},4}}{T_{\text{er}}} & 0 & 0 & \frac{X_{\text{er},1}}{T_{\text{er}}} & 0 & \frac{X_{\text{er},2}}{T_{\text{er}}}\\
0 & 0 & 0 & 0 & \frac{X_{\text{ew},1}}{T_{\text{ew}}} & \frac{X_{\text{ew},2}}{T_{\text{ew}}}\\
0 & 0 & 0 & \frac{X_{\text{eg},1}}{T_{\text{eg}}} & \frac{X_{\text{eg},2}}{T_{\text{eg}}} & \frac{X_{\text{eg},3}}{T_{\text{eg}}}
\end{bmatrix};$$

$$B_{\text{chiller}} = \begin{bmatrix}
0 & 0 & 0 & 0 & \frac{X_{\text{cr},3}}{T_{\text{cr}}}\\
\frac{X_{\text{cw},3}}{T_{\text{cw}}} & \frac{X_{\text{cw},4}}{T_{\text{cw}}} & 0 & 0 & 0\\
\frac{X_{\text{cg},4}}{T_{\text{cg}}} & \frac{X_{\text{cg},5}}{T_{\text{cg}}} & 0 & 0 & 0\\
0 & 0 & 0 & 0 & \frac{X_{\text{er},3}}{T_{\text{er}}}\\
0 & 0 & \frac{X_{\text{ew},3}}{T_{\text{ew}}} & \frac{X_{\text{ew},4}}{T_{\text{ew}}} & 0\\
0 & 0 & \frac{X_{\text{eg},4}}{T_{\text{eg}}} & \frac{X_{\text{eg},5}}{T_{\text{eg}}} & 0
\end{bmatrix};$$

$$C_{\text{chiller}} = \begin{bmatrix}
0 & 1 & 0 & 0 & 0 & 0\\
0 & 0 & 0 & 0 & 0 & 0\\
0 & 0 & 0 & 0 & 1 & 0\\
0 & 0 & 0 & 0 & 0 & 0\\
X_{\text{Ew},1} & 0 & 0 & X_{\text{Ew},2} & 0 & 0\\
X_{\text{Q},1} & 0 & 0 & X_{\text{Q},2} & 0 & 0\\
X_{\text{cop},1} & 0 & 0 & X_{\text{cop},2} & 0 & 0
\end{bmatrix}; \quad D_{\text{chiller}} = \begin{bmatrix}
0 & 0 & 0 & 0 & 0\\
0 & 1 & 0 & 0 & 0\\
0 & 0 & 0 & 0 & 0\\
0 & 0 & 0 & 1 & 0\\
0 & 0 & 0 & 0 & X_{\text{Ew},3}\\
0 & 0 & 0 & 0 & X_{\text{Q},3}\\
0 & 0 & 0 & 0 & 0
\end{bmatrix}.$$

The directed graph for chiller's state-space model is shown in Fig. 4.2. Known from Fig. 4.2, the inlet parameters of condenser-side cooling water (i.e., $\Delta G_{\text{cw,E}}$ and $\Delta t_{\text{cw,E}}$) produce direct influence on the exit water temperature of condenser

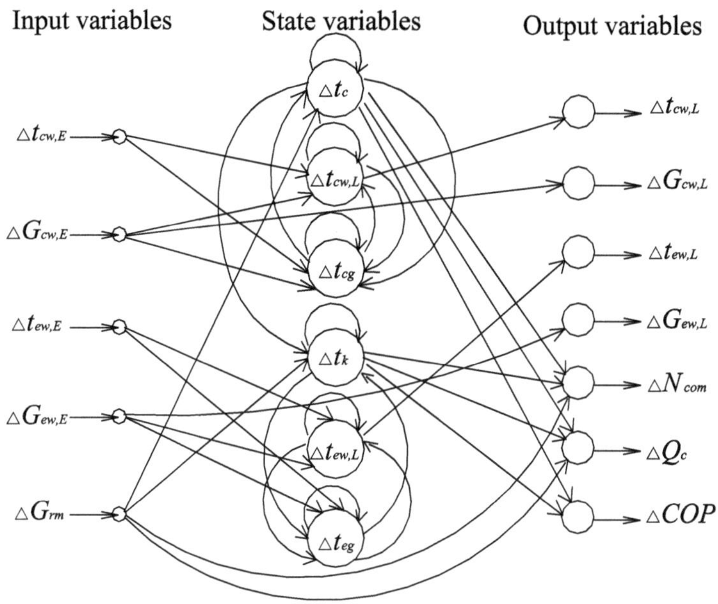

Fig. 4.2 Directed graph for chiller model

$(\Delta t_{\text{cw,L}})$, and so do the inlet parameters of evaporator-side chilled water (i.e., $\Delta G_{\text{ew,E}}$ and $\Delta t_{\text{ew,E}}$) on the exit water temperature of evaporator $(\Delta t_{\text{ew,L}})$. It can be known as well from Fig. 4.2 that the condenser-side cooling water inlet parameters (i.e., $\Delta G_{\text{cw,E}}$ and $\Delta t_{\text{cw,E}}$) and evaporator-side chilled water inlet parameters (i.e., $\Delta G_{\text{ew,E}}$ and $\Delta t_{\text{ew,E}}$) produce indirect influence on chiller's power (ΔN_{com}), cooling capacity, (ΔQ_{c}) and performance (ΔCOP) through the condensing temperature (Δt_{c}) and the evaporating temperature (Δt_{k}) of chiller, while the influence of refrigerant flow rate (ΔG_{rm}) on the two output variables, i.e., ΔN_{com} and ΔQ_{c}, is in a direct way.

4.1.2.3 Cooling Tower

The state-space model for cooling tower can be written as follows:

$$\dot{X}_{\text{tower}} = A_{\text{tower}} X_{\text{tower}} + B_{\text{tower}} u_{\text{tower}} \tag{4.7}$$

$$y_{\text{tower}} = C_{\text{tower}} X_{\text{tower}} + D_{\text{tower}} u_{\text{tower}} - C_{\text{tower}} A_{\text{tower}}^{-1} \xi_{\text{tower}} \tag{4.8}$$

where

$$X_{\text{tower}} = x_{\text{tower}} + A_{\text{tower}}^{-1} \xi_{\text{tower}}$$
$$x_{\text{tower}} = [\Delta t_{\text{a,L}}, \Delta W_{\text{a,L}}, \Delta t_{\text{w,L}}, \Delta t'_{\text{w,L}}]^{\text{T}}$$
$$y_{\text{tower}} = [\Delta t_{\text{a,L}}, \Delta W_{\text{a,L}}, \Delta G_{\text{a,L}}, \Delta t'_{\text{w,L}}, \Delta G_{\text{w,L}}]^{\text{T}}$$
$$u_{\text{tower}} = [\Delta t_{\text{a,E}}, \Delta W_{\text{a,E}}, \Delta G_{\text{a,E}}, \Delta t_{\text{w,E}}, \Delta G_{\text{w,E}}]^{\text{T}}$$
$$\xi_{\text{tower}} = \left[\xi_{t_{\text{a,L}}}, \xi_{W_{\text{a,L}}}, \xi_{t_{\text{w,L}}}, 0\right]^{\text{T}}$$

$$A_{\text{tower}} = \begin{bmatrix} \frac{X_{\text{t,1}}}{T_{\text{t,ta}}} & \frac{X_{\text{t,2}}}{T_{\text{t,ta}}} & \frac{X_{\text{t,3}}}{T_{\text{t,ta}}} & 0 \\ 0 & \frac{Y_{\text{t,1}}}{T_{\text{t,wa}}} & \frac{Y_{\text{t,2}}}{T_{\text{t,wa}}} & 0 \\ \frac{Z_{\text{t,1}}}{T_{\text{t,tw}}} & \frac{Z_{\text{t,2}}}{T_{\text{t,tw}}} & \frac{Z_{\text{t,3}}}{T_{\text{t,tw}}} & 0 \\ 0 & 0 & \frac{M_{\text{t,1}}}{M_{\text{sump}}} & \frac{M_{\text{t,2}}}{M_{\text{sump}}} \end{bmatrix}; \quad B_{\text{tower,a}} = \begin{bmatrix} \frac{X_{\text{t,4}}}{T_{\text{t,ta}}} & \frac{X_{\text{t,6}}}{T_{\text{t,ta}}} & \frac{X_{\text{t,7}}}{T_{\text{t,ta}}} & \frac{X_{\text{t,5}}}{T_{\text{t,ta}}} & \frac{X_{\text{t,8}}}{T_{\text{t,ta}}} \\ 0 & \frac{Y_{\text{t,4}}}{T_{\text{t,wa}}} & \frac{Y_{\text{t,5}}}{T_{\text{t,wa}}} & \frac{Y_{\text{t,3}}}{T_{\text{t,wa}}} & \frac{Y_{\text{t,6}}}{T_{\text{t,wa}}} \\ \frac{Z_{\text{t,4}}}{T_{\text{t,tw}}} & \frac{Z_{\text{t,6}}}{T_{\text{t,tw}}} & \frac{Z_{\text{t,7}}}{T_{\text{t,tw}}} & \frac{Z_{\text{t,5}}}{T_{\text{t,tw}}} & \frac{Z_{\text{t,8}}}{T_{\text{t,tw}}} \\ 0 & 0 & 0 & 0 & \frac{M_{\text{t,3}}}{M_{\text{sump}}} \end{bmatrix};$$

$$C_{\text{tower}} = \begin{bmatrix} 1 & 0 & 0 & 0 \\ 0 & 1 & 0 & 0 \\ 0 & 0 & 0 & 0 \\ 0 & 0 & 0 & 1 \\ 0 & 0 & 0 & 0 \end{bmatrix}; \quad D_{\text{tower}} = \begin{bmatrix} 0 & 0 & 0 & 0 & 0 \\ 0 & 0 & 0 & 0 & 0 \\ 0 & 0 & 1 & 0 & 0 \\ 0 & 0 & 0 & 0 & 0 \\ 0 & 0 & 0 & 0 & 1 \end{bmatrix}.$$

The directed graph for the state-space model of cooling tower is presented in Fig. 4.3. From the directed graph, the direct or indirect influence of input variables on the output ones of cooling tower can be easily judged. For example, we can see from Fig. 4.3 that the exit air temperature of cooling tower $(\Delta t_{\text{a,L}})$ is under the direct influence of inlet air temperature $(\Delta t_{\text{a,E}})$, inlet air humidity $(\Delta W_{\text{a,E}})$, air flow rate $(\Delta G_{\text{a,E}})$, and inlet cooling water temperature $(\Delta t_{\text{w,E}})$, while the influence of cooling water flow rate $(\Delta G_{\text{w,E}})$ on the exit air temperature of cooling tower $(\Delta t_{\text{a,L}})$ is indirect. The exit water temperature of the tower's sink $\left(\Delta t'_{\text{w,L}}\right)$ is directly influenced by the flow rate of cooling water $(\Delta G_{\text{w,E}})$ and is indirectly influenced by the other input variables.

4.1.2.4 Air-Conditioned Room

The state-space model for air-conditioned room is given by the following:

$$\dot{x}_{\text{room}} = A_{\text{room}} \cdot x_{\text{room}} + B_{\text{room}} \cdot u_{\text{room}} \tag{4.9}$$

$$y_{\text{room}} = C_{\text{room}} x_{\text{room}} \tag{4.10}$$

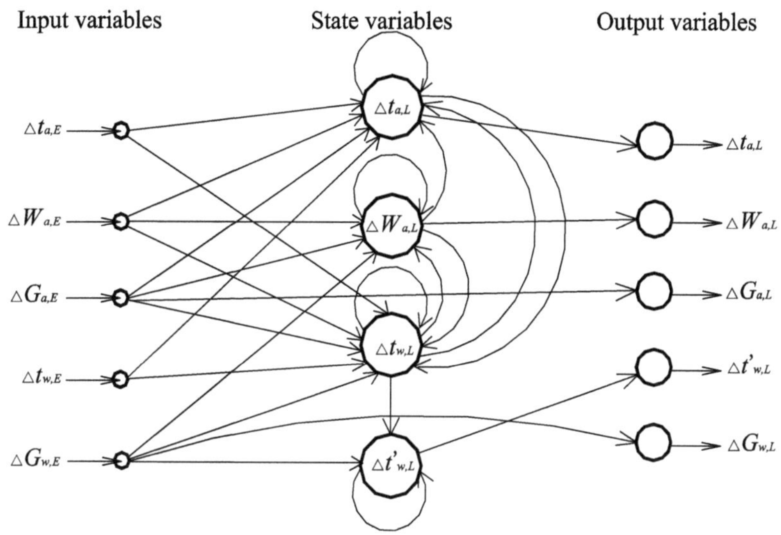

Fig. 4.3 Directed graph for cooling tower model

where

$$x_{\text{room}} = [\,\Delta t_{\text{a,s}} \quad \Delta W_{\text{a,s}} \quad \Delta t_{\text{riw,s}} \quad \Delta t_{\text{a},n} \quad \Delta W_{\text{a},n} \quad \Delta t_{\text{riw},n} \quad \Delta t_{\text{rew},n} \quad \Delta t_{\text{a,r}} \quad \Delta W_{\text{a,r}} \quad \Delta t_{\text{riw,r}}\,]^{\text{T}};$$

$$y_{\text{room}} = [\,\Delta t_{\text{a,s}}, \quad \Delta W_{\text{a,s}}, \quad \Delta t_{\text{a},n}, \quad \Delta W_{\text{a},n}, \quad \Delta t_{\text{a,r}}, \quad \Delta W_{\text{a,r}}\,]^{\text{T}};$$

$$u_{\text{room}} = \left[\,\Delta t_{\text{a},i} \quad \Delta W_{\text{a},i} \quad \Delta G_{\text{a},i} \quad \Delta A^{(1)}_{\text{rq},n} \quad \Delta A^{(2)}_{\text{rq},n} \quad \Delta A^{(i)}_{\text{rq},n} \quad \Delta t_{\text{a,out}} \quad \Delta I_{\text{sol}}\,\right]^{\text{T}};$$

$$A_{\text{room}} = \begin{bmatrix} X_{\text{tas},1}/T_{\text{tas}} & 0 & X_{\text{tas},3}/T_{\text{tas}} & X_{\text{tas},4}/T_{\text{tas}} & 0 & 0 & 0 & 0 & 0 & 0 \\ 0 & X_{\text{was},1}/T_{\text{was}} & 0 & 0 & 0 & 0 & 0 & 0 & 0 & 0 \\ X_{\text{triws},1}/T_{\text{triws}} & 0 & X_{\text{triws},2}/T_{\text{triws}} & 0 & 0 & 0 & 0 & 0 & 0 & 0 \\ X_{\text{tan},1}/T_{\text{tan}} & 0 & 0 & X_{\text{tan},2}/T_{\text{tan}} & 0 & X_{\text{tan},3}/T_{\text{tan}} & X_{\text{tan},4}/T_{\text{tan}} & X_{\text{tan},5}/T_{\text{tan}} & 0 & 0 \\ 0 & X_{\text{wan},1}/T_{\text{wan}} & 0 & 0 & X_{\text{wan},2}/T_{\text{wan}} & 0 & 0 & 0 & 0 & 0 \\ 0 & 0 & 0 & X_{\text{triwn},1}/T_{\text{triwn}} & 0 & X_{\text{triwn},2}/T_{\text{triwn}} & 0 & 0 & 0 & 0 \\ 0 & 0 & 0 & X_{\text{trewn},1}/T_{\text{trewn}} & 0 & 0 & X_{\text{trewn},2}/T_{\text{trewn}} & 0 & 0 & 0 \\ 0 & 0 & 0 & X_{\text{tar},1}/T_{\text{tar}} & 0 & 0 & 0 & X_{\text{tar},2}/T_{\text{tar}} & 0 & X_{\text{tar},3}/T_{\text{tar}} \\ 0 & 0 & 0 & 0 & X_{\text{war},1}/T_{\text{war}} & 0 & 0 & 0 & X_{\text{war},2}/T_{\text{war}} & 0 \\ 0 & 0 & 0 & 0 & 0 & 0 & 0 & X_{\text{triwr},1}/T_{\text{triwr}} & 0 & X_{\text{triwr},2}/T_{\text{triwr}} \end{bmatrix};$$

$$B_{\text{room}} = \begin{bmatrix} X_{\text{tas},2}/T_{\text{tas}} & 0 & X_{\text{tas},5}/T_{\text{tas}} & 0 & 0 & 0 & 0 & 0 & 0 \\ 0 & X_{\text{was},2}/T_{\text{was}} & X_{\text{was},3}/T_{\text{was}} & 0 & 0 & 0 & 0 & 0 & 0 \\ 0 & 0 & 0 & 0 & 0 & 0 & 0 & 0 & 0 \\ 0 & 0 & X_{\text{tan},6}/T_{\text{tan}} & X^{(1)}_{\text{tan},7}/T_{\text{tan}} & X^{(2)}_{\text{tan},7}/T_{\text{tan}} & \cdots & X^{(i)}_{\text{tan},7}/T_{\text{tan}} & 0 & 0 \\ 0 & 0 & X_{\text{wan},3}/T_{\text{wan}} & X_{\text{wan},4}/T_{\text{wan}} & 0 & 0 & 0 & 0 & 0 \\ 0 & 0 & 0 & 0 & 0 & 0 & 0 & 0 & 0 \\ 0 & 0 & 0 & 0 & 0 & 0 & 0 & X_{\text{trewn},3}/T_{\text{trewn}} & X_{\text{trewn},4}/T_{\text{trewn}} \\ 0 & 0 & X_{\text{tar},4}/T_{\text{tar}} & 0 & 0 & 0 & 0 & 0 & 0 \\ 0 & 0 & X_{\text{tar},4}/T_{\text{tar}} & 0 & 0 & 0 & 0 & 0 & 0 \\ 0 & 0 & 0 & 0 & 0 & 0 & 0 & 0 & 0 \end{bmatrix};$$

$$C_{\text{room}} = \begin{bmatrix} 1 & 0 & 0 & 0 & 0 & 0 & 0 & 0 & 0 & 0 \\ 0 & 1 & 0 & 0 & 0 & 0 & 0 & 0 & 0 & 0 \\ 0 & 0 & 0 & 1 & 0 & 0 & 0 & 0 & 0 & 0 \\ 0 & 0 & 0 & 0 & 1 & 0 & 0 & 0 & 0 & 0 \\ 0 & 0 & 0 & 0 & 0 & 0 & 0 & 1 & 0 & 0 \\ 0 & 0 & 0 & 0 & 0 & 0 & 0 & 0 & 1 & 0 \end{bmatrix}.$$

The directed graph of the state-space model of air-conditioned room is shown in Fig. 4.4. From Fig. 4.4, we can clearly see that the influence of supply air temperature $(\Delta t_{\text{a,s}})$ on the air temperature of supply air zone is direct, while its influence on that of the other two zones (i.e., the working zone and return air zone) is indirect. So is the influence of supply air humidity $(\Delta W_{\text{a,s}})$ on the indoor air humidity of different zones. The influence of supply air flow rate (ΔG_{a}) on the air

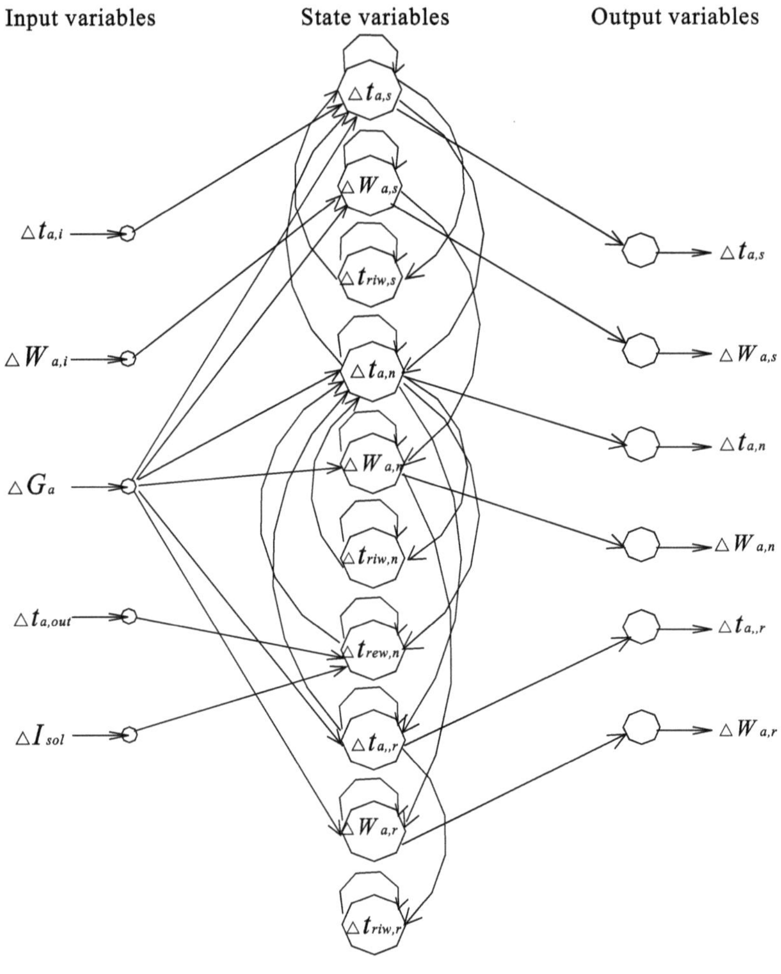

Fig. 4.4 Directed graph for air-conditioned room model

conditions of all indoor zones is direct. The ambient perturbations $(\Delta t_{\mathrm{a,out}}, \Delta I_{\mathrm{sol}})$ produce direct influence on the working zone, while they produce indirect influence on the other zones.

4.1.2.5 Duct (Pipe) and Fan (Pump)

(1) Graph description for duct (pipe)

The state-space model of straight-through duct is given by,

$$X_{\mathrm{zduct}} = A_{\mathrm{zduct}} X_{\mathrm{zduct}} + B_{\mathrm{zduct}} u_{\mathrm{zduct}} \tag{4.11}$$

$$y_{\mathrm{zduct}} = C_{\mathrm{zduct}} X_{\mathrm{zduct}} + D_{\mathrm{zduct}} u_{\mathrm{zduct}} - C_{\mathrm{zduct}} A_{\mathrm{zduct}}^{-1} \xi_{\mathrm{zduct}} \tag{4.12}$$

where

$$X_{\mathrm{zduct}} = x_{\mathrm{zduct}} + A_{\mathrm{zduct}}^{-1} \xi_{\mathrm{zduct}};$$

For the case of condensation on the interior surface of duct,

$$x_{\mathrm{zduct}} = [\Delta t_{\mathrm{da,L}}, \Delta W_{\mathrm{da,L}}, \Delta t_{\mathrm{dg}}]^{\mathrm{T}}; \quad y_{\mathrm{zduct}} = [\Delta t_{\mathrm{da,L}}, \Delta W_{\mathrm{da,L}}, \Delta G_{\mathrm{da,L}}]^{\mathrm{T}};$$
$$u_{\mathrm{zduct}} = [\Delta t_{\mathrm{da,E}}, \Delta W_{\mathrm{da,E}}, \Delta G_{\mathrm{da,E}}]^{\mathrm{T}}; \quad \xi_{\mathrm{zduct}} = [\xi_{\mathrm{t,dweta}}, \xi_{\mathrm{w,dweta}}, 0]^{\mathrm{T}};$$

$$\Lambda_{\mathrm{zduct}} = \begin{bmatrix} \frac{X_{\mathrm{dwet,1}}}{T_{\mathrm{d,ta}}} & \frac{X_{\mathrm{dwet,2}}}{T_{\mathrm{d,ta}}} & \frac{X_{\mathrm{dwet,3}}}{T_{\mathrm{d,ta}}} \\ 0 & \frac{Y_{\mathrm{dwet,1}}}{T_{\mathrm{d,wa}}} & \frac{Y_{\mathrm{dwet,2}}}{T_{\mathrm{d,wa}}} \\ \frac{Z_{\mathrm{dwet,1}}}{T_{\mathrm{d,dg}}} & \frac{Z_{\mathrm{dwet,2}}}{T_{\mathrm{d,dg}}} & \frac{Z_{\mathrm{dwet,3}}}{T_{\mathrm{d,dg}}} \end{bmatrix}; B_{\mathrm{zduct}} = \begin{bmatrix} \frac{X_{\mathrm{dwet,4}}}{T_{\mathrm{d,ta}}} & \frac{X_{\mathrm{dwet,5}}}{T_{\mathrm{d,ta}}} & \frac{X_{\mathrm{dwet,6}}}{T_{\mathrm{d,ta}}} \\ 0 & \frac{Y_{\mathrm{dwet,3}}}{T_{\mathrm{d,wa}}} & \frac{Y_{\mathrm{dwet,4}}}{T_{\mathrm{d,wa}}} \\ \frac{Z_{\mathrm{dwet,4}}}{T_{\mathrm{d,dg}}} & \frac{Z_{\mathrm{dwet,5}}}{T_{\mathrm{d,dg}}} & \frac{Z_{\mathrm{dwet,6}}}{T_{\mathrm{d,dg}}} \end{bmatrix};$$

$$C_{\mathrm{zduct}} = \begin{bmatrix} 1 & 0 & 0 \\ 0 & 1 & 0 \\ 0 & 0 & 0 \end{bmatrix}; \quad D_{\mathrm{zduct}} = \begin{bmatrix} 0 & 0 & 0 \\ 0 & 0 & 0 \\ 0 & 0 & 1 \end{bmatrix}.$$

For the case of non-condensation on the interior surface of duct,

$$x_{\mathrm{zduct}} = [\Delta t_{\mathrm{da,L}}, \Delta t_{\mathrm{dg}}]^{\mathrm{T}}; y_{\mathrm{zduct}} = [\Delta t_{\mathrm{da,L}}, \Delta W_{\mathrm{da,L}}, \Delta G_{\mathrm{da,L}}]^{\mathrm{T}};$$
$$u_{\mathrm{zduct}} = [\Delta t_{\mathrm{da,E}}, \Delta W_{\mathrm{da,E}}, \Delta G_{da,E}]^{T}; \xi_{\mathrm{zduct}} = [\xi_{\mathrm{t,ddrya}} 0]^{\mathrm{T}};$$

$$A_{\mathrm{zduct}} = \begin{bmatrix} \frac{X_{\mathrm{ddry,1}}}{T_{\mathrm{d,ta}}} & \frac{X_{\mathrm{ddry,2}}}{T_{\mathrm{d,ta}}} \\ \frac{Z_{\mathrm{ddry,1}}}{T_{\mathrm{d,dg}}} & \frac{Z_{\mathrm{ddry,2}}}{T_{\mathrm{d,dg}}} \end{bmatrix}; \quad B_{\mathrm{zduct}} = \begin{bmatrix} \frac{X_{\mathrm{ddry,3}}}{T_{\mathrm{d,ta}}} & 0 & \frac{X_{\mathrm{ddry,4}}}{T_{\mathrm{d,ta}}} \\ \frac{Z_{\mathrm{ddry,3}}}{T_{\mathrm{d,dg}}} & 0 & \frac{Z_{\mathrm{ddry,4}}}{T_{\mathrm{d,dg}}} \end{bmatrix};$$

$$C_{\mathrm{zduct}} = \begin{bmatrix} 1 & 0 \\ 0 & 0 \\ 0 & 0 \end{bmatrix}; D_{\mathrm{zduct}} = \begin{bmatrix} 0 & 0 & 0 \\ 0 & 1 & 0 \\ 0 & 0 & 1 \end{bmatrix}.$$

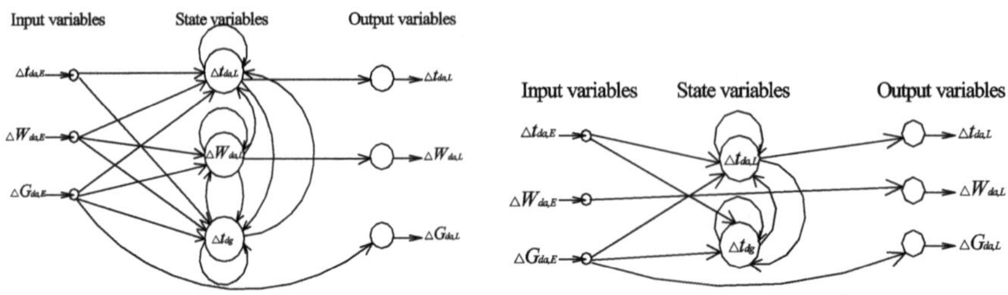

(a) Moisture-condensation case (b) None-condensation case

Fig. 4.5 Directed graph for straight-through duct model. **a** Moisture-condensation case, **b** Non-condensation case

The directed graph for straight-through duct model is shown in Fig. 4.5. It can be seen from Fig. 4.5a that under the moisture condensation (when the duct wall temperature is lower than air dew-point temperature), the exit air temperature $\left(\Delta t_{\text{da,L}}\right)$ is directly influenced by all input variables, i.e., inlet air temperature $\left(\Delta t_{\text{da,E}}\right)$, inlet air humidity $\left(\Delta W_{\text{da,E}}\right)$, and air flow rate $\left(\Delta G_{\text{da,E}}\right)$; the exit air humidity $\left(\Delta W_{\text{da,L}}\right)$ is directly influenced by air flow rate $\left(\Delta G_{\text{da,E}}\right)$ and inlet air humidity $\left(\Delta W_{\text{da,E}}\right)$ and indirectly influenced by inlet air temperature $\left(\Delta t_{\text{da,E}}\right)$. As the duct wall temperature is higher than air dew-point temperature (non-condensation case as shown in Fig. 4.5b), the exiting air temperature $\left(\Delta t_{\text{da,L}}\right)$ is no longer influenced by inlet air humidity $\left(\Delta W_{\text{da,E}}\right)$, and the exiting air humidity $\left(\Delta W_{\text{da,L}}\right)$ is no longer influenced by inlet air temperature $\left(\Delta t_{\text{da,E}}\right)$ and air flow rate. $\left(\Delta G_{\text{da,E}}\right)$

Likewise, the state-space model of straight-through pipe is given by,

$$\dot{X}_{\text{zpipe}} = A_{\text{zpipe}} X_{\text{zpipe}} + B_{\text{zpipe}} u_{\text{zpipe}} \tag{4.13}$$

$$y_{\text{zpipe}} = C_{\text{zpipe}} X_{\text{zpipe}} + D_{\text{zpipe}} u_{\text{zpipe}} - C_{\text{zpipe}} A_{\text{zpipe}}^{-1} \xi_{\text{zpipe}} \tag{4.14}$$

where

$$X_{\text{zpipe}} = x_{\text{zpipe}} + A_{\text{zpipe}}^{-1} \xi_{\text{zpipe}}; \quad x_{\text{zpipe}} = \left[\Delta t_{\text{pw,L}}, \Delta t_{\text{pg}}\right]^{\text{T}};$$

$$y_{\text{zpipe}} = \left[\Delta t_{\text{pw,L}}, \Delta G_{\text{pw,L}}\right]^{\text{T}}; \quad u_{\text{zpipe}} = \left[\Delta t_{\text{pw,E}}, \Delta G_{\text{pw,E}}\right]^{\text{T}}; \xi_{\text{zpipe}} = \left[\xi_{\text{t pw,L}}, 0\right]^{\text{T}};$$

$$A_{\text{zpipe}} = \begin{bmatrix} \frac{X_{\text{p,1}}}{T_{\text{p,tw}}} & \frac{X_{\text{p,2}}}{T_{\text{p,tw}}} \\ \frac{Y_{\text{p,1}}}{T_{\text{p,pg}}} & \frac{Y_{\text{p,2}}}{T_{\text{p,pg}}} \end{bmatrix}; \quad B_{\text{zpipe}} = \begin{bmatrix} \frac{X_{\text{p,3}}}{T_{\text{p,tw}}} & \frac{X_{\text{p,4}}}{T_{\text{p,tw}}} \\ \frac{Y_{\text{p,3}}}{T_{\text{p,pg}}} & \frac{Y_{\text{p,4}}}{T_{\text{p,pg}}} \end{bmatrix}; \quad C_{\text{zpipe}} = \begin{bmatrix} 1 & 0 \\ 0 & 0 \end{bmatrix};$$

$$D_{\text{zpipe}} = \begin{bmatrix} 0 & 0 \\ 0 & 1 \end{bmatrix}.$$

The directed graph for the state-space model of straight-through pipe can be obtained as shown in Fig. 4.6.

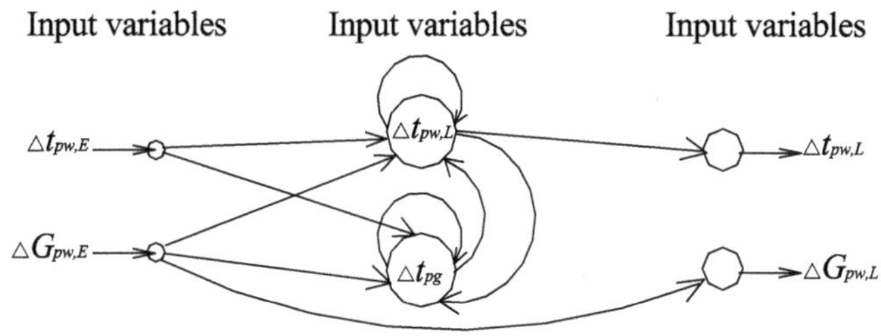

Fig. 4.6 Directed graph for straight-through pipe model

For the three-way duct (or pipe), the state variables are normally ignored, and there only exist dynamic relationships between the input and output variables.

The model of the three-way air duct is given by,

$$y_{\text{sduct}} = D_{\text{sduct}} \cdot u_{\text{sduct}} \tag{4.15}$$

For the confluent flow,

$$\begin{aligned}
y_{\text{sduct}} &= \left[\Delta t_{\text{da3,L}}, \Delta W_{\text{da3,L}}, \Delta G_{\text{da3,L}}\right]^{\text{T}}; \quad u_{\text{sduct}} = \left[u_{\text{sduct,1}}, u_{\text{sduct,2}}\right]^{\text{T}};\\
u_{\text{sduct,1}} &= \left[\Delta t_{\text{da1,E}}, \Delta W_{\text{da1,E}}, \Delta G_{\text{da1},E}\right]^{T}; u_{\text{sduct,2}} = \left[\Delta t_{\text{da2,E}}, \Delta W_{\text{da2,E}}, \Delta G_{\text{da2,E}}\right]^{\text{T}};\\
D_{\text{sduct}} &= \left[D_{\text{da1,E}}, D_{\text{da2,E}}\right]
\end{aligned}$$

$$D_{\text{da1,E}} = \begin{bmatrix} \frac{(G_{\text{da1,E}})_{\text{o}}}{(G_{\text{da3,L}})_{\text{o}}} & 0 & \frac{[(t_{\text{da1,E}})_o - (t_{\text{da3,L}})_{\text{o}}]}{(G_{\text{da3,L}})_{\text{o}}} \\ 0 & \frac{(G_{\text{da1,E}})_{\text{o}}}{(G_{\text{da3,L}})_{\text{o}}} & \frac{[(W_{\text{da1,E}})_o - (W_{\text{da3,L}})_{\text{o}}]}{(G_{\text{da3,L}})_{\text{o}}} \\ 0 & 0 & 1 \end{bmatrix};$$

$$D_{\text{da2,E}} = \begin{bmatrix} \frac{(G_{\text{da2,E}})_{\text{o}}}{(G_{\text{da3,L}})_{\text{o}}} & 0 & \frac{[(t_{\text{da2,E}})_{\text{o}} - (t_{\text{da3},L})_{\text{o}}]}{(G_{\text{da3,L}})_{\text{o}}} \\ 0 & \frac{(G_{\text{da2,E}})_{\text{o}}}{(G_{\text{da3,L}})_{\text{o}}} & \frac{[(W_{\text{da2,E}})_o - (W_{\text{da3,L}})_{\text{o}}]}{(G_{\text{da3,L}})_{\text{o}}} \\ 0 & 0 & 1 \end{bmatrix};$$

For the split flow,

$$\begin{aligned}
y_{\text{sduct}} &= \left[y_{\text{sduct,2}}, y_{\text{sduct,3}}\right]^{T}; \quad u_{\text{sduct}} = \left[\Delta t_{\text{da1,E}}, \Delta W_{\text{da1,E}}, \Delta G_{\text{da1,E}}\right]^{\text{T}};\\
y_{\text{sduct,2}} &= \left[\Delta t_{\text{da2,L}}, \Delta W_{\text{da2},L}, \Delta G_{\text{da2,L}}\right]^{T}; \quad y_{\text{sduct,3}} = \left[\Delta t_{\text{da3},L}, \Delta W_{\text{da3,L}}, \Delta G_{\text{da3,L}}\right]^{\text{T}};
\end{aligned}$$

$$D_{\text{sduct}} = \begin{bmatrix} D_{\text{fda12,E}} \\ D_{\text{fda13,E}} \end{bmatrix}; D_{\text{fda12,E}} = \begin{bmatrix} 1 & 0 & 0 \\ 0 & 1 & 0 \\ 0 & 0 & \kappa_a \end{bmatrix}; \quad D_{\text{fda13},E} = \begin{bmatrix} 1 & 0 & 0 \\ 0 & 1 & 0 \\ 0 & 0 & 1-\kappa_{\text{a}} \end{bmatrix}.$$

The dynamic relationships between the input and output variables of the three-way water pipe are given by,

$$y_{\text{spipe}} = D_{\text{spipe}} \cdot u_{\text{spipe}} \tag{4.16}$$

For the confluent flow,

$$y_{\text{spipe}} = \left[\Delta t_{\text{pw3,L}}, \Delta G_{\text{pw3,L}}\right]^{\text{T}}; \quad u_{\text{spipe}} = \left[\Delta t_{\text{pw1,E}}, \Delta G_{\text{pw1,E}}, \Delta t_{\text{pw2,E}}, \Delta G_{\text{pw2,E}}\right]^{\text{T}};$$

$$D_{\text{spipe}} = \begin{bmatrix} \frac{(G_{\text{pw1,E}})_o}{(G_{\text{pw3,L}})_o} & \frac{[(t_{\text{pw1,E}})_o - (t_{\text{pw3,L}})_o]}{(G_{\text{pw3,L}})_o} & \frac{(G_{\text{pw2,E}})_o}{(G_{\text{pw3,L}})_o} & \frac{[(t_{\text{pw2,E}})_o - (t_{\text{pw3,L}})_o]}{(G_{\text{pw3,L}})_o} \\ 0 & 1 & 0 & 1 \end{bmatrix};$$

For the split flow,

$$y_{\text{spipe}} = \left[\Delta t_{\text{pw2,L}}, \Delta G_{\text{pw2,L}}, \Delta t_{\text{pw3,L}}, \Delta G_{\text{pw3,L}}\right]^{\text{T}}; u_{\text{sduct}} = \left[\Delta t_{\text{pw1,E}}, \Delta G_{\text{pw1,E}}\right]^{\text{T}};$$

$$D_{\text{spipe}} = \begin{bmatrix} 1 & 0 \\ 0 & \kappa_{\text{w}} \\ 1 & 0 \\ 0 & 1-\kappa_{\text{w}} \end{bmatrix}.$$

The directed graphs for three-way duct and three-way pipe are shown in Figs. 4.7 and 4.8, respectively. Taking Fig. 4.7 (directed graph of the three-way duct model), for example, we can see that the output air temperature $(\Delta t_{\text{da3,L}})$ and humidity $(\Delta W_{\text{da3,L}})$ of confluent-flow duct are directly influenced by input air temperature and humidity as well as air flow rate of incoming two branches. For the split-flow duct, the outlet air temperature and humidity of the two branches are only influenced by input air temperature and humidity of the main duct.

(2) Graph description for fan (pump)

Figures 4.9 and 4.10 are directed graph corresponding to the state-space model of fan and pump from which we can see that the outlet fluid temperature of

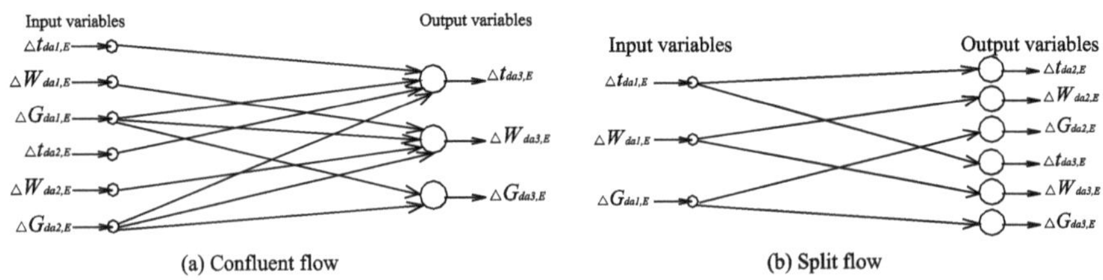

Fig. 4.7 Directed graph for three-way duct model. **a** Confluent flow, **b** Split flow

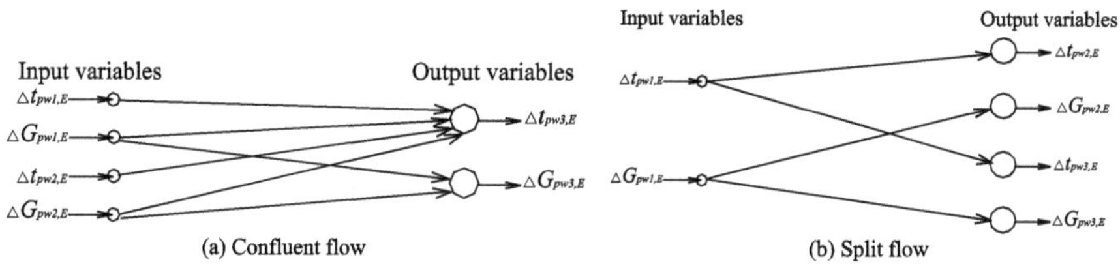

Fig. 4.8 Directed graph for three-way pipe model. **a** Confluent flow, **b** Split flow

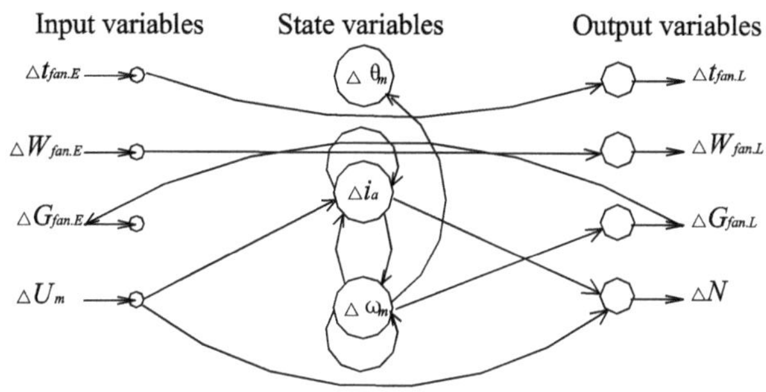

Fig. 4.9 Directed graph for fan model

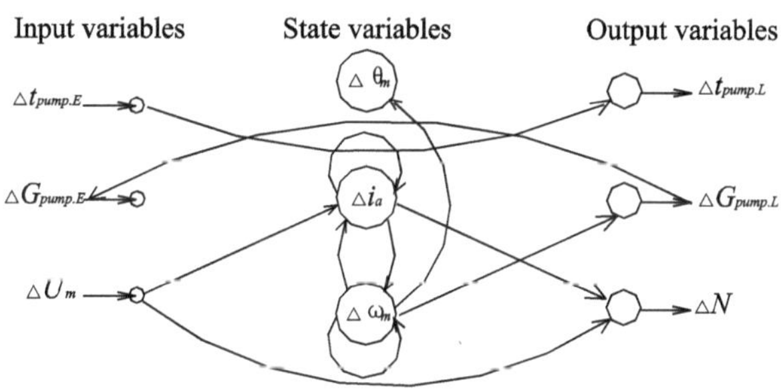

Fig. 4.10 Directed graph for pump model

fan/water pump $\left(\Delta t_{\mathrm{fan,L}}/\Delta t_{\mathrm{pump,L}}\right)$ is only directly influenced by inlet fluid temperature $\Delta t_{\mathrm{fan,E}}/\Delta t_{\mathrm{pump,E}}$ and the fan/water pump output power ΔN and the fluid flow rate follow the change of motor input electric voltage ΔU_{m}. In Figs. 4.9 and 4.10, i_{a} is motor input current (A); θ_{m} is angular position of motor (rads); and ω_{m} is angular speed of motor (rad/s).

4.2 Graph-Theory Modeling for HVAC System

4.2.1 Basic Method

Graphs are not only suitable for describing relationships among the input, the state, and the output variables of dynamic component models, but also suitable for describing link relationships and information flow among parts of a complex system. It helps us have an overall understanding on the complex system and analyze interactions among the system's components.

According to the definition of graphs [1], a graph is made up of branches of nodes and arcs connecting the nodes. Arcs are also known as branch lines, straight lines, or edges. Taking the graph in Fig. 4.11, for example, it consists of nodes $X_1 - X_7$ and arcs $L_1 - L_7$ which can be expressed by Eq. (4.17)

$$G = (X, L) \tag{4.17}$$

where

$$X = \{X_1, X_2, \ldots, X_n\}; \quad L = \{L_1, L_2, \ldots, L_n\}.$$

In the graph representation for a system, the graph nodes denote component units of the system, and the directed arcs for mass and energy flows among unit components. The directed graph can offer us an overall structure information of a system and assist us to develop the system model.

In order to facilitate computer recognition of the system structure, the directed graph is often rewritten into a matrix form also known as structural matrix. Suppose a graph is composed of m nodes, then the system structural matrix is defined as below:

$$S = [s_{i,j}]_{m \times m} \quad i = 1, 2, \ldots, m; j = 1, 2, \ldots, m \tag{4.18}$$

If $s_{i,j} = d_{i,j}$, it means there exists a directed arc flow from node X_i to X_j, and the flow dimension is $d_{i,j}$; and if $s_{i,j} = 0$, it means there is no arc connection between node X_i and X_j.

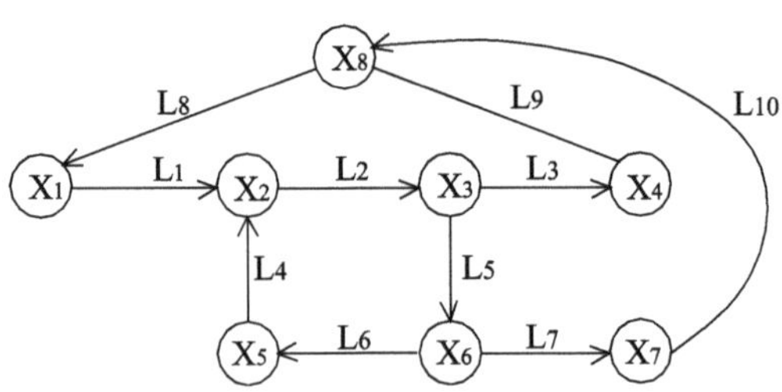

Fig. 4.11 Graph representation for a system

Assuming that each arc in the system (Fig. 4.11) is of one flow dimension, the structure matrix of the system can be written as below:

$$S = \begin{array}{c} \\ X_1 \\ X_2 \\ X_3 \\ X_4 \\ X_5 \\ X_6 \\ X_7 \\ X_8 \end{array} \begin{array}{c} \begin{array}{cccccccc} X_1 & X_2 & X_3 & X_4 & X_5 & X_6 & X_7 & X_8 \end{array} \\ \begin{bmatrix} 0 & 1 & 0 & 0 & 0 & 0 & 0 & 0 \\ 0 & 0 & 1 & 0 & 0 & 0 & 0 & 0 \\ 0 & 0 & 0 & 1 & 0 & 1 & 0 & 0 \\ 0 & 0 & 0 & 0 & 0 & 0 & 0 & 1 \\ 0 & 1 & 0 & 0 & 0 & 0 & 0 & 0 \\ 0 & 0 & 0 & 0 & 1 & 0 & 1 & 0 \\ 0 & 0 & 0 & 0 & 0 & 0 & 0 & 1 \\ 1 & 0 & 0 & 0 & 0 & 0 & 0 & 0 \end{bmatrix} \end{array} \tag{4.19}$$

From the above structure matrix, we can get the structural characteristics of the system as below:

(1) For a system consisting of m nodes, its structural matrix consists of at least $(m - 1)$ nonzero elements (the simple mode of connection between the nodes is series connection);
(2) If all the elements in the *i*th column are zero (0), the *j*th row must contain at least one nonzero element and vice versa;
(3) The nodes corresponding to the rows or columns whose elements are all zero represent a unit known as system terminal unit. The all-zero row corresponds to output terminal unit, and the all-zero column corresponds to the input terminal unit;
(4) Series unit is represented by the nonzero elements above the main diagonal of the matrix;
(5) The parallel output terminal is represented by multiple nonzero elements in a row, and the nonzero elements equal to the number of parallel branches. Likewise, the parallel input terminal is represented by multiple nonzero elements in a column;
(6) The nonzero elements below the main diagonal of the structure matrix represent the system feedback.

To sum up, a system structure matrix can clearly illustrate the position of components and their mutual relationship in the system. Meanwhile, it can reflect the system hierarchy as well.

4.2.2 Case Study

Taking the HVAC system in Fig. 2.40, for example, the graphic description for the system can be obtained according to the graph-theory modeling approach, as shown in Fig. 4.12, where 'S_{refri}' stands for refrigeration system, 'S_{conw}' for cooling coolant system which consists of cooling coolant pump, cooling tower, and pipes,

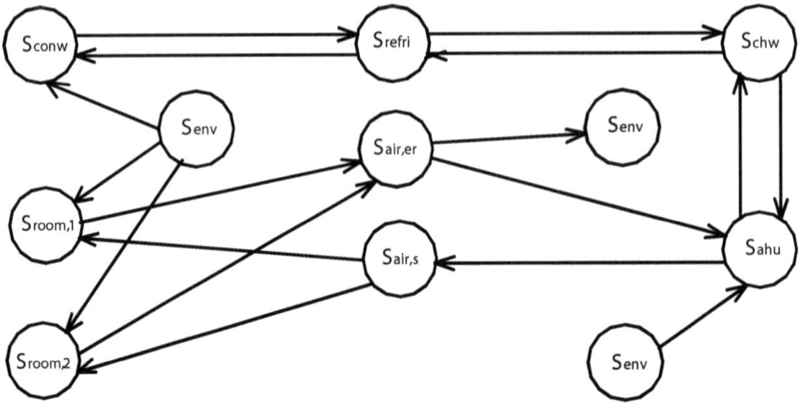

Fig. 4.12 Directed graph for the HVAC system presented in Fig. 2.40

'S_{chw}' for chilled coolant system which consists of chilled coolant pump and pipes, 'S_{ahu}' for air-handling system which consists of surface heat exchanger and supply air fan, '$S_{\text{air,s}}$' for supply air system, '$S_{\text{air,er}}$' for exhaust-return air system, and 'S_{room}' for air-conditioned room.

Based on the directed graph in Fig. 4.12, the HAVC system can be expressed by structural matrix as below:

$$S = \begin{array}{c} S_{conw} \\ S_{refri} \\ S_{chw} \\ S_{ahu} \\ S_{air,er} \\ S_{air,s} \\ S_{room,1} \\ S_{room,2} \\ S_{env} \end{array} \overset{\begin{array}{ccccccccc} S_{conw} & S_{refri} & S_{chw} & S_{ahu} & S_{air,er} & S_{air,s} & S_{room,1} & S_{room,2} & S_{env} \end{array}}{\begin{bmatrix} 0 & 2 & 0 & 0 & 0 & 0 & 0 & 0 & 0 \\ 2 & 0 & 2 & 0 & 0 & 0 & 0 & 0 & 0 \\ 0 & 2 & 0 & 2 & 0 & 0 & 0 & 0 & 0 \\ 0 & 0 & 2 & 0 & 0 & 3 & 0 & 0 & 0 \\ 0 & 0 & 0 & 3 & 0 & 0 & 0 & 0 & 3 \\ 0 & 0 & 0 & 0 & 0 & 0 & 3 & 3 & 0 \\ 0 & 0 & 0 & 0 & 3 & 0 & 0 & 0 & 0 \\ 0 & 0 & 0 & 0 & 3 & 0 & 0 & 0 & 0 \\ 3 & 0 & 0 & 3 & 0 & 0 & 4 & 4 & 0 \end{bmatrix}} \tag{4.20}$$

In Eq. (4.20), $S_{12} = S_{21} = 2$ represents from cooling coolant system (S_{conw}) to cold supply system (S_{refri}) and from cold supply system (S_{refri}) to cooling coolant system (S_{conw}), respectively, and both have two variables i.e., cooling coolant flow rate and cooling coolant temperature. $S_{23} = S_{32} = 2$ represents from cold supply system (S_{refri}) to chilled coolant system (S_{chw}) and from chilled coolant system (S_{chw}) to cold supply system(S_{refri}), respectively, and both have two variables i.e., chilled coolant flow rate and chilled coolant temperature. $S_{44} = S_{46} = S_{54} = S_{59} = S_{67} = S_{68} = S_{75} = S_{85} = S_{91} = S_{94} = 3$ represents from air-handling system (S_{ahu}) to air supply system ($S_{\text{air,s}}$), from exhaust-return air system ($S_{\text{air,er}}$) to air-handling system (S_{ahu}) and external environment (S_{env}), from air supply system ($S_{air,s}$) to conditioned room (S_{room}), from air-conditioned room (S_{room}) to ambient environment (S_{env}), and from ambient environment (S_{env}) to cooling coolant system (S_{conw}) and air-handling system (S_{ahu}), respectively, and all have three variables i.e., air flow rate, air temperature, and air humidity; $S_{97} = S_{98} = 4$ represents from ambient environment (S_{env}) to air-conditioned room (S_{room}), respectively, and both have four variables, i.e., air flow rate, air temperature, air humidity, and solar radiation.

Figure 4.13 shows directed graph for subsystems in the HVAC system, where X_{coil} stands for surface heat exchanger; X_{fan} for fan; X_{pump} for pump (the subscripts '1' and '2' denote cooling coolant pump and chilled coolant pump, respectively); X_{tower} for cooling tower; X_{sduct} for three-way duct (the subscripts '1' and '3' denote split-flow duct in exhaust-return air system and supply air system, respectively, and the subscripts '2' and '4' denote confluent-flow duct in exhaust-return air system and supply air system, respectively.); X_{zduct} for straight-through duct (the subscripts '1' and '2' denote duct from the air-handling unit to $X_{sduct,1}$ and that from $X_{sduct,1}$ to air-conditioned room, respectively, and the subscript '3' denotes duct from $X_{sduct,2}$ to $X_{sduct,3}$); X_{zpipe} for straight-through pipe (the subscripts '1' and '2' denote pipe from cooling tower to chiller's condenser and that from chiller's condenser to cooling tower, respectively, and the subscripts '3' and '4' denote pipe from chiller's evaporator to surface heat exchanger and that from surface heat exchanger to chiller's evaporator); X_{spipe} for three-way pipe; X_{fa} for air valve; X_{fw} for pipe valve; and X_{room} for air-conditioned room.

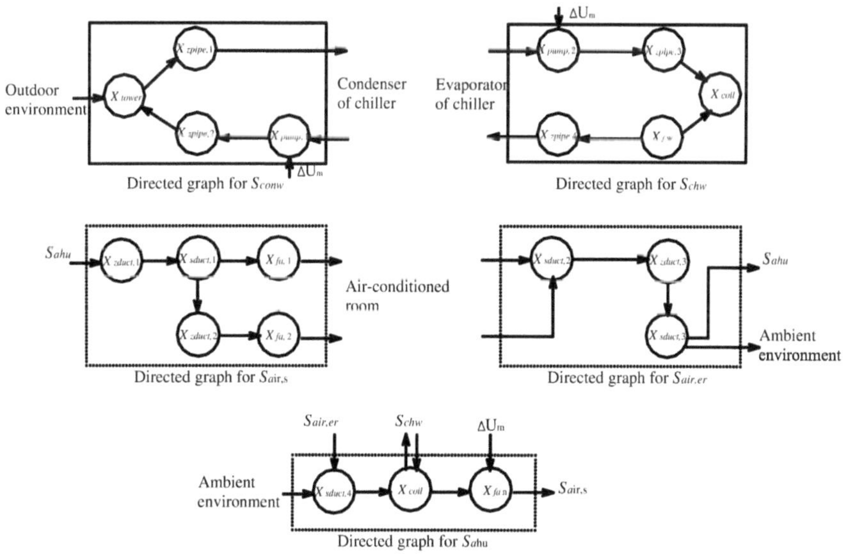

Fig. 4.13 Directed graph for subsystems in the HVAC system

4.3 Structure-Matrix Analysis Approach

Graph-theory description of the model can intuitively get the nature of influence of the input variables on the state and the output ones of a dynamic model. To have a deeper understanding of the dynamic models, the structure-matrix method is employed for the model analysis to provide important guidance for the feasibility analysis of system control scheme.

4.3.1 Model Structural Matrix

Consider a linear time-invariant system:

$$\begin{cases} \dot{x}(\tau) = Ax(\tau) + Bu(\tau) \\ \dot{y}(\tau) = Cx(\tau) + Du(\tau) \end{cases} \tag{4.21}$$

where A, B, C, and D are quantity matrices.

For such kind of mathematical description of (A, B, C, D), there must be a corresponding structural description of $(\bar{A}, \bar{B}, \bar{C}, \bar{D})$ i.e., structural matrix among which $\bar{A}, \bar{B}, \bar{C}, \bar{D}$ are corresponding matrix parameters with zero elements '0' and other nonzero elements marked by '$\times$'. It can be seen that suppose two linear systems (A_1, B_1, C_1, D_1) and (A_2, B_2, C_2, D_2) have different mathematical descriptions but corresponding position of the zero elements in the matrix is the same, then they have the same structural system $(\bar{A}, \bar{B}, \bar{C}, \bar{D})$. In such case, the two systems have structural equivalence.

Concerning a structural matrix, since there is no quantitative description of its elements, we cannot identify its rank. But the rank of a structural matrix can be obtained by maximum rank of a numerical matrix corresponding to the structural matrix. In order to differentiate the rank of the numerical matrix, we name the rank of a structural matrix as 'generic rank' as below:

$$\mathrm{gr}(\bar{S}) = \max\,(\mathrm{rank}\,(\mathrm{S})) \tag{4.22}$$

For details on calculation of generic rank of a structural matrix, please refer to Johnson [4].

4.3.2 Reachability Analysis of Model Input–Output

In a dynamic system model, there are basically three types of variables: i.e., the state (x), the input (u), and the output (y). To fully understand the mutual influence among them, especially whether or not the state variables can be reachable by the input ones, or whether or not the state variables can affect the output ones. This is extremely important during the structural analysis of a dynamic model. Suppose every state variable of the system model (x_i) can be reachable by at least one input variable (u_j), then the system model is of input reachability. Suppose every state variable of the system model (x_i) can reach at least one of the output ones (y_k), then the system model is of output reachability. The input reachability and output reachability reflect whether or not the state variables of the model system can be changed by the input ones and whether or not they can be reflected by the output ones of the model system.

Introducing adjacency matrix to describe the logical relationship of the system model [5] in accordance with the dynamic model described by Eq. (4.21), the basic elements are the state, the input, and the output variables of the model. Suppose these variables are the basic units and are arranged in form of block combination, then the adjacency matrix of the system can be written as follows:

$$A_{\mathrm{a}} = \begin{array}{c} \\ x \\ u \\ y \end{array} \begin{array}{c} \begin{array}{ccc} x & u & y \end{array} \\ \begin{bmatrix} \bar{A}^{\mathrm{T}} & 0 & \bar{C}^{\mathrm{T}} \\ \bar{B}^{\mathrm{T}} & 0 & \bar{D}^{\mathrm{T}} \\ 0 & 0 & 0 \end{bmatrix} \end{array} \tag{4.23}$$

From Eq. (4.23), we can easily see the structural matrix that appears in A_{a} i.e., $\bar{A}, \bar{B}, \bar{C}, \bar{D}$ are all transposed. Taking the elements in matrix $\bar{A}$, for example, suppose an element in the matrix $\bar{A}$ a_{ij} = '×' suggests that x_i would be affected by x_j, then such kind of influence means that there should be a_{ji} = '×' in A_{a}. In the adjacency matrix, the all-zero column corresponds to the input vector of system model, and the all-zero row corresponds to the output vector of system model.

The adjacency matrix only shows the direct influence among system units. In practice, our main concern is whether or not some system units can impact other units within the system through others (i.e., indirect influence). In such case, we can use kth power of A_a (A_a^k) to represent the kth heavy indirect relationship among system units, e.g., one element a_{ij}^k = '×' in A_a^k means the unit S_i affects the unit S_j through the other $(k-1)$ units. The power operation here can be made based on the general rules of matrix multiplication. Since A_a is structural matrix, the addition and multiplication operation between the elements should be the logic addition and multiplication.

For a system consisting of N elements, suppose there exists reachable relationship between any of its two elements S_i and S_j, we can always get a reflection in the corresponding units $A_a, A_a^2, \ldots, A_a^N$. If each unit can reach itself, we can get the following matrix to describe every reachable relationship:

$$R = I \cup A_{\mathrm{a}} \cup A_{\mathrm{a}}^2 \cup \ldots \cup A_{\mathrm{a}}^N \tag{4.24}$$

The matrix R in Eq. (4.24) is called system reachability matrix. It is also a form of expression of structural matrix in which the element r_{ij} = '×' means that the unit

S_i can directly or indirectly reach the unit S_j, and the element r_{ij} = '0' means that S_i can not reach the unit S_j. According to the logic algorithm, formula (4.24) can be rewritten as follows:

$$R = (I \cup A_{\mathrm{a}})^N = \begin{bmatrix} (\bar{A}^{\mathrm{T}} + I)^N & 0 & (\bar{A}^{\mathrm{T}} + I)^{N-1}\bar{C}^{\mathrm{T}} \\ \bar{B}^{\mathrm{T}}(\bar{A}^{\mathrm{T}} + I)^{N-1} & I & \bar{B}^{\mathrm{T}}(\bar{A}^{\mathrm{T}} + I)^{N-2}\bar{C}^{\mathrm{T}} \\ 0 & 0 & I \end{bmatrix} \tag{4.25}$$

Meanwhile,

$$R = \begin{matrix} & \begin{matrix} x & u & y \end{matrix} \\ \begin{matrix} x \\ u \\ y \end{matrix} & \begin{bmatrix} R_{xx} & R_{xu} & R_{xy} \\ R_{ux} & R_{uu} & R_{uy} \\ R_{yx} & R_{yu} & R_{yy} \end{bmatrix} \end{matrix} \tag{4.26}$$

Thus, the input reachability and output reachability can be described by use of structural matrices R_{ux} and R_{xy}, respectively, as below:

$$R_{ux} = \bar{B}^{\mathrm{T}}(\bar{A}^{\mathrm{T}} + I)^{n-1} \tag{4.27}$$

$$R_{xy} = (\bar{A}^{\mathrm{T}} + I)^{n-1}\bar{C}^{\mathrm{T}} \tag{4.28}$$

If there does not exist all-zero column in the matrix R_{ux}, then the system is of input reachability; and if there does not exist all-zero row in the matrix R_{xy}, then the system is of output reachability. If there exsits all-zero row in the matrix R_{ux}, then it means the input variables corresponding to the all-zero row do not affect any of the state variables of model system. By using the method of structure-matrix analysis, the input and output reachability of dynamic component models (state-space model) can be investigated.

4.3.3 Controllability/Observability Analysis of Model

In order to get/find a feasible control scheme, we must discuss the system's controllability and observability in advance. Using the previously described system structural matrix, we can carry out structural analysis of system controllability. This is very meaningful since it helps to understand the influence of each control action on the system and provide us global guiding information prior to the design of control system. Although the study of system structural controllability cannot characterize it quantitatively, and so cannot make a detailed analysis of strength of controllability, etc., but we can carry out qualitative analysis of the system controllability based on logic.

Likewise, the linear time-invariant system (Eq. 4.21) is considered for the discussion. If $(\bar{A}, \bar{B})$ has an equivalent system of controllable structure in the common sense, then $(\bar{A}, \bar{B})$ is structurally controllable. Otherwise, $(\bar{A}, \bar{B})$ is structurally uncontrollable.

Davidson and Morari started from system input reachability and combined it with judgment of the generic rank to get the necessary and sufficient conditions for controllability of a system structure as below [5]:

(a) The system is input reachable;
(b) $\mathrm{gr}([\bar{A}, \bar{B}]) = n$.

The condition (a) points out that all system states can be affected by changes of input; and the condition (b) points out that inputs have sufficient freedom to change the system state to a desired state.

In the control of a dynamic system, there often exist many several possibilities for variable selection. In system design, we should decide which control actions are necessary, and the rest of control actions can be omitted or may be used to implement an auxiliary control target. For a feasible control scheme, apart from need to consider system constraint conditions given by process conditions, the most important point is that all the selected control variables must guarantee system controllability. Normally, with regard to system structural matrix $(\bar{A}, \bar{B})$, suppose the generic rank of the structural matrix does not change after deleting a column (which corresponds to a control variable) in the matrix, then it is considered that the control variable corresponding to the deleted column does not affect the system controllability. Otherwise, the control variable corresponding to the deleted column does affect the system controllability, and it must be considered when choosing controllable variables. It is admitted that the controllability usually describes the ability of input variables to control the state of system.

The observability means to what degree the state of system can be reflected by the input and the output variables, i.e., provided the system model is fully known, the system states can be got by the input and the output information from measurements. Like the controllability analysis, the observability analysis is of equally important significance to the state feedback control design. The necessary and sufficient conditions for system structural observability are as follows [5]:

(a) Output reachable system
(b) $gr\left(\begin{bmatrix} \bar{A} \\ \bar{C} \end{bmatrix}\right) = n.$

4.3.4 *Case Study*

In this section, the dynamic characteristics of models of various parts of air-conditioning system are discussed with the method of structure-matrix analysis. By using the structure-matrix method, the reachability, controllability, and observability of the models can be analyzed.

4.3.4.1 Air-Cooling Coil

(1) Input–output reachability analysis

According to the structure-matrix theory, the coefficient matrix of air-cooling coil model can be structured as below:

For dry-condition model:

$$\bar{A}_{\text{coil}} = \begin{bmatrix} \times & 0 & \times \\ 0 & \times & \times \\ \times & \times & \times \end{bmatrix}; \quad \bar{B}_{\text{coil}} = \begin{bmatrix} \times & \times & 0 & 0 & 0 \\ 0 & 0 & \times & 0 & \times \\ \times & \times & \times & 0 & \times \end{bmatrix};$$

$$\bar{C}_{\text{coil}} = \begin{bmatrix} \times & 0 & 0 \\ 0 & 0 & 0 \\ 0 & \times & 0 \\ 0 & 0 & 0 \\ 0 & 0 & 0 \end{bmatrix}; \quad \bar{D}_{\text{coil}} = \begin{bmatrix} 0 & 0 & 0 & 0 & 0 \\ 0 & \times & 0 & 0 & 0 \\ 0 & 0 & 0 & 0 & 0 \\ 0 & 0 & 0 & \times & 0 \\ 0 & 0 & 0 & 0 & \times \end{bmatrix}.$$

For wet-condition model:

$$\bar{A}_{\text{coil}} = \begin{bmatrix} \times & 0 & 0 & \times \\ 0 & \times & \times & \times \\ 0 & \times & \times & \times \\ \times & \times & 0 & \times \end{bmatrix}; \quad \bar{B}_{\text{coil}} = \begin{bmatrix} \times & \times & 0 & 0 & 0 \\ 0 & 0 & \times & \times & \times \\ 0 & 0 & \times & \times & \times \\ \times & \times & \times & 0 & \times \end{bmatrix};$$

$$\bar{C}_{\text{coil}} = \begin{bmatrix} \times & 0 & 0 & 0 \\ 0 & 0 & 0 & 0 \\ 0 & \times & 0 & 0 \\ 0 & 0 & \times & 0 \\ 0 & 0 & 0 & 0 \end{bmatrix}; \quad \bar{D}_{\text{coil}} = \begin{bmatrix} 0 & 0 & 0 & 0 & 0 & 0 \\ 0 & \times & 0 & 0 & 0 & 0 \\ 0 & 0 & 0 & 0 & 0 & 0 \\ 0 & 0 & 0 & 0 & 0 & 0 \\ 0 & 0 & 0 & 0 & 0 & \times \end{bmatrix}.$$

For the case of dry condition, the model dimension of air-cooling coil is three (i.e., $n = 3$). By using Eq. (4.27) and (4.28), we have

$$R_{\text{ux,coil}} = \bar{B}^{\text{T}}_{\text{coil}}(\bar{A}^{\text{T}}_{\text{coil}} + I)^2 = \begin{bmatrix} \times & \times & \times \\ \times & \times & \times \\ \times & \times & \times \\ 0 & 0 & 0 \\ \times & \times & \times \end{bmatrix} \tag{4.29}$$

$$R_{\text{xy,coil}} = (\bar{A}^{\text{T}}_{\text{coil}} + I)^2 \bar{C}^{\text{T}}_{\text{coil}} = \begin{bmatrix} \times & 0 & \times & 0 & 0 \\ \times & 0 & \times & 0 & 0 \\ \times & 0 & \times & 0 & 0 \end{bmatrix} \tag{4.30}$$

The result in Eq. (4.29) shows that apart from the input variable $\Delta W_{\text{a,E}}$, the other input variables can affect all the state ones. Since there does not exist all-zero column in the matrix $R_{\text{ux,coil}}$ (Eq. 4.29), we can conclude that the air-cooling coil is an input reachability system under dry working condition. The result in Eq. (4.30) signifies that each state variable can affect the exit water temperature $(\Delta t_{\text{w,L}})$ and the exit air temperature $(\Delta t_{\text{a,L}})$ of coil, and this is consistent with the previous graph description. Since there does not exist all-zero row in the matrix $R_{\text{xy,coil}}$ (Eq. 4.30), we can conclude that the air-cooling coil is an output reachability system under dry working condition.

For wet working condition, the model dimension of air-cooling coil is four (i.e., $n = 4$). The corresponding structure matrixes can be obtained as below:

$$R_{\text{ux,coil}} = \bar{B}^{\text{T}}_{\text{coil}}(\bar{A}^{T}_{\text{coil}} + I)^3 = \begin{bmatrix} \times & \times & \times & \times \\ \times & \times & \times & \times \\ \times & \times & \times & \times \\ \times & \times & \times & \times \\ \times & \times & \times & \times \end{bmatrix} \tag{4.31}$$

$$R_{\text{xy,coil}} = (\bar{A}^{\text{T}}_{\text{coil}} + I)^3 \bar{C}^{\text{T}}_{\text{coil}} = \begin{bmatrix} \times & 0 & \times & \times & 0 \\ \times & 0 & \times & \times & 0 \\ \times & 0 & \times & \times & 0 \\ \times & 0 & \times & \times & 0 \end{bmatrix} \tag{4.32}$$

Since there is no all-zero column in the matrix $R_{\text{ux,coil}}$ (Eq. 4.31) and no all-zero row in the matrix $R_{\text{xy,coil}}$ (Eq. 4.32), the air-cooling coil model under wet condition is of both input and output reachability. The matrix in Eq. (4.31) manifests that all input variables can attain all state ones, which means each input variable can produce direct or indirect influence on any state variable of air-cooling coil under wet condition. Likewise, it can be seen from the matrix in Eq. (4.32) that each state variable can only produce influence on the three output variables, i.e., $\Delta t_{\text{w,L}}$, $\Delta t_{\text{a,L}}$, and $\Delta W_{\text{a,L}}$.

(2) Controllability/observability analysis

From the previous analysis, we know that regardless of whether it is dry working condition or wet working condition, structural models of air-cooling coil are all input reachable, thus meets the first necessary and sufficient condition of controllable model structure. Now test the generic rank of the matrix $[\bar{A}_{\text{coil}}\bar{B}_{\text{coil}}]$ based on the condition (b) in the theorem.

For the dry working condition:

$$\text{gr}([\bar{A}_{\text{coil}}\bar{B}_{\text{coil}}]) = \text{gr}\left(\begin{bmatrix} \times & 0 & \times & \times & \times & 0 & 0 & 0 \\ 0 & \times & \times & 0 & 0 & \times & 0 & \times \\ \times & \times & \times & \times & \times & \times & 0 & \times \end{bmatrix}\right) = 3 \quad (4.33)$$

For the wet working condition:

$$\text{gr}(\bar{A}_{\text{coil}}\bar{B}_{\text{coil}}) = \text{gr}\left(\begin{bmatrix} \times & 0 & 0 & \times & \times & \times & 0 & 0 & 0 \\ 0 & \times & \times & \times & 0 & 0 & \times & \times & \times \\ 0 & \times & \times & \times & 0 & 0 & \times & \times & \times \\ \times & \times & 0 & \times & \times & \times & \times & 0 & \times \end{bmatrix}\right) = 4 \quad (4.34)$$

From Eq. (4.33) and (4.34), we know that the generic rank of structural model of air-cooling coil under both dry and wet working condition meets the second necessary and sufficient condition for controllable model structure. Therefore, we can conclude the dynamic model of air-cooling coil has a controllable structure. The generic rank of matrix $[\bar{A}_{\text{coil}}\bar{B}_{\text{coil}}]$ does not change after deleting any column, which indicates that we can choose any one or more among the input variables (i.e., $\Delta t_{\text{w,E}}, \Delta G_{\text{w,E}}, \Delta t_{\text{a,E}}, \Delta W_{\text{a,E}},$ and $\Delta G_{\text{a,E}}$) as the control variables. In real situations, the control variables of the air-cooling coil are normally $\Delta t_{\text{w,E}}$, $\Delta G_{\text{w,E}}$, or $\Delta G_{\text{a,E}}$, and the other two input variables, i.e., $\Delta t_{\text{a,E}}$ and $\Delta W_{\text{a,E}}$, are often disturbance inputs.

Similarly,

$$\text{gr}\left(\begin{bmatrix} \bar{A}_{\text{coil}} \\ \bar{C}_{\text{coil}} \end{bmatrix}\right) = \begin{cases} 3, \text{dry conditions} \\ 4, \text{wet conditions} \end{cases} \quad (4.35)$$

According to necessary and sufficient conditions for observable system structure, the structural model of the air-cooling coil under both dry and wet working conditions is of structural observability.

4.3.4.2 Chiller

(1) Input–output reachability analysis

The coefficient matrix of the chiller's state-space model is structured as below:

$$\bar{A}_{\text{chiller}} = \begin{bmatrix} \times & 0 & \times & 0 & 0 & 0 \\ 0 & \times & \times & 0 & 0 & 0 \\ \times & \times & \times & 0 & 0 & 0 \\ 0 & 0 & 0 & \times & 0 & \times \\ 0 & 0 & 0 & 0 & \times & \times \\ 0 & 0 & 0 & \times & \times & \times \end{bmatrix}; \quad \bar{B}_{\text{chiller}} = \begin{bmatrix} 0 & 0 & 0 & 0 & \times \\ \times & \times & 0 & 0 & 0 \\ \times & \times & 0 & 0 & 0 \\ 0 & 0 & 0 & 0 & \times \\ 0 & 0 & \times & \times & 0 \\ 0 & 0 & \times & \times & 0 \end{bmatrix};$$

$$\bar{C}_{\text{chiller}} = \begin{bmatrix} 0 & \times & 0 & 0 & 0 & 0 \\ 0 & 0 & 0 & 0 & 0 & 0 \\ 0 & 0 & 0 & 0 & \times & 0 \\ 0 & 0 & 0 & 0 & 0 & 0 \\ \times & 0 & 0 & \times & 0 & 0 \\ \times & 0 & 0 & \times & 0 & 0 \\ \times & 0 & 0 & \times & 0 & 0 \end{bmatrix}; \quad \bar{D}_{\text{chiller}} = \begin{bmatrix} 0 & 0 & 0 & 0 & 0 \\ 0 & \times & 0 & 0 & 0 \\ 0 & 0 & 0 & 0 & 0 \\ 0 & 0 & 0 & \times & 0 \\ 0 & 0 & 0 & 0 & \times \\ 0 & 0 & 0 & 0 & \times \\ 0 & 0 & 0 & 0 & 0 \end{bmatrix}.$$

Dimension of chiller's model is 6 (i.e., $n = 6$). Thus, by using Eq. (4.27) and (4.28), we have

$$R_{\text{ux,chiller}} = \bar{B}^{\text{T}}_{\text{chiller}}(\bar{A}^{\text{T}}_{\text{chiller}} + I)^5 = \begin{bmatrix} \times & \times & \times & 0 & 0 & 0 \\ \times & \times & \times & 0 & 0 & 0 \\ 0 & 0 & 0 & \times & \times & \times \\ 0 & 0 & 0 & \times & \times & \times \\ \times & \times & \times & \times & \times & \times \end{bmatrix} \tag{4.36}$$

$$R_{\text{xy,chiller}} = (\bar{A}^{\text{T}}_{\text{chiller}} + I)^5 \bar{C}^{\text{T}}_{\text{chiller}} = \begin{bmatrix} \times & 0 & 0 & 0 & \times & \times & \times \\ \times & 0 & 0 & 0 & \times & \times & \times \\ \times & 0 & 0 & 0 & \times & \times & \times \\ 0 & 0 & \times & 0 & \times & \times & \times \\ 0 & 0 & \times & 0 & \times & \times & \times \\ 0 & 0 & \times & 0 & \times & \times & \times \end{bmatrix} \tag{4.37}$$

We can easily conclude from Eq. (4.36) that the chiller's dynamic model is input reachable in which the input variables $\Delta t_{\text{cw,E}}$ and $\Delta G_{\text{cw,E}}$ can reach the state variables Δt_{c}, $\Delta t_{\text{cw,L}}$, and Δt_{cg}; the input variables $\Delta t_{\text{ew,E}}$ and $\Delta G_{\text{ew,E}}$ can reach the state variables Δt_{k}, $\Delta t_{\text{ew,L}}$, and Δt_{eg}; and the input variable ΔG_{rm} can reach all state variables of chiller. The matrix in Eq. (4.37) tells us the chiller's dynamic model is output reachable in which the state variables Δt_{c}, $\Delta t_{\text{cw,L}}$, and Δt_{cg} can reach the output variables $\Delta t_{\text{cw,L}}$, $\Delta E_{\text{w,com}}$, ΔQ_{c}, and ΔCOP; and the state variables Δt_{k}, $\Delta t_{\text{ew,L}}$, and Δt_{eg} can reach output variables $\Delta t_{\text{ew,L}}$, $\Delta E_{\text{w,com}}$, ΔQ_{c}, and ΔCOP .

(2) Controllability/observability analysis

From above analysis, we know that the structure-matrix model of chiller is input reachable, and

$$\operatorname{gr}([\bar{A}_{\text{chiller}}\ \bar{B}_{\text{chiller}}]) = \operatorname{gr}\left(\begin{bmatrix} \times & 0 & \times & 0 & 0 & 0 & 0 & 0 & 0 & 0 & \times \\ 0 & \times & \times & 0 & 0 & 0 & \times & \times & 0 & 0 & 0 \\ \times & \times & \times & 0 & 0 & 0 & \times & \times & 0 & 0 & 0 \\ 0 & 0 & 0 & \times & 0 & \times & 0 & 0 & 0 & 0 & \times \\ 0 & 0 & 0 & 0 & \times & \times & 0 & 0 & \times & \times & 0 \\ 0 & 0 & 0 & \times & \times & \times & 0 & 0 & \times & \times & 0 \end{bmatrix}\right) = 6 \tag{4.38}$$

which meets the necessary and sufficient conditions for controllable system structure. Thus, the dynamic model of chiller is structurally controllable.

Since the generic rank of matrix $[\bar{A}_{\text{tower}}\bar{B}_{\text{tower}}]$ does not change after deleting any column corresponding to input variables, we can choose any one or more among $[\Delta t_{\text{cw,E}}, \Delta G_{\text{cw,E}}, \Delta t_{\text{ew,E}}, \Delta G_{\text{ew,E}}, \Delta G_{\text{rm}}]$ as the control variables. In practice, the control variables for chiller are often $\Delta G_{\text{cw,E}}$, $\Delta G_{\text{ew,E}}$, or ΔG_{rm}, and the other two input variables (i.e., $\Delta t_{\text{cw,E}}$ and $\Delta t_{\text{ew,E}}$,) are taken as disturbance inputs.

Similarly, because

$$\operatorname{gr}\left(\begin{bmatrix} \bar{A}_{\text{chiller}} \\ \bar{C}_{\text{chiller}} \end{bmatrix}\right) = 6. \tag{4.39}$$

and the structural model of the chiller is output reachable, the dynamic model of chiller is structurally observable.

4.3.4.3 Cooling Tower

(1) Input–output reachability analysis

The coefficient matrix of the cooling tower model is structured as below:

$$\bar{A}_{\text{tower}} = \begin{bmatrix} \times & \times & \times & 0 \\ 0 & \times & \times & 0 \\ \times & \times & \times & 0 \\ 0 & 0 & \times & \times \end{bmatrix}; \quad \bar{B}_{\text{tower}} = \begin{bmatrix} \times & \times & \times & \times & 0 \\ 0 & \times & \times & \times & 0 \\ \times & \times & \times & \times & \times \\ 0 & 0 & 0 & 0 & \times \end{bmatrix};$$

$$\bar{C}_{\text{tower}} = \begin{bmatrix} \times & 0 & 0 & 0 \\ 0 & \times & 0 & 0 \\ 0 & 0 & 0 & 0 \\ 0 & 0 & 0 & 1 \\ 0 & 0 & 0 & 0 \end{bmatrix}; \quad \bar{D}_{\text{tower}} = \begin{bmatrix} 0 & 0 & 0 & 0 & 0 \\ 0 & 0 & 0 & 0 & 0 \\ 0 & 0 & \times & 0 & 0 \\ 0 & 0 & 0 & 0 & 0 \\ 0 & 0 & 0 & 0 & \times \end{bmatrix}.$$

Dimension of the cooling tower model is 4 (i.e., n=4). According to Eq. (4-27) and (4-28), we have

$$R_{\mathrm{ux,tower}} = \bar{B}^{\mathrm{T}}_{\mathrm{tower}}(\bar{A}^{\mathrm{T}}_{\mathrm{tower}} + I)^3 = \begin{bmatrix} \times & \times & \times & \times \\ \times & \times & \times & \times \\ \times & \times & \times & \times \\ \times & \times & \times & \times \\ \times & \times & \times & \times \end{bmatrix} \tag{4.40}$$

$$R_{xy,\mathrm{tower}} = (\bar{A}^{\mathrm{T}}_{\mathrm{tower}} + I)^3 \bar{C}^{\mathrm{T}}_{\mathrm{tower}} = \begin{bmatrix} \times & \times & 0 & \times & 0 \\ \times & \times & 0 & \times & 0 \\ \times & \times & 0 & \times & 0 \\ 0 & 0 & 0 & \times & 0 \end{bmatrix} \tag{4.41}$$

The structure matrix in Eq. (4.40) signifies that all input variables can affect all state ones, and this manifests that the cooling tower model is input reachable. The structure matrix in Eq. (4.41) signifies that the three state variables, i.e., $\Delta t_{\mathrm{a,L}}$, $\Delta W_{\mathrm{a,L}}$, and $\Delta t_{\mathrm{w,L}}$, can affect the exit water temperature of tower's sink $\left(\Delta t'_{\mathrm{w,L}}\right)$, and this manifests that the cooling tower model is output reachable.

(2) Controllability/observability analysis

Known from above analysis, the structural model of cooling tower is input reachable, and meanwhile,

$$\mathrm{gr}([\bar{A}_{\mathrm{tower}}\ \bar{B}_{\mathrm{tower}}]) = \mathrm{gr}\left(\begin{bmatrix} \times & \times & \times & 0 & \times & \times & \times & \times & 0 \\ 0 & \times & \times & 0 & 0 & \times & \times & \times & 0 \\ \times & \times & \times & 0 & \times & \times & \times & \times & \times \\ 0 & 0 & \times & \times & 0 & 0 & 0 & 0 & \times \end{bmatrix}\right) = 4 \tag{4.42}$$

Therefore, the dynamic model of cooling tower is structurally controllable.

Since the generic rank of matrix $[\bar{A}_{\mathrm{tower}}\ \bar{B}_{\mathrm{tower}}]$ does not change after deleting any column corresponding to input variables, we can choose any one or more among the input variables (i.e.,$\Delta t_{\mathrm{a,E}}$, $\Delta W_{\mathrm{a,E}}$, $\Delta G_{\mathrm{a,E}}$, $\Delta t_{\mathrm{w,E}}$, and $\Delta G_{\mathrm{w,E}}$) as the control variables. In real situations, the control variable is $\Delta G_{\mathrm{w,E}}$ or $\Delta G_{\mathrm{a,E}}$, and the other input variables (i.e., $\Delta t_{\mathrm{a,E}}$, $\Delta W_{\mathrm{a,E}}$, and $\Delta t_{\mathrm{w,E}}$) are disturbance inputs.

Similarly,

$$\mathrm{gr}\left(\begin{bmatrix} \bar{A}_{\mathrm{tower}} \\ \bar{C}_{\mathrm{tower}} \end{bmatrix}\right) = 4 \tag{4.43}$$

and the structural model of cooling tower is output controllable, thus the cooling tower model is structurally observable.

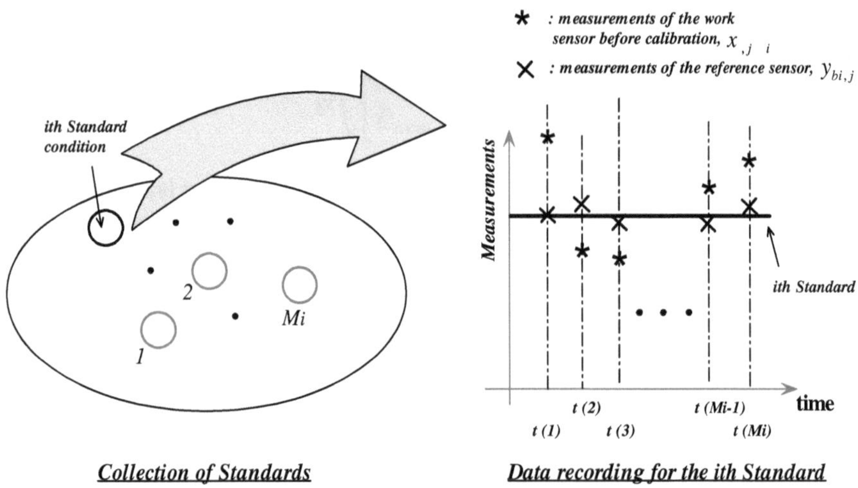

Fig. 5.1 Illustration of a conventional calibration

comparatively reliable information. Sometimes, a measurement does not come directly from one measured quantity; for example, an orifice flow rate meter gives the value of flow rate based on the measurement of fluid pressure differences, area and temperature, and correcting coefficients. Under this kind of scenario, a combined standard uncertainty should be used to attribute the propagation of measurement uncertainties. It is defined as the positive square root of a sum of terms, involving all the variances or covariance of these quantities.

In a conventional calibration, standards are generally predefined and maintained constant. The comparison between a working sensor and a known benchmark or reference sensor can be carried out either using the sensors' signals or the reading instrument. The process of a conventional calibration of a sensor based on a sole quantity is illustrated as in Fig. 5.1.

Both the working sensor and the reference sensor are placed in the uniform standards to produce measurements simultaneously. The number and value of selected standards should be adequate to cover the working range of the working sensor for an effective calibration. For illustration purposes, the number of standards is denoted as N and the values are indicated as i-th standard, from 1 to n, circled in the left figure. Along the horizontal axis of the right figure, $t(1)$ to $t(M_i)$ are the time stamps throughout the calibration process. For each standard, M_1 to M_n times of measurements from both the working sensor and the reference sensor are repeated and recorded. As shown in Fig. 5.1, the readings from both sensors corresponding to i-th standard are recorded for the number of M_i instances.

Measurements from the working sensor are denoted as $x_{i,\ j}$ for i-th standard and j-th instance. Correspondingly, measurements from the reference sensor are denoted as $y_{bi,\ j}$. Subscript bi stands for i-th benchmark.

Dimension of the cooling tower model is 4 (i.e., n=4). According to Eq. (4-27) and (4-28), we have

$$R_{ux,\text{tower}} = \bar{B}^{\text{T}}_{\text{tower}}(\bar{A}^{\text{T}}_{\text{tower}} + I)^3 = \begin{bmatrix} \times & \times & \times & \times \\ \times & \times & \times & \times \\ \times & \times & \times & \times \\ \times & \times & \times & \times \\ \times & \times & \times & \times \end{bmatrix} \tag{4.40}$$

$$R_{xy,\text{tower}} = (\bar{A}^{\text{T}}_{\text{tower}} + I)^3 \bar{C}^{\text{T}}_{\text{tower}} = \begin{bmatrix} \times & \times & 0 & \times & 0 \\ \times & \times & 0 & \times & 0 \\ \times & \times & 0 & \times & 0 \\ 0 & 0 & 0 & \times & 0 \end{bmatrix} \tag{4.41}$$

The structure matrix in Eq. (4.40) signifies that all input variables can affect all state ones, and this manifests that the cooling tower model is input reachable. The structure matrix in Eq. (4.41) signifies that the three state variables, i.e., $\Delta t_{\text{a,L}}$, $\Delta W_{\text{a,L}}$, and $\Delta t_{\text{w,L}}$, can affect the exit water temperature of tower's sink $\left(\Delta t'_{\text{w,L}}\right)$, and this manifests that the cooling tower model is output reachable.

(2) Controllability/observability analysis

Known from above analysis, the structural model of cooling tower is input reachable, and meanwhile,

$$\text{gr}([\bar{A}_{\text{tower}}\ \bar{B}_{\text{tower}}]) = \text{gr}\left(\begin{bmatrix} \times & \times & \times & 0 & \times & \times & \times & \times & 0 \\ 0 & \times & \times & 0 & 0 & \times & \times & \times & 0 \\ \times & \times & \times & 0 & \times & \times & \times & \times & \times \\ 0 & 0 & \times & \times & 0 & 0 & 0 & 0 & \times \end{bmatrix}\right) = 4 \tag{4.42}$$

Therefore, the dynamic model of cooling tower is structurally controllable.

Since the generic rank of matrix $[\bar{A}_{\text{tower}}\ \bar{B}_{\text{tower}}]$ does not change after deleting any column corresponding to input variables, we can choose any one or more among the input variables (i.e.,$\Delta t_{\text{a,E}}$, $\Delta W_{\text{a,E}}$, $\Delta G_{\text{a,E}}$, $\Delta t_{\text{w,E}}$, and $\Delta G_{\text{w,E}}$) as the control variables. In real situations, the control variable is $\Delta G_{\text{w,E}}$ or $\Delta G_{\text{a,E}}$, and the other input variables (i.e., $\Delta t_{\text{a,E}}$, $\Delta W_{\text{a,E}}$, and $\Delta t_{\text{w,E}}$) are disturbance inputs.

Similarly,

$$\text{gr}\left(\begin{bmatrix} \bar{A}_{\text{tower}} \\ \bar{C}_{\text{tower}} \end{bmatrix}\right) = 4 \tag{4.43}$$

and the structural model of cooling tower is output controllable, thus the cooling tower model is structurally observable.

4.3.4.4 Air-Conditioned Room

(1) Input–output reachability analysis

The coefficient matrix of the state-space model of conditioned room is structured as below:

$$\bar{A}_{\text{room}} = \begin{bmatrix} \times & 0 & \times & \times & 0 & 0 & 0 & 0 & 0 & 0 \\ 0 & \times & 0 & 0 & 0 & 0 & 0 & 0 & 0 & 0 \\ \times & 0 & \times & 0 & 0 & 0 & 0 & 0 & 0 & 0 \\ \times & 0 & 0 & \times & 0 & \times & \times & \times & 0 & 0 \\ 0 & \times & 0 & 0 & \times & 0 & 0 & 0 & 0 & 0 \\ 0 & 0 & 0 & \times & 0 & \times & 0 & 0 & 0 & 0 \\ 0 & 0 & 0 & \times & 0 & 0 & \times & 0 & 0 & 0 \\ 0 & 0 & 0 & \times & 0 & 0 & 0 & \times & 0 & \times \\ 0 & 0 & 0 & 0 & \times & 0 & 0 & 0 & \times & 0 \\ 0 & 0 & 0 & 0 & 0 & 0 & 0 & \times & 0 & \times \end{bmatrix};$$

$$\bar{B}_{\text{room}} = \begin{bmatrix} \times & 0 & \times & 0 & 0 \\ 0 & \times & \times & 0 & 0 \\ 0 & 0 & 0 & 0 & 0 \\ 0 & 0 & \times & 0 & 0 \\ 0 & 0 & \times & 0 & 0 \\ 0 & 0 & 0 & 0 & 0 \\ 0 & 0 & 0 & \times & \times \\ 0 & 0 & \times & 0 & 0 \\ 0 & 0 & \times & 0 & 0 \\ 0 & 0 & 0 & 0 & 0 \end{bmatrix};$$

$$\bar{C}_{\text{room}} = \begin{bmatrix} \times & 0 & 0 & 0 & 0 & 0 & 0 & 0 & 0 & 0 \\ 0 & \times & 0 & 0 & 0 & 0 & 0 & 0 & 0 & 0 \\ 0 & 0 & 0 & \times & 0 & 0 & 0 & 0 & 0 & 0 \\ 0 & 0 & 0 & 0 & \times & 0 & 0 & 0 & 0 & 0 \\ 0 & 0 & 0 & 0 & 0 & 0 & 0 & \times & 0 & 0 \\ 0 & 0 & 0 & 0 & 0 & 0 & 0 & 0 & \times & 0 \end{bmatrix};$$

$$\bar{D}_{\text{room}} = \begin{bmatrix} 0 & 0 & 0 & 0 & 0 \\ 0 & 0 & 0 & 0 & 0 \\ 0 & 0 & 0 & 0 & 0 \\ 0 & 0 & 0 & 0 & 0 \\ 0 & 0 & 0 & 0 & 0 \\ 0 & 0 & 0 & 0 & 0 \end{bmatrix}.$$

Dimension of air-conditioned room model is 10 (i.e., $n = 10$). By using Eq. (4.27) and (4.28), we get

$$R_{\text{ux,room}} = \bar{B}_{\text{room}}^{\text{T}}(\bar{A}_{\text{room}}^{\text{T}} + I)^9 = \begin{bmatrix} \times & 0 & \times & \times & 0 & \times & \times & \times & 0 & \times \\ 0 & \times & 0 & 0 & \times & 0 & 0 & 0 & \times & 0 \\ \times & \times & \times & \times & \times & \times & \times & \times & \times & \times \\ \times & 0 & \times & \times & 0 & \times & \times & \times & 0 & \times \\ \times & 0 & \times & \times & 0 & \times & \times & \times & 0 & \times \end{bmatrix} \tag{4.44}$$

$$R_{\text{xy,room}} = (\bar{A}_{\text{room}}^{\text{T}} + I)^9 \bar{C}_{\text{room}}^{\text{T}} = \begin{bmatrix} \times & 0 & \times & 0 & \times & 0 \\ 0 & \times & 0 & \times & 0 & \times \\ \times & 0 & \times & 0 & \times & 0 \\ \times & 0 & \times & 0 & \times & 0 \\ 0 & 0 & 0 & \times & 0 & \times \\ \times & 0 & \times & 0 & \times & 0 \\ \times & 0 & \times & 0 & \times & 0 \\ \times & 0 & \times & 0 & \times & 0 \\ 0 & 0 & 0 & 0 & 0 & \times \\ \times & 0 & \times & 0 & \times & 0 \end{bmatrix} \tag{4.45}$$

From Eq. (4.44), we can easily judge that the air-conditioned room model is input reachable in which the input variable $\Delta G_{\text{a},i}$ can reach all state variables of room model; the input variables $\Delta G_{\text{a},i}$, $\Delta t_{\text{a,out}}$, and ΔI_{sol} can reach air temperature of all indoor zones; and the input variable $\Delta W_{\text{a},i}$ can reach air humidity of all indoor zones. The matrix in Eq. (4.45) manifests that the air-conditioned room model is output reachable, and it can clearly reflect the output reachability from state variables to output ones.

(2) Controllability/observability analysis

From above analysis, we know that the structural model of the conditioned room is input reachable, and meanwhile,

$$\text{gr}([\bar{A}_{\text{room}}\ \bar{B}_{\text{room}}]) = \text{gr}\left(\begin{bmatrix} \times & 0 & \times & \times & 0 & 0 & 0 & 0 & 0 & 0 & \times & 0 & \times & 0 & 0 \\ 0 & \times & 0 & 0 & 0 & 0 & 0 & 0 & 0 & 0 & 0 & \times & \times & 0 & 0 \\ \times & 0 & \times & 0 & 0 & 0 & 0 & 0 & 0 & 0 & 0 & 0 & 0 & 0 & 0 \\ \times & 0 & 0 & \times & 0 & \times & \times & \times & 0 & 0 & 0 & 0 & \times & 0 & 0 \\ 0 & \times & 0 & 0 & \times & 0 & 0 & 0 & 0 & 0 & 0 & 0 & \times & 0 & 0 \\ 0 & 0 & 0 & \times & 0 & \times & 0 & 0 & 0 & 0 & 0 & 0 & 0 & 0 & 0 \\ 0 & 0 & 0 & \times & 0 & 0 & \times & 0 & 0 & 0 & 0 & 0 & 0 & \times & \times \\ 0 & 0 & 0 & \times & 0 & 0 & 0 & \times & 0 & \times & 0 & 0 & \times & 0 & 0 \\ 0 & 0 & 0 & 0 & \times & 0 & 0 & 0 & \times & 0 & 0 & 0 & \times & 0 & 0 \\ 0 & 0 & 0 & 0 & 0 & 0 & 0 & \times & 0 & \times & 0 & 0 & 0 & 0 & 0 \end{bmatrix}\right) = 10 \tag{4.46}$$

Thus, the dynamic model of conditioned room is structurally controllable. According to the real conditions, we generally choose $[\Delta t_{\text{a},i}, \Delta W_{\text{a},i}]$, $[\Delta G_{\text{a},i}]$, or $[\Delta t_{\text{a},i}, \Delta W_{\text{a},i}, \Delta G_{\text{a},i}]$ $[\Delta t_{\text{a},i}, \Delta W_{\text{a},i}, \Delta G_{\text{a},i}]$ as control variables and $[\Delta Q_n, \Delta W_n, \Delta G_{\text{a},i}, \Delta t_{\text{out}}, \Delta I_{\text{sol}}]$ as disturbance inputs.

Similarly,

$$\mathrm{gr}\left(\begin{bmatrix} \bar{A}_{\mathrm{room}} \\ \bar{C}_{\mathrm{room}} \end{bmatrix}\right) = 10. \tag{4.47}$$

and the structural model of the conditioned room is output reachable; therefore, the dynamic model of the conditioned room is structurally observable.

References

1. Balakrishnan VK.: Graph theory. Shaum's Outline Series, New York (1997)
2. Chatrrand, G., Oellermann, O.: Applied and algorithmic graph theory. McGraw-Hill, New York (1993)
3. Heckmann, T., Schwanghart, W., Phillips, J.D.: Graph theory—recent developments of its application in geomorphology. Geomorphology **243**(1), 130–146 (2015)
4. Johnston, B.D.: Determination of the generic rank of structural matrices. Int. J. Control **40**(2), 257–264 (1984)
5. Chen, S., Han, Z., Hu, Q.: The control theory of linear system. East China Normal University Press, Shanghai (China) (2000)

Chapter 5
Virtual Measurement Modeling

5.1 Virtual Calibration

5.1.1 Conventional Calibration

Generally, a measurement is only an approximation of the 'true' value of a measurand. Because of the involved errors, i.e., systematic errors and random errors, such a measurement should always be accompanied by a statement of uncertainty [1]. Measurement quality, which defines our knowledge about factors that lead to the difference between a measurement and a measurand, can only be maintained with confidence through a calibration process. Conventionally, a calibration refers to '*the set of operations that establish, under specified conditions, the relationship between values of quantities indicated by a measuring instrument, a measuring system or values represented by a material measure, and the corresponding known values of a measurand*' [2]. Essentially, a calibration is a comparison and correction process. Standards are needed in the process as '*a material measure or physical property that defines or reproduces the unit of measurement of a base or derived quantity*' [1]. They refer to the known values of a measurand in the given scenarios. For different accuracy and applications, different levels of standards are used, including fundamental or absolute standards, international standards, national or primary standards, secondary standards, working standards, and alternative standards. Alternative standards can consist of a reference standard, a reference material, and a certified reference material.

Validity of measurements from calibrated sensors is reflected by the associated uncertainty [1, 2]. It characterizes the dispersion of measurements attributed to a measurand [1]. Generally, it is expressed as a standard deviation. According to the new international standard, two types of uncertainties, Type A and Type B, are defined. Type A evaluation of uncertainty is based on the statistical analysis of series of observations. Type B evaluation of uncertainty is by means other than the statistical analysis of series of observations [1, 3]. It is usually based on a pool of

Y. Yao and Y. Yu, *Modeling and Control in Air-conditioning Systems*,
Energy and Environment Research in China, DOI 10.1007/978-3-662-53313-0_5

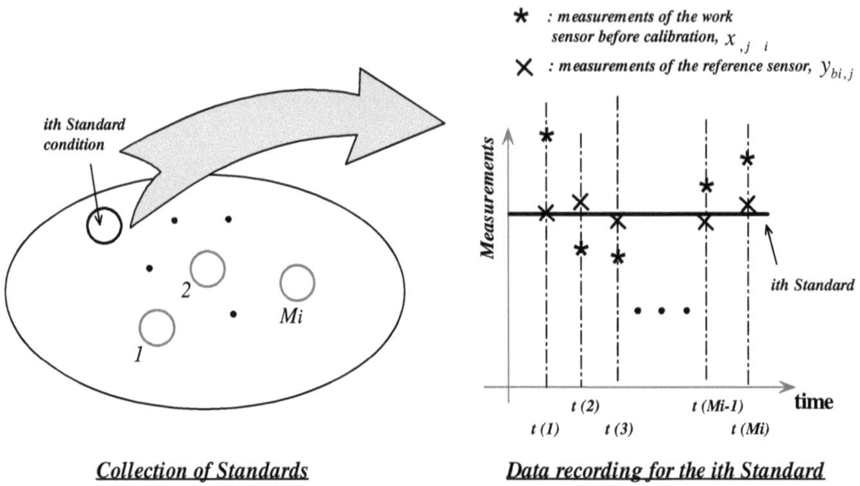

Fig. 5.1 Illustration of a conventional calibration

comparatively reliable information. Sometimes, a measurement does not come directly from one measured quantity; for example, an orifice flow rate meter gives the value of flow rate based on the measurement of fluid pressure differences, area and temperature, and correcting coefficients. Under this kind of scenario, a combined standard uncertainty should be used to attribute the propagation of measurement uncertainties. It is defined as the positive square root of a sum of terms, involving all the variances or covariance of these quantities.

In a conventional calibration, standards are generally predefined and maintained constant. The comparison between a working sensor and a known benchmark or reference sensor can be carried out either using the sensors' signals or the reading instrument. The process of a conventional calibration of a sensor based on a sole quantity is illustrated as in Fig. 5.1.

Both the working sensor and the reference sensor are placed in the uniform standards to produce measurements simultaneously. The number and value of selected standards should be adequate to cover the working range of the working sensor for an effective calibration. For illustration purposes, the number of standards is denoted as N and the values are indicated as i-th standard, from 1 to n, circled in the left figure. Along the horizontal axis of the right figure, $t(1)$ to $t(M_i)$ are the time stamps throughout the calibration process. For each standard, M_1 to M_n times of measurements from both the working sensor and the reference sensor are repeated and recorded. As shown in Fig. 5.1, the readings from both sensors corresponding to i-th standard are recorded for the number of M_i instances.

Measurements from the working sensor are denoted as $x_{i,\ j}$ for i-th standard and j-th instance. Correspondingly, measurements from the reference sensor are denoted as $y_{bi,\ j}$. Subscript bi stands for i-th benchmark.

To ease the presentation hereafter, we do not differentiate the reading signal from the working sensor and the reading from additional instrument on the working sensor. It means for any measurement x, it can be either of these two:

$$x = \left(x_{\text{signal}} \vee x_{\text{reading}}\right) \tag{5.1}$$

where x represents the instance of a measurement from the to-be-calibrated working sensor; subscript *signal* denotes the transmission signal from the working sensor and *reading* denotes the reading from an additional instrument upon the working sensor, respectively.

Define the multiple readings (measurements) from the working sensor recorded at different time stamps (instances) associated with a given i-th standard condition as a vector X_i:

$$X_i = \left(x_{i,1}, x_{i,2}, x_{i,3}, \ldots, x_{i,j}, \ldots, x_{i,M_i}\right)'_{1\times M_i} \tag{5.2}$$

where subscript i stands for i-th condition, and M_i for the total number of measurements recorded for the given condition. Entry $x_{i,j}$ represents a reading instance for i-th condition at time stamp j (j is a number from 1 to M_i).

In the meantime, the set of measurements from the reference sensor recorded at different time stamps under i-th standard condition is a vector Y_{bi}:

$$Y_{bi} = \left(y_{bi,1}, y_{bi,2}, y_{bi,3}, \ldots, y_{bi,j}, \ldots, y_{bi,M_i}\right)'_{1\times M_i} \tag{5.3}$$

where subscript bi denotes the benchmark value for i-th condition. Entry $y_{bi,j}$ represents an instance of the benchmark at different time stamp j (j is a number from 1 to M_i).

When known standards are directly used rather than the measurements from the reference sensor, the entries in Y_{bi} are replaced with a known standard value y_{si}. Therefore, the benchmark vector in Eq. (5.3) becomes:

$$Y_{bi} = \left(y_{si}, y_{si}, y_{si}, \ldots, y_{si}\right)'_{1\times M_i} \tag{5.4}$$

With the obtained measurements from the working sensor and the reference sensor, a comparison can be applied to correct the working sensor. The mathematic formulation of a conventional calibration is to find the function, f, which can minimize the distance between the two data series.

$$\min_f J = \left|\bar{Y}_b - f(\bar{X})\right| \tag{5.5}$$

In Eq. (5.5), $\bar{Y}_b = \left(\bar{Y}_{b1}, \bar{Y}_{b2}, \bar{Y}_{b3}, \ldots, \bar{Y}_{bn}\right)'_{1\times n}$ is the processed standard or benchmark vector, including n number of standards $\left(\bar{Y}_{bi}\right)$, which is the average value of readings from the reference sensor in calibration conditions. It can be written as:

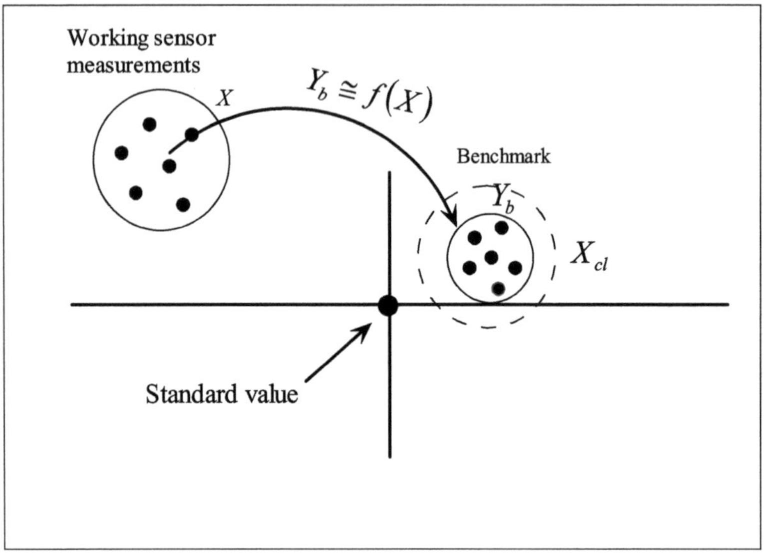

Fig. 5.2 Sensor calibration based on comparison method

$$\bar{Y}_{bi} = \frac{\sum_{j=1}^{M_i} y_{bi,j}}{M_i} \tag{5.6}$$

Similarly, $\bar{X} = (\bar{X}_1, \bar{X}_2, \bar{X}_3, \ldots, \bar{X}_n)'_{1\times n}$ is the processed measurement vector from the working sensor, including n number of average values ($\bar{X}_i$) from the working sensor in the n number of calibration conditions; f is the correction function to be identified in the calibration process.

As shown in Fig. 5.2, a function f, which is identified after the calibration, should map the original measurements from the working sensor to a better approximation of the benchmark. The benchmark is very close, but can be barely equal, to the standard value of the measurement.

Correspondingly, the new reading from the working sensor is corrected by function f as:

$$x_{\mathrm{cl}} = f(x) \tag{5.7}$$

where subscript *cl* stands for *calibrated*.

5.1.2 Methodology of Virtual In Situ Calibration

Reliable sensing serves as the base of understanding, monitoring, and control of building energy systems. However, as stated in Chap. 1 of this book, the conventional calibration method has practical problems with sensors in building systems: (1) time and monetary cost for reinstalling sensors; (2) disruption to a normal operation; (3) difficulty in accessing various sensors in pipelines, hidden spaces, etc.; and (4) large quantity of sensors in the system. Meanwhile, a conventional calibration cannot eliminate the potential systematic error due to the change of working environment, which is so hard to identify. A manufacture calibrated sensor does not necessary mean it can provide reliable measurement when it is placed in the working environment. Thus, an alternative calibration method is needed that can be implemented continuously and broadly to reduce the uncertainty and improve the confidence of measurements. Despite the findings in the related work, there is no study that systematically establishes a generic methodology and addresses the required conditions for conducting a virtual in situ calibration.

The methodology we propose here is built upon but significantly different from a conventional calibration. The framework is obtained by mathematically extracting the characteristics of essential aspects involved in a conventional calibration, including mainly the calibration environment assessment, benchmark establishment, and uncertainty quantification. Unlike a conventional calibration, where the standard or benchmark is controlled and readily available, the measurand as the standard used in the virtual in situ calibration needs to be approximated by the benchmark value that is either statistically established or mathematically modeled. Correspondingly, the two approaches for establishing the needed standards are termed as statistical-based method and modeling-based method, which are detailed in the following sections.

The procedure in both methods includes four essential components:

1. Evaluating the calibration environment;
2. Establishing benchmarks;
3. Fulfilling the calibration with benchmarks, and
4. Assessing the overall uncertainty.

In the following sections, the mathematic framework of these four components is introduced.

5.1.2.1 Environment Evaluation

A calibration is in fact a comparison and correction process, either against the known standard value or the benchmark value measured by reference sensors. In a conventional calibration, the calibration environments are controlled to be acceptable in terms of uniformity and stability for conducting such a comparison. However, it is different in a virtual calibration process. With an in situ calibration, the environment is by nature the working environment of the target sensor, including the dynamics and continuous changes. It could be an advantage and a disadvantage. The advantage is that any systematic errors that will be imposed on the working sensor from the environment are exposed and can be handled during the live calibration process. The disadvantage is that an additional and continuous environment assessment is needed to establish reliable benchmarks. In some circumstances, it could be long to acquire all the needed benchmarks to represent the working range of the sensor.

Evaluation of the environment should be performed accordingly:

1. With the statistical-based approach, the environment is preferred to be uniform and steady, satisfying a set of predefined criteria for the sensors as in a conventional calibration;
2. With the modeling-based approach, the environment should be steady with only slow or no change as in a conventional calibration.

A uniform environment may be defined mathematically as:

$$\left| x_{si,j} - \frac{\sum_{si=1}^{p} x_{si,j}}{p} \right| \leq Th_{\text{uniform}} \tag{5.8}$$

where x is the estimate (reading) of the measurand from the multiple sensors observing the same phenomenon. Subscript si denotes i-th sensor, and j represents the time stamp (instance). p is the size of the measurements. Th is the defined threshold, and subscript uniform denotes the uniform condition.

Assuming, when p is large, p-number of readings $X_j = \left(x_{s1}, x_{s2}, \ldots, x_{sp}\right)_j$ from sensor 1 to p obey a normal distribution with $X_j \sim N(\mu, \sigma^2)$ because of the Central Limit Theorem [1, 4]; then, with limited p samples, t as:

$$t = \frac{\bar{X}_j - \mu}{S/\sqrt{p}} \tag{5.9}$$

gives a student T-distribution with $p - 1$ degree of freedom, where μ is the expected average of the population, $\bar{X}_j$ is the mean value of the readings from the sensors, S is the standard deviation of the readings.

With a confidence interval as $(1 - \alpha)$, the expected average should fall in the range between $\left[\bar{X}_j - t_{\alpha/2}\left(S/\sqrt{p}\right), \bar{X}_j + t_{\alpha/2}\left(S/\sqrt{p}\right)\right]$, where $t_{\alpha/2}$ is the corresponding critical value of T-distribution.

The uniform condition threshold can be given as:

$$Th_{\text{uniform}} = t_{\alpha/2} \cdot \frac{S}{\sqrt{p}} \tag{5.10}$$

In addition to a uniform condition, the environment should also be steady for at least a period of time in order to establish reliable benchmarks during the calibration. It can be quantified based on the standard deviations of the measurements from the involved sensors in a moving window.

$$\left| x_{si,j} - \frac{\sum_{j=k}^{k+m-1} x_{si,j}}{m} \right| \leq Th_{\text{steady}} \tag{5.11}$$

where m is the number of time stamps from k to $k+m-1$. Th is the defined threshold and subscript *steady* denotes the steady condition.

Similar to the uniform environment, the m readings $X_{si} = (x_k, x_{k+1}, \ldots, x_{k+m-1})_{si}$ within the giving time window for each involved sensor i are assumed to obey normal distribution with $X_{si} \sim N(\mu, \sigma^2)$ when m is large; then:

$$t = \frac{\bar{X}_{si} - \mu}{S/\sqrt{m}} \tag{5.12}$$

gives a student T-distribution with $m-1$ degree of freedom for the m samples. The expected average of the population should fall in $\left[\bar{X}_{si} - t_{\alpha/2}(S/\sqrt{m}), \bar{X}_{si} + t_{\alpha/2}(S/\sqrt{m})\right]$, where $t_{\alpha/2}$ is the corresponding critical value of T-distribution.

The stability condition threshold (Th_{steady}) can be given as:

$$Th_{\text{steady}} = t_{\alpha/2} \cdot \frac{S}{\sqrt{m}} \tag{5.13}$$

In the previous criteria for a uniform and steady condition, the confidence level should be justifiable based on the real situation and applications. Asking for a high confidence such as 99 % could be extremely difficult and consequently rule out the conditions where a calibration can actually be conducted to improve the measurement accuracy.

5.1.2.2 Establishing Benchmarks

1. Statistical-based method

With a statistical method, it is assumed that a redundancy of sensors exists in the system for observing the same phenomenon or measuring the same measurand. The benchmarks are established based on the assumption that, after removing the outliers, sensors' measurements at a given time stamp obey a normal distribution $X_j \sim N(\mu, \sigma^2)$ converging around the measurand. The mean of samples can be used to estimate the mean of the population. As the sample size m increases, the sample mean is likely to be closer to the mean of the population, since the variance of sample mean has a relationship with the variance of population:

$$\sigma_{\bar{X}}^2 = \frac{m \cdot \sigma^2}{m-1} \tag{5.14}$$

where $\sigma_{\bar{X}}^2$ is the variance of the samples, σ^2 is the variance of the population, and m is the sample size.

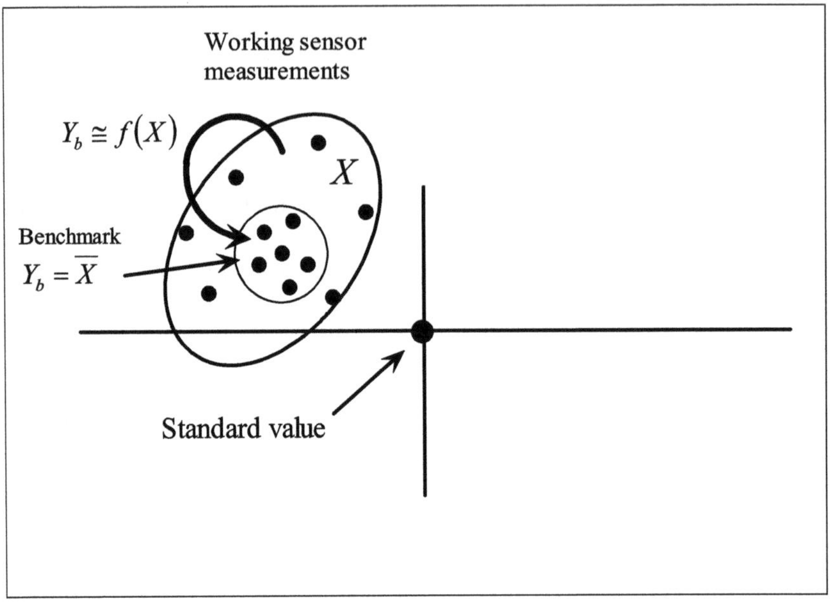

Fig. 5.3 Virtual in situ calibration with statistical-based method

The statistical virtual in situ calibration method can be illustrated as in Fig. 5.3.

Taking X_{bi} as the measurements from the working sensors for conducting the statistical-based calibration regarding benchmark value Y_{bi}, it can be generalized as:

$$X_{bi} = \begin{bmatrix} x_{s1,k} & \cdot & x_{s1,k+j} & \cdot & x_{s1,k+m-1} \\ \cdot & \cdot & \cdot & \cdot & \cdot \\ x_{si,k} & \cdot & x_{si,k+j} & \cdot & x_{si,k+m-1} \\ \cdot & \cdot & \cdot & \cdot & \cdot \\ x_{sp,k} & \cdot & x_{sp,k+j} & \cdot & x_{sp,k+m-1} \end{bmatrix}_{p \times m} \tag{5.15}$$

where the first subscript of the entries x in the matrix indicates the sequence of the working sensors, from 1 to p, and the second subscript denotes time stamps (sets) from k to $k+m-1$.

If p sensors from the population measure the same standard value at a time stamp (instance) j to establish the benchmark Y_{bi}, the benchmark can be estimated as:

$$Y_{bi,j} = \frac{\sum_{si=1}^{p} x_{si,j}}{p} \tag{5.16}$$

If the process is repeatable from time stamp k to $k+m-1$ for the same fixed standard, the benchmark mean value $\bar{Y}_{bi}$ is established as:

$$\bar{Y}_{bi} = \frac{\sum_{si=1}^{p} \sum_{j=k}^{k+m-1} x_{si,j}}{p \times m} \tag{5.17}$$

In general, with m instances corresponding to a benchmark, the variance of the samples is given as:

$$\sigma_{\bar{x}}^2 = \frac{1}{m-1} \sum_{h=1}^{m} (x_h - Y_{bi})^2 \tag{5.18}$$

where x_h is a reading from the selected group of sensors for benchmark Y_{bi}, and h is a counter.

It should be noticed that if the subset of measurements from the working sensors at a given time stamp has a big deviation from the mean, the subset should be trimmed off before calculating the benchmark. It is possible that such a benchmark could not be established if the overall deviation from the sensors is not acceptable. Under this circumstance, either the environment should be re-evaluated or the method should be changed.

2. Modeling-based method

Under the circumstance where the redundancy of sensors in a system for observing the same phenomenon is not available, it will not be possible to directly conduct a statistical-based virtual in situ calibration. The second method termed as modeling-based calibration can be used.

The method is illustrated as in Fig. 5.4. With the modeling-based calibration method, the benchmark is not obtained from the parallel working sensors, but from some mathematical models constructed upon relative variables as $(v_1, v_2, \ldots, v_p)$ in the system. Based on our previous studies on packaged air-conditioning units [5], the calibration function f in a modeling-based calibration is more general than the one used in a statistical-based virtual calibration method. Variables, including measurements from physical sensors, control commands, and status information, might be included to enhance the accuracy and robustness of the virtual calibration. Variables $v_1, v_2, \ldots, v_p$ used for establishing the benchmarks might not be the same as the ones in the calibration function of the working sensor. They could be replaced by other variables $z_1, z_2, \ldots, z_p$, which are easier to be accurately measured or obtained.

As shown in Fig. 5.4, assuming the involved relative variables for obtaining the benchmark Y_{bi} are $v_1, v_2, \ldots, v_p$, measurements of the corresponding sensors in a steady environment with m time stamps (instances) can be written as a matrix V_{bi}:

$$V_{bi} = \begin{bmatrix} v_{1,k} & \cdot & v_{1,k+j} & \cdot & v_{1,k+m-1} \\ \cdot & \cdot & \cdot & \cdot & \cdot \\ v_{i,k} & \cdot & v_{i,k+j} & \cdot & v_{i,k+m-1} \\ \cdot & \cdot & \cdot & \cdot & \cdot \\ v_{p,k} & \cdot & v_{p,k+j} & \cdot & v_{p,k+m-1} \end{bmatrix}_{p \times m} \tag{5.19}$$

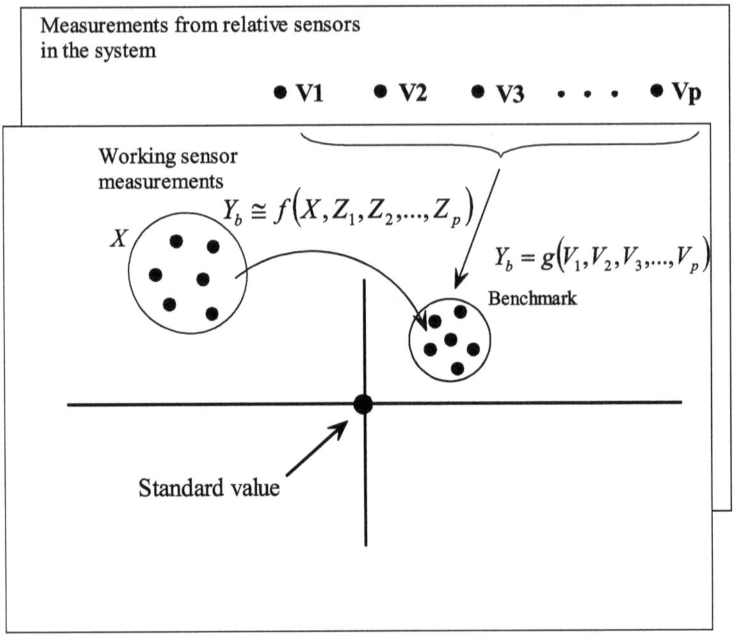

Fig. 5.4 Virtual in situ calibration with modeling-based method

where v is the reliable measurement from the involved individual sensor for constructing the benchmark, the first subscript indicates the p variables, and the second subscript denotes time stamps (instances) from k to $k+m-1$.

The benchmark for i-th standard can be modeled with the measurements from the involved sensors at j-th time stamp:

$$Y_{bi} = g\left(\mathrm{v}_{1,j}, \mathrm{v}_{2,j}, \ldots, \mathrm{v}_{p,j}\right) \tag{5.20}$$

If the process is repeated from time stamp k to $k+m-1$ for the same fixed standard, the benchmark mean value $\bar{Y}_{bi}$ is established as:

$$\bar{Y}_{bi} = \frac{\sum_{i=1}^{p} \sum_{j=k}^{k+m-1} g\left(\mathrm{v}_{i,j}\right)}{p \times m} \tag{5.21}$$

5.1.2.3 Calibration

A full calibration can be conducted when the benchmarks in the whole working range of the sensor are collected. The established benchmarks, using either the statistical-based method or modeling-based method, can be written as a vector:

$$\bar{Y}_b = \begin{pmatrix} \bar{Y}_{b1} \\ \cdot \\ \bar{Y}_{bi} \\ \cdot \\ \bar{Y}_{bn} \end{pmatrix} \tag{5.22}$$

$\bar{Y}_{b1}$ to $\bar{Y}_{bn}$ in Eq. (5.22) are selected correspondingly as the upper and lower limits, respectively, to represent the sensors' working range. Due to the changing external conditions, the collection of benchmarks might across a certain time period for the desired conditions. For N number of benchmarks, the overall matrix of sensor readings is:

$$X = \begin{pmatrix} X_{b1} \\ \cdot \\ X_{bi} \\ \cdot \\ X_{bn} \end{pmatrix} \text{ or } V = \begin{pmatrix} V_{b1} \\ \cdot \\ V_{bi} \\ \cdot \\ V_{bn} \end{pmatrix} \tag{5.23}$$

Each of the components in the X matrix is a matrix given by Eq. (5.15) and each component in the V matrix is given by Eq. (5.19).

The distance from the working sensor to the benchmarks should be evaluated with Eq. (5.5) as in a conventional calibration for finding the mapping function f. After the in situ calibration, a sensor's reading can be expressed as Eq. (5.7) for a statistical-based method. Or, it can be written as:

$$X_{\text{cl}} = f(X, Z_1, Z_2, \ldots, Z_p) \tag{5.24}$$

for a modeling-based method, where $Z_1, Z_2, \ldots, Z_p$ are the selected variables in the virtual in situ calibration model that provides the insight of the driving forces leading to the measurement errors.

5.1.2.4 Uncertainty Analysis

A critical process of the virtual in situ calibration is to evaluate the combined total uncertainty [1]. Once a calibration is finished, this combined total uncertainty becomes the uncertainty of the calibrated working sensor.

The combined total uncertainty can be influenced by factors in the aforementioned processes, including: ① uncertainty of the model; ② uncertainty of the benchmarks; ③ uniformity and stability of the environment.

1. Uncertainty of the model

   The combined standard uncertainty for uncorrelated input quantities in any calibration model f is given as:

$$e_f^2(y) = \sum_{i=1}^{n} \left(\frac{\partial f}{\partial x_i}\right)^2 e^2(x_i) \tag{5.25}$$

where e_f is the combined uncertainty of the model, y is the dependent measurement, f is the mapping function for y, x is the uncorrelated input quantity, and e is the associated standard uncertainty. n and subscript i are counters.

When the input quantities are correlated, the appropriate expression for the combined variance associated with the result of a measurement is:

$$\begin{aligned} e_f^2(Q) &= \sum_{i=1}^{n}\sum_{j=1}^{n} \frac{\partial f}{\partial x_i}\frac{\partial f}{\partial x_j} e(x_i, x_j) \\ &= \sum_{i=1}^{n} \left(\frac{\partial f}{\partial x_i}\right)^2 e^2(x_i) + 2\sum_{i=1}^{n-1}\sum_{j=i+1}^{n} \frac{\partial f}{\partial x_i}\frac{\partial f}{\partial x_j} e(x_i, x_j) \end{aligned} \tag{5.26}$$

where subscript i and j are counters.

2. Uncertainty of benchmarks

   The uncertainty of individual benchmark Y_{bi} is calculated based on the sample variance. The variance of the observations' average, rather than the variance of the samples, is the proper measure of the uncertainty of a measurement result [1]:

$$\sigma^2(\bar{X}) = \frac{1}{m \cdot (m-1)} \sum_{h=1}^{m} (x_h - Y_{bi})^2 \tag{5.27}$$

   where the subscript h is the counter of the samples, and Y_{bi} is the established benchmark.

3. Uncertainty of the calibration environment

   The uncertainty associated with the environment uniformity and stability is given as Eqs. (5.10) and (5.13), respectively. Eventually, the combined total uncertainty with the proposed virtual in situ calibration method for any specific benchmark Y_{bi} is:

$$e_{\mathrm{cl},i} = \sqrt{e_f^2 + \sigma(\bar{X})^2 + Th_{\mathrm{uniform}}^2 + Th_{\mathrm{steady}}^2} \tag{5.28}$$

5.1.3 Case Study

We conducted a modeling-based in situ virtual calibration to the supply air temperature sensor in packaged rooftop units (RTUs) with the proposed methodology. RTUs have generally four sensors installed in the air loop, including the outdoor air temperature, return air temperature, supply air temperature, and zone air temperature. As a key component, the supply air temperature sensor is directly included in the control logic for energy conservation. It is also utilized as an important input to some unmeasured variables, such as the supply air fan temperature rise, mixed air temperature, supply air humidity ratio, and cooling capacity, to support fault detection and diagnosis of RTUs. Therefore, the accuracy of the supply air temperature sensor has a great impact on the operation performance of RTUs and the environment they serve. However, through a series of tests, it was found that the supply air temperature sensor, which is typically mounted after the gas-fired heating coil, can barely provide reliable readings of the supply air temperature when the unit is operated in a heating mode. The error, varying from 11.7 to 19.2 °C, is systematic and associated with the working environment. Because of the feature, the error cannot be eliminated with a conventional calibration. In the meantime, due to a cost consideration, there is normally no redundancy of the air-side sensors in RTUs. A virtual in situ calibration with modeling method serves as a great option for correcting the measurements. The essential process and the results are presented below.

For the modeling-based virtual calibration of this supply air temperature sensor, the relative variables v indicated in Eq. (5.19) are the heating capacity Q_{H}, specific heat c_{p}, measured supply airflow rate $G_{\mathrm{a,sup,meas}}$, specific volume of air v_{a}, mixed air temperature MAT, and supply fan temperature rise ΔT_{fan}. Benchmarks of the supply air temperature were established with the following model:

$$\begin{aligned}\mathrm{SAT_b} &= g\left(Q_{\mathrm{H}}, c_{\mathrm{p}}, G_{\mathrm{a,sup,meas}}, v_{\mathrm{a}}, \mathrm{MAT}, \Delta T_{\mathrm{fan}}\right)\\ &= \frac{Q_{\mathrm{H}}}{c_{\mathrm{p}} \times G_{\mathrm{a,sup,meas}}} v_{\mathrm{a}} + \mathrm{MAT} + \Delta T_{\mathrm{fan}}\end{aligned} \tag{5.29}$$

Among the variables, c_{p} and v_{a}, as the properties of air, are regarded constant in the sensor working range and do not impact the uncertainty of the supply air sensor. To cover the full working range of the supply air sensor in heating mode, we varied the supply air temperature from 27 to 44.6 °C [as $\bar{Y}_{b1}$ and $\bar{Y}_{bn}$ in Eq. (5.22)]. Different outside air temperature, heating stage, and outside air damper position were applied to emulate the reality. There are totally 16 benchmarks established in the range.

For the modeling-based calibration, a steady-state condition should be applied in order to alleviate impacts from the dynamics and obtain reliable data for establishing the benchmarks. A threshold of 0.2 °C [as Th_{steady} in Eq. (5.11)] was adopted to detect the steady-state condition. When the system reached the required steady-state condition, data were collected at the sampling intervals to establish the benchmarks, three times a minute. The sampling size for each benchmark is around 30 times within 10 min. The summarized readings are given in Table 5.1, where $\mathrm{SAT_b}$ is the benchmark obtained with Eq. (5.29) and SAT is the direct measurement from the supply air temperature sensor.

Table 5.1 Modeling-based calibration of the supply air temperature sensor

Benchmark #	$G_{a,\ sup,\ meas}$ (m^3/s)	MAT (°C)	Q_H (kW)	ΔT_{fan} (°C)	SAT_b (°C)	SAT (°C)
H-1	0.87	12.3	38.1	0.9	42.2	48.9
H-2	0.97	15.3	38.1	0.9	42.4	48.9
H-3	0.88	13.0	38.1	0.9	42.7	49.7
H-4	0.86	14.1	38.1	0.9	44.3	51.1
H-5	0.98	7.2	38.1	0.9	33.9	37.7
H-6	1.07	11.4	38.1	0.9	35.9	40.3
H-7	0.96	10.1	38.1	0.9	37.3	41.2
H-8	0.96	12.1	38.1	0.9	39.3	43.3
H-9	0.87	15.2	24.8	0.9	34.8	37.4
H-10	0.97	16.1	24.8	0.9	34.0	36.7
H-11	0.87	16.1	24.8	0.9	35.8	38.2
H-12	0.86	17.6	24.8	0.9	37.6	39.2
H-13	0.98	9.2	24.8	0.9	26.9	28.1
H-14	1.07	11.6	24.8	0.9	27.8	28.4
H-15	0.97	11.9	24.8	0.9	29.7	38.3
H-16	0.97	13.6	24.8	0.9	31.5	32.7

To offset the systematic error associated with the working environment where the sensor is placed, the inaccurate supply air temperature sensor was calibrated as:

$$\mathrm{SAT_c} = f\left(\mathrm{SAT_{meas}}, H_{\mathrm{stage}}, \mathrm{OAD_{st}}\right) \tag{5.30}$$

where $\mathrm{SAT_c}$ is the supply air temperature after the calibration, f is the function for correcting the original readings from the physical sensor, $\mathrm{SAT_{meas}}$ is the reading from the physical sensor, H_{stage} is the heating stage command (1 or 2), $\mathrm{OAD_{st}}$ is the outdoor air damper setpoint (0 or 30 %). The chosen independent variables are corresponding to the calibration variables in Eq. (5.24). Compared to the independent variables in Eq. (5.29), variables in Eq. (5.30) are much easier to obtain.

The final calibration function of the supply air temperature sensor was obtained as:

$$\mathrm{SAT_c} = \mathrm{SAT_{meas}} - \begin{bmatrix} 0.6056 + 0.3836 \times H_{\mathrm{stage}} + 1.3444 \times H_{\mathrm{stage}}^2 + \\ 0.3047 \times \mathrm{OAD_{st}} + 0.0648 \times \mathrm{OAD_{st}^2} \\ -4.7222 \times H_{\mathrm{stage}} \times \mathrm{OAD_{st}} \end{bmatrix} \tag{5.31}$$

Uncertainties of the other four variables in the calibration function g as in Eq. (5.29) were evaluated to obtain the combined standard uncertainty e_f of the supply air sensor with Eq. (5.28).

As a calculated variable, the heating capacity Q_H has a combined uncertainty from the independent variables of air-side flow and gas-heated-side flow.

Table 5.2 Uncertainty analysis of SAT

Variables		Uncertainty
Dependent ones	Gas heating capacity, Q_H	±2 %
	Measured supply airflow rate, $G_{a,\ sup,\ meas}$	±1 %
	MAT	±0.6 °C
	Supply fan temperature rise, ΔT_{fan}	±0.1 °C
	Environment steady state, Th_{steady}	±0.2 °C
Calibrated SAT	42.2 °C	±0.7 °C
	42.4 °C	±0.6 °C
	42.7 °C	±0.7 °C
	44.3 °C	±0.7 °C
	33.9 °C	±0.6 °C
	35.9 °C	±0.6 °C
	37.3 °C	±0.6 °C
	39.3 °C	±0.6 °C
	34.8 °C	±0.6 °C
	34.0 °C	±0.6 °C
	35.8 °C	±0.6 °C
	37.6 °C	±0.6 °C
	26.9 °C	±0.6 °C
	27.8 °C	±0.6 °C
	29.7 °C	±0.6 °C
	31.5 °C	±0.6 °C

By applying Eq. (5.25), it was found this variable has an uncertainty about ±2 % across the working range. The full deduction can be found in the appendix of [5]. The uncertainties of the dependent variables and the final results of the total uncertainty are listed in Table 5.2. The final results showed that, after the calibration, the erratic errors were significantly alleviated and the accuracy of the physical supply air temperature sensor was improved from up to 19.2 to 0.7 °C.

5.2 Virtual Sensing

The availability of a rich set of high value sensor information would enable a level of building optimization and improvement not previously possible. Virtual sensors for capacity and power consumption at the device level would allow real-time monitoring of device, sub-system, and whole-building efficiencies. This information could be used along with other virtual sensors to enable diagnosis, tracking, and economic impact evaluation of specific faults. The virtual sensor outputs and embedded models could also be employed within a control optimizer that determines set points that minimize operating costs at each operating condition. If all devices incorporated embedded intelligence, then there would undoubtedly be a significant amount of redundant information that could used to diagnose sensor faults (e.g., faults within devices providing virtual sensor outputs). Compared with the physical sensors, virtual sensors have several advantages in addition to lower cost. For example, virtual sensors could be more easily added as retrofits in a number of important applications, such as measurement of refrigerant flow rate or pressure. A physical sensor for refrigerant mass flow would require opening up the system, recovering the refrigerant, installing the sensors, and then recharging the system. The installation of refrigerant pressure sensors would require access to threaded service ports on the equipment, which can cause refrigerant leakage over time [6]. The same refrigerant-side sensor information could be obtained using noninvasive surface-mounted temperature sensors along with models. In some occasions, it is very difficult to install physical sensors that can accurately measure a desired quantity. For example, it is very difficult to obtain accurate mixed air temperature measurements at the inlet of cooling coils or evaporators because the compactness of the mixing chamber creates highly non-uniform flow and temperature characteristics. However, an accurate effective mixed air temperature is readily obtained using a model and measurements obtained at other more uniform locations (e.g., coil outlet and/or return and ventilation air streams). Virtual sensors have been developed in other fields to obtain measurements indirectly in a cost-effective, noninvasive or/and practical manner, but are only recently the subject of development for building systems. There is no widely accepted definition of virtual sensing. In the context of this section, virtual sensing is considered to include any indirect method of determining a measureable quantity that utilizes outputs from other physical and/or virtual sensors along with process models and/or property relations.

5.2.1 Development Methodology for Virtual Sensing

Virtual sensors are the embodiment of virtual sensing techniques. For the sake of simplicity, the term virtual sensor is used interchangeably with virtual sensing to present development methodology. Although many different types of virtual sensors have been developed, there is no widely accepted definition and no systematic virtual sensing development methodology. It is meaningful to categorize virtual sensors before attempting to describe general approaches for their development.

5.2.1.1 Categorization of Virtual Sensors

Virtual sensors can be categorized, as shown in Fig. 5.5, according to three interrelated criteria that affect development approaches: ① measurement characteristics, ② modeling methods, and ③ application.

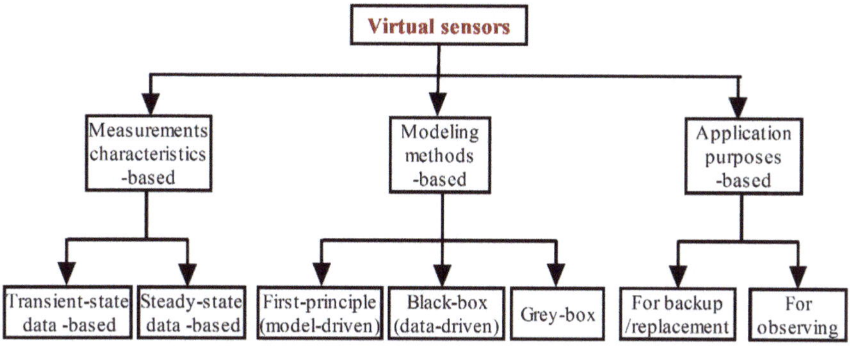

Fig. 5.5 Categorization for virtual sensors

The measurement characteristic category refers to whether the desired virtual sensor outputs are transient or steady-state variables. A transient virtual sensor incorporates a transient model to predict the transient behavior of an unmeasured variable in response to measured transient inputs. This type of sensor would be necessary for feedback control or if transient information was useful in fault identification. For performance monitoring or fault identification, it is often adequate and/or desirable to assume that the modeled process is quasi-static. In this case, steady-state models (i.e., with no dynamic terms) are utilized. This modeling is appropriate when the measured input quantities change slowly or the modeled process responds very 'quickly' to changes in inputs. Many processes in the food and biochemistry industry, such as fermentation processes, utilize transient-state virtual sensors [7]. Transient-state virtual sensors are also very common in the specialty chemistry field [8], while steady-state virtual sensors represent the majority of the applications in different fields [9–16].

With respect to modeling methods, virtual sensors can be divided into three types: first-principle (model-driven), black-box (data-driven), and gray-box virtual sensors. First-principle (physical or white-box) virtual sensors are most commonly derived from fundamental physical laws and have parameters with some physical significance. For example, DeWolf et al. [17] developed a virtual slurry polymerization reactor sensor based on a Kalman filter and Prasad et al. [18] applied a multi-rate Kalman filter to the control of a polymerization process. For the same application, Doyle [19] utilized a nonlinear observer method. In contrast to first-principle virtual sensors, black-box (data-driven) approaches utilize empirical correlations without any knowledge of the physical process. Examples include multivariate principle component analysis [20, 21], partial least squares [22, 23], artificial neural networks [24, 25], and so on. A gray-box virtual sensor utilizes a combination of physical and empirical models in estimating the output of an unmeasured process [10, 13, 16].

According to application, virtual sensors can be divided into backup/replacement and observing virtual sensors. Backup/replacement virtual sensors are used either to back up or replace existing physical sensors. A backup virtual sensor can provide a check on the accuracy of an installed sensor and even enable virtual calibration. For example, the reliability of temperature sensors is affected by incorrect installation, hostile environmental conditions, or natural drift [26]. A replacement application is dictated by cost and reliability considerations, such as the virtual tire–air pressure sensor studied for automobiles [27–29]. Either backup/replacement or observer sensors can be used for a variety of end-use applications, including performance monitoring, control, and fault detection and diagnostics (FDD). FDD virtual sensors allow flagging of a fault and 'measurement' of the fault severity. For instance, a virtual airflow sensor could be a cost-effective replacement for a physical sensor and be used to directly identify a fan or heat exchanger fouling fault. Alternatively, diagnostic fault indicators could be energy rates or other variables that have no physical sensor counterpart. Since the early 1990s, a number of FDD virtual sensors have been studied, such as FDD virtual sensors for a turbo generator [30] and an ethylene cracking process [7].

5.2.1.2 General Steps in Developing Virtual Sensors

In general, the process can be summarized as (i) data collection and preprocessing, (ii) model selection and training, and (iii) sensor implementation and validation, which is illustrated in Fig. 5.6.

Proper data collection and preprocessing (prefiltering or data outliers) are fundamentals in the development of accurate and reliable virtual sensor models. The type and range of test data depends on the modeling approach. Transient sensors require transient test data, whereas transient data should be filtered for steady-state modeling approaches. A 'steady-state detector' may be used as a preprocessor to eliminate transient data [31, 32]. For black-box models, principle component analysis is a popular approach for preprocessing data in order to aid in the model selection [33–35].

Model selection and training are the most difficult and critical steps in the process of developing a virtual sensor. There are many model types to choose from and each requires a process of determining the proper model order, estimating parameters, and then redefining the model selection/order. The previous section

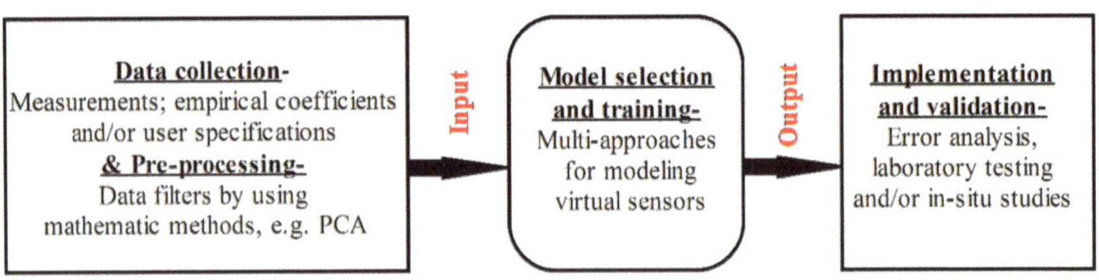

Fig. 5.6 General steps in developing virtual sensors

provided an overview of possible modeling approaches. However, there is a bit of an art involved in identifying an appropriate model.

A virtual sensor could be implemented as part of a control or monitoring system or as a standalone sensor with its own hardware, embedded software, and input/output channels. In either case, the virtual sensor implementation needs to be tested in both laboratory and in situ studies to validate performance and evaluate robustness (e.g., error analysis). Statistical approaches (e.g., student's t test [36]) can be used to validate accuracy. It is important to assess the performance using independent data [37].

5.2.2 *Case Study*

A proper amount of supply airflow, as a key component of the control system for light commercial RTUs (roof-top units), is critical to maintain desired control effectiveness, energy efficiency, and indoor air quality. For example, ASHRAE Standard 62.1-2007 specifies ventilation and circulation airflow rate requirements according to occupancy and floor area. Improper supply airflow rate (SCFM) could directly impair temperature distribution and cause poor indoor air quality [38]. Typically, low SCFM across the RTUs could lead to intensive heating cycling and energy losses. On the other hand, SCFM measurement plays a remarkable role in automated FDD technology for performance monitoring, diagnostics, and optimization of RTUs. It is typically used to detect RTU air-side faults such as the dirty indoor filter and slipping supply fan belt, which lead to decreased SCFM in RTUs. Systematically, it is used as an input in determining the actual heating capacity, which can serve for system monitoring and observing. It also could be compared with heating datasheet to predict heating coil efficiency for performance fault diagnostics (e.g., an indication of fault levels) and impact evaluation [39, 40].

Despite the importance of SCFM measurement, over the last decades, a number of studies have focused on finding good solutions for measuring SCFM with physical airflow meters [41–47]. The single largest focus of techniques is based on air dynamic pressure measurements using a pitot traverse or on air velocity by vane anemometer. However, in general, using a physical airflow meter absolutely exposes a series of weaknesses. For instance, credibility of SCFM measurement would be compromised dramatically after long-term use in adverse duct work surroundings. Periodical calibration is required but rarely followed practically. In the meantime, physical airflow meters are very costly in terms of procurement and installation, ranging from hundreds to thousands of dollars. Many more expenses emerge along with maintenance (e.g., repair or rebuild). Besides, particularly, installing physical airflow meters in RTUs is even more unrealistic. RTUs usually have a compact structure and short duct work. Such efficient configuration leaves barely any space for a physical meter, let alone measures the true value of SCFM. Moreover, the relative price of physical airflow meters over RTUs is high. A decent

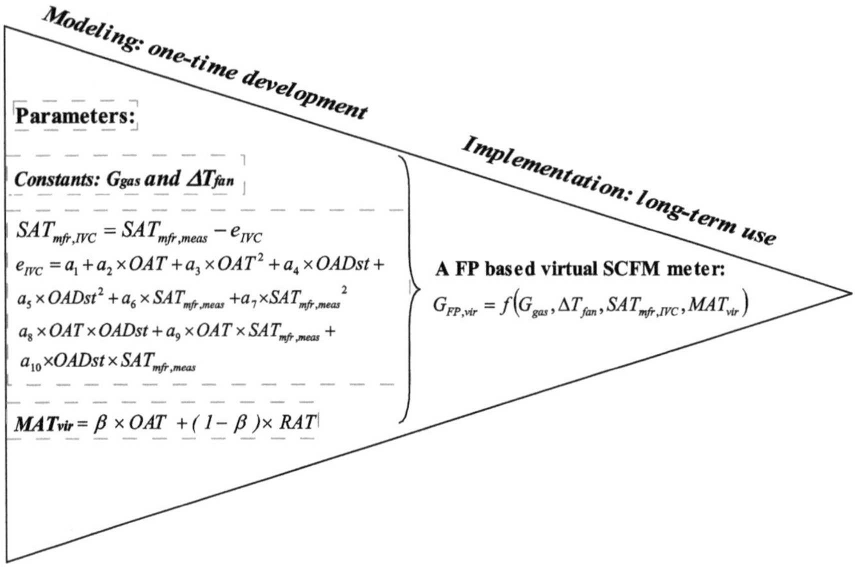

Fig. 5.7 Modeling and implementation procedures for a first-principle virtual SCFM meter in RTUs

physical airflow meter plus installation could cost up to one thousand dollars for one unit and eat up the cost advantage of economic RTUs.

To resolve the disadvantages of conventional measuring techniques in general, Li et al. [48] proposed a virtual sensing technology using low-cost measurements and mathematical models to estimate the desired quantities. Within the past two decades, virtual meters/sensors have been widely adopted in a number of fields and have enabled many intelligent features that would otherwise not be possible or economical [49–51]. In building systems, Li and Braun launched about eleven virtual sensors for a vapor compression cycle equipment to economically and systematically improve the real-time monitoring, control, and diagnosis [6, 39, 40]. Recently, as shown in Fig. 5.7, Yu et al. proposed a first-principle-based virtual SCFM meter for RTUs by correlating a first-principle model in combination with low-cost virtual or virtually calibrated temperature sensors (an improved virtually calibrated manufacturer-installed supply air temperature sensor $SAT_{mfr,\ IVC}$ (calibrated SAT for the manufacturer-installed temperature sensor by using the improved virtual calibration method), a virtual mixed air temperature sensor MAT_{vir} (virtual mixed air temperature) [52–55]. Compared with physical airflow meters, the first-principle-based virtual SCFM meter obtains a number of merits, including high cost-effectiveness, good accuracy, and long-term availability after one-time development and generic characteristics for all RTUs.

We further define the following variables to facilitate the discussion of the supply airflow rate sensor (Table 5.3).

Table 5.3 Variables used in the virtual supply airflow rate development

a_1, a_2,…, a_{10}: correlated coefficient	OAT: outside air temperature, °F (°C)
C_a: air capacity rate, Btu/(hr °F) (kW/K)	OAD_{st}: outside air damper position
C_g: gas capacity rate, Btu/(hr °F) (kW/K)	OAT_{VC}: virtually calibrated OAT, °F (°C)
e_{IVC}: offset error for calibration of the manufactured installed supply air temperature sensor by using the improved virtual calibration method, °F (°C)	Q_c: cooling capacity, Btu/hr (kJ/s)
e_{val}: related error between $G_{GB,\ vir}$ and $G_{a,\ sup,\ meas}$	Q_H: heating capacity, Btu/hr (kJ/s)
h_1: enthalpy of air before the thermal equipment Btu/lbm (kJ/kg)	RAT: return air temperature, °F (°C)
h_2: enthalpy of air after the thermal equipment Btu/lbm (kJ/kg)	$G_{GB,\ vir}$: gray-box-based virtual supply airflow rate, cfm (m^3/s)
Hstage: gas heating stage	$G_{a,\ sup,\ meas}$: measured supply airflow rate, cfm (m^3/s)
MAT_{vir}: virtual mixed air temperature, °F (°C)	ΔT_{fan}: temperature rise across the supply fan, °F (°C)
N_{Hstg}: number of gas heating stage	β: outside fresh air ratio

In this case study, we present a gray-box-based virtual SCFM meter, which notably enhances some deficiencies of the first-principle-based virtual SCFM meter; for instance,

- *It is much easier for implementation and maintenance*
  A first-principle-based virtual SCFM meter relies heavily on indirect measurements of a virtually calibrated supply air temperature sensor and a virtual mixed air temperature sensor. Owing to the drifts and other changes of data during routine operations and during system renovation (e.g., serious damages of physical sensors), there is often no objective measurement to reassess the virtual or virtually calibrated sensors. Moreover, most parameters of the above two sensors (e.g., OAT and return air temperature [RAT]) require periodically traditional calibration or virtual calibration before they can be used, since the accuracy and reliability of the fragile physical sensors often cannot be guaranteed (e.g., bath in hostile work environment for a long time) [56]. In contrast, the gray-box-based virtual SCFM meter mainly uses direct measurement and only requires virtual calibration of an OAT sensor prior to its use.
- *It has more capability for facilitating fault diagnostics*
  Since the first-principle-based virtual SCFM meter is proposed by utilizing an energy balance in combination with accurate $SAT_{mfr,\ IVC}$ and MAT_{vir} measurements across heating coils, it cannot be used to develop a virtual heating capacity meter inversely for further automated FDD applications. However, the measurement of a gray-box based virtual SCFM meter could be gathered for determining virtual heating capacity and within diagnosing algorithms for performance fault detection (e.g., low heating coil efficiency detection).

Additionally, the gray-box-based virtual SCFM meter also inherits good characteristics of the first-principle-based virtual SCFM meter, such as high cost-effectiveness, good robustness against variations in multi-variable operating conditions, and applicability to similar RTUs. This innovative virtual meter could serve as a permanent monitoring tool to indicate real-time SCFM measurement and/or to automatically detect and diagnose improper quantity of SCFM for RTUs.

The study begins with addressing carefully designed heating experiments and investigation by using acquired experimental results associated with the heating process. It is found that a gray-box-based virtual SCFM meter in RTUs can be generated by using three parameters for any number of heating stage (N_{Hstg}): $\mathrm{SAT}_{\mathrm{mfr,\ meas}}$, $\mathrm{OAD}_{\mathrm{st}}$ and OAT. Then, algorithm development, model validation, and uncertainly analysis for the virtual meter are provided. After that, detailed implementation issues including measuring and processing parameters as well as an implementation flowchart of the meter are presented. The study concludes that this non-intrusive virtual meter can accurately predict SCFM values for RTUs with high robustness.

5.2.3 *Model Development*

Experimental settings

As shown in Table 5.4, fifty-six sets of tests for heating stage (Hstage) 1 and forty sets for Hstage 2 are performed in a laboratory with artificial climate chambers to develop a virtual SCFM meter in a 7.5 ton RTU. An additional thirty heating experiments (eighteen sets for Hstage1 and twelve sets for Hstage 2) are carried out for validation purposes. Heating process is adopted in this study, since it is much superior to cooling process regarding the simplicity and reliability during developing and implementing a virtual SCFM meter. A wide range of OAT and direct

Table 5.4 Experimental design

Number of tests for training model		56 for Hstage 1; 40 for Hstage 2		
Number of tests of model validation		18 for Hstage 1; 12 for Hstage 2		
Parameters	Comments	Minimum	Maximum	Interval
	Number of heating stage, N_{Hstg}	1	2	No interval
	Outside air damper position, $\mathrm{OAD}_{\mathrm{st}}$ (%)	0	40	10
	Outside air temperature, OAT, °F (°C)	25(−3.9)	50(10.0)	About 5(2.8)
	Measured SCFM, $G_{\mathrm{a,\ sup,\ meas}}$, cfm($m^3$/s)	1200 (0.57)	2200 (1.04)	About 200 (0.09)

SCFM measurement $G_{a,\ sup,\ meas}$, including multi-combinations of OAD_{st} and N_{Hstg} information, is covered throughout the test [55]:

1. In an attempt to generalize gas Hstage, both the stages of gas heating are operated.
2. OAD_{st} is modulated from 0 to 40 % with an interval of 10 %. According to ventilation requirements with heating on [38], OAD_{st} should be switched to a minimum position for offsetting heating energy losses through a reduction of outdoor air intake. 40 % could definitely be the upper limit of a minimum damper position.
3. OAT measurement is acquired from 25 °F (−3.9 °C) to 50.0 °F (10.0 °C), which is a typical temperature range that requires the operating of mechanical heating.
4. To consider the validity of the generic SCFM model under fault conditions, the uncertainty of SCFM is extended as high as 50 % with an interval of 10 %, which means the SCFM is reduced from 100 to 50 % of design SCFM level [2400 cfm (1.13 m^3/s)].
5. Besides, for building applications, to some extent, RAT is reckoned as quasi-constant with small variations. By this reasoning, RAT, as a direct measurement, is generally collected but not 'controlled' in experiments.

Experimental results investigation

Observed from the indispensable measurements above, $SAT_{mfr,\ meas}$, OAT, RAT as well as the system operating information OAD_{st} and N_{Hstg} are easy to be directly measured. But RAT varies in a relatively small range in a typical space served by a light commercial RTU. Thus, variables of $SAT_{mfr,\ meas}$, OAT, and OAD_{st} as well as N_{Hstg} acquired from the carefully designed experiments are reinvestigated to develop a virtual SCFM meter.

In Fig. 5.8, nine linear regressions by using available results are depicted with dependent variable $G_{a,\ sup,\ meas}$ and independent variables $SAT_{mfr,\ meas}$, OAD_{st} and OAT. Each data point of the regressions denotes a set of testing result. We can see:

① ***In principle, there exists a significantly positive linear relationship between*** $G_{a,\ sup,\ meas}$ ***and*** $\mathbf{SAT_{mfr,\ meas}}$.

Generally, all linear regressions (L1 ~ L9) point to the fact that with a given N_{Hstg}, fixed OAT and OAD_{st}, $G_{a,\ sup,\ meas}$ is a strong function of $SAT_{mfr,\ meas}$. $G_{a,\ sup,\ meas}$ increases while $SAT_{mfr,\ meas}$ raises.

② ***The linear relationship is influenced by*** $\mathbf{OAD_{st}}$.

From the results in Group 1-a and Group 1-b, with a fixed OAT and N_{Hstg}, $G_{a,\ sup,\ meas}$ increases while OAD_{st} climbs. The linear line translates along with different OAD_{st}. R-square values of them are higher than 95 %.

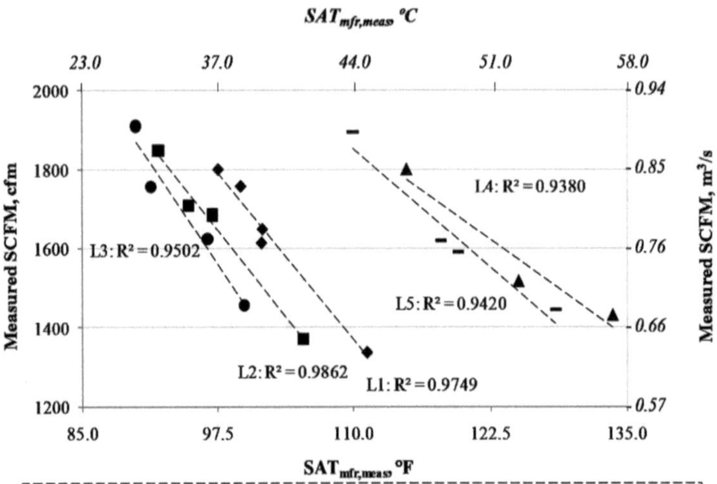

Group#	Regressions	Definite values	Variable
Group 1 -a:	L1:	Hstage1, OAT40°F (4.4°C)	**OADst0%**
	L2:	Hstage1, OAT40°F(4.4°C)	**OADst10%**
	L3:	Hstage1, OAT40°F(4.4°C)	**OADst20%**
Group 1 -b:	L4:	Hstage2, OAT40°F(4.4°C)	**OADst10%**
	L5:	Hstage2, OAT40°F(4.4°C)	**OADst20%**
Group 3 -a:	L2:	OAT40°F(4.4°C), OADst10%	**Hstage1**
	L4:	OAT40°F(4.4°C), OADst10%	**Hstage2**
Group 3 -b:	L3:	OAT40°F(4.4°C), OADst20%	**Hstage1**
	L5:	OAT40°F(4.4°C), OADst20%	**Hstage2**

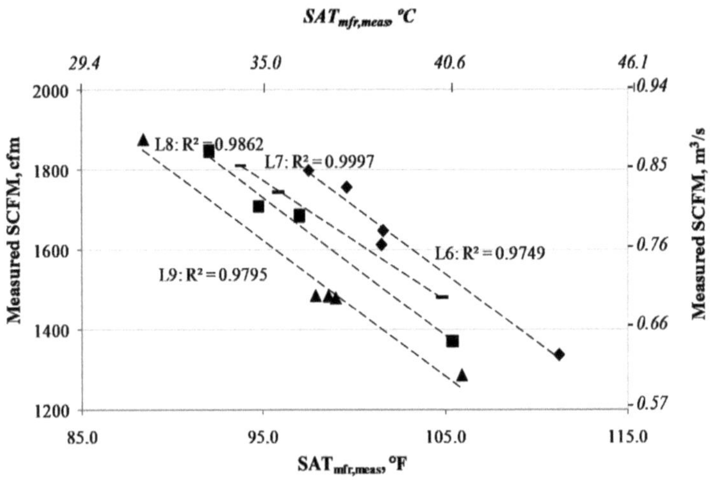

Group#	Regressions	Definite values	Variable
Group 2 -a:	L6:	Hstage1, OADst0%	**OAT40°F(4.4°C)**
	L7:	Hstage1, OADst0%	**OAT30°F(-1.1°C)**
Group 2 -b:	L8:	Hstage1, OADst10%	**OAT40°F(4.4°C)**
	L9:	Hstage1, OADst10%	**OAT32°F(0°C)**

Fig. 5.8 Experimental results investigation

③ ***The linear relationship is also influenced by* OAT.**

From the data in Group 2-a and Group 2-b, when OAD_{st} and N_{Hstg} keep constant, $G_{a,\ sup,\ meas}$ increases while OAT increases. The linear line also translates along with different OAT. R-square values of them are up to 99 % or more.

④ ***The linear relationship can be further categorized according to N_{Hstg}.***

From the data in Group 3-a and Group 3-b, with definite values of OAT and OAD_{st}, the dependency of $G_{a,\ sup,\ meas}$ on N_{Hstg} is not obvious, but $SAT_{mfr,\ meas}$ increases while N_{Hstg} increases. In Hstage 1, $SAT_{mfr,\ meas}$ varies within the range of 85 °F (29.4 °C) to 110 °F (43.3 °C), while in Hstage 2, the scope is 110 °F (43.3 °C) to 125 °F (51.7 °C). Instead of being treated as a regression variable, N_{Hstg} groups the lines of $G_{a,\ sup,\ meas}$ along with $SAT_{mfr,\ meas}$, OAD_{st} and OAT.

In summary, the above observations have shown that a gray-box-based virtual SCFM meter in RTUs can be generated by using three parameters for a given N_{Hstg}: $SAT_{mfr,\ meas}$, OAD_{st} and OAT.

Algorithm development

On the basis of investigations presented in the last subsection, a gray-box-based virtual SCFM meter is proposed accordingly. Adequately, fifty-six sets of experimental results for Hstage1 and forty sets of experimental results for Hstage 2 with strict experimental settings described in Table 5.4 are used for correlation of a virtual SCFM meter. Excellent correlation by using the linear correlation method is obtained into the following form:

$$\begin{aligned} G_{GB,vir} = {} & a_1 + a_2 \times OAT + a_3 \times OAT^2 + a_4 \times OAD_{st} + a_5 \\ & \times OAD_{st}^2 + a_6 \times SAT_{mfr,meas} + a_7 \times SAT_{mfr,meas}^2 \\ & + a_8 \times OAT \times OAD_{st} + a_9 \times OAT \times SAT_{mfr,meas} \\ & + a_{10} \times OAD_{st} \times SAT_{mfr,meas} \end{aligned} \tag{5.32}$$

where $G_{GB,\ vir}$ is the gray-box-based virtual supply airflow rate in cfm (m^3/s); a_1, $a_2,\ldots,\ a_{10}$ are the correlated coefficients.

By Hstage categories, coefficients of the virtual SCFM meter in a 7.5 Ton RTU are listed in Table 5.5. The R-square values are 96.37 % for Hstage 1 and 98.38 % for Hstage 2. Validation of this meter is presented in the next subsection.

Model Validation

Eighteen sets of experimental results for Hstage 1 and twelve sets for Hstage 2, covering comprehensive operation conditions, are used for validation. The error e_{val}, related to $G_{GB,\ vir}$ and $G_{a,\ sup,\ meas,}$ is:

$$e_{val} = \frac{G_{a,sup,vir} - G_{GB,vir}}{G_{a,sup,vir}} \tag{5.33}$$

For a better view of the results, e_{val} are plotted in Fig. 5.9 for each N_{Hstg}. As it can be seen, for Hstage 1, the absolute error e_{val} is lower than 5.8 %. For Hstage 2,

Table 5.5 SCFM model coefficients for the experimental RTU

IP units	**Coefficient**	a_1	a_2	a_3	a_4	a_5
	Hstage 1	4791	5.882	0.9841	−435.6	701.9
	Hstage 2	3620	39.92	1.0671	−304.2	847.7
	Coefficient	a_6	a_7	a_8	a_9	a_{10}
	Hstage 1	−40.44	0.1736	21.85	−0.6493	−18.27
	Hstage 2	−26.81	0.1636	−28.47	−0.8938	3.201
SI units	**Coefficient**	a_1	a_2	a_3	a_4	a_5
	Hstage 1	2.001	0.03985	0.00144	−0.1757	0.3381
	Hstage 2	2.043	0.06533	0.001585	−0.4316	0.3410
	Coefficient	a_6	a_7	a_8	a_9	a_{10}
	Hstage 1	−0.04351	0.0002782	0.01807	−0.0009571	−0.01459
	Hstage 2	−0.03729	0.0002415	−0.02293	−0.001309	0.001179

e_{val} is lower than ±4.3 %. This demonstrates that the gray-box-based virtual SCFM meter could accurately predict the true value of SCFM in RTUs.

Uncertainty Analysis

The eighteen sets of experimental results for Hstage1 and twelve sets for Hstage 2 are used for uncertainty analysis of the gray-box-based virtual SCFM meter in relative to $\text{SAT}_{\text{mfr, meas}}$, OAD_{st}, and OAT in this section. The root sum square method of uncertainty calculation is applied to the variables. The random uncertainty is expressed as Eq. 5.34:

$$e(G_{\text{GB,vir}}) = \left[\left(\frac{\partial G_{\text{GB,vir}}}{\partial \text{SAT}_{\text{mfr,meas}}} e(\text{SAT}_{\text{mfr,meas}})\right)^2 + \left(\frac{\partial G_{\text{GB,vir}}}{\partial \text{OAD}_{\text{st}}} e(\text{OAD}_{\text{st}})\right)^2 + \left(\frac{\partial G_{\text{GB,vir}}}{\partial \text{OAT}} e(\text{OAT})\right)^2\right]^{1/2} \tag{5.34}$$

where $e(\text{SAT}_{\text{mfr, meas}})$, $e(\text{OAD}_{\text{st}})$, and e(OAT) are uncertainties.

Table 5.6 summarizes uncertainties of the three independent input variables, as well as the calculated uncertainties of $\dot{V}_{\text{GB,vir}}$. As it can be seen, the absolute uncertainty of the virtual SCFM meter is lower than 5.9 % for Hstage 1 and lower than 3.4 % for Hstage 2. It demonstrates that the SCFM meter is very accurate.

Implementation procedure

This part is dedicated to provide an instruction of deploying this gray-box-based virtual SCFM meter in the field including measuring and processing parameters as well as a graphical flowchart as in Fig. 5.10.

Measuring and processing parameters

Measurements of OAD_{st} and N_{Hstg} are straightforward and readily available, while the measurements of $\text{SAT}_{\text{mfr, meas}}$ and OAT are not. Measuring and processing of these two temperature measurements are described in this subsection. To ensure the robustness and reliability of this virtual meter, first a steady-state detector is used to filter out the transient data of all temperature measurements involved in Eq. 5.34 [31].

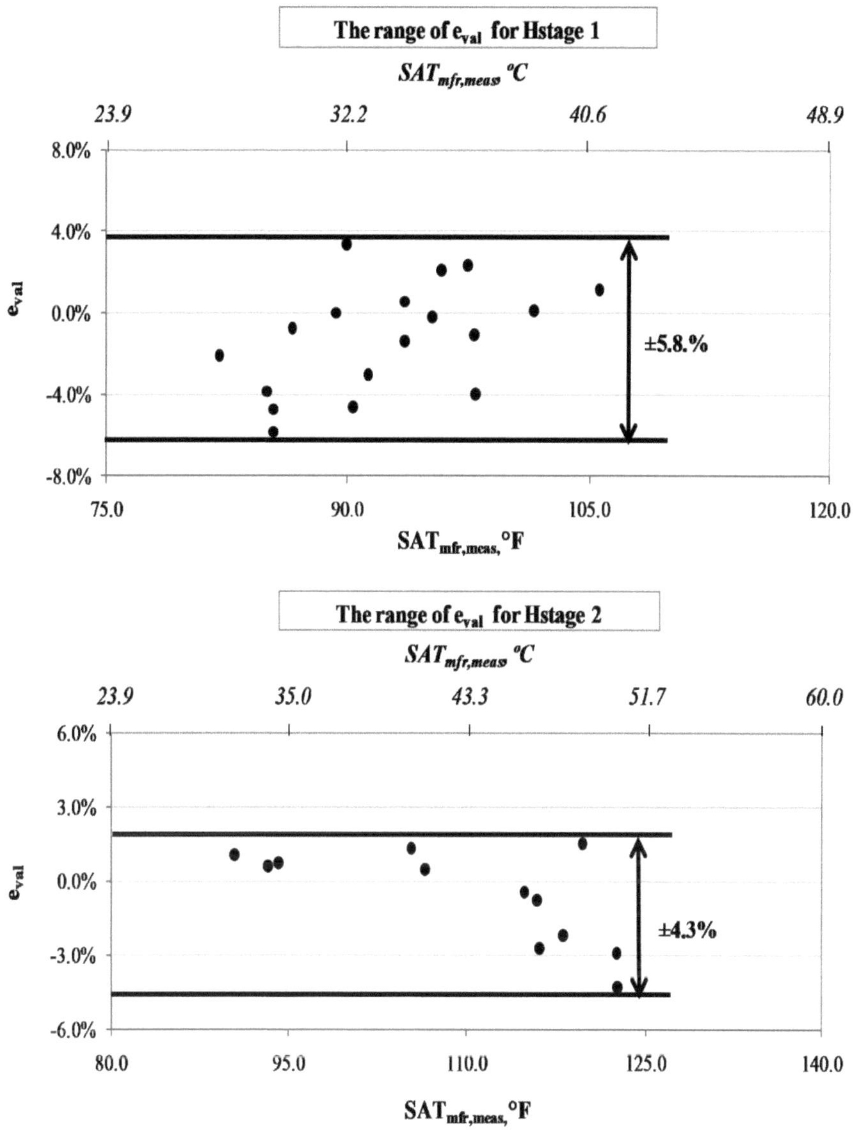

Fig. 5.9 Range of e_{val} for Hstage 1 and Hstage 2

Direct measurement of $SAT_{mfr,\ meas}$ Under steady-state condition, to get reasonable $SAT_{mfr,\ meas}$ for the virtual meter, sensor failure detection is necessary. It can be determined when $SAT_{mfr,\ meas}$ measurement is out of the normal temperature scales [26] for a certain amount of time. For example, a manufacturer-installed SAT sensor is considered to be failed under conditions where $SAT_{mfr,\ meas}$ measurement keeps lower than −50 °F (−45.6 °C) or higher than 200 °F (93.3 °C) for more than

Table 5.6 Uncertainty analysis of the gray-box-based virtual SCFM meter

Independent variables	Inputs	Uncertainty
Measurements of the manufacturer-installed SAT sensor, $SAT_{mfr, meas}$, °F (°C)	18 sets of experimental results for Hstage 1; 12 sets for Hstage 2	±1.0 (0.6)
Outside air damper position, OAD_{st}		5 %
Outside air temperature, OAT, °F (°C)		±1.0 (0.6)

Dependent variable		A gray-box-based virtual SCFM meter, $G_{GB, vir}$								
Hstage 1	Scenario ID	H1_1	H1_2	H1_3	H1_4	H1_5	H1_6	H1_7	H1_8	H1_9
	Uncertainty (%)	5.9	4.8	4.7	4.4	5.1	4.4	4.1	5.2	4.2
	Scenario ID	H1_10	H1_11	H1_12	H1_13	H1_14	H1_15	H1_16	H1_17	H1_18
	Uncertainty (%)	3.8	3.4	3.7	3.7	3.6	3.9	3.5	4.2	3.3
Hstage 2	Scenario ID	H2_1	H2_2	H2_3	H2_4	H2_5	H2_6			
	Uncertainty (%)	1.0	2.1	3.0	2.8	1.7	1.2			
	Scenario ID	H2_7	H2_8	H2_9	H2_10	H2_11	H2_12			
	Uncertainty (%)	2.2	2.5	3.4	2.6	1.0	2.5			

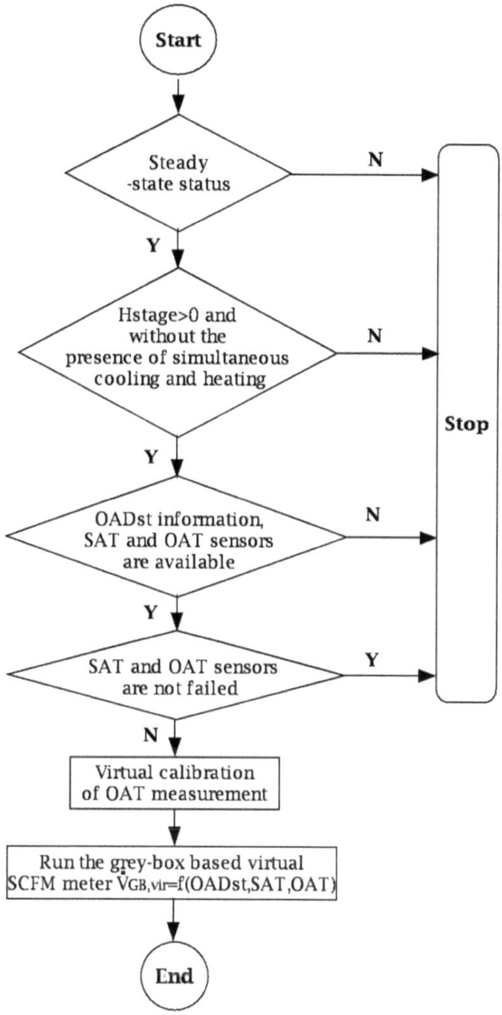

Fig. 5.10 An implementation flowchart of a gray-box-based virtual SCFM meter

eight hours. If so, the sensor has to be replaced prior to its use for the virtual SCFM meter.

Virtual calibration of OAT Despite the sensor failure detection for an OAT sensor, there is an additional great challenge by using direct OAT measurement. The fact is that direct measurement of an OAT sensor is notoriously difficult to attain due to a number of practical dilemmas. The original manufacturer preinstalled OAT sensor is easily worn out since it totally bathes in a harsh outdoor environment. In another direction, the way of installing a global OAT sensor is very costly and more than often infeasible. An accurate OAT sensor requires high initial cost and further maintenance expense (e.g., manual periodical calibration). What's more, the global OAT sensor is installed on and continuously exposed to external wall, which is easily affected by the presence of strong solar radiation.

In order to 'qualify' direct measurement of a flimsy OAT sensor, Yu and Li [56] proposed a virtual calibration methodology based on statistical or modeling methods. Compared with traditional calibration, this innovative calibration technology surpasses with a number of merits, such as high cost-effectiveness, achieving real-time, self-calibration, and handling a notably huge amount of measurements acquired from physical and virtual sensors. A statistical method to virtually calibrate direct measurement of an OAT sensor was applied in the following form:

$$\mathrm{OAT}_{\mathrm{VC}} = a_1 \times \mathrm{OAT} + a_2 \tag{5.35}$$

where a and b are correlated coefficients; $\mathrm{OAT}_{\mathrm{VC}}$ is virtually calibrated OAT measurement in °F (°C).

With accurate parameters accomplished, a graphical flowchart is explored accordingly to graphically sum up the implementation procedures of a gray-box-based virtual SCFM meter.

Implementation flowchart

A gray-box-based virtual SCFM meter can be implemented through the following procedures:

Step 1: Check the steady-state status;
Step 2: Check the heating and cooling status;
Step 3: Check the availability of $\mathrm{OAD}_{\mathrm{st}}$ information, SAT, and OAT sensors;
Step 4: Check that SAT and OAT sensors are not failed;
Step 5: Virtually calibrate OAT measurement;
Step 6: Run the gray-box-based virtual SCFM meter $G_{\mathrm{GB,vir}}$.

References

1. ISO/IEC 2008. Uncertainty of measurement—Part 2: guide to the expression of uncertainty in measurement (2008)
2. International Organization for Standardization. International Vocabulary of Basic and General Terms in Metrology (1993)
3. Ray, S.: Error propogation and uncertainty evaluation for automatic control—a neglected part of engineering education. Proc. Technol. **4**, 629–635 (2012)
4. Dudley, R.M.: Uniform Central Limit Theorems. Cambridge University Press, Cambridge (1999)
5. Yu, D., Li, H., Yang, M.: A virtual supply airflow rate meter for rooftop air-conditioning units. Build. Environ. **46**(6), 1292–1302 (2011)
6. Li, H., Braun, J.E.: Development, evaluation, and demonstration of a virtual refrigerant charge sensor. HVAC&R Res. **15**(1), 117–136 (2009)
7. Kampjarvi, P., Sourander, M., Komulainen, T., Vatanski, N., Nikus, M., Jms-Jounela, S.L.: Fault detection and isolation of an on-line analyzer for an ethylene cracking process. Control Eng. Pract. **16**(1), 1–13 (2008)

8. Bonne, D., Jorgensen, S.B.: Data-driven modeling of batch processes. In: Proceedings of 7th International Symposium on Advanced Control of Chemical Processes, ADCHEM (2004)
9. Qin, S.J.: Neural networks for intelligent sensors and control—practical issues and some solutions. Neural Syst. Control, 213–234 (1997)
10. Casali, A., Gonzalez, G., Torres, F., Vallebuona, G., Castelli, L., Gimenez, P.: Particle size distribution soft-sensor for a grinding circuit. Powder Technol. **99**(1), 15–21 (1998)
11. Park, S., Han, C.: A nonlinear soft sensor based on multivariate smoothing procedure for quality estimation in distillation columns. Comput. Chem. Eng. **24**(2–7), 871–877 (2000)
12. Assis, J.D., Filho, M.: Soft sensors development for on-line bioreactor state estimation. Comput. Chem. Eng. **24**(2), 1099–1103 (2000)
13. Meleiro, L.A.C., Finho, R.M.: A self-tuning adaptive control applied to an industrial large scale ethanol production. Comput. Chem. Eng. **24**(2–7), 925–930 (2000)
14. Radhakrishnan, V.R., Mohamed, A.R.: Neural networks for the identification and control of blast furnace hot metal quality. J. Process Control **10**(6), 509–524 (2000)
15. Devogelaere, D., Rijckaert, M., Leon, O.G., Lemus, G.C.: Application of feedforward neural networks for soft sensors in the sugar industry. In: Proceedings of the VII Brazilian Symposium on Neural Networks, pp. 2–6 (2002)
16. James, S., Legge, R., Budman, H.: Comparative study of black-box and hybrid estimation methods in fed-batch fermentation. J. Process Control **12**(1), 113–121 (2002)
17. DeWolf, S., Cuypers, R.L.E., Zullo, L.C., Vos, B.J., Bax, B.J.: Model predictive control of a slurry polymerisation reactor. Comput. Chem. Eng. **20**, 955–961 (1996)
18. Prasad, V., Schley, M., Russo, L.P., Wayne Bequette, B.: Product property and production rate control of styrene polymerization. J. Process Control **12**(3), 353–372 (2002)
19. Doyle, F.J.: Nonlinear inferential control for process applications. J. Process Control **8**(5–6), 339–353 (1998)
20. Gonzalez, G.D.: Soft sensors for processing plants. In: Proceedings of the Second International Conference on Intelligent Processing and Manufacturing of Materials (1999)
21. Warne, K., Prasad, G., Rezvani, S., Maguire, L.: Statistical and computational intelligence techniques for inferential model development: a comparative evaluation and a novel proposition for fusion. Eng. Appl. Artif. Intell. **17**(8), 871–885 (2004)
22. Frank, I.E., Friedman, J.H.: A statistical view of some chemometrics regression tools. Technometrics **35**(2), 109–135 (1993)
23. Kourti, T.: Process analysis and abnormal situation detection: from theory to practice. IEEE Control Syst. Mag. **22**(5), 10–25 (2002)
24. Poggio, T., Girosi, F.: Regularization algorithms for learning that are equivalent to multilayer networks. Science **247**(4945), 978–982 (1990)
25. Bishop, C.M.: Neural Networks for Pattern Recognition. Oxford University Press, USA (1995)
26. ASHRAE. ASHRAE Handbook—Fundamentals. American Society of Heating, Refrigerating and Air-conditioning Engineers, Inc., Atlanta (2009)
27. Gustafsson, M.D., Forssell, U., Löfgren, M., Persson, N., Quicklund, H.: Virtual Sensors of Tire Pressure and Road Friction. Society of Automotive Engineers (SAE), Inc. (2001)
28. Nishikawa, Y.: Tire Air Pressure Detecting Device. EP832768, US 5982279. Denso Corp, Nippon Soken (1998)
29. Taguchi, T.: Tire Air Pressure Detecting Device. EP783982. Nippon Denso Co (1997)
30. Gomez, E., Unbehauen, H., Kortmann, P., Peters, S.: Fault detection and diagnosis with the help of fuzzy-logic and with application to a laboratory turbogenerator. In: Proceedings of the 13th IFAC World Congress, vol. N, pp. 175–180 (1996)
31. Li, H., Braun, J.E.: An improved method for fault detection and diagnosis applied to packaged air conditioners. ASHRAE Trans. **109**(2), 683–692 (2003)
32. Wichman, A., Braun, J.E.: A smart mixed-air temperature sensor. HVAR&R Res. **15**(1), 101–115 (2009)
33. Serneels, S., Verdonck, T.: Principal component analysis for data containing outliers and missing elements. Comput. Stat. Data Anal. **52**(3), 1712–1727 (2008)

34. Stanimirova, I., Daszykowski, M., Walczak, B.: Dealing with missing values and outliers in principal component analysis. Talanta **72**(1), 172–178 (2007)
35. Walczak, B., Massart, D.L.: Robust principal components regression as a detection tool for outliers. Chemometr. Intell. Lab. Syst. **27**(1), 41–54 (1995)
36. Gosset, W.S.: The probable error of a mean. Biometrika **6**(1), 1–25 (1908)
37. Hastie, T., Tibshirani, R., Friedman, J.: The Elements of Statistical Learning: Data Mining, Inference, and Prediction. Springer, Berlin (2001)
38. ASHRAE. ASHRAE Standard 62.1-2007, "Ventilation for Acceptable Indoor Air Quality". American Society of Heating, Refrigerating and Air-conditioning Engineers, Inc., Atlanta (2007)
39. Li, H., Braun, J.E.: A methodology for diagnosing multiple-simultaneous faults in vapor compression air conditioners. HVAC&R Res. **13**(2), 369–395 (2007)
40. Li, H., Braun, J.E.: Decoupling features and virtual sensors for diagnosis of faults in vapor compression air conditioners. Int. J. Refrig. **30**(3), 546–564 (2007)
41. ASHRAE. ASHRAE Standard 41.2-1987, "Standard Methods for Laboratory Airflow Measurement". American Society of Heating, Refrigerating and Air-conditioning Engineers, Inc., Atlanta (1987)
42. Howell, R.H., Sauer, H.J.: Field measurements of air velocity: pitot traverse or vane anemometer. ASHRAE J. **32**(3), 46–52 (1990)
43. Howell, R.H., Sauer, H.J.: Airflow measurements at coil faces with vane anemometers. ASHRAE Trans. **96**(1), 502–511 (1990)
44. Riffat, S.B.: Turbulent flow in a duct: measurement by a tracer gas technique. Build. Serv. Eng. Res. Technol. **11**(1), 21–26 (1990)
45. Riffat, S.B.: Airflow rate through a heat-exchanger coil. Appl. Energy **38**(3), 231–238 (1991)
46. Palmiter, L., Francisco, P.: "Development of a Simple Device for Field Airflow Measurement of Residential Air Handling Equipment, Phase I and Phase II". Ecotope Final Report for DOE 2000 (2000)
47. ASHRAE. ASHRAE Standard 110-1995, Method of Testing Performance of Laboratory Fume Hoods. American Society of Heating, Refrigerating and Air-conditioning Engineers, Inc., Atlanta (1995)
48. Li, H., Yu, D., Braun, J.E.: A review of virtual sensing technology and application in building systems. HVAC&R Res. (under review, 2010)
49. Hardy, N., Maroof, A.A.: ViSIAr—a virtual sensor integration architecture. Robotica **17**(6), 635–647 (1999)
50. Oosterom, M., Babuška, R.: Virtual sensor for fault detection and isolation in flight control systems-fuzzy modeling approach. In: Proceedings of the IEEE Conference on Decision and Control, vol. 3, pp. 2645–2650 (2000)
51. Kadlec, P., Gabrys, B., Strandt, S.: Data-driven soft sensors in the process industry. Comput. Chem. Eng. **33**(4), 795–814 (2009)
52. Yu, D., Li, H., Yang, M.: A virtual supply airflow meter in rooftop air conditioning units. Build. Environ. **46**(6), 1292–1302 (2010)
53. Yu, D., Li, H., Yu, Y. An improved virtual calibration of a supply air temperature sensor in rooftop air conditioning units. HVAC &R Res. (In press, 2010)
54. Yu, D., Li, H., Yu, Y.: A gray-box based virtual supply airflow meter in rooftop air-conditioning units. J. Therm. Sci. Eng. Appl. **3**(1), 011005.1-7 (2011)
55. Yu, D., Li, H., Yu, Y., Xiong, J.: Virtual calibration of a supply air temperature sensor in rooftop air conditioning units. HVAC & R Res. **17**(1), 31 (2010)
56. Yu, Y., Li, H.: Virtual in-situ calibration methodology for building systems. Autom. Constr. (2015)

Chapter 6
Control Design Based on State-Space Model

In this chapter, the methodology of MPC in terms of the formulation, model structure, and solving method is first briefly reviewed. Then, Wiener, Hammerstein, and Hammerstein–Weiner structures are introduced, which can be used to handle the nonlinearity in a building system. Thereafter, they are applied to a hydraulic heating system to evaluate the performance. Meanwhile, design principle of state-feedback control system based on state-space model is illustrated, and corresponding examples are given.

6.1 Model-Based Predictive Control (MPC)

6.1.1 Introduction of MPC

Model-based predictive control (MPC) technology provides us with opportunities for improving the building energy system performance by incorporating the system models and utilizing future information. With over 2000 industrial installations, MPC is currently the most widely implemented advanced control technology for process plants [1]. A lot of literatures summarized the MPC theory and recent studies in both industry and academia [2–12]. Although MPC paradigm includes many different variants with all kinds of features, all MPC systems involve the basic elements to generate the control trajectory for a process: an internal model, an optimization solver, and a moving horizon feature, as shown in Fig. 6.1. Different MPCs have the varieties of the model format, disturbance estimation method, objective, solver, etc. Many techniques are utilized in addition to improve the performance of MPC for given circumstances, for instance, moving block, reference look-ahead, time-variant weighting, and time-variant constraints.

Y. Yao and Y. Yu, *Modeling and Control in Air-conditioning Systems*,
Energy and Environment Research in China, DOI 10.1007/978-3-662-53313-0_6

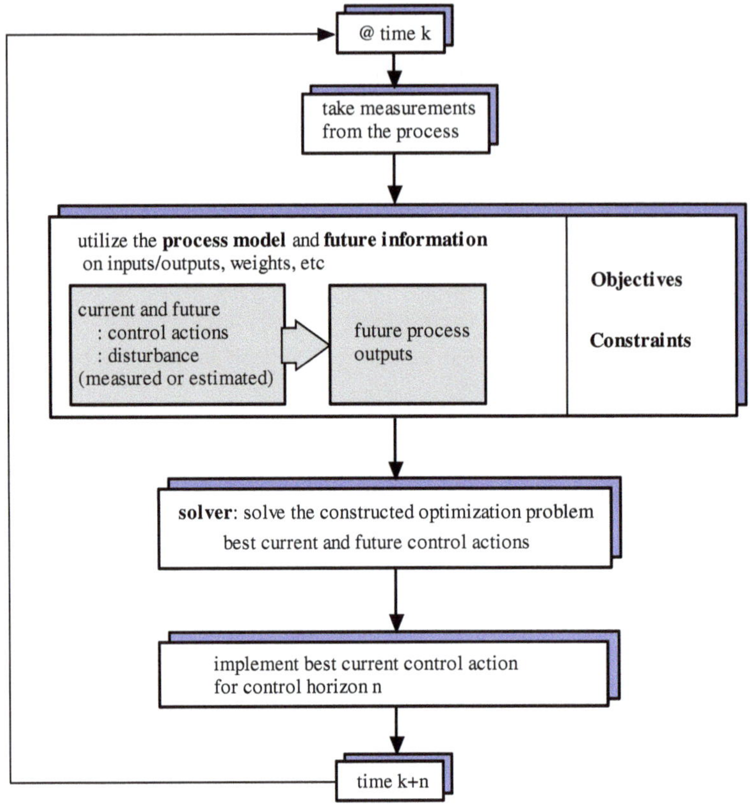

Fig. 6.1 Flowchart for model-based predictive control scheme

6.1.2 *MPC in Broad Definition*

From a broad point of view, any control strategy with the aforementioned three elements can be considered as an MPC. For every move, an open-loop optimal control problem is formulated and solved over a finite horizon. The solver evaluates the cost function based on the system simulation and the given constraints. The controller picks the first set of optimal control variables for implementation and the whole process repeats.

6.1.2.1 Controller Formulation

For a general MPC, the objective function as given in Eq. (6.1) can be loosely defined as almost anything, as long as it has a clear mathematic expression to represent the cost of the system. The system model is also a general model, which can be, for example, a group of differential algebraic equations, neural network, or model bank. The system is subject to a set of constraints on the states, outputs, or control variables.

$$\min J = f(u, x, y) \tag{6.1}$$

where

$$\dot{x} = f(x, u, w)$$
$$0 = h(x, u)$$

$$x^L \leq x \leq x^U$$
$$y^L \leq y \leq y^U$$
$$u^L \leq u \leq u^U$$

where J is the cost function, x is the vector of state variables, y is the vector of system output, u is the control input, w is the vector of disturbance, $\dot{x}$ is the derivative of state variables. Subscripts L and U mean the lower bound and upper bound of the corresponding variables.

6.1.2.2 Solving Methods

The dynamic control problem defined by Eq. (6.1) is very often casted as a numerical optimization programming problem. The control inputs, or both the control inputs and the states in the modeling equations are parameterized along the prediction horizon. When only control inputs are parameterized, it is referred as a sequential approach. The system model is remained as a simulator to obtain the cost and evaluate the constraints for different control actions. The second approach is termed collocation or simultaneous approach, where both the control and system model are converted into polynomial form that satisfies the system differential algebraic equations. With the later approach, models in general need to be expressed in a set of explicit equations and follow some strict syntax in order to perform automatic numeric processing. After that, many methods and optimization packages can be applied to solve the optimization programming problems. The solution approaches vary a lot in terms of involved mathematic techniques/principles and computational cost. The selection of a proper solver can be difficult, depending on the problem features and the required performance of the control.

In general, a numerical solver is usually designed for a particular problem type. No optimization solver is universally applicable, or it works at an extra cost. Meanwhile, the solution of a general optimization problem (e.g., constrained, nonlinear, and mixed integer) can be computationally expensive and no global optimization is guaranteed. The satisfaction of conditions is only ensured at the selected evaluation points. Findeisen categorized three principle approaches to solving optimal control problems: Hamilton–Jacobi–Bellman equations, Euler–

Lagrange differential equations, and finite parameterization on the control inputs and constraints [13]. A few literatures available discussed the related issues involved in general MPC from different points of view (e.g., [14–16]).

6.1.2.3 Linear Quadratic MPC

Linear classical MPC is one of the most mature optimal control strategies. It refers to a group of optimization problem where: (1) the internal dynamic model is linear and time-invariant (LTI); (2) the objective function is a quadratic function in terms of the control inputs, measureable outputs, and rates of control inputs; and (3) the constraints are on the control inputs, measureable outputs, and rates of control inputs.

(1) Controller formulation

A classical MPC has the following quadratic programming formulation, including cost function, linear model, and boxed constraints:

$$\min \sum_{i=1}^{k} \|e^y\|_Q^2 + \|\Delta u\|_S^2 + \|u\|_R^2 \tag{6.2}$$

where

$$\begin{aligned} \dot{x} &= Ax + Bu \\ y &= Cx \end{aligned}$$

$$\begin{aligned} & x^L \le x \le x^u \\ & y^L \le y \le y^U \\ & u^L \le u \le u^U \\ & \Delta u^L \le \Delta u \le \Delta u^U \end{aligned}$$

where e^y is the output errors, Δu is the rates of control inputs, A is the matrix of states, B is the matrix of inputs, and C is the matrix of outputs. The subscripts Q, S, and R denote the weighting matrixes on the reference errors, rates of control inputs, and control inputs, respectively.

The common model types are finite impulse and step response (SR) models, autoregressive moving average with exogenous input (ARMAX)/controlled autoregressive integrated moving average (CARIMA) models, transfer function models, and more general, state-space models.

A finite impulse response (FIR) or SR model is widely used in system control since it can fit arbitrarily complex stable linear dynamics based on experimental data. No or less advance knowledge about the system is required to develop the model. Methods, such as ordinary least square or optimization based on minimization of prediction and measured errors, can be used to process the data and obtain the model parameters. Equations (6.3) and (6.4) give the general expression of an FIR and a SR model, respectively.

$$y(k) = \sum_{i=1}^{n} a_i u(k-i) \tag{6.3}$$

$$y(k) = \sum_{i=1}^{n} a_i \Delta u(k-i) + a_n u(k-n-i) \tag{6.4}$$

where u is the input signal, y is the output signal, n is the number of parameter terms, a is the filter or model coefficient, Δu is the unit input change, and k is the current time stamp.

For a multiple-input-multiple-output (MIMO) system, the coefficients become a matrix instead of a vector to represent the mapping relationship between the multiple inputs and multiple outputs. As seen obviously from the expression, a large number of data are generally needed to identify the parameters used in an FIR or SR model for a complex system. The parameters can be up to hundreds for a complex system. For a MIMO system, the disadvantage of great number of parameters becomes troublesome. Since the model does not include any structure information from the system, data over-fitting and high modeling uncertainty may exist.

Equations (6.5) and (6.6) show the structure of an ARMAX and a CARIMA model, respectively. In Eq. (6.5), from left to right, the terms are autoregressive part, moving average part, and exogenous inputs. In Eq. (6.6), the moving average term includes an integral part.

$$Ay(k) = Ce(k) + Bu(k) \tag{6.5}$$

$$Ay(k) = \frac{C}{\Delta} e(k) + Bu(k) \tag{6.6}$$

Both models can be deduced based on measurements and in an adaptive (predictive) manner. With CARIMA model, good output predictions and future control sequence can be obtained alternatively to minimize the cost function. The derivation of optimal prediction of CARIMA model can be obtained by recursion of Diophantine equation [17].

A transfer function model is applicable to both stable and unstable plants. It can be deduced based on regression or from first principles. As the term implied, the model maps the direct relationship between inputs and outputs. The advantage of using a transfer function model also includes its compact expression, which requires less parameter than in an FIR model. Equation (6.7) gives the structure for a transfer function model. Like an FIR model, a transfer function is considered less effective for MIMO plants.

$$Ay(k) = Bu(k) \tag{6.7}$$

Any aforementioned linear dynamic model for classical MPC can be transformed to a state-space model. A state-space model provides a nice structure for MPC, and it can easily handle MIMO system. It has uniform treatment of stable, integrating, and unstable processes.

(2) Solving methods

Since the model and the constraints are linear, and the cost function is formed in quadratic, classical linear MPC can be eventually casted as a quadratic programming problem. A quadratic programming problem is a convex optimization problem with a unique optimal solution. It can generally be solved with low computational cost, which makes it suitable for online implementation. To offset nonzero values on the variables, an integral term of manipulated variables based on augmented model is generally used in MPC. For most plants in building systems, there are hard constraints and/or soft constraints on the inputs, outputs, and/or changing rates. For instance, the total input to a fan cannot be more than the maximum capacity. The supply air temperature needs to stay in an acceptable range. A constraint imposed on compressor speed changing rate can prevent the compressor from frequent and large-scale speed modulation. A wider constraints during unoccupied hours and narrower constraints during occupied hours on the indoor air temperature can lead to more energy conservative operation. Constraints can also be utilized to guide MPC toward the desired operation direction. Techniques, such as Lagrange multipliers, Kuhn–Tucker conditions, active set methods, primal-dual method, interior points, are the solution candidates for constrained quadratic programming problems.

6.1.2.4 Handling of Nonlinear Elements in MPC

While classical linear MPC runs fast, it cannot directly handle nonlinear problems. Most building energy systems are nonlinear in nature. Additional considerations are needed in order to utilize it. For conventional buildings, there are generally single mechanical heating or cooling sources deployed in the space; the controlled variable is typically the space air temperature. For instance, in a VAV conditioned building, the supply air temperature, and the airflow delivered into the individual room are the manipulated variables. In a fan-coil-conditioned space, the supply water temperature and the airflow through each fan coil are the manipulated variables. In a residential building, the power input to the air-conditioner or furnace can be considered as the manipulated variable. Under these circumstances, it is beneficial to design an MPC with the building treated as a linear time-invariant system. Figure 6.2 illustrates the layout for a linear MPC structure with heat flux as the system inputs and direct measurable variables as outputs.

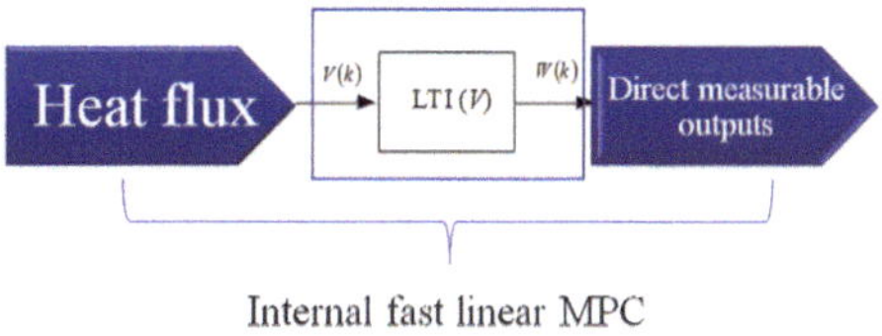

Fig. 6.2 Pure linear classical MPC for building systems

This approach has the following benefits: (1) by using a linear model and quadratic cost function, the problem becomes a highly structured convex problem and can be solved fast. There are mature solution and analysis methodologies; (2) the space system can be decoupled from the mechanical system. The space model can be reutilized for other study purposes; and (3) the manipulated variables can be set as the heat flux and the cost function can directly reflect the overall energy consumption.

While the building system dynamics can be reasonably simplified into a linear time-invariant model as a resistance–capacitance (RC) thermal network, the non-linearity still exists. Figure 6.3 illustrates the typical model structure in reality. The two ends of the linear time-invariant (LTI) system are static nonlinear mappings. A system with only the static nonlinearity before the LTI is termed a Hammerstein nonlinear system. A system with only the static nonlinearity after the LTI is a Wiener nonlinear system. A system with the nonlinearity at both ends is termed Hammerstein–Wiener (HW) nonlinear system [18]. The nonlinearity in a Hammerstein system comes from the fact that the heat flux into the air space is usually not the direct manipulated variable. For example, with a VAV box we control the damper position instead of the amount of heat into the space. In a Wiener system, the direct measurable variables are not the target control variables. For instance, we want to control thermal comfort, but we usually only have measurements on the air temperature and humidity.

To utilize the advantages of linear MPC for online implementation and solve the nonlinear system with static nonlinearity, the system inputs and outputs need to be processed accordingly. For a Wiener model, as shown in Fig. 6.4, the outputs from the LTI are mapped into the desired outputs by using the inverse function of the mapping. The signal is then compared to the inversed reference set point. The errors are fed into the linear MPC to form a closed-loop feedback control. In a thermal comfort-oriented control, the direct map is from the room air temperature, room air humidity, air velocity, metabolic rate, external work, clothing insulation to the thermal comfort index. In a Hammerstein model, as illustrated in Fig. 6.5, the control action generated by the linear MPC first goes through a static nonlinear

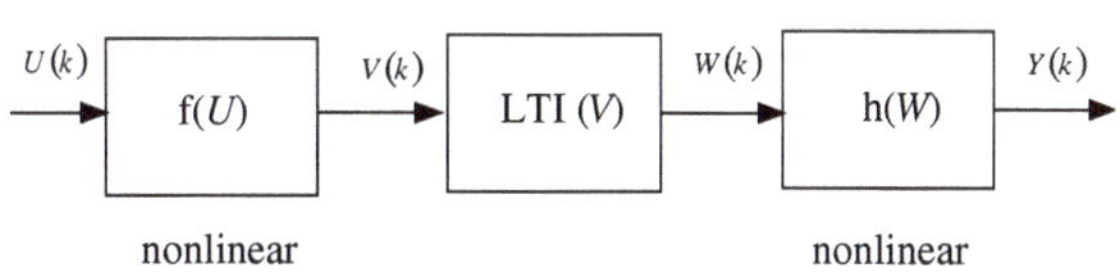

Fig. 6.3 Hammerstein–Wiener nonlinear model

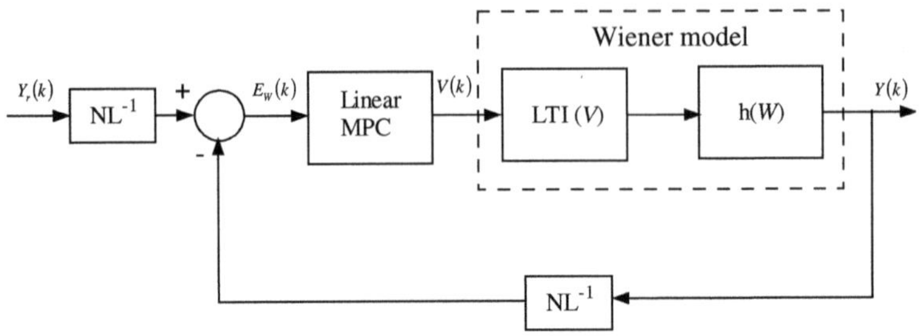

Fig. 6.4 Control structure for Wiener model plant

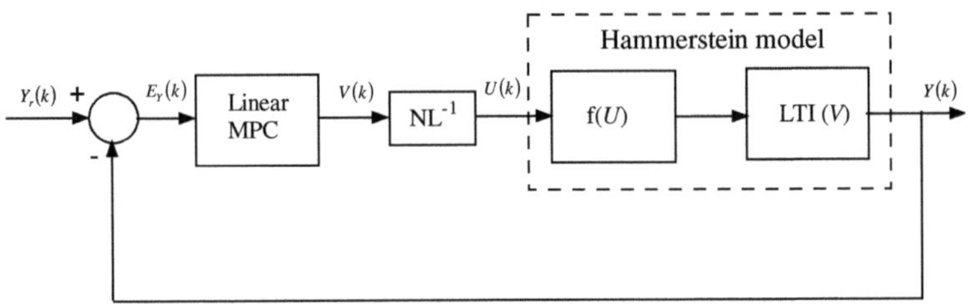

Fig. 6.5 Control structure for Hammerstein model plant

inverse function, e.g., from the heat flux to the water flow rate. The actual signal is then obtained and fed into the system after the inverse map.

Nonlinear Hammerstein–Wiener (HW) model consists of a linear dynamic block and two static output nonlinearity functions on both sides. This structure is more realistic in representing space conditioning for thermal comfort in most building systems. The nonlinear HW system can be described by the following equations:

$$x(k+1) = Ax(k) + Bu(k) \tag{6.8}$$

$$y(k) = Cx(k) \tag{6.9}$$

$$z(k) = h(y(k)) \tag{6.10}$$

$$v(k) = f(u(k)) \tag{6.11}$$

where A, B, and C are the state, input, and output matrix, respectively, x is the state variable vector in the LTI, y is the output vector in the LTI, u is the input vector in the LTI, z is the overall system output vector, v is the overall system input vector, h and f are the two static mapping functions, and k is the time step.

For the hydronic heating and cooling system in this study, the HW model and the linear MPC-based control loop is illustrated in Fig. 6.6.

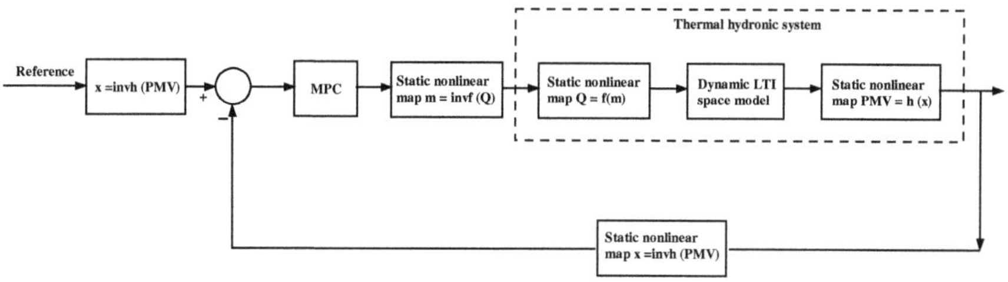

Fig. 6.6 Hydronic system in Hammerstein–Wiener structure

6.2 Applications of MPC in HVAC Field

6.2.1 Control of a Hybrid Ventilation Unit

The hybrid ventilation unit includes three individually functional components: the enthalpy recovery unit (ERV), vapor compression unit, and active desiccant dehumidifier. It provides necessary ventilation air to the space and maintains the space humidity level during summer to avoid condensation on the surface of the water-based hydraulic thermal system. The supply airflow rate and supply air temperature set points are given by the central workstation. The space sensible load is mainly assumed by the water-based hydraulic system while the hybrid unit takes care of the latent load. To avoid a large fluctuation of the indoor air conditions and as a matter of fact, the operation of the hybrid unit, in terms of supply airflow rate and supply air temperature, is constrained in a narrow range.

With a conventional cooling-based dehumidification method, the dehumidification capacity is coupled with the sensible cooling process and limited by the evaporator surface temperature. Dehumidification is mainly a result of cooling rather than a control variable. The supply air can only be dehumidified to the extent where the supply airflow rate and supply air temperature match the space sensible cooling load. With the application of variable speed drivers, the coupling issue can be mitigated with improved control performance. For example, reducing supply airflow rate can decrease the air-side sensible heat transfer coefficient and therefore increase the latent cooling capacity. The cooling capacity delivered by a vapor compression unit is largely determined by the control of the compressor. According to the vapor cycle chart and the model deduction presented before, the refrigerant side mass flow and enthalpy change dominate the cooling output. Increasing the compressor speed, with the condenser fan modulated accordingly, leads to lower evaporator surface temperature and higher sensible and latent cooling capacity.

To achieve a desired control objective, different control strategies and actions may be taken. An optimal may only be achieved when the action of the individual components is coordinated. Traditional PID-controlled multi-SISO approach does not consider the strong coupling among and within the individual component in equipment as the hybrid ventilation unit. Jakobsen [19] proposed a MIMO control strategy to identify the optimal settings for individual components. They pointed out that adjusting the compressor speed, modulating the condenser fan speed, or controlling the supply fan speed could achieve the same marginal cooling capacity increment expectation, but only one optimum existed. Their study showed the energy consumption penalty in a conventional multiple SISO control where the set points for the components are fixed.

Denis et al. [20] studied the application of nonlinear predictive control of a water-cooled vapor compression unit. The multi-objectives, including cooling capacity, energy efficiency, smooth command, were weighted to form the quadratic cost function. The compressor speed, expansion valve openness, and pump speeds for evaporator and condenser were manipulated variables. A 20-s sampling period and four steps prediction were adopted after trial and error. The simulation results showed a higher COP by about 20 % could be achieved with optimal control. Since a nonlinear optimization algorithm is utilized, the computational cost was reported higher in their study.

In reality, an optimal operation of an equipment is almost always subjected to a set of constraints. For example, the compressor speed can only be lower than the maximum but no less than a value (e.g., 30 %) for motor overheat protection. The optimization is usually also a compromise among the multiple objectives, e.g., additional dehumidification versus energy consumption. With a desired low energy consumption, the cooling or heating capacity has to be satisfied first. In addition to the obvious cost, the changing rate of the key components, such as the compressor, shall also be considered. A frequent and large speed change results in an early burnout. While the overall unit may not cost big money, the fault diagnosis, installation, component replacement, and the discomfort during the maintenance all count into the cost.

In this section, the classical MPC with linearized internal model is investigated for the optimal control of the hybrid unit. The base structure and rule-based multi-SISO control are first introduced. The controllable variables in the unit and their modulating range are stated. Three model linearization methods are briefly described to obtain the internal linear model for the MPC controllers. The performance of the proposed MPC for the three components in the hybrid unit is presented. The simulation results and findings are discussed.

6.2.1.1 Basic Structure and Control of the Hybrid Ventilation Unit

Figure 6.7 illustrates the base structure of the hybrid ventilation unit. There are five variable speed drives installed, four of them modulate fans' speed and one modulates the compressor speed. The enthalpy recovery wheel and the active desiccant wheel are equipped with speed adjustable motors. A gas burner is fed with natural gas through a modulating valve to control the regeneration air temperature. The binary points provided by the manufacturer include the following: system on/off, supply fan on/off, compressor on/off, compressor VFD status, condenser fan status, condenser fan VFD status, regeneration fan on/off, burner power on/off. The airflow, temperature, dew point, relative humidity and enthalpy of the outdoor air,

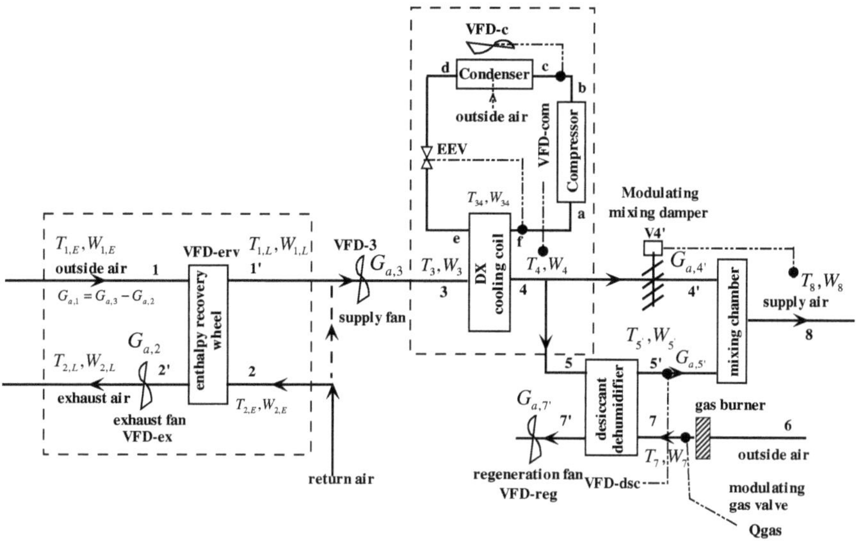

Fig. 6.7 Illustration of the hybrid ventilation unit with denotes

return air, air entering the coil, air leaving the coil, etc., are monitored through the manufacturer control panel. In this illustration, we keep the notes in consistence, as much as possible, with the nomenclature of this book. To ease the discussion, numbers as well as lower-cased letters are used to denote the variables at different locations in the system. This convention continues in the following figures of Simulink models associated with the components in the hybrid ventilation unit.

Tables 6.1 and 6.2 collect the parameter of the hybrid ventilation unit. The efficiency of design conditions are taken from the manufacture datasheet.

The original control of the hybrid unit is provided by the manufacturer with an integrated direct digital control system to enhance the control capabilities. The four types of control methods are listed in Table 6.3.

As with a typical rule-based control, the hybrid unit is divided into multiple loops in the original control algorithm: the outdoor air intake loop, the heat pump loop, the active dehumidification loop, and the mixing air loop. The outdoor air intake loop modulates the VFD on the module to maintain the outdoor airflow rate set point. The enthalpy recovery wheel is controlled at a full rotation speed for most of the time to recycle the heat and moisture from the exhaust air. The vapor compression unit loop modulates the compressor speed and coordinates the VFD on the condenser to provide a predefined supply air temperature. The active dehumidification loop controls the active desiccant wheel and the natural gas burner to maintain the return air humidity ratio. Currently, with control logic provided by the manufacturer, the supply airflow and the outside airflow are constantly fixed at arbitrary values.

As aforementioned, conventional rule-based control can hardly deal with the strong coupling within and among the elements. Meanwhile, the control law is usually set for a fixed reference point, without considering the control efforts. In an

Table 6.1 Design parameters of the desiccant wheels

	Enthalpy recovery wheel	Active desiccant wheel
Nominal airflow rate (cfm)	2000	660
Wheel diameter (mm)	787	787
Wheel depth (mm)	152	102
Wheel split ratio	1/2	2/3
Rotary speed (rpm)	30	0.42
Channel size (mm)	1.8*4.2	1.5*3.4
Desiccant thickness (microns)	25	65
Substrate thickness	15	75
Desiccant material	3A molecular sieves	Silica gel
Desiccant density (kg/m^3)	760	700
Desiccant specific heat (J/kg K)	1000	1000
Desiccant conductivity (W/m K)	0	0
Separation factor	0.1	1
Maximum moisture loading (kg/kg)	0.2	0.36
Heat of adsorption (J/kg)	2,791,000	2,791,000
Substrate material	Aluminum	Glass fiber paper
Substrate density (kg/m^3)	2700	500
Substrate specific heat (J/kg K)	900	900
Substrate conductivity (W/m K)	237	0
Convective heat transfer coefficient (W/m^2 K)	36.2	43.3
Convective mass transfer coefficient (m/s)	0.03	0.036

Table 6.2 Facts of the hybrid desiccant ventilation unit

DX unit + active desiccant unit
Single variable speed compressor, pressure: 144–278 PSI, total refrigerant charge: 15–21 lbs (6.8–9.53 kg), maximum cooling capacity: 12.5 tons (44 kW), maximum heating capacity: 150,000 BTU/h (44 kW), maximum burner capacity: 74,592 BTU/h (21 kW)
Supply fan motor: 3 HP (2.23 kW), supply airflow: 2250–3600 cfm (1.062–1.7 kg/s), condenser fan motor: 1 HP (0.74 kW), condenser fan airflow rate: 9375 cfm (4.42 kg/s), Regen fan motor: ½ HP (0.37), regeneration fan maximum airflow: 660 cfm (0.311 kg/s). Sensible heating ratio: 0.38–0.6 (with 40 % OA), 0.24–0.5 (with 100 % OA)

MPC optimal control, the control law for a MIMO is naturally included through the internal model. The cost function can be defined to represent the control efforts and the tracking errors. The constraints on the inputs and outputs are naturally embedded as well. It is suitable for an equipment control to reduce the energy consumption for the given set points. By incorporating multiple criteria and/or constraints, the hardware life cycle can also be extended.

Table 6.3 The typical control modes of the hybrid unit

Mode	Description	Notes
A	Space temperature and humidity control	The air temperature and relative humidity are monitored in a space and used to determine the supply air temperature and humidity set points. Temperature dominant control
B	Space temperature and return humidity control	The air temperature and relative humidity are monitored in the return air duct to determine the set point of supply humidity to maintain humidity set point. Temperature dominates
C	Supply temperature and space (return) humidity control	Used where other air-conditioning units are deployed to maintain the space temperature. This unit is controlled to provide a predetermined moderate or neutral supply air temperature. Humidity dominates
D	Building automation variable integration with supply temperature and return humidity control	Used to condition multiple spaces or feeding VAV boxes. The temperature and relative humidity are monitored in the return air duct to determine what supply humidity is needed to maintain humidity set points. Temperature dominates

6.2.1.2 Model Linearization and State-Space Realization (Design)

The internal model for the low-level MPC requires a linear format. This section handles the nonlinear model linearization for the direct expansion (DX) unit. Jordan [21] provided an excellent overall summary of linearization methods of nonlinear state equations. The main methods widely used in applications are Taylor's series expansion, optimal linearization method, and global linearization method.

A nonlinear system can be expressed in the following form:

$$\dot{x} = f(x, u, \tau), x(0) = x_0 \tag{6.12}$$

where x is the states vector, u is the control inputs, τ is time, and f is the nonlinear dynamic mapping.

In many circumstance, numerical methods can be applied to directly solve the set of equations. The analysis of a nonlinear system imposes challenges. It can also be approximated by using linearization and put in the state-space form as follows:

$$\dot{x} = Ax(\tau) + Bu(\tau), x(0) = x_0 \tag{6.13}$$

where A and B are the state matrix and input matrix.

This linear system has analytical solution as follows:

$$x(\tau) = e^{A(\tau-\tau_0)}x_0 + \int_{\tau_0}^{\tau} e^{A(\tau-\tau_0)}Bu(\tau)d\tau, \quad x(0) = x_0 \tag{6.14}$$

Taylor's series expansion utilizes the following approximation about an equilibrium point:

$$f(x) = \sum_{n=0}^{\infty} \frac{f^{(n)}(a)}{n!}(x-a)^n \tag{6.15}$$

By taking only the terms before the second order, the linearization in incremental form can be obtained as follows:

$$\Delta\dot{x} = \begin{bmatrix} \frac{\partial f_1}{\partial x_1} & \cdots & \frac{\partial f_1}{\partial x_n} \\ \vdots & \vdots & \vdots \\ \frac{\partial f_m}{\partial x_1} & \cdots & \frac{\partial f_m}{\partial x_n} \end{bmatrix} \Delta x + \begin{bmatrix} \frac{\partial f_1}{\partial u_1} & \cdots & \frac{\partial f_1}{\partial u_n} \\ \vdots & \vdots & \vdots \\ \frac{\partial f_m}{\partial u_i} & \cdots & \frac{\partial f_m}{\partial u_n} \end{bmatrix} \Delta u \tag{6.16}$$

Optimal linearization utilizes the least squares method to find the linear expression for the original nonlinear system. The goal is to minimize the inner product of the error as follows:

$$I = (e, e) \tag{6.17}$$

$$e = Ax + Bu - f(x(\tau), u(\tau), \tau) \tag{6.18}$$

A global linearization uses global diffeomorphism through the state variables transformation to solve the nonlinear system equations [22]. A set of new variables, k, are defined so that the new system in terms of the new variables has a linear format:

$$\dot{k} = A\dot{k} + Bv \tag{6.19}$$

$$k = f(x) \tag{6.20}$$

The new input v is a nonlinear map of the original inputs u and the function defined by the states x. With the solution to the new linear system about k, the original system can be obtained via the inverse transformation:

$$x = f^{-1}(k) \tag{6.21}$$

A local linearization through Taylor's series is generally adopted for engineering analysis. Choosing a right operating point is critical to obtain an accurate linear model for a nonlinear system. When a system works in a wide range with multiple equilibrium points, a constant linearization is needed to ensure the controller performance.

6.2.1.3 MPC Architecture for the Hybrid Ventilation Unit

The hybrid ventilation unit has eight variables that can be continuously manipulated, as collected in Table 6.4. The condenser fan, compressor rotation speed, and the main unit fan jointly form the control of the vapor compression unit. The operation of this unit dominates the performance of the hybrid unit since it provides both cooling and dehumidification to the air stream. The control of the vapor compression unit aims to provide the required supply air temperature with optimal energy consumption and less control efforts. The enthalpy recovery wheel is a non-source system. It does not convert electrical energy to thermal energy. The control is to maximize the heat and moisture exchange between the outdoor air and exhaust air when the exchange is beneficial to the space, and minimize the exchange when it is not. The control of the active desiccant unit has narrow margin since the performance of the dehumidification is largely determined by the predefined configuration and operation. The hybrid ventilation unit mainly controls the space humidity. The supply air temperature is also controlled to avoid a large change and disturbance to the space air which is thermally conditioned by the water-based system. The controlled variables are listed in Table 6.5.

Three MPCs can be structured for the three components individually, as illustrated in Fig. 6.8. With this structure, the supervisory controller supplies the reference set points to the MPCs. The function of the three components in the hybrid unit is clear:

Table 6.4 Manipulated variables of the hybrid ventilation unit

Variables	Notes	Purpose	Maximum	Minimum	Unit
Vapor compression unit					
VFD-c	Condenser airflow rate	Maintain <278 PSI (19 bar) condenser pressure	4.42	0	kg/s
VFD-com	Compressor rotation speed	Control the outlet air temp of evaporator	30	100	%
VFD_3	Main unit fan speed	Maintain total supply airflow rate	1.7	0	kg/s
Enthalpy recovery unit					
VFD_{erv}	Recovery wheel rotation speed	Constant speed	30	0	rpm
VFD_{ex}	Exhaust airflow rate	Match outdoor airflow rate	1.04	0	cfm
Active desiccant unit					
Q_g	Gas burner	Maintain the regeneration air temperature	21	0	kW
VFD_{reg}	Regeneration airflow rate	Maintain the	0.31	0	kg/s
VFD_{dsc}	Desiccant wheel rotation speed	Constant speed	0.42	0	rpm

Table 6.5 Controlled variables of the hybrid ventilation unit

Variables	Notes	Purpose	Maximum	Minimum	Unit
P_{com}, T_{com}	Compressor outlet line pressure	Maintain <278 PSI condenser pressure	19	9	Bar
T_4	Cooling coil outlet air temperature	Common control practice	26 + 273	9 + 273	K
T_7	Regeneration air temperature	Avoid overheating of the regeneration air	80 + 273	Outdoor air temp	K
T_8	Supply air temperature	Maintain the space ventilation and humidity	26 + 273	13 + 273	K
w_8	Supply air humidity	Maintain the space humidity	From supervisor	–	g/kg

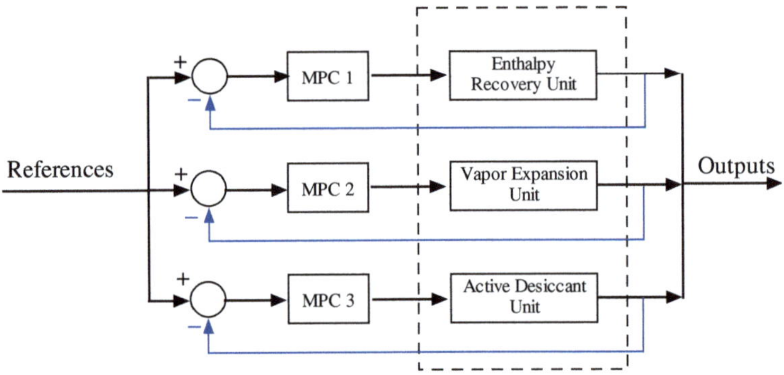

Fig. 6.8 MPC architecture—Example 1

- The ERV controls the outdoor air intake. For example, it maximizes the outdoor air intake for free cooling and minimizes the outdoor air intake for mechanical cooling and heating;
- The vapor expansion unit tracks the supply air temperature; it is desirable and it uses less electricity or with less control efforts for the same load condition;
- The active desiccant unit tracks the supply air humidity; it runs only when the humidity after the vapor expansion unit is high.

The decentralized structure has better reliability since the computational load is divided into three. Another approach could be an integrated MPC for the whole unit, as shown in Fig. 6.9. The computation load is higher with more manipulated variables and constrains. The tuning effort for the centralized MPC control is also higher. The benefit of using a distributed MPC structure is more obvious when the equipment complexity grows.

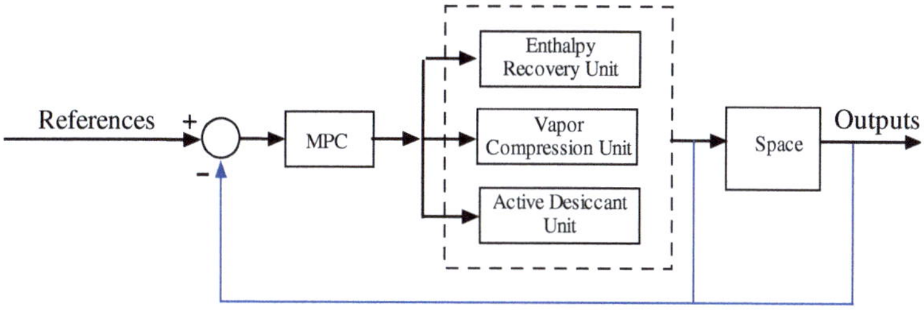

Fig. 6.9 MPC architecture—Example 2

(1) Enthalpy recovery wheel

The enthalpy recovery wheel is a non-source unit for outdoor air intake preprocessing. The function is to recover the heat and moisture from the space exhaust air into the outdoor air intake in mechanical heating seasons. In mechanical cooling seasons, the process is reversed with heat and moisture transferred from the outdoor air intake into the exhaust air. Figure 6.10 shows the variables involved in the MPC of the ERV.

As shown in Fig. 6.10, the outdoor air temperature, humidity, return air temperature, humidity are measured disturbances on the first-order system. The enthalpy recovery wheel unit is a first-order system along with varying gain, varying delay, and disturbances. The inputs on the exhaust fan airflow rate and the wheel rotation speed are constrained. All these conditions remark that MPC is a good choice for the control.

The model of the enthalpy recovery wheel is programmed in MATLAB as an S-function. The S-function block of the model is wrapped in a sub-system with additional inputs and outputs as plotted in Fig. 6.11. To compare the performance of the MPC with conventional PIDs in terms of the reference tracking, two models with PID controllers are also built for the unit. As a SISO-oriented controller, the two manipulated variables, the wheel rotating speed and the exhaust fan speed, need to be assigned to two references separately to minimize the coupling issue. The sign of the PI gains in the two PID controllers shall be adjusted when the process switches from direct action to inverse action, e.g., from winter to summer. ERV model with PID controllers is shown in Fig. 6.12.

The control cost function of MPC is defined as the summation of the weighted quadratic function of the tracking errors, manipulated variables, and changing rates.

Fig. 6.10 MPC structure for enthalpy recovery unit (ERV)

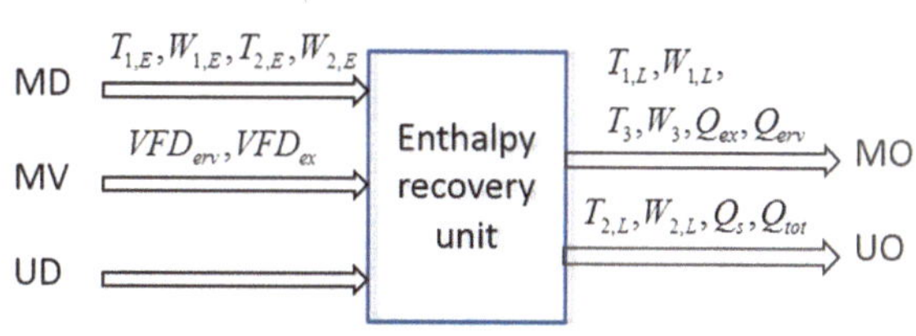

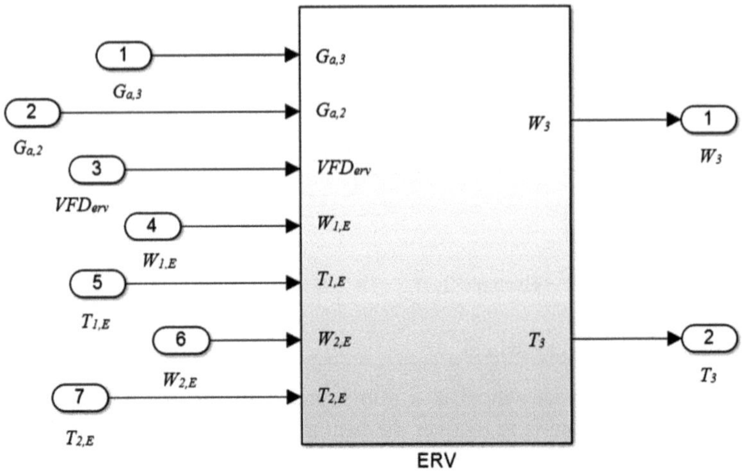

Fig. 6.11 ERV model wrapped as a block

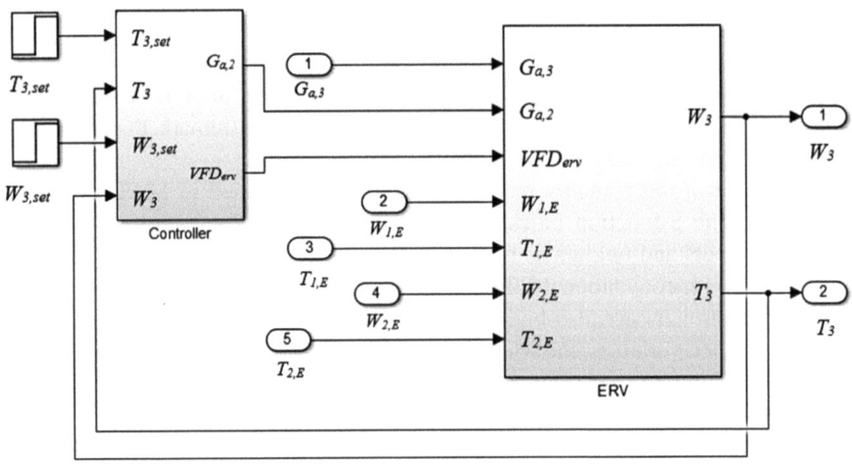

Fig. 6.12 ERV model with PID controllers

$$\min \sum_{i=1}^{k} \|e^y\|_Q^2 + \|\Delta u\|_S^2 + \|e^u\|_R^2 \tag{6.22}$$

As a non-source equipment, the accurate control of the ERV to achieve both the required supply air temperature and humidity is extremely difficult. The disturbances come from outside air and return air, including the return air quantity. As SISO-based controller, setting and tuning PIDs for the coupled thermal and mass transfer process can be challenging. We pick one-hour data in winter and control the unit with both PIDs and MPC to compare the performance. The controllers only track the air temperature since air humidity is not a concern in the space in winter and there is no humidifier equipped. Handling the control with PID requires the tuning of the controller for both the enthalpy recovery wheel rotation speed and the exhaust fan. With MPC control, it is naturally considered and the different control performances are achieved by adjusting the weighting matrix for the reference tracking, manipulated variables, the rates, etc.

For the enthalpy recovery wheel, when the system is operated at the point of $G_3 = 1.13$ kg/s, $G_2 = 0.65$ kg/s, $\omega_{\text{erv}} = 20$ rpm, $w_{\text{e}} = 11.05$ g/kg, $T_{\text{e}} = 302\,\text{K}$, $w_{\text{r}} = 8.4$ g/kg, $T_{\text{r}} = 298\,\text{K}$, the system matrices A and B are calculated as follows:

$$A_{\text{m}} = \begin{bmatrix} -0.0565 & 0 & 0.0325 & 0 \\ 0 & -0.0565 & 0 & 0 \\ 0 & 0 & -0.0105 & 0 \\ 0 & 0 & 0 & -0.0124 \end{bmatrix} \tag{6.23}$$

The eigenvalue of the characteristic matrix is as follows:

$$\lambda_{\text{m}} = \begin{bmatrix} 5.650e-002 \\ 5.650e-002 \\ 1.050e-002 \\ 1.240e-002 \end{bmatrix} \tag{6.24}$$

The matrices all have full rank, which indicates the system is both observable and controllable. Meanwhile, the eigenvalues of the linearized dynamic model are shown in Eq. (6.24) for the operating point. All negative real parts are found for the values, which means the system is asymptotically stable [23, 24]. The MPC controller is programmed in MATLAB.

Figure 6.13 plots the comparison of the results between two PID controls and MPC control. PID one (PID-1) has the two controllers both track the supply air temperature. PID two (PID-2) has the two PID controllers track the set points of the supply air temperature and the humidity separately. The controllers are tuned for winter season and winter weather data is used. All the controllers work well to track the supply air temperature set point. With the weight 1 on the supply air temperature tracking, MPC responses a little bit slower than the two PID controllers. The tracking error of MPC is less than the two PIDs as zoomed-in in Fig. 6.14, with the response from MPC almost overlap the set points. The control actions of MPC are plotted in Fig. 6.15.

The same controllers are applied with the weather data changed from winter to summer. It is found that the PID controllers do not function well since the action of the unit is complex and determined by four disturbances. PID-1 drives the outdoor air intake to the maximum and turns off the enthalpy recovery wheel, while PID-2 drives both the outdoor air intake and the enthalpy recovery wheel rotation speed to the maximum. This demonstrates that PID controller, as a reference tracking only controller, is not a good candidate for the control of the ERV unit. With the same

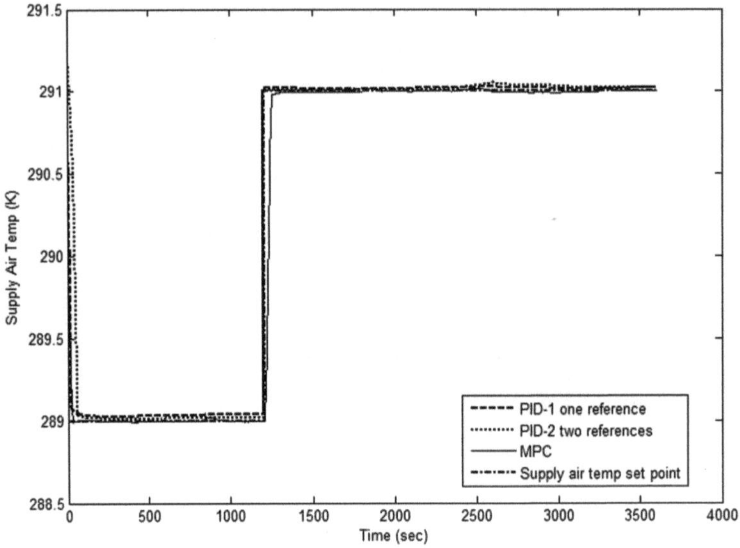

Fig. 6.13 Comparison of PIDs and MPC for ERV control—winter 1 h

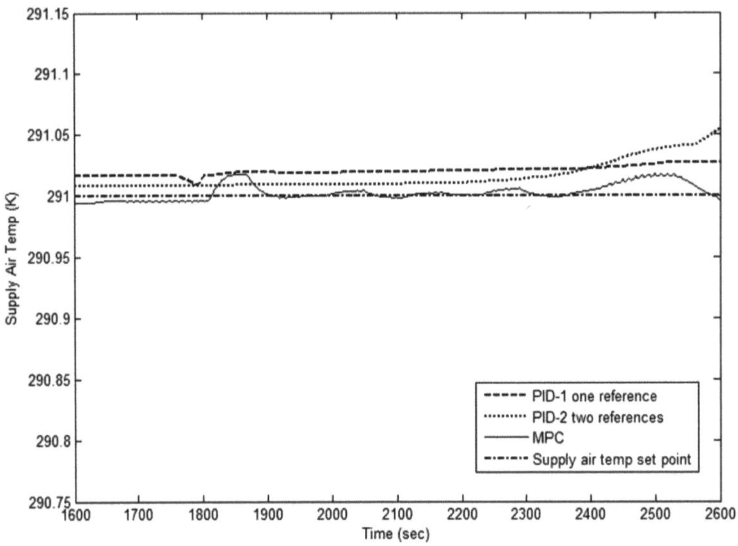

Fig. 6.14 Comparison zoom in—winter 1 h

settings, although the set points cannot be maintained due to the feasibility, MPC drives the outdoor air intake to the minimum intake almost immediately and turns the ERV rotation speed to the maximum. The control efforts of MPC for a given target can also be adjusted by changing the weights on the manipulated variables.

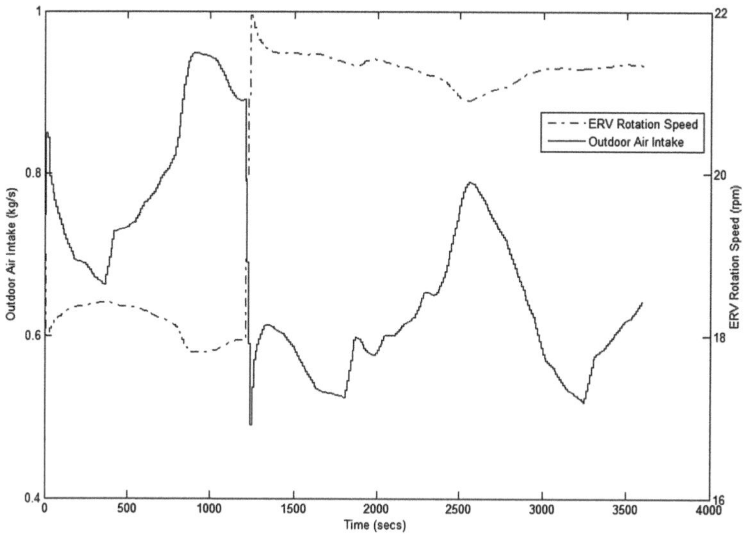

Fig. 6.15 Outdoor air intake and ERV rotation speed of MPC

The control weights on the inputs and outputs are the main variables to drive the MPC toward the desired direction. Four additional scenarios as listed in Table 6.6 are further studied for the MPC control of the ERV unit.

In Scenario 1, the mixed air humidity set point is picked as the most important variable. A weight of 0.5 is also set for the outdoor air intake. With the settings, MPC drives the outdoor air to a low level as long as the humidity set point can be satisfied. Figure 6.16 plots the outdoor air intake and the ERV rotation speed. The ERV rotates at high speed for most of the time since the weight of the control is zero. The supply air humidity is maintained well as shown in Fig. 6.17. Except the initial shoot, the supply air humidity almost overlaps with the set point. The moisture is transferred from the outdoor air to the exhaust air. Since the supply air temperature gets zero weight, the response of the supply air temperature, as shown in Fig. 6.18, is not controlled.

Table 6.6 MPC control scenarios for summer and swing weather conditions

Scenario	Weather	Humidity set point (g/kg)	Temp set point (K)	Weight on humidity	Weight on temp	Weight on OA intake	Weight on ERV speed
1	Summer	10	299–299.5	1	0	0.5	0
2	Summer	10	299–299.5	0	1	0	0
3	Summer	8	290	1	1	0	0
4	Swing	8	290	0	Variable	Variable	0

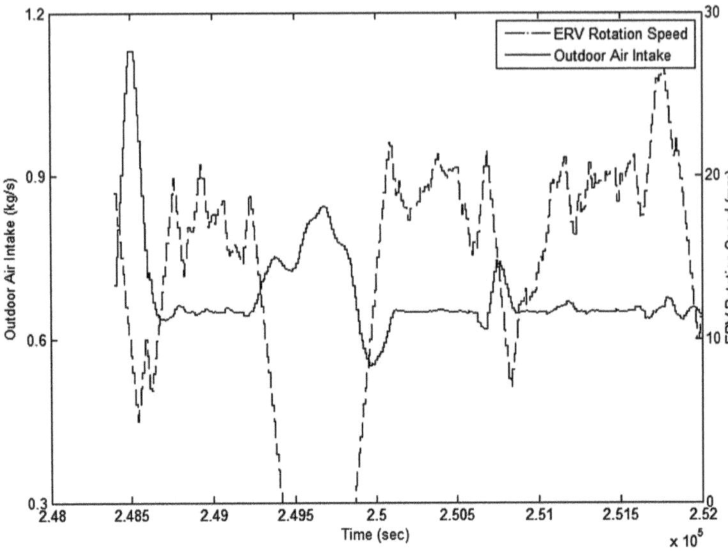

Fig. 6.16 Outdoor air intake and ERV rotation speed in summer (Scenario 1)

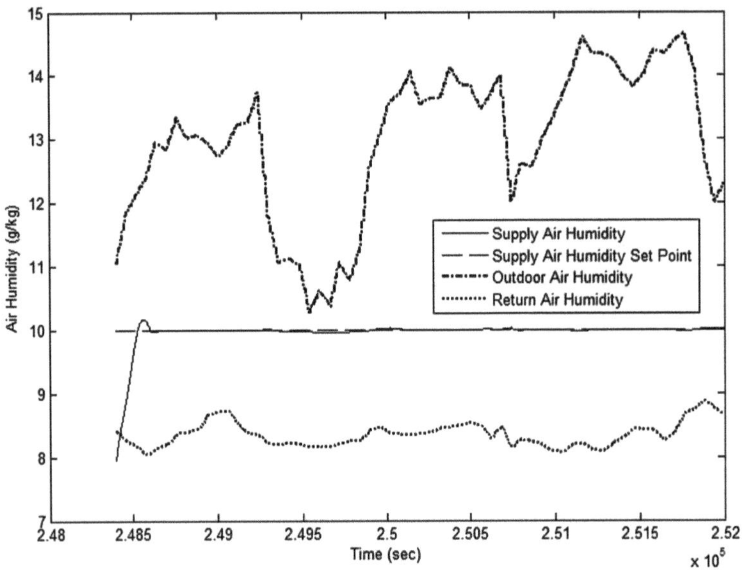

Fig. 6.17 ERV supply air humidity control in summer (Scenario 1)

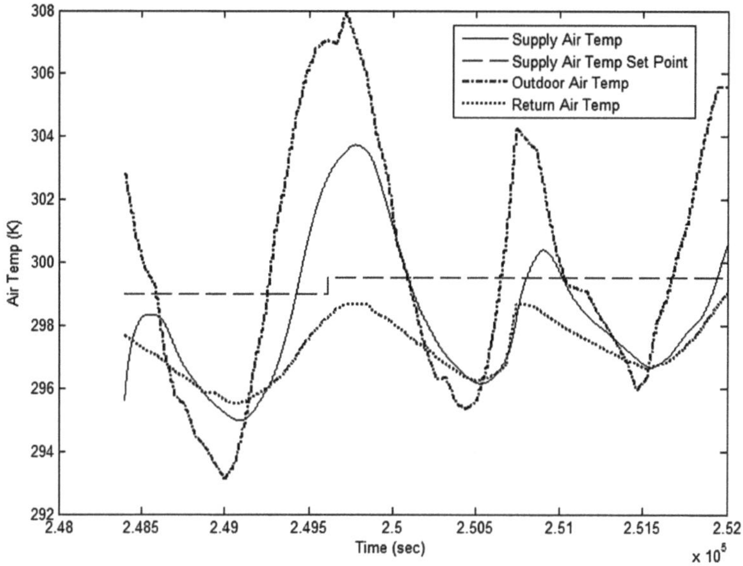

Fig. 6.18 ERV air temperature control in summer (Scenario 1)

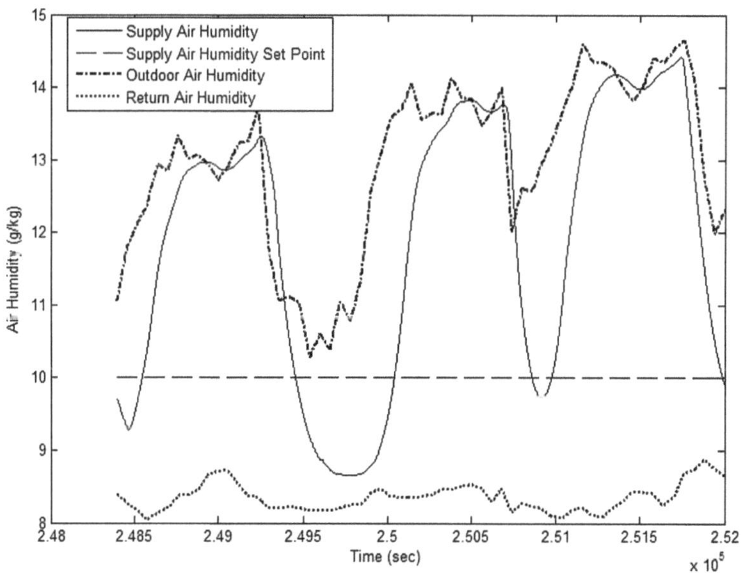

Fig. 6.19 ERV air humidity control in summer (Scenario 2)

For the same time period and working conditions, Scenario 2 gives a weight as 1 to the supply air temperature and 0 to the supply air humidity. The weight on both the outdoor air intake and the ERV rotation speed are set to zero. The MPC mainly

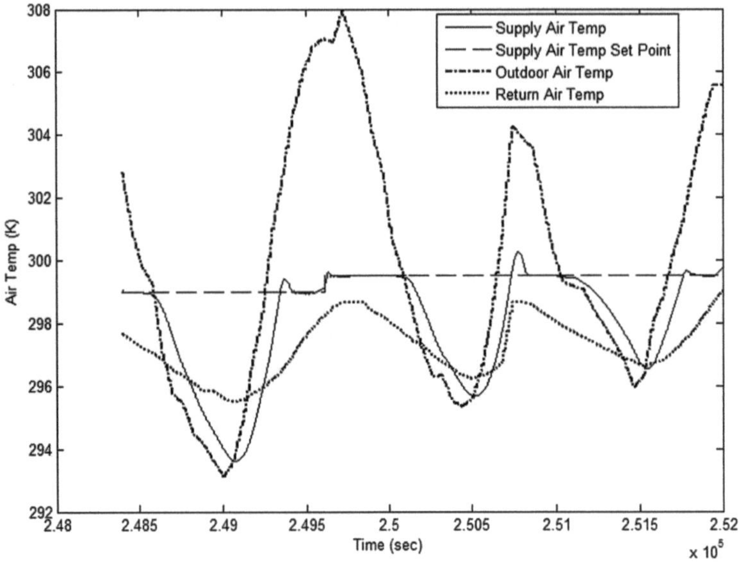

Fig. 6.20 ERV air temperature control in summer (Scenario 2)

concerns the supply air temperature. In Fig. 6.19, the humidity set point is ignored by the MPC controller. The humidity responses naturally to the mass transfer process. Although the time period is in summer, there are several short periods when the supply air temperature can possibly be met and MPC makes it as shown in Fig. 6.20. Figure 6.21 plots the corresponding modulation of the outdoor air intake and the ERV rotation speed under the weighting settings. The outdoor air intake is maximized when the outdoor air temperature is lower than the supply air temperature set point, which is desirable for energy savings.

In Scenario 3, the supply air humidity set point is set to 8 g/kg and the temperature set point is set to 290 K. As in Scenario 3, the effort of outdoor air intake and ERV rotation speed are ignored by setting the weights to zero. However, the set points of the supply air temperature and humidity are reduced to 290 K and 8 g/kg, respectively. During the time period, the outdoor air humidity is constantly above 10 g/kg, and high to about 15 g/kg. As plotted in Fig. 6.22, the MPC automatically reduces the outdoor air intake and modulates the enthalpy recovery wheel up to the maximum speed to recycle the heat from the exhaust air. The minimum outdoor air intake to maintain the ventilation air during the mechanical cooling season is achieved. This indicates, in summer, a more realistic set points can drive the MPC toward minimum outdoor intake. The supervisory-level controller can put a low limit on the outdoor air intake to ensure the ventilation air intake.

It is desired that the control of the outdoor air processing unit can bring in more outdoor air when the space is in cooling mode and the outdoor air condition is favorable to maximize the free cooling benefit. When the outdoor air is warm, the intake is reduced to the minimum. To investigate the control of MPC, a time slot is

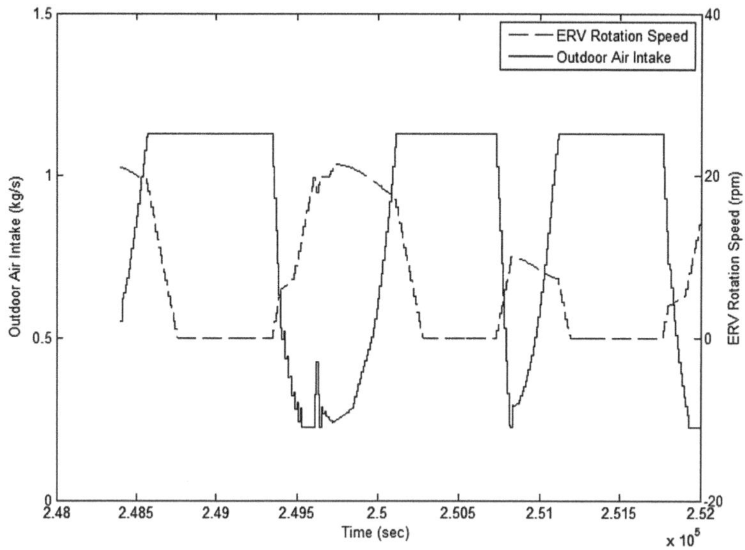

Fig. 6.21 ERV manipulated variables in summer (Scenario 2)

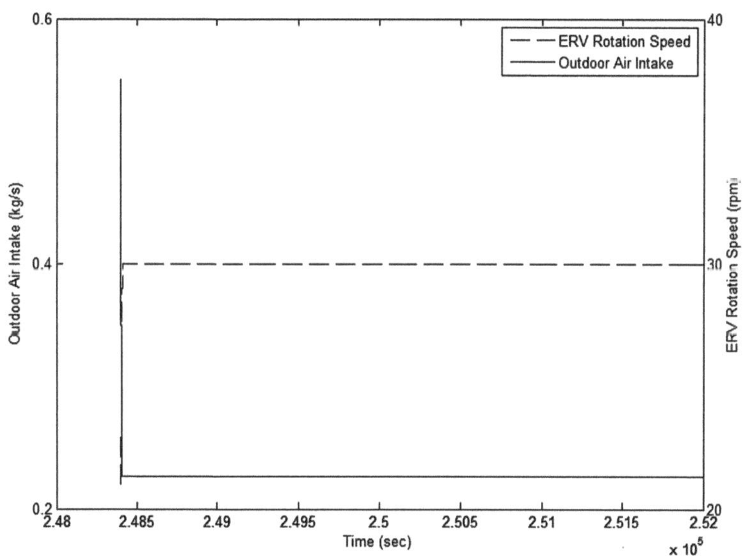

Fig. 6.22 ERV manipulated variables in summer (Scenario 3)

picked as Scenario 4. As shown in Figs. 6.23 and 6.24, the outdoor air temperature fluctuates between 284 and 302 K and the outdoor air humidity is around 5 g/kg for most of the time. A dynamic linearization or weighting adjustment is needed to ensure the linear model represents the original nonlinear model well and the MPC

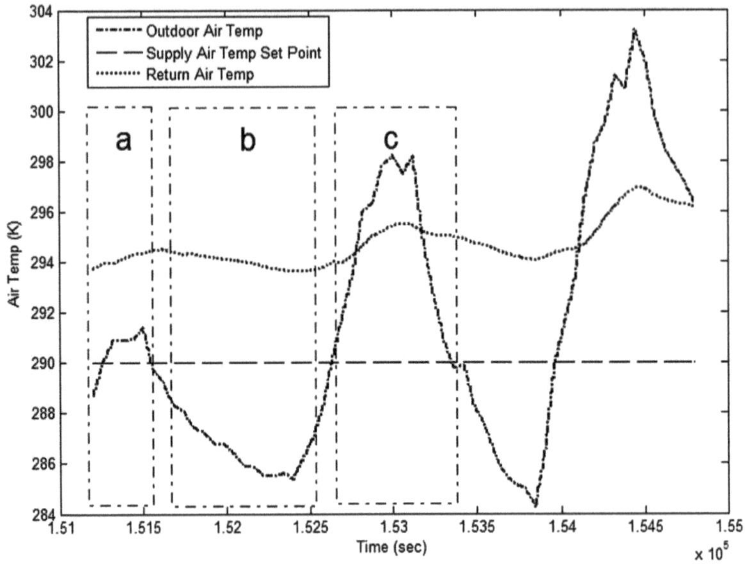

Fig. 6.23 Disturbance conditions (Scenario 4)

functions properly, due to the following three different modes in the time period as marked in Fig. 6.23.

- For the initial time period, condition 'a' in Fig. 6.23, the outdoor air temperature and the return air temperature are close to each other and the set point;
- For condition 'b', the outdoor air temperature is lower than the set point while the return air temperature is higher than the supply air temperature set point;
- For condition 'c', both the return air and the outdoor air temperature are higher than the supply air temperature set point.

With a rule-based operation, for condition 'a' and with space in cooling mode, maximum outdoor air intake shall be encouraged since the outdoor air is favorable. For condition 'b', a modulation is probably needed to maintain the supply air temperature set point. For condition 'c', a minimum outdoor air intake is considered to avoid penalty since the outdoor air temperature is higher than the return air temperature.

With the MPC, a conditionally consecutive linearization and dynamic weighting are used to accommodate the large switches. When the outdoor air temperature is lower than the return air temperature, the system is linearized around the operation point. The supply air temperature gets a weight as 1. To prevent the unit from over-tuning, a weight of 0.5 is imposed on the outdoor air intake modulation rate. When the outdoor air temperature changes to be higher than the room air temperature and the supply air temperature set point, a weight as 1 is added to the outdoor air intake and the supply air temperature output. The simulation results are plotted in Figs. 6.24, 6.25, 6.26, and 6.27.

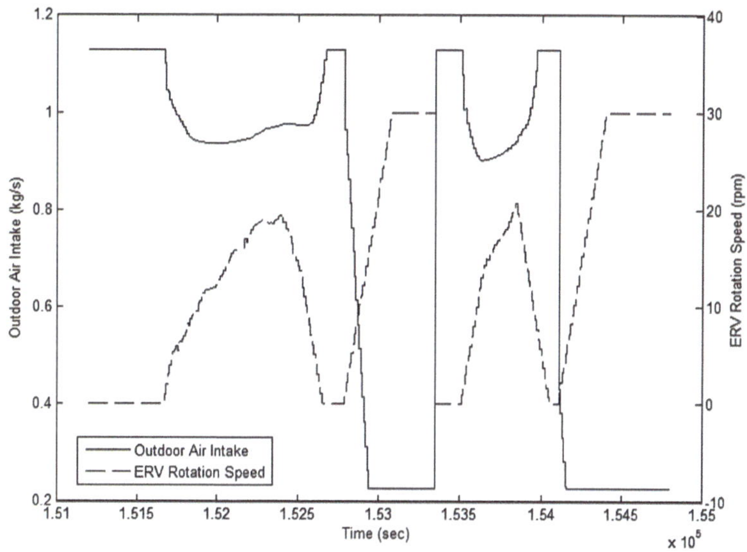

Fig. 6.24 ERV manipulated variables in swing season (Scenario 4)

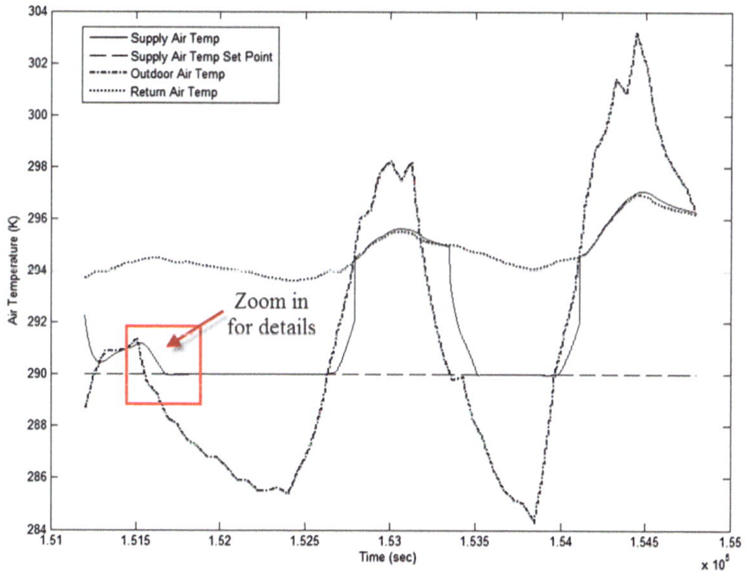

Fig. 6.25 ERV air temperature control in swing season (Scenario 4)

It is shown in Fig. 6.24 that the outdoor air intake is kept to maximum when the outdoor air temperature is lower than the return air temperature and close to the supply air temperature. An economizer with maximum outdoor air intake is

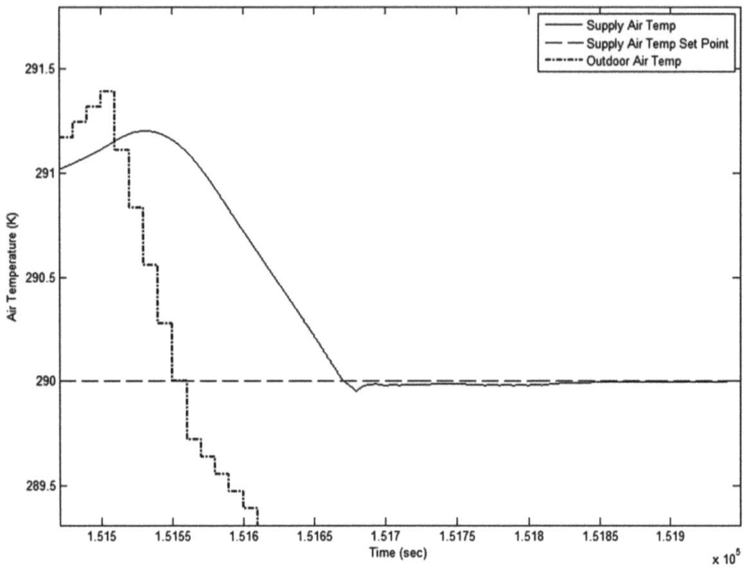

Fig. 6.26 ERV air temperature tracking zoom in as in Fig. 6.25 (Scenario 4)

achieved. When the outdoor air temperature drops to be lower than the supply air temperature, a modulation action is taken by the MPC. From Figs. 6.25 and 6.26, it is shown that the supply air temperature is controlled to 290 K as the set point. This strategy is similar to free cooling operation as in rule-based control. 290 K is selected to prevent over-cooling of the space and it is adjustable if more free cooing is desired. When the outdoor air rises to be higher than the return air temperature, it is a typical cooling season and a minimum outdoor air intake is desired. As illustrated in Fig. 6.27, MPC reduces the outdoor air intake to the minimum and turns the ERV rotation speed to full to recycle the heat. The warm outdoor air intake is cooled down to close to the return air temperature. The outdoor humidity set point is not weighed since the outdoor air in the season is mild or dry. Dehumidification is not a concern with temperature control dominates.

(2) Vapor compression unit

The vapor compression unit itself is a highly complex nonlinear system. The components include the compressor, condenser, thermal expansion valve, evaporator, and main fan. The thermal expansion valve is a self-regulating device with proportional control. On the evaporator side, a heat and moisture-coupled transfer process happens. To provide dehumidified air to the space, the compressor varies the speed and coordinates with the condenser fan for wider cooling capacity modulation. As pointed out in the introduction, a same marginal cooling capacity can be achieved via the operation on the compressor, supply air fan, or condenser fan. The goal of utilizing an MPC for this unit is to bring in the benefits of MIMO

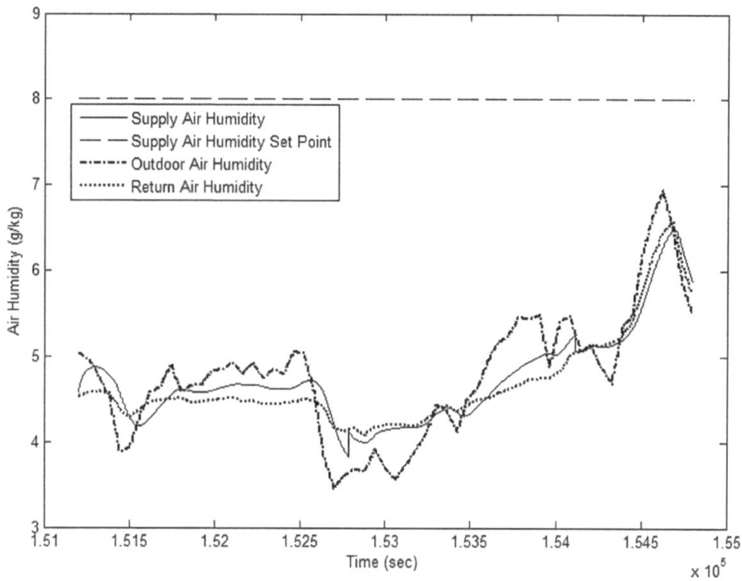

Fig. 6.27 ERV air humidity control in swing season (Scenario 4)

Fig. 6.28 Abstracted DX unit MPC structure

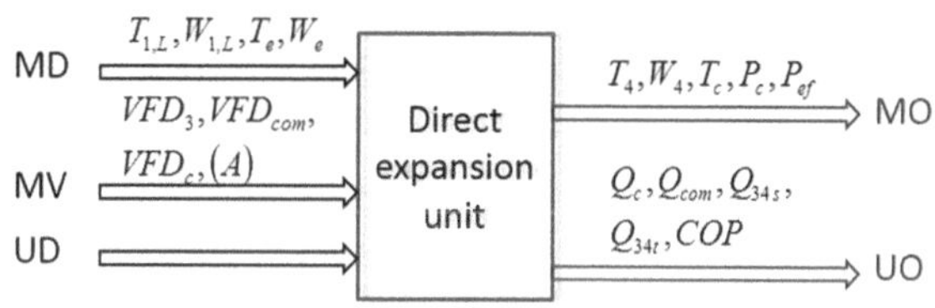

control along with energy consumption, equipment life cycle cost, and handling of dynamic coupling and constraints. The system monitors the supply air temperature and humidity as the controlled variables. The disturbances include the varying outside air temperature, return air temperature, and humidity. To extend the life span of the unit, we also need to penalize the compressor speed modulation rate from frequent changing. The structure is illustrated in Fig. 6.28 for the vapor compression unit.

Figure 6.29 shows the integrated model built in Simulink. The model of the DX unit includes the four blocks: the condenser, evaporator, compressor, and expansion valve. The inputs and outputs of the components are marked with notes. Figures 6.30, 6.31, 6.32, and 6.33 plot the individual component model in Simulink. The combined nonlinear model serves as a system simulator for the MPC controller. Another linearized model around the operation point is utilized in the MPC for control.

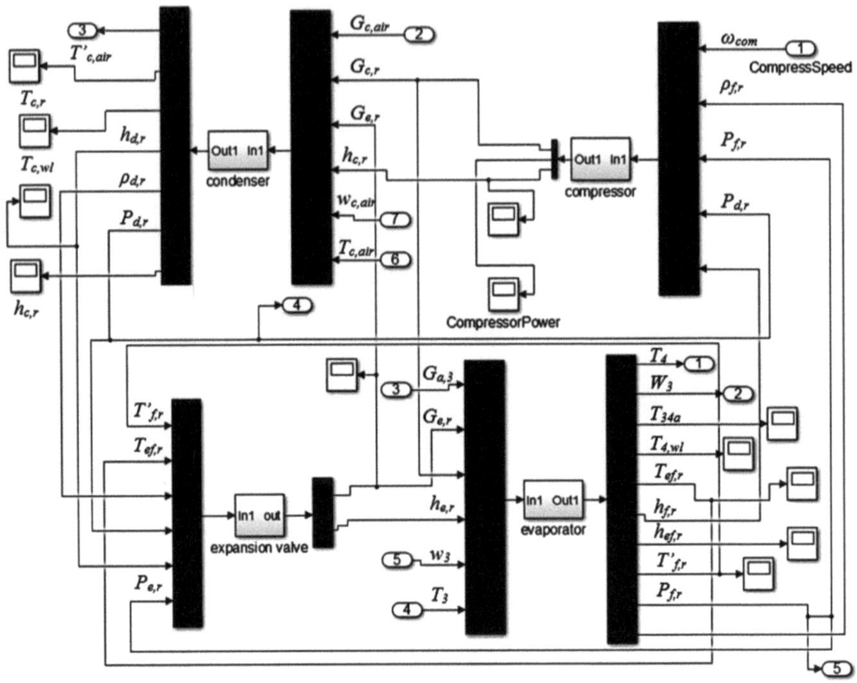

Fig. 6.29 DX unit model in Simulink

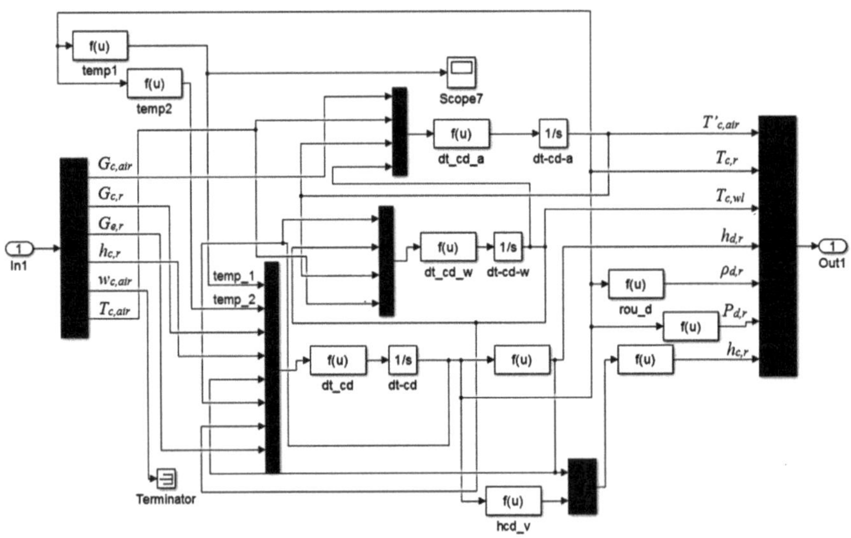

Fig. 6.30 Condenser model in Simulink

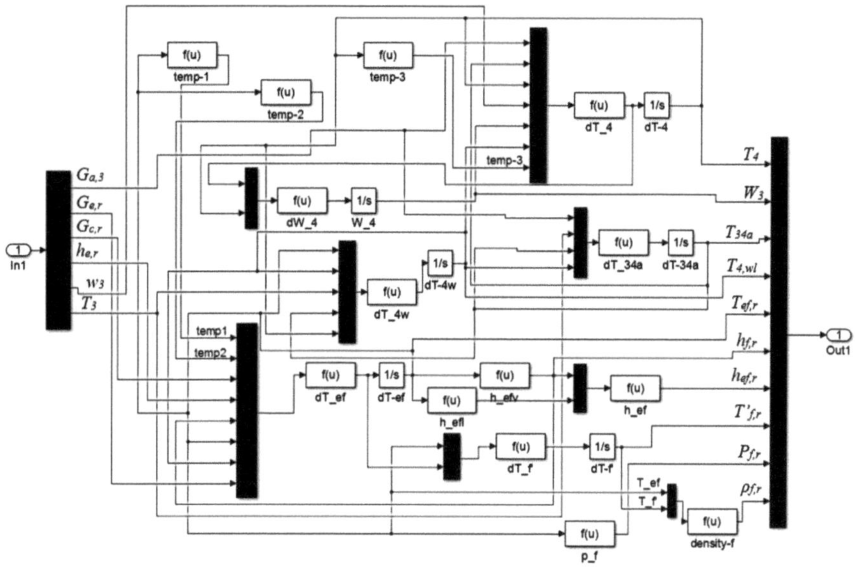

Fig. 6.31 Evaporator model in Simulink

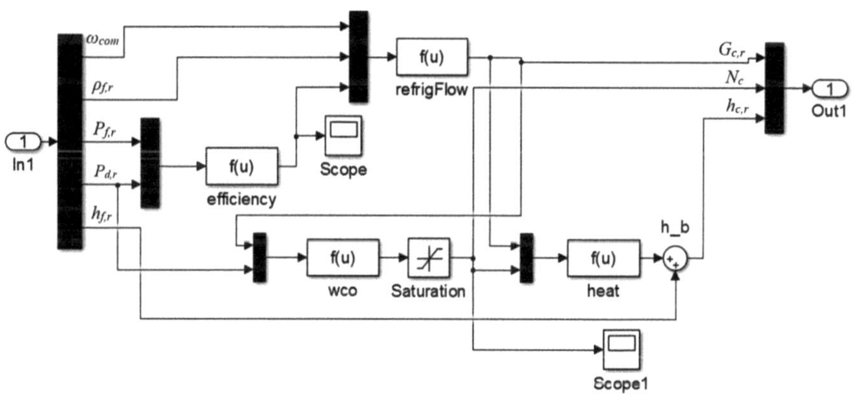

Fig. 6.32 Compressor model in Simulink

As a cooling unit, the typical operation is to maintain a predefined supply air temperature and allow the supply air humidity to drift around. This control can cause the adverse sensible and latent load coupling issue when the space sensible and latent load does not match the sensible and latent capacity of the unit. Under

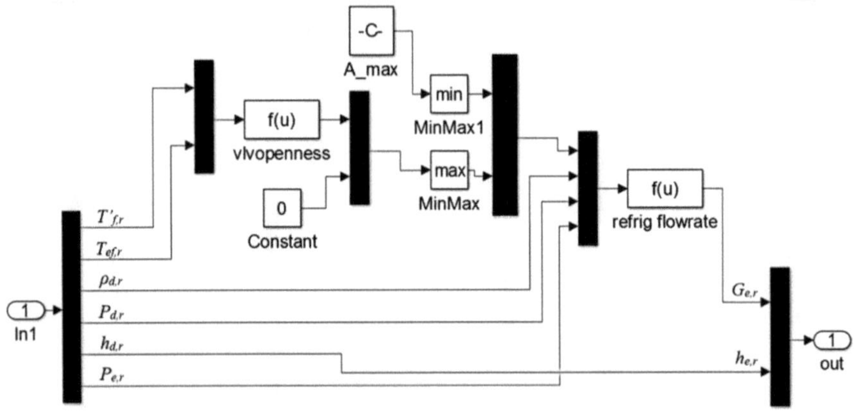

Fig. 6.33 Expansion valve model in Simulink

most circumstances, the space humidity is allowed to fluctuate within a range (e.g., 45–65 %), instead of a fixed set point. Another possible solution is to allow the supply airflow rate to be modulated accordingly when the sensible capacity and the latent capacity do not match. For the ventilation unit deployed, the control is decoupled by adding an additional dehumidification unit. The DX unit only controls the supply air temperature within a predefined range, allowing the humidity to fluctuate. The additional dehumidification unit monitors the room air humidity.

With conventional PID control, the modulation of the fan, condenser, and compressor needs to be put in a set of rules to ensure a good control on the supply air temperature and avoid interest conflicts. The component wearing, maintenance and overall unit life span can hardly be described in the control rules. With optimization-based MPC, the modulation of the compressor and fans can be integrated in the control law by penalizing the frequent modulation. The coupling among the components and variables is evaluated by the internal model and the optimization.

In reality, a DX system is designed to operate in the vicinity of a set point for a given load that does not change significantly. The model of the DX unit built in MATLAB is linearized for the space conditioning in summer around the typical operation point. Here, we pick the following: T_3=24 °C, W_3 = 10 g/kg, T_3 = 30 °C, W_e = 14 g/kg, and G_3 = 1.4 kg/s. The reduced-order system has eight states and seven inputs. The compressor rotation speed, the condenser fan airflow rate, and the main supply fan airflow rate are the three manipulated variables. The variable we does not have any effect on the states since the air humidity at the condenser side is not of concern. The supply air temperature, supply air humidity, condenser air temperature, condenser inline pressure, and evaporator inline pressure are the five measured outputs.

The system matrices of this DX unit are obtained as follows for the operation condition:

$$A_{\mathrm{m}} = \begin{bmatrix} -0.8255 & -0.4288 & 0 & 0 & 0 & -0.4825 & 1.3080 & 0 \\ -0.3701 & -0.1922 & 0 & 0 & 0 & -0.2163 & 0.5864 & 0 \\ 0 & 0 & -1.8466 & 0 & 0 & 0 & 0 & 2.1007 \\ 0 & 0 & 0 & -7.3625 & 0.6589 & 0 & 0 & 7.2909 \\ 0 & 0 & 0 & 0.2524 & -9.0039 & 0 & 8.2293 & 0 \\ 0 & 0 & 0 & 0 & 0 & -0.5985 & 0.4696 & 0 \\ 0.2134 & 0 & 0 & 0 & 0.8077 & 0.2495 & -1.3068 & 0 \\ 0 & 0 & 0.2101 & 0.8956 & 0 & 0 & 0 & -1.3157 \end{bmatrix} \tag{6.25}$$

$$B_{\mathrm{m}} = \begin{bmatrix} 0 & 0 & 0.4580 & 0 & 0.4288 & 0 & 0 \\ 0 & 0 & 0.2053 & 0 & 0.1922 & 0 & 0 \\ 0 & -0.3638 & 0 & 0 & 0 & -0.2541 & 0 \\ 0.01 & 0 & 0 & 0 & 0 & 0 & 0 \\ -0.0116 & 0 & 0 & 0 & 0 & 0 & 0 \\ 0 & 0 & 0.5481 & 0.1289 & 0 & 0 & 0 \\ 0 & 0 & 2.3751 & 0.0361 & 0 & 0 & 0 \\ 0 & -0.7871 & 0 & 0 & 0 & 0.2101 & 0 \end{bmatrix} \tag{6.26}$$

The eigenvalue of the characteristic matrix is as follows:

$$\lambda_{\mathrm{m}} = \begin{bmatrix} -9.8785e+000 \\ -8.2224e+000 \\ -2.0686e+000 \\ -1.4176e+000 \\ -6.0030e-001 \\ -1.8185e-014 \\ -9.8400e-002 \\ -1.6580e-001 \end{bmatrix} \tag{6.27}$$

Table 6.7 presents two scenarios for studying performance of the proposed MPC. As in the study of MPC for ERV unit, the trended real data is condensed from seconds to minutes to strengthen the disturbances.

Table 6.7 MPC control scenarios for the DX unit

Scenario	Humidity set point (g/kg)	Temp set point (K)	Weight on humidity	Weight on temp	Others
1	7	285–282	0	300	Constraints on the manipulated variables and changing rates. Step disturbance on the supply air temperature set point
2	7	283	0	300	As Scenario 1, with step disturbance on the inlet air temperature. Supply airflow rate is further constrained between 1.1 and 1.5 kg/s

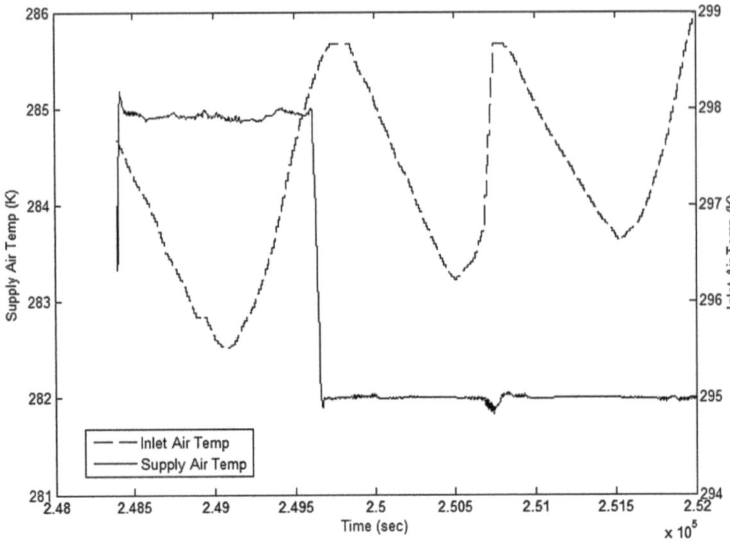

Fig. 6.34 Air temperature, Scenario 1

In Scenario 1, a step input disturbance is imposed on the supply air temperature set point to investigate the controller's performance. A weight of 300 is assigned to the temperature set point. From Fig. 6.34, it is shown that the MPC controls the system to the predefined set point. The system responses closely to the set point step change. For the first twenty minutes, when the evaporator inlet temperature is low to 295 K, the MPC drives the compressor to about 1400 rpm and the condenser fan to around 1.5 kg/s. When the inlet air temperature increases to 298 K, and the set point steps down to 282 K, the compressor speed is driven up to 2000 rpm. The condenser fan speed also increases correspondingly to the load change. The control efforts are plotted in Fig. 6.35 for the compressor speed and condenser fan airflow rate. To accommodate the input step disturbance, the supply fan airflow rate is reduced from about 1.6 to 1.2 kg/s as shown in Fig. 6.36.

Figure 6.37 shows that the inlet air temperature and the outdoor air temperature swing with similar trend. The condenser fan airflow rate drops when the outdoor air temperature decreases.

In Scenario 2, a step disturbance is introduced to the inlet air temperature. The supply air temperature response and inlet air temperature are plotted in Fig. 6.38. When the inlet air temperature drops suddenly from 295 to 293 K, the supply air temperature decreases by up to 0.5 K. The controller then drives the unit back to the set point quickly. Figure 6.39 shows the sharp speed drop of the compressor. With the decreased load, the compressor speed keeps at a medium speed. As in Scenario 1, the condenser fan responses to the outdoor air temperature changing trend.

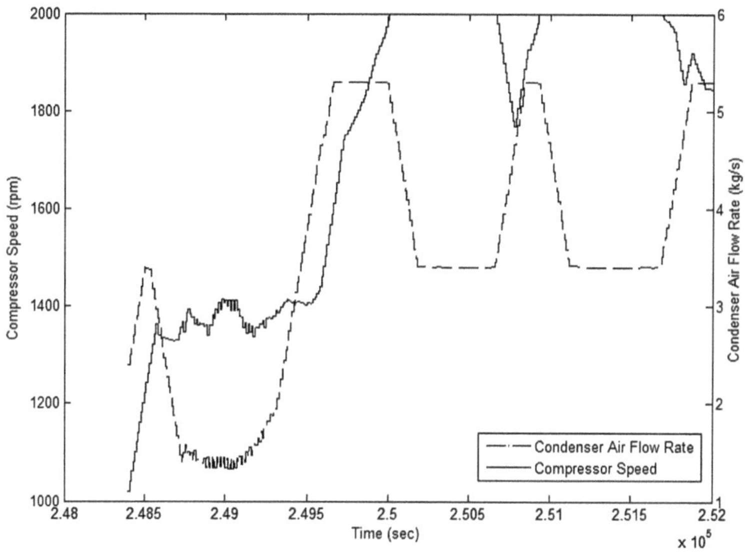

Fig. 6.35 Control variables 1 (Scenario 1)

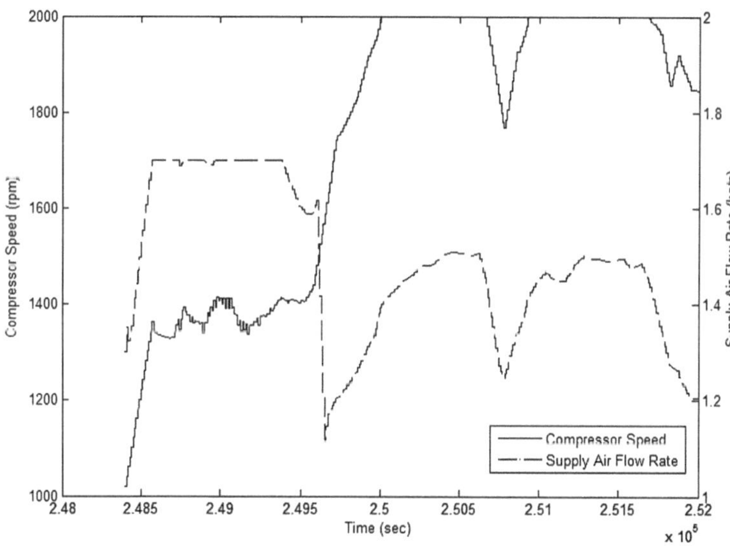

Fig. 6.36 Control variables 2 (Scenario 1)

Figure 6.39 illustrates that the condenser fan airflow rate decreases when the outdoor air temperature is at the valley. In this scenario, the supply fan airflow rate decreases when the compressor speed decreases as shown in Fig. 6.40. The cooling capacity is picked up by the increased condenser fan speed.

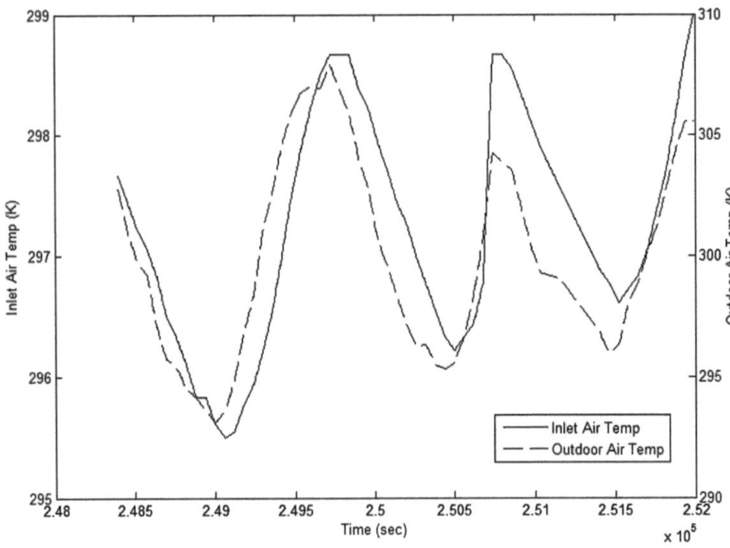

Fig. 6.37 Air temperature (Scenario 1)

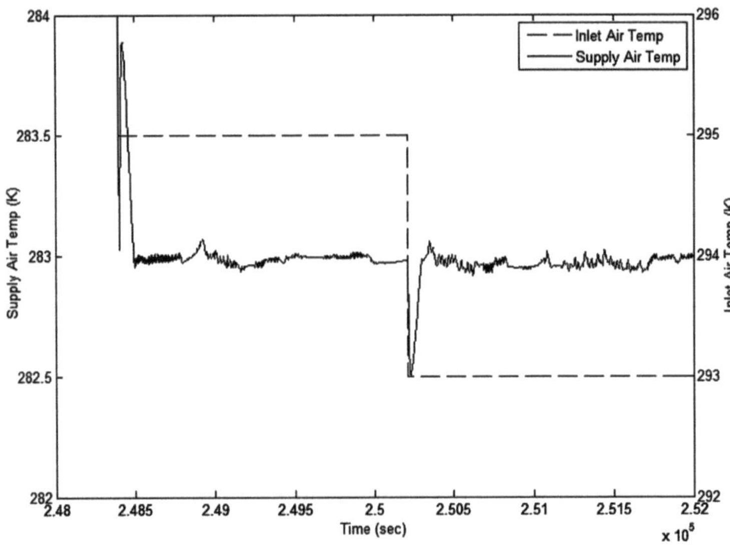

Fig. 6.38 Air temperature (Scenario 2)

(3) Active desiccant unit

An active desiccant unit has a similar structure as an enthalpy recovery wheel (i.e., passive desiccant wheel) in terms of the matrix. The moisture and heat transfer differ because of the rotation speed and the air stream ratio. This unit alone has

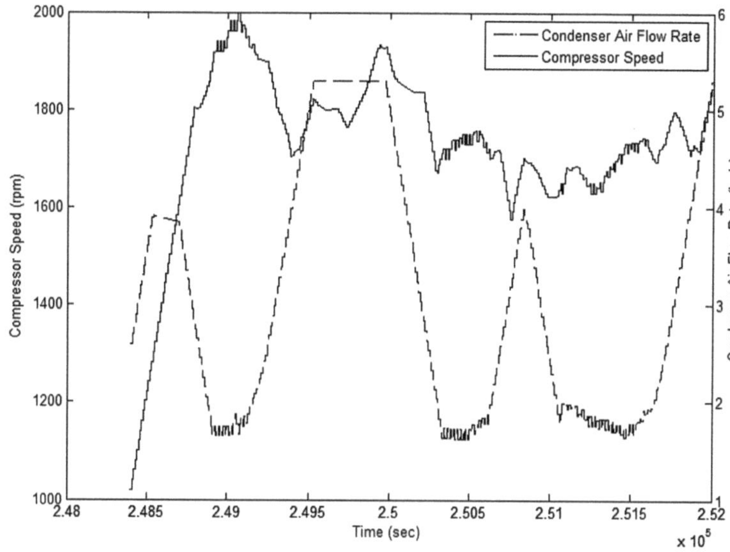

Fig. 6.39 Control variables 1 (Scenario 2)

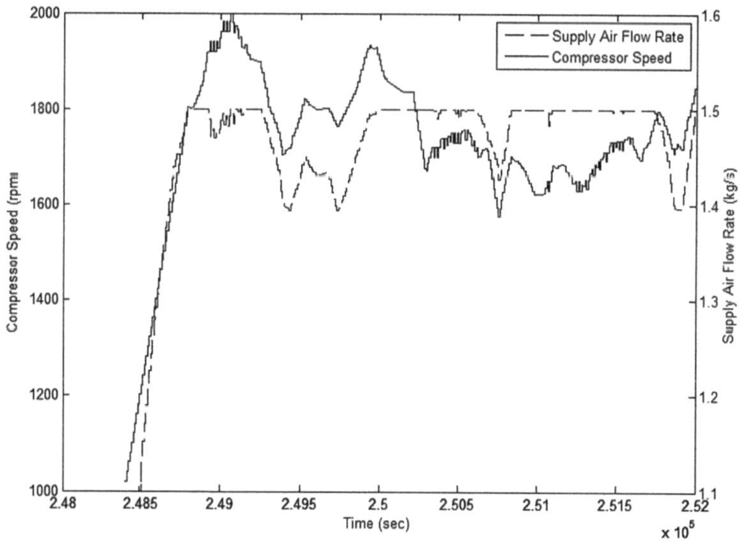

Fig. 6.40 Control variables 2 (Scenario 2)

multiple constraints on the manipulated variables, e.g., the regeneration air temperature and the bypass airflow rate. The desiccant wheel and the heat exchanger are also coupled together to provide proper supply air humidity. Because of the coexistence of heat and mass transfer, the combined desiccant unit does not just dry

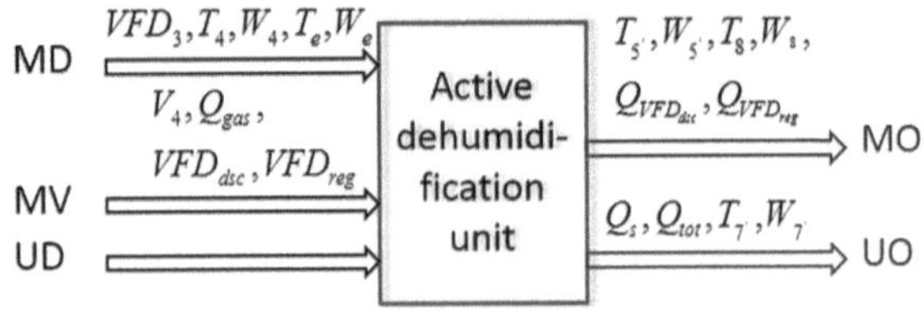

Fig. 6.41 Schematic diagram for the active dehumidification unit

the air, but also add unneeded heat into the supply air. The strong coupled MIMO features make it difficult to apply PID controllers without violating some of the physics. For example, to dehumidify a humid air stream, the regeneration air might have to be heated up to more than 100 °C degree because of the dehumidification efficiency limitation. To avoid this from happening, an additional algorithm (i.e., trip off) is needed. As a heating-driven device, the temperature and humidity control of the unit can only be possibly achieved in a very narrow range. In this section, the performance of an MPC is evaluated.

The existing control set by the manufacturer uses the active desiccant unit to follow the supply air humidity and indirectly maintain the return air humidity. The temperature is monitored by the cooling DX unit. Figure 6.41 illustrates the involved inputs and outputs of the device.

The model is built in Simulink as illustrated in Fig. 6.42. The MPC for the active desiccant unit is designed with the linearized nonlinear model. The sensible and latent conditioning effectiveness is transferred into the effectiveness for enthalpy and relative humidity. The effectiveness is near constant and changes only in a very narrow range for a given active desiccant wheel. The main drive for dehumidification of an active desiccant wheel is the temperature of the regeneration air [25]. The linear model is obtained around the operation point: T_4 = 18 °C, w_4 = 8 g/kg,

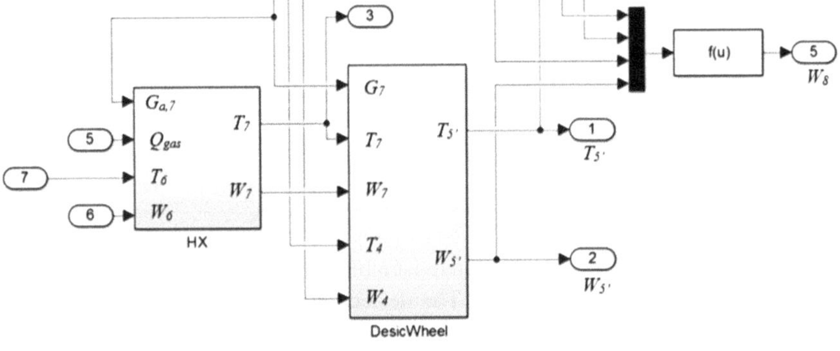

Fig. 6.42 Active desiccant dehumidification model in Simulink

$T_e = 25$ °C, $w_e = 14$ g/kg, $G_7 = 0.31$ kg/s, and $q_g = 10\,\text{kW}$. The system has 5 states and 2 manipulated variables. The system matrices of this active desiccant unit are obtained as follows:

$$A_m = \begin{bmatrix} -0.0167 & 0 & 0.0064 & 0 & 0 \\ 0 & -0.0167 & -0.0015 & 0 & 0 \\ 0 & 0 & -0.31 & 0 & 0 \\ 0.155 & 0 & 0 & -0.5649 & 0 \\ 0 & 0.155 & 0 & 0 & -0.5649 \end{bmatrix} \tag{6.28}$$

$$B_m = \begin{bmatrix} 0 & -0.2928 & 0.0186 & 0.0122 & 0 & -0.0084 & -0.0003 \\ 0 & -0.0077 & 0.0073 & 0.0017 & 0 & 0.0042 & 0.0002 \\ 0 & -32.2581 & 0 & 0 & 1 & 0 & 0.31 \\ -1.8509 & 6.7448 & 0 & 0.4099 & 0 & 0 & 0 \\ 0.4819 & -1.7562 & 0.4099 & 0 & 0 & 0 & 0 \end{bmatrix} \tag{6.29}$$

The system is stable with all the eigenvalues negative:

$$\lambda_m = \begin{bmatrix} -5.649e-001 \\ -1.670e-002 \\ -5.649e-001 \\ -1.670e-002 \\ -3.100e-001 \end{bmatrix} \tag{6.30}$$

Three scenarios as collected in Table 6.8 are conducted to test the performance of using MPC for the unit control.

To force a humidity tracking, a high penalty is imposed on the tracking error. As shown in Fig. 6.43, both the regeneration airflow and the heating input are set to the maximum when the outdoor air humidity is high. A minimum 16 kW is needed from the heat exchanger to maintain the supply air humidity at 5.5 g/kg level. The

Table 6.8 MPC control scenarios for the active desiccant unit

Scenario	Humidity set point (g/kg)	Temperature set point (K)	Weight on humidity	Weight on temperature	Others
1	5.5	292–291	1000	0	Normal constraints
2	6	291–289	0	50	Supply air humidity between [0, 7]
3	6	292	0.01	0.001	Rate constraints on the gas burner, supply air humidity [0 6], supply air temp [290 295]

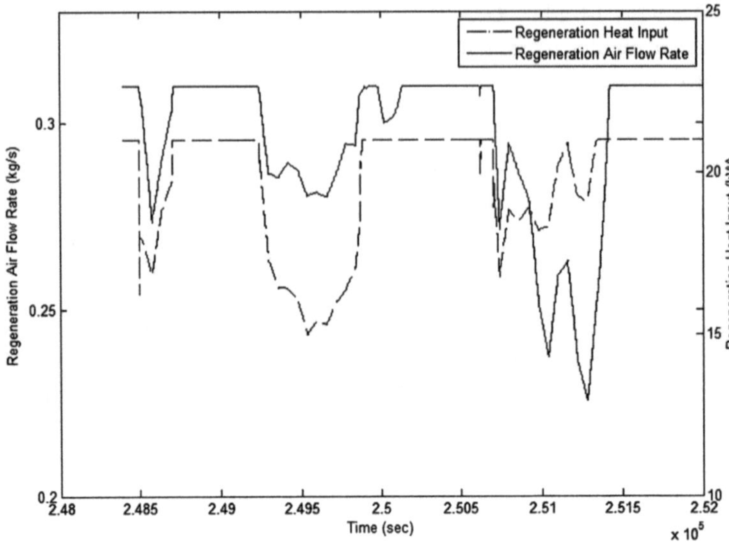

Fig. 6.43 Manipulated variables (Scenario 1)

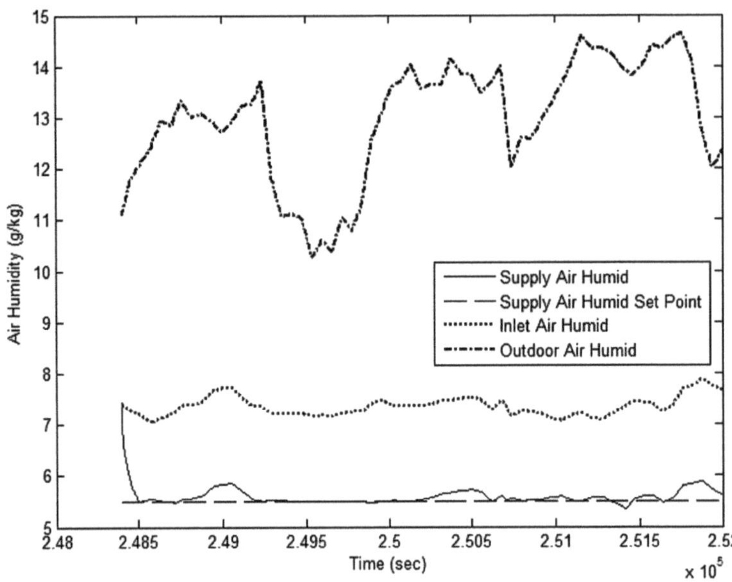

Fig. 6.44 Air humidity (Scenario 1)

supply air humidity as shown in Fig. 6.44 follows the set point well. Since the air temperature gets zero weight, there is no trade-off between the additional heat and dehumidification with the MPC control. As shown in Fig. 6.45, the supply air

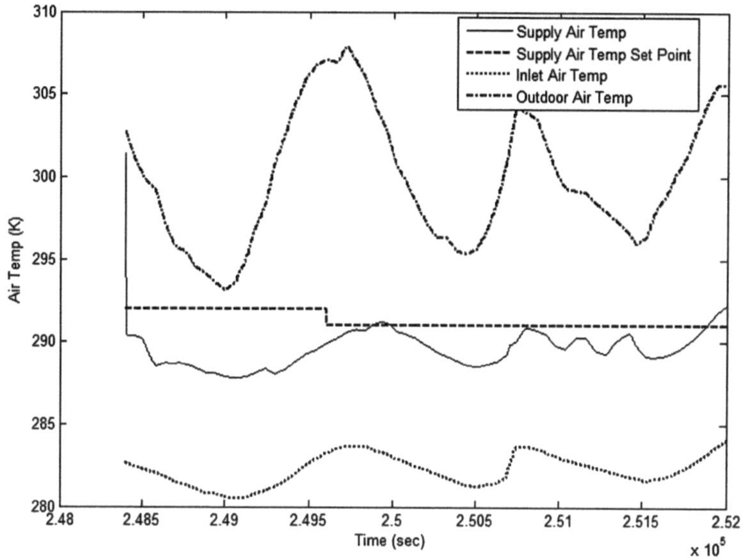

Fig. 6.45 Air temperature (Scenario 1)

temperature fluctuates around 290 K when the inlet air is around 280 K. The unneeded heat shall be handled by the control of the DX unit.

In Scenario 2, the condition is switched with penalty weight added on the air temperature. A constraint is specified on the supply air humidity to prevent the supply air humidity from going beyond 7 g/kg. As shown in Fig. 6.46, large modulations on the heat input exist in the middle and at the end of the period. As shown in Figs. 6.47 and 6.48, the supply air humidity close to 7 g/kg triggers the modulation, and the supply air temperature follows the set point for most of the time. There are circumstances where the control of the supply air temperature cannot be satisfied with only the unit.

The modulating rate constraints are imposed on the airflow rate in Scenario 3. The low and high limits on the supply air humidity and temperature are specified as well in the control law. The simulation results are plotted in Figs. 6.49, 6.50, and 6.51. As shown in Fig. 6.51, the MPC responses with high heat input when the inlet air humidity is high. Since the weight on the supply air humidity is higher than that on the supply air temperature, the supply air humidity follows the set point better than the supply air temperature. However, there are some circumstances where the control causes overshoot on the air temperature and humidity with sudden increase on the regeneration airflow rate as plotted in Figs. 6.50 and 6.51.

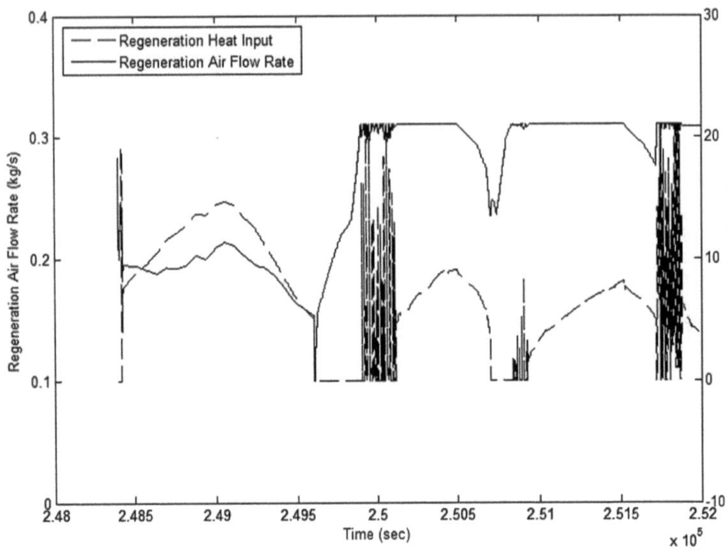

Fig. 6.46 Manipulated variables (Scenario 2)

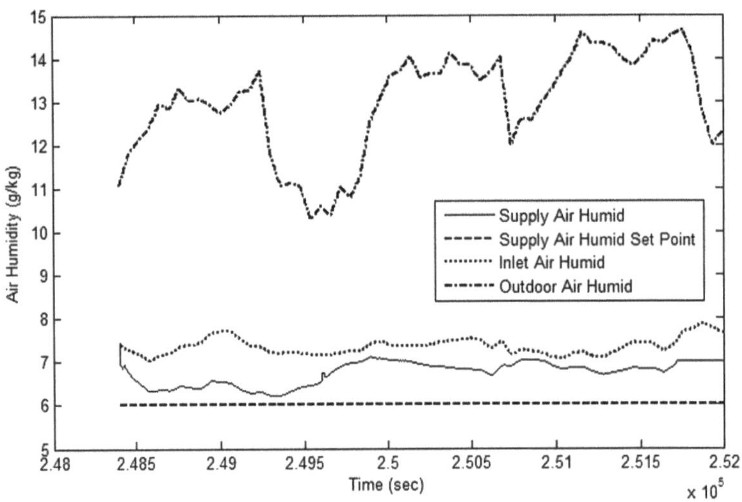

Fig. 6.47 Air humidity (Scenario 2)

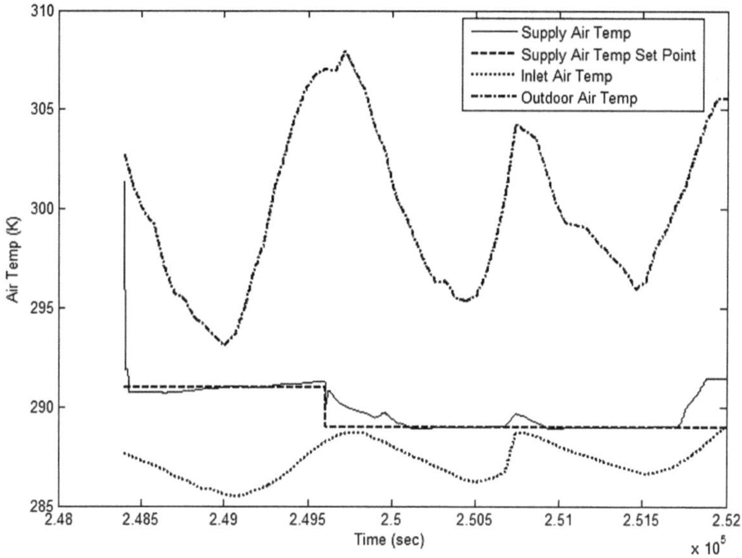

Fig. 6.48 Air temperature (Scenario 2)

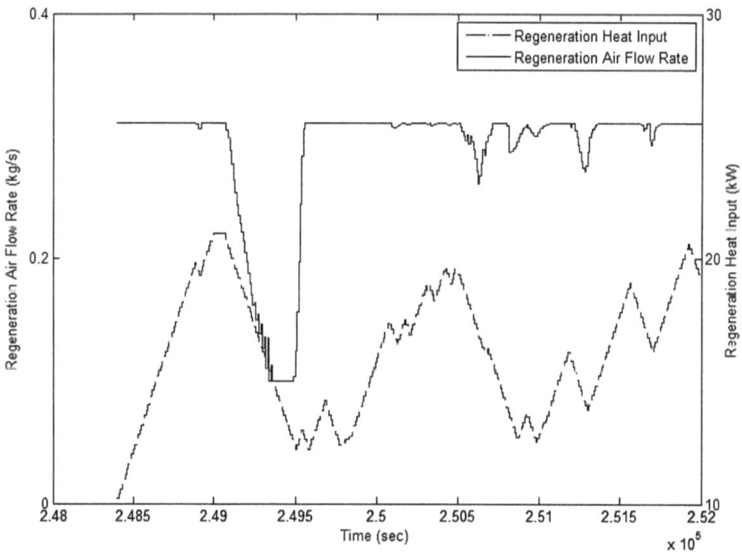

Fig. 6.49 Manipulated variables (Scenario 3)

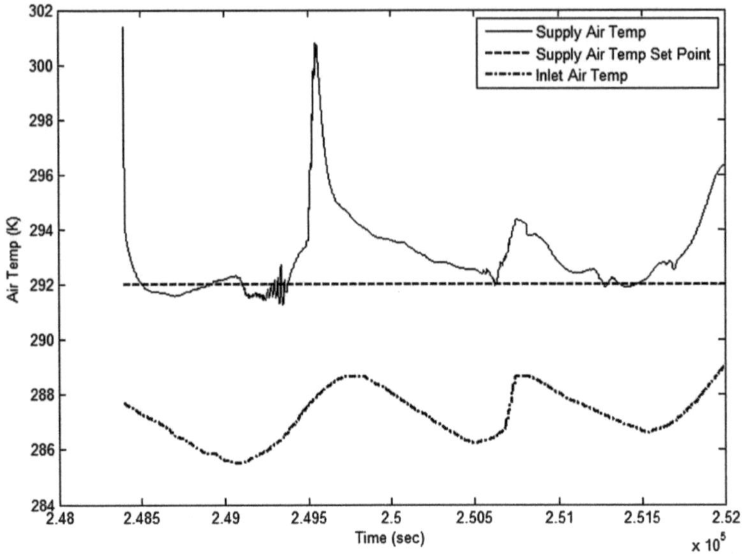

Fig. 6.50 Air temperature (Scenario 3)

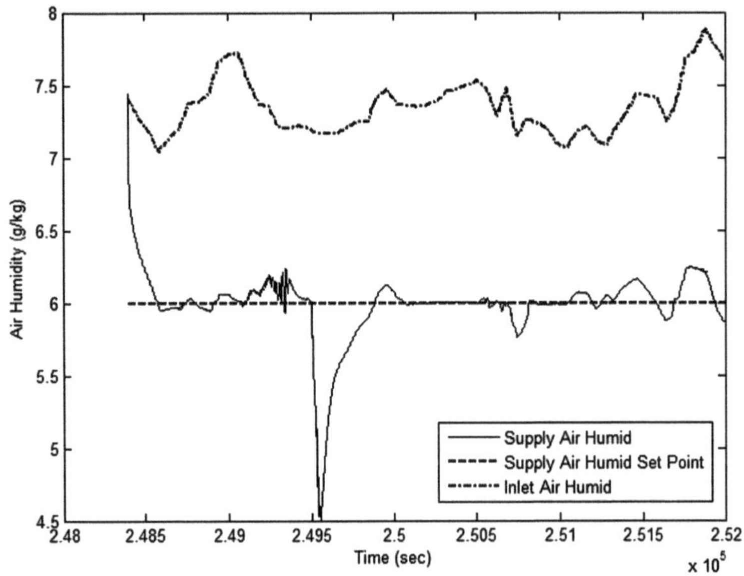

Fig. 6.51 Air humidity (Scenario 3)

6.2.1.4 Small Summary

The MIMO nature of the complex device implies that an MPC can be a good candidate for the control to handle the strong coupling effects and incorporate the constraints on the inputs and states/controlled variables. The dynamic models for the three units are established in MATLAB/Simulink in this chapter to investigate the transient behavior and the control performance of MPC. A distributed structure is adopted since the function of the hybrid unit can be decomposed into three elements:

(1) The ERV controls the outdoor air intake. For example, it maximizes the outdoor air intake for free cooling and minimizes the outdoor air intake for mechanical cooling and heating;
(2) The vapor expansion unit tracks the supply air temperature; it is desirable and it uses less electricity for the same load condition;
(3) The active desiccant unit tracks the supply air humidity; it runs only when the humidity after the vapor expansion unit is high.

Compared to a centralized MPC, the distributed structure has higher reliability and lower computational cost. In an MPC controller, the main tuning parameters are the weights on the manipulated variables, reference tracking errors, and manipulated variable changing rates. The relative value of the weights greatly determines the performance of the controller. For example, to prevent the compressor from frequent ramping up and down, e.g., to extend the life span of the compressor, a weight can be added on the changing rate. For an MIMO system, since the manipulated variables are in different units, scaling is very important in practical. The weights should be scaled based on the expected magnitudes of disturbances and reference changes, on the allowed magnitude of each input signal, and on the allowed deviation of outputs.

Several scenarios are designed to investigate the performance of classical MPC for the three components in the hybrid ventilation unit. The nonlinear models are linearized around the operating points to form the internal linear model for future dynamic prediction. The following points are identified with the study:

(1) The MPC controllers on the ERV and vapor compression unit have good disturbance rejection capability. The system response to the step changes on the inputs and the set points with acceptable stability.
(2) The MPC controller on the enthalpy recovery wheel can integrate the algorithm to maximize the use of free cooling air when the outdoor air is favorable and minimize the outdoor air intake for ventilation.
(3) The MPC controller on the vapor compression unit can effectively coordinate the multiple inputs and multiple outputs for a given set point. A tuning on the weights may guide the unit toward high energy efficiency for a given load.
(4) The MPC controller on the active desiccant wheel can manipulate the heat input and regeneration airflow to follow the humidity set point for most of the time and avoid large penalty on the unwanted heat.

6.2.2 Control of Space Thermal Conditioning

6.2.2.1 System Description

The Intelligent Workplace north integrated building energy system is a living laboratory for the research on high-performance buildings and conditioning techniques as shown in Fig. 6.52. It is located on the campus of Carnegie Mellon University, Pittsburgh, USA. The building features a reconfigurable open space and accommodates offices for some thirty graduate students and faculty members. Approximately 600 m^2 area is conditioned. Twenty-six sets of water mullions are deployed along the facade to provide heating and a moderate level of cooling to the space. Each set of water mullion consists of four vertical embedded pipes and is equipped with a modulating valve on the inlet pipe. A surface temperature sensor is also installed on the outlet point of each mullion. The water is supplied from the campus grid via a tertiary pump. A heat exchanger is installed in the basement to covert the steam from the campus grid into hot water and feed the space in winter. In the heating season, the primary supply water temperature is adjusted by modulating the steam valve. In the cooling season, the primary supply water temperature is determined by the campus grid and not adjustable locally. To avoid condensation on the surface of the water-based conditioning terminals and also provide addition control flexibility, a mixing valve is mounted after the tertiary pump. The mixing valve controls the supply water temperature circulated to the space. The other water-based terminals for local environment control include suspended radiant ceilings and ductless radiant chilled beams. When additional or individual conditioning is needed, these terminals can be controlled individually to provide extra heating and cooling.

As a living and lived-in laboratory, the space has been constantly used for experimental research projects. The control algorithm and logic of the sub-systems and equipment are not always the same. The basic structure of the control architecture includes a layer of rule-based supervisory control and the local SISO-based on/off and PID control. The space is designed for easy reconfiguration and

Fig. 6.52 Internal view of the hydronic building system

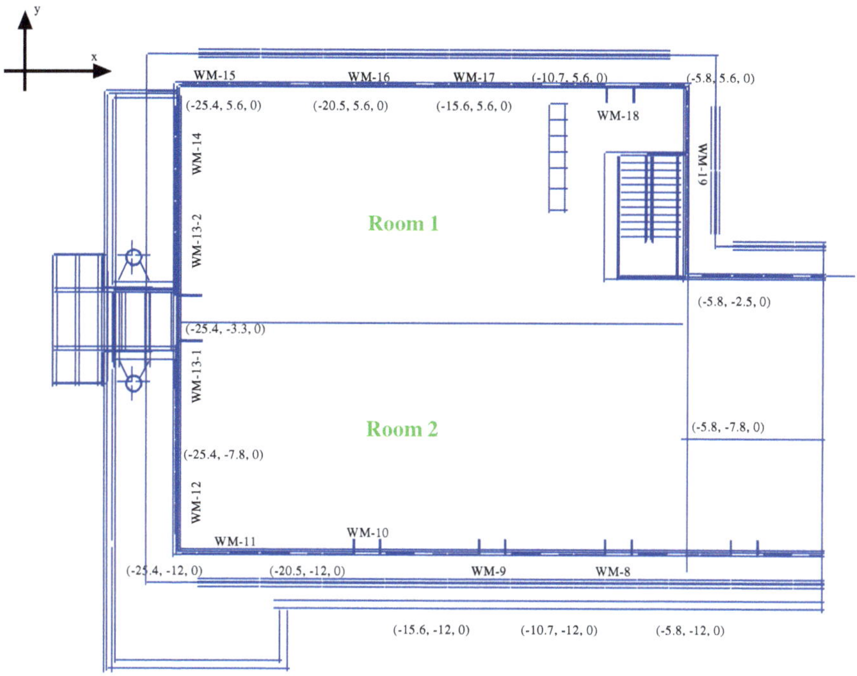

Fig. 6.53 Illustration of the space layout

accommodates different experiments. For this study, the space is divided into two zones: one along the east facade, and the other along the west facade. Figure 6.53 illustrates the space layout.

6.2.2.2 Model Development

The thermal dynamics of a building is determined mainly by the thermal mass of the building materials and the inside air. The time constant of a heavy building can be up to hours. To predict the future evolution of indoor air condition for acceptable thermal comfort, the dynamics need to be represented accurately in the model. Calculating instantaneous space load is critical in dynamic modeling of building thermal systems. Building envelop includes the external walls, roof, and floor. They are the boundary between the indoor and the outdoor environment. For energy consumption and thermal comfort evaluation, the dynamics and interaction of the indoor air and the envelope are of great importance since they determine the actual thermal load and evolution of indoor air temperature and influence the thermal comfort. The time constant of a model tells how fast the building system can response to the change in the mechanical system while the model provides some

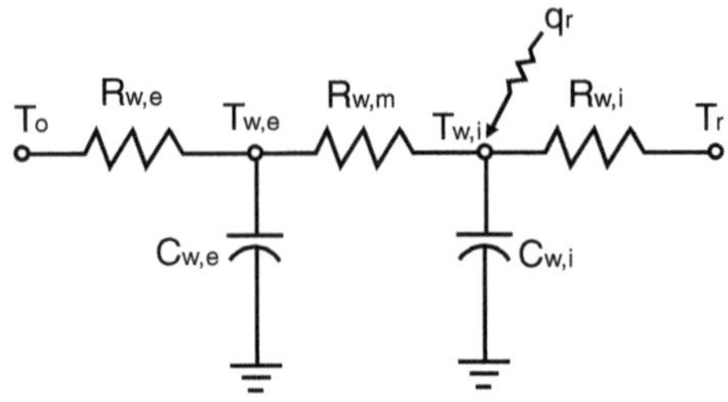

Fig. 6.54 3R2C analogue heat transfer model

insights on what corresponding actions shall be taken for a given indoor temperature goal.

The deduction of the transient space model is based on the energy conservation, the first law of thermodynamics. The thermal capacitance and resistance involved in the heat balance of a building thermal system is analogue to the capacitance and resistance as in an electric network. Similar to the electric network, the thermal network has the linearity and can be put in the state-space format. This approach is also used in the whole building system simulation tools (EnergyPlus, Trnsys, etc.). Figure 6.54 depicts such an analogue of a wall divided into two layers with two capacitances and three resistances. A heat flux, such as radiation, can be regarded as current source acting on the inside or outside capacitance.

Heat transfer through an opaque wall as shown in Fig. 6.54 can be modeled as follows:

$$C_{\mathrm{w,e}}\frac{\mathrm{d}T_{\mathrm{w,e}}}{\mathrm{d}t}=\frac{\left(T_{\mathrm{oa}}-T_{\mathrm{w,e}}\right)}{R_{\mathrm{w,e}}}-\frac{\left(T_{\mathrm{w,e}}-T_{\mathrm{w,i}}\right)}{R_{\mathrm{w,m}}} \tag{6.31}$$

$$C_{\mathrm{w,i}}\frac{\mathrm{d}T_{\mathrm{w,i}}}{\mathrm{d}t}=q_{\mathrm{r}}+\frac{\left(T_{\mathrm{w,e}}-T_{\mathrm{w,i}}\right)}{R_{\mathrm{w,m}}}-\frac{\left(T_{\mathrm{w,i}}-T_{\mathrm{ra}}\right)}{R_{\mathrm{w,i}}} \tag{6.32}$$

where C is the total thermal capacitance of the given layer (kJ/K), T_{w} is the representative mass temperature of the given layer wall (K), T_{o} is the outdoor air temperature (K), R_{w} is the thermal resistance between the nodes (K/kW), and q_{r} is the heat flux due to the radiation through the fenestration on the internal surface (kW). The subscripts, e, m, and i denote the external, middle, and internal layers of the construction, respectively.

Figure 6.55 illustrates the thermal network of one construction with an opaque wall and a transparent window for a building with four rooms.

Convection is the main heat transfer mechanism between the air node, the internal heat gains, and the enclosures. Based on the heat balance, a first-order equation can be obtained to model the evolution of room air temperature:

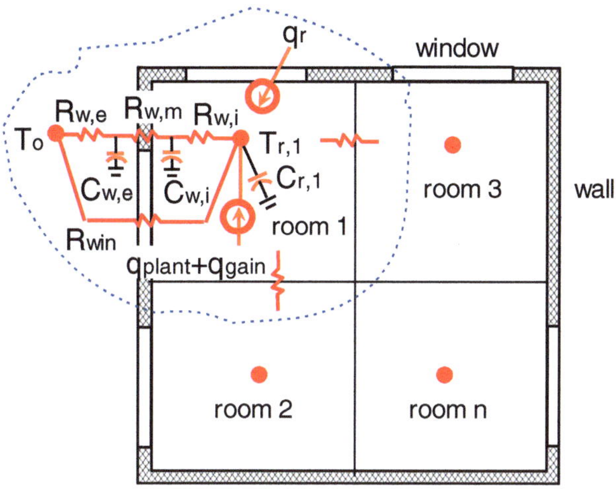

Fig. 6.55 Thermal network in a multi-zone building

$$\rho V_{\mathrm{ra},j} c_{\mathrm{p}} \frac{\mathrm{d}T_{\mathrm{ra},j}}{\mathrm{d}t} = q_{\mathrm{plant},j} + q_{\mathrm{g},j} + q_{\mathrm{w},j} + q_{\mathrm{gain},j} + q_{\mathrm{inf},j} \tag{6.33}$$

where ρ is the air density (kg/m^3), $V_{\mathrm{r},j}$ is the total air volume in the room (m^3), c_{p} is the specific heat of room air (kJ/kg K), $T_{\mathrm{r},j}$ is the room air temperature (K), $q_{\mathrm{plant},j}$ is the total heat flux from the mechanical system into the air node (kW), $q_{\mathrm{g},j}$ is the total heat flux into the space from the fenestration (kW), $q_{\mathrm{w},j}$ is the total heat flux into the space from the opaque construction (kW), $q_{\mathrm{gain},j}$ is the total internal heat gain from the occupants, equipment, lighting, and so forth into the room air (kW), and $q_{\mathrm{inf},j}$ is the total heat flux into the room air due to infiltration (kW). Subscript j denotes the jth room.

The thermal components can be further decomposed to the driven factors, e.g., temperature differences. For simplicity, some of the subscripts for enumeration are omitted in the following text.

$$q_{\mathrm{g},j} = \sum A_{\mathrm{g}} \cdot K_{\mathrm{g}} \left(T_{\mathrm{ra},j} - T_{\mathrm{oa}}\right) \tag{6.34}$$

$$q_{\mathrm{w},j} = \sum A_{\mathrm{w}} \cdot K_{\mathrm{w}} \left(T_{\mathrm{ra},j} - T_{\mathrm{w},i}\right) \tag{6.35}$$

$$q_{\mathrm{inf},j} = \mathrm{ACH} \cdot V_{\mathrm{ra},j} \left(T_{\mathrm{ra},j} - T_{\mathrm{oa}}\right)/3 \tag{6.36}$$

where A_{g} is the surface area of the fenestration (m^2), K_{g} is the lumped average heat transfer coefficient of the fenestration (kW/m^2 K), T_{e} is the outdoor air temperature (K), A_{w} is the surface area of the opaque construction (m^2), Kw is the lumped average heat transfer coefficient of the opaque construction (kW/m^2 K), T_{w} is the inside surface temperature of the opaque construction (K), and ACH is the air change rate due to infiltration (/h).

The complexity of temperature evolution in a building comes from the thermal interaction among the zones and between the indoor and outdoor air through the enclosures. For a large building, it could be a challenge to deduce and denote the physical model equations. The number of governing equations grows exponentially as the number of zones and surfaces increases. The following deduction could be utilized to automatically obtain the A matrix for the building state-space model [26]:

$$A_{ij} = 0, \quad \text{if} \quad i \neq j, \quad (i,j) \notin \varepsilon \tag{6.37}$$

$$A_{ij} = 1/\left(C_i R_{ij}\right), \quad \text{if} \quad i \neq j, \quad (i,j) \in \varepsilon \tag{6.38}$$

$$A_{ij} = -\sum\nolimits_{j \neq i} A_{ij}, \quad \text{if} \quad i = j, \quad (i,j) \in \varepsilon \tag{6.39}$$

where A is the state matrix, ε here denotes the adjacent relationship, C is the total thermal capacitance of the given node (kJ/K), and R is the total thermal resistance between two nodes (K/kW). The subscripts, i and j, denote the nodes.

The moisture balance in the air node can be similarly modeled. Unlike the air temperature modeling, the overall space humidity balance is typically lumped as one node without differentiating the spaces:

$$\rho V_{\mathrm{ra}} \frac{\mathrm{d}W_{\mathrm{ra}}}{\mathrm{d}\tau} = G_{\mathrm{plant}} \cdot \left(W_{\mathrm{plant}} - W_{\mathrm{ra}}\right) + 1.2 \cdot \mathrm{ACH} \cdot V_{\mathrm{ra}}\left(W_{\mathrm{oa}} - W_{\mathrm{ra}}\right)/3600 + \frac{q_{\mathrm{l}}}{h_{\mathrm{w}}} \tag{6.40}$$

where W is the average humidity ratio (kg/kg), G_{plant} is the total supply airflow rate into the space from the plant (kg/s), V_{ra} is total air volume in the space (m^3), q_{l} is the total latent heat gain in the space (kW), and h_{w} is the specific latent heat of water vaporization (kJ/kg). Subscripts ra and oa denote room air and outdoor air, respectively.

The heat gain from solar radiation through fenestrations is treated differently due to the two-step heat transfer mechanism. The heat flux of solar radiation first penetrates fenestrations, hits the floor, and then is reflected to all other surfaces. The surfaces are warmed up by the heat flux then gradually transfer the heat into the air node mainly via convection. The algorithm proposed by Skartveit and Oleseth [27, 28] based on the sky cleanness index is used to process the hourly horizontal solar radiation. Normal solar radiation is then corrected into the plane of building surface at any azimuth and tilt angle with the method established by Liu and Jordan [29]. The model of the heat flux through a fenestration has the following relationship:

$$q_{\rm r} = A_{\rm g} \cdot \alpha \cdot \varepsilon_{\rm g} \cdot q_{\rm r-surf} \tag{6.41}$$

where $q_{\rm r}$ is the solar radiation transmitted into the room (W/m^2), α is the average solar absorptivity of the surfaces, $\varepsilon_{\rm g}$ is transmissivity of the fenestration, and $q_{\rm r-surf}$ is the total solar radiation corrected to the plane of the window (W/m^2).

The internal heat gains can be regarded as instant load and simplified as a function of total load intensity:

$$q_{\rm s} = f(I_{\rm s}) \tag{6.42}$$

$$q_{\rm l} = f(I_{\rm l}) \tag{6.43}$$

where q is the total load from the occupants, equipment, and lighting, etc. (kW), and I is the total load intensity (kW). The subscripts s and l denote sensible and latent parts, respectively, and f maps the corresponding load based on the schedule.

6.2.2.3 Thermal Comfort Model

Thermal comfort is defined as the 'condition of mind which expresses satisfaction with the thermal environment and is assessed by subjective evaluation' [30]. As a subjective thing, an absolute satisfaction is almost impossible to achieve in a conditioned space with more than one occupant. However, a relatively acceptable thermal comfort to the occupants can be evaluated based on the existing findings. Predictive mean vote (PMV) and predicted percentage of dissatisfaction (PPD) are the indexes utilized to assess the thermal comfort condition of an indoor environment.

PMV gives the mean value of the votes of a large group of people. Values from −3 to 3 are utilized to express the conditions from very cold to very hot. A value, between −0.5 and 0.5, indicates a good thermal comfort indoor environment. PPD is another index in the quantitative evaluation of thermal comfort. A PPD value higher than 10 % indicates the indoor environment may not be acceptable. Models of PMV and PPD are quasi-steady state and can be expressed as the following equations [30, 31]:

$$\begin{aligned} \text{PMV} = {} & \left(0.028 + 0.3033\text{e}^{-0.036M}\right) \cdot \\ & \left\{ \begin{array}{l} (E-N) - 3.05 \cdot 10^{-3}[5733 - 6.99(E-N) - \text{Pa}] - \\ 0.42[(E-N) - 58.15] - 1.7 \cdot 10^{-5}E(5867 - \text{Pa}) - \\ 0.0014E(34 - T_{\rm a}) - 3.96 \cdot 10^{-8} \cdot \text{fcl} \cdot \left[(T_{\rm clo} + 273)^4 - (T_{\rm mrt} + 273)^4\right] - \\ \text{fcl} \cdot a_{\rm c}(T_{\rm clo} - T_{\rm a}) \end{array} \right\} \end{aligned} \tag{6.44}$$

$$\text{PPD} = 100 - 95 \cdot \text{e}^{-0.03353 \cdot \text{PMV}^4 - 0.2179 \cdot \text{PMV}^2} \tag{6.45}$$

where E is metabolism (W/m^2), N is external work (W/m^2), Pa is partial water vapor pressure (Pa), fcl is ratio of clothed body surface area to nude body surface area, T_{clo} is surface temperature of clothing, and ac is convectional heat transfer coefficient (W/m^2 K).

The surface temperature of clothing is given by the following:

$$T_{\mathrm{clo}} = 35.7 - 0.028(E - N) - 0.155 \cdot R_{\mathrm{clo}} \left\{ \begin{array}{c} 3.96 \cdot 10^{-8} \cdot \mathrm{fcl} \left[\begin{array}{c} (T_{\mathrm{clo}} + 273)^4 - \\ (T_{\mathrm{mrt}} + 273)^4 \end{array} \right] \\ -\mathrm{fcl} \cdot a_{\mathrm{c}} (T_{\mathrm{clo}} - T_{\mathrm{a}}) \end{array} \right\} \tag{6.46}$$

$$a_{\mathrm{c}} = \begin{cases} 2.38 \cdot (T_{\mathrm{clo}} - T_{\mathrm{a}})^{0.25} & \text{for} \quad 2.38(T_{\mathrm{clo}} - T_{\mathrm{a}})^{0.25} \geq 12.1\sqrt{V_{\mathrm{a}}} \\ 12.1\sqrt{V_{\mathrm{a}}} & \text{for} \quad 2.38(T_{\mathrm{clo}} - T_{\mathrm{a}})^{0.25} \leq 12.1\sqrt{V_{\mathrm{a}}} \end{cases} \tag{6.47}$$

where T_{mrt} is the mean radiant temperature (K) and V_{a} is the air velocity (m/s).

6.2.2.4 Hydraulic Thermal System

The test bed space is mainly conditioned by a group of radiant pipes deployed along the envelope. Additional radiant panels and cool beams are installed in the offices to provide extra thermal conditioning capacity during extreme days. Lumped parameter models are adopted in this study to simulate the heat transfer of the water-based terminals. The average representative temperature of the terminal mass can be found by solving:

$$M_{\mathrm{ter}} c_{\mathrm{p,ter}} \frac{\mathrm{d}T_{\mathrm{ter}}}{\mathrm{d}\tau} = q_{\mathrm{e}} - q_{\mathrm{l}} \tag{6.48}$$

where M_{ter} is lumped mass of the terminal (W/m^2), T_{ter} is the representative temperature of the terminal (K), $c_{\mathrm{p,ter}}$ is the specific heat of the terminal (kJ/kg K), q_{e} is heat transferred from the fluid to the mass (kW), and q_{l} is heat dissipated from the terminal to the space (kW).

The total heat getting into the fluid residing in the terminals is simply as follows:

$$q_{\mathrm{e}} = G \cdot c_{\mathrm{p}} \left(T_{\mathrm{w,s}} - T_{\mathrm{ter}} \right) \tag{6.49}$$

where G is the fluid mass flow rate (kg/s), and $T_{w,s}$ is the supply water temperature (K).

The heat transferred from the terminals to the air node can be modeled by the following:

$$q_{\text{out}} = K_{\text{ter}} A_{\text{ter}} (T_{\text{ter}} - T_{\text{r}})^n \tag{6.50}$$

where K_{ter} is the overall heat transfer coefficient (kW/m^2 K), A_{ter} is the representative surface area (m^2), and n is the adjusting coefficient.

6.2.2.5 MPC Formulation

An MPC works to achieve a minimum cost based on existing knowledge and future prediction. With the above first-principle model for the building system, a cost function needs to be determined. For a classical MPC, this cost function needs to be defined as a quadratic one. It may include reference tracking errors, control efforts, and actuator movement efforts. Air temperature, thermal comfort, and even monetary cost can be wrapped into it as well. Also, a band of thermal comfort value may be imposed as constraints to differentiate the operation variations due to occupancy changes.

As shown in Fig. 6.56, two heat fluxes from the terminals are the direct inputs to the air nodes. The outdoor air temperature, boundary surface temperature, and solar radiation heat flux are the external inputs that affect the space temperature through the construction surfaces. In addition, the sensible heat gain, latent heat gain, infiltration airflow rates, etc., are the disturbances to the system. The indoor air temperature and average air humidity are the direct outputs from the LTI space system. The mathematic expression of the LTI system for the two room layout 2R1C space can be put as follows:

$$x(k+1) = Ax(k) + Bu(k) \tag{6.51}$$

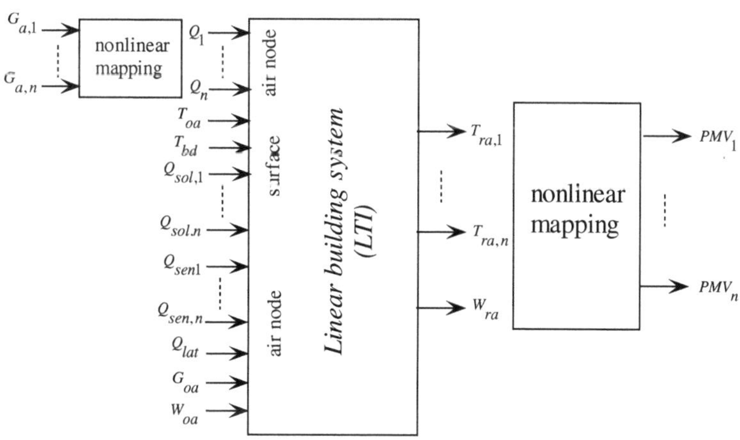

Fig. 6.56 Illustration of the variables in HW nonlinear MPC

$$x = \left[\, T_{\mathrm{ra},i=1,2} \quad T_{m,i=1\ldots11} \quad W_{\mathrm{ra}} \,\right]^{\mathrm{T}}_{14\times1} \tag{6.52}$$

$$u = \left[\dot{Q}_{i=1..2}\right]^{\mathrm{T}}_{2\times1} \tag{6.53}$$

$$d = \left[\, \dot{Q}_{\mathrm{sen},i=1,2} \quad \dot{Q}_{\mathrm{lat}} \quad \dot{Q}_{\mathrm{sol},i=1,2} \,\right]^{\mathrm{T}}_{5\times1} \tag{6.54}$$

$$w = \left[\, T_{\mathrm{oa},i=1\ldots6} \quad T_{\mathrm{og}} \quad G_{\mathrm{oa}} \quad W_{\mathrm{oa}} \,\right]^{\mathrm{T}}_{9\times1} \tag{6.55}$$

where u, d, and w are the manipulated variable vector, disturbance vector, and external disturbance vector, respectively, $\mathrm{G_{oa}}$ is the infiltration airflow rate (kg/s), and $\dot{Q}$ is the heat flux (kW). The subscripts ra, oa, m, sen, lat, sol, og denote the room air, outdoor air, material, sensible heat, latent heat, solar heat, and ground condition, respectively.

In the nonlinear HW-MPC used in this study, water flow rates to the space terminals are considered as the manipulated variables for the space thermal conditioning. Instead, the supply water temperature may be regarded as one manipulated variable if the whole space is treated as one zone. In the prediction horizon, e.g., 100 min, twenty variables in total need to be optimized by the MPC controller. Constraints on the inputs are the minimum and maximum water flow rates. With 10–20 min set as one control interval, which is long enough to achieve a stable water distribution among the terminals, the water flow changing rates are not constrained. Figure 6.57 illustrates the model built in MATLAB/Simulink. The nonlinear map from the air temperature and humidity to thermal comfort is calculated by a function programmed in MATLAB.

6.2.2.6 Results and Analysis

The scenarios collected in Table 6.9 are considered in this study to evaluate the performance of classical MPC and proposed HW nonlinear MPC for the hydraulic system, in regard of energy, thermal comfort, and computational cost. The MPC objectives are selected as the room air temperature, thermal comfort, and energy consumption, respectively. Both Scenarios SC-1 and SC-1' are temperature reference oriented control.

The experiments were conducted for five days. Figure 6.58 plots the outdoor air temperature, relative humidity, solar radiation to the two rooms for the five days. The outdoor air fluctuates between 2 and 18 °C during the simulation period, which represents a typical winter.

A conventional space thermal conditioning is to track a room air temperature reference. It is typically a fixed set point, e.g., 24 °C. A base case is set as a 24/7 occupied schedule with the fixed room air temperature set point at 24 °C. Figure 6.59 plots the room air temperature, relative humidity, PMV value, and heat inputs. The MPC modulates the heat inputs properly and maintained well the room air temperature close to the reference. The overall energy consumption for the 24/7

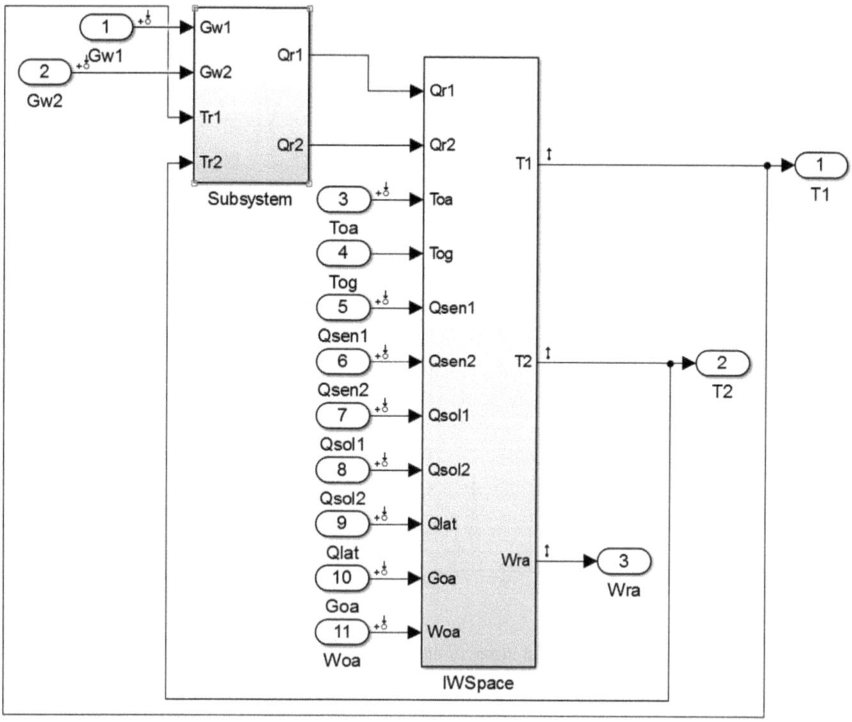

Fig. 6.57 MATLAB modeling for nonlinear hydronic system

Table 6.9 Control scenarios for classical MPC and HW nonlinear MPC

Scenario	Problem description
SC-1	Classical MPC, direct heat flux inputs, temperature-oriented control. 24/7 schedule
SC-1′	Classical MPC, direct heat flux inputs, temperature oriented control. Office schedule
SC-2	Wiener nonlinear MPC, thermal comfort-oriented control
SC-3	Wiener nonlinear MPC, energy consumption and thermal comfort-oriented control. Thermal comfort serves as constraints, PMV = [−0.5, 0.5]
SC-4	HW nonlinear MPC, thermal comfort-oriented control
SC-5	HW nonlinear MPC, energy consumption, and thermal comfort-oriented control. Thermal comfort serves as constraints, PMV = [−0.5, 0.5]
SC-6-1	As case 5, with variable occupancy schedule. When unoccupied, PMV = [−3, 3]; when occupied, PMV = [−0.5, 0.5]. Band control
SC-6-2	As case 6-1, dynamic constraints based on outdoor air temperature
SC-7	As case 4, with variable occupancy schedule and full look-ahead ability. When unoccupied, PMV = [−3, 3]; when occupied, PMV = 0

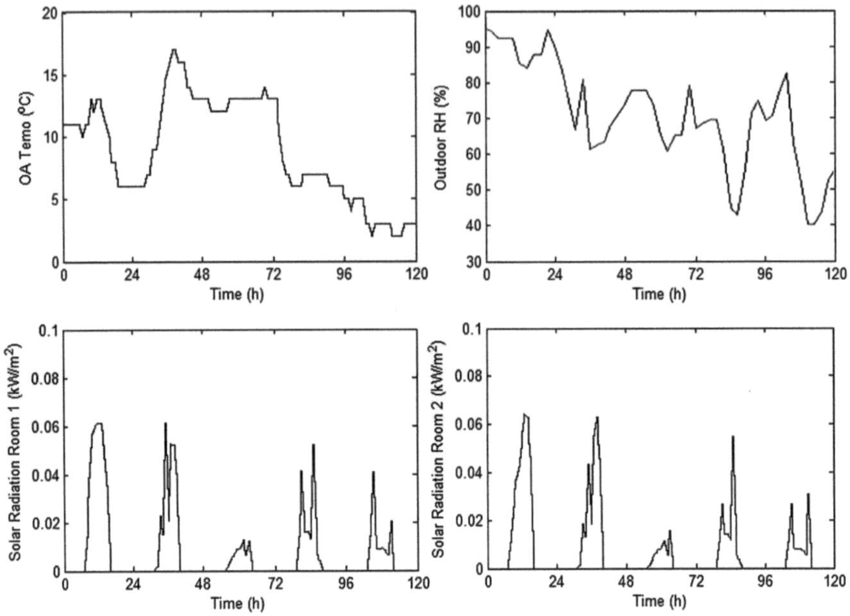

Fig. 6.58 Outdoor air condition and solar radiation

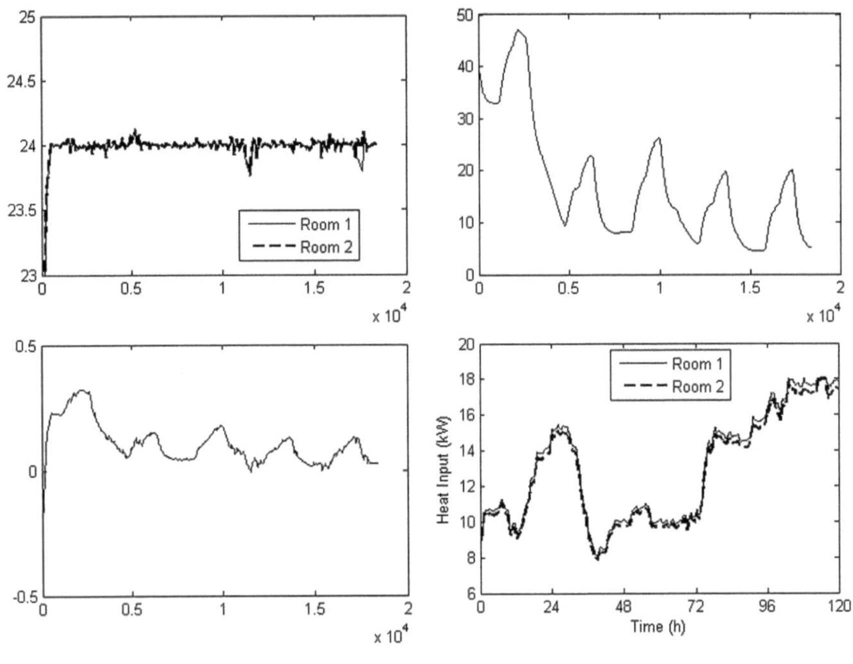

Fig. 6.59 Simulation results, SC-1

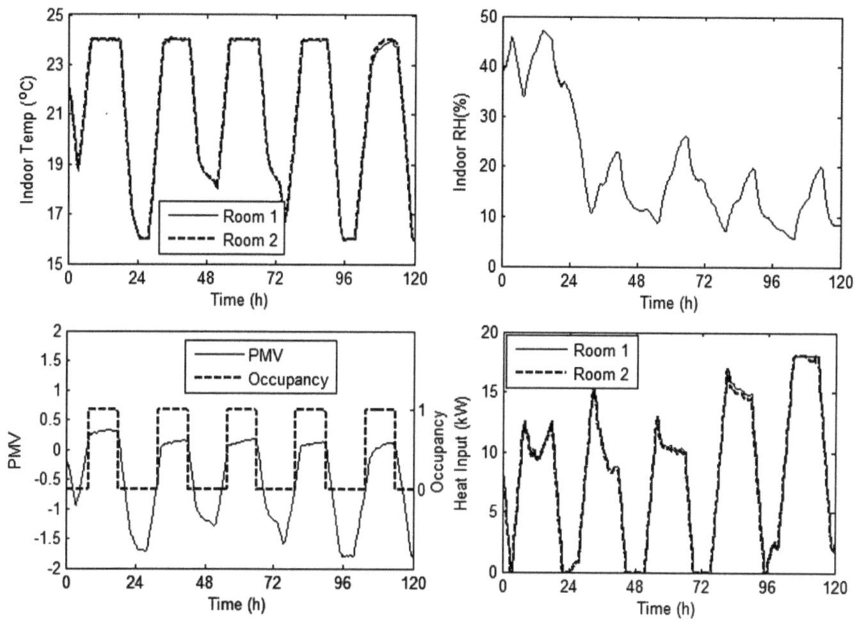

Fig. 6.60 Simulation results, SC-1′

base case in the five days is 3091 kWh. The PMV calculation shows that, with a fixed room air temperature, the PMV value fluctuates around 0.2. Energy savings can be achieved if the thermal comfort can be used as the reference instead of the air temperature.

An office occupancy-based air temperature control is also simulated for the five-day period serving as an additional base case. The results are collected in Fig. 6.60. The system is configured to have a morning warm-up operation based on rule of thumb to ensure a smooth transition in the space from unoccupied status in the night to occupied status in the day. The overall energy consumption for this base case is 1941 kWh, reduced by about 37 % from SC-1′.

Scenario SC-2 is constructed to investigate the thermal comfort-based MPC control. The nonlinearity on the energy consumption side is ignored in this control. With a Wiener structure for the thermal comfort mapping, the internal classical MPC works well with the overall hourly summation on square PMV equal to 0.07. The PMV tracking output is plotted in Fig. 6.61, which is really close to the set point 0 for neutral thermal comfort. The manipulated variables in this scenario are the heat fluxes to the space air. Compared to Scenario SC-1, the room air temperature drops to be below 24 °C and fluctuates along with the room air humidity. The overall energy consumption is reduced by around 6 % from SC-1 to 2895 kWh. With different room air temperature, the indoor air relative humidity increases slightly.

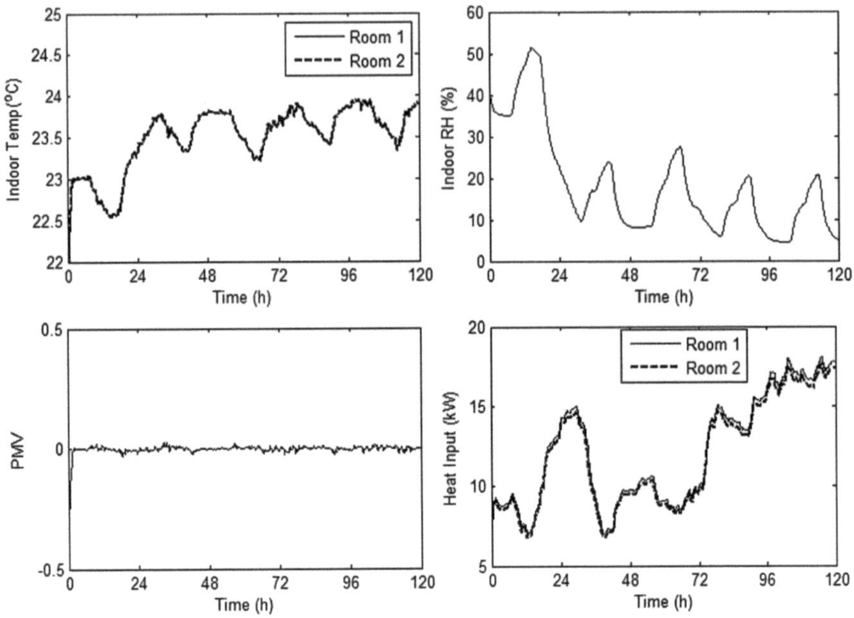

Fig. 6.61 Simulation results, SC-2

While setting PMV to 0 in MPC can maintain a very good indoor thermal comfort, a slightly cool indoor air condition in winter can bring in more energy savings. Less than 10 % occupants vote dissatisfaction when PMV drops to −0.5 according to the existing study in [30]. In Scenario SC-3, the MPC is configured to minimize the energy consumption with PMV constrained between −0.5 and 0.5. As shown in Fig. 6.64, the MPC successfully drives the system to the lower limit since it causes the lowest energy consumption in winter. Compared to Scenario SC-2, with PMV equal to −0.5 and the corresponding room air humidity, the indoor air temperature is reduced by about 2 °C. The overall energy consumption decreases to 1977 kWh, which is about 36 % less than the base condition.

From Scenario SC-2 and SC-3, we see that the proposed Wiener nonlinear MPC can make full use of the advantages of classical linear MPC. It turns the room air temperature-based space air-conditioning into a thermal comfort-oriented control while achieving additional energy savings. The overall run time for 5 days takes only 76–176 s to identify the optimums. The nonlinearity is successfully handled with the Wiener MPC approach. In reality, a heat flux is usually not directly controllable from the controller. The corresponding nonlinearity from the actual manipulated variables, e.g., water flow rates, to the heat fluxes needs to be considered. Scenarios SC-4–SC-7 are designed to investigate the performance of HW nonlinear MPC, including the aspects of reference tracking, disturbance rejection, cost function minimization, and computational cost.

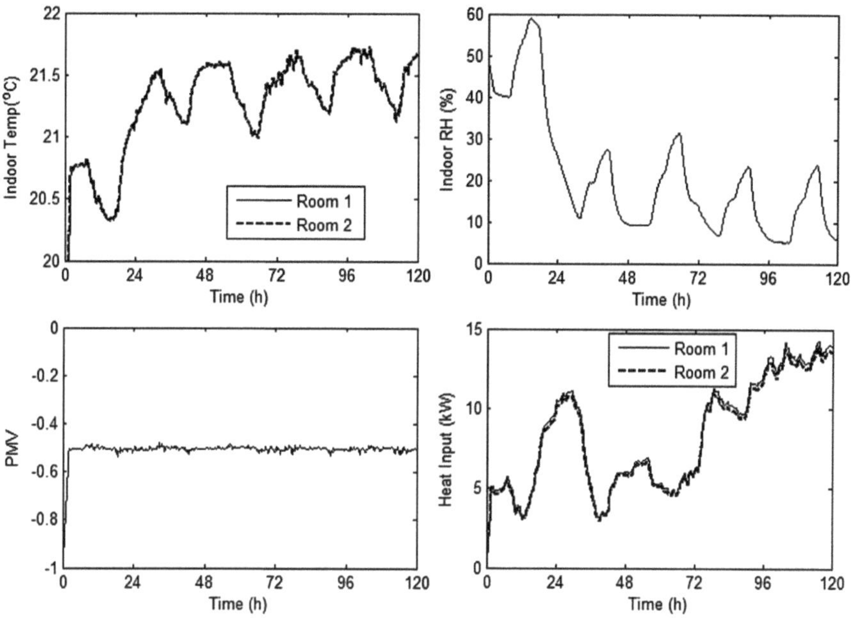

Fig. 6.62 Simulation results, SC-3

In Scenario SC-4, the water flow rates are used as the manipulated variables in the HW nonlinear MPC. The controller is designed to minimize the discomfort from the occupants by letting PMV track neutral comfort value. From Fig. 6.63, it is shown that, compared to Scenario SC-2, the HW-MPC handles the system well and the PMV tracking error is close to that of a Wiener nonlinear MPC where the heat fluxes are used as the manipulated variables. With this control system layout, the computational cost is increased by only 2 s than Scenario SC-2. Similarly, Scenario SC-5 is included to evaluate the performance of HW nonlinear MPC with thermal comfort constrained. The controller brings the system to the lowest acceptable PMV in order to minimize the energy consumption. The system performance in Fig. 6.64 is very similar to Scenario SC-3 as plotted in Fig. 6.62 in terms of PMV value, except that the driving force becomes the hot water flow rates in SC-5. In the following scenarios, HW nonlinear MPC is utilized to respect the reality of nonlinearity.

Commercial office buildings may have variable occupancy. The change of set point is also a disturbance to the controller in addition to the unmeasured solar heat flux and internal heat gains. In Scenario SC-6-1, a variable occupancy schedule is studied and a band control is imposed. When the space is unoccupied, the constraint on PMV is loosened to [−3, 3]. They are reduced to [−0.5, 0.5] when the space is occupied to maintain acceptable thermal comfort. Under those conditions, the controller is set to minimize the energy consumption during the 5 days. As plot in Fig. 6.65, the controller drives the system between the two different constrains with

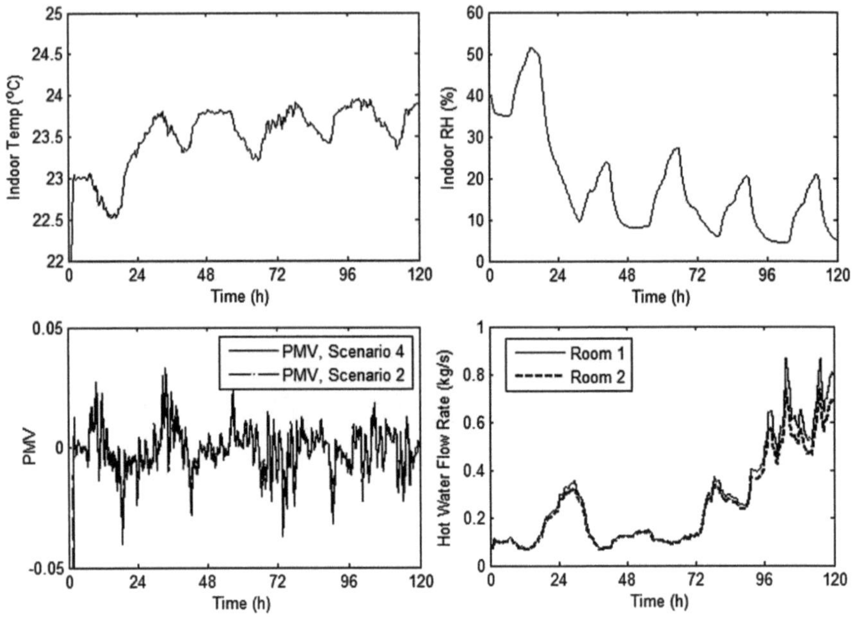

Fig. 6.63 Simulation results, SC-4

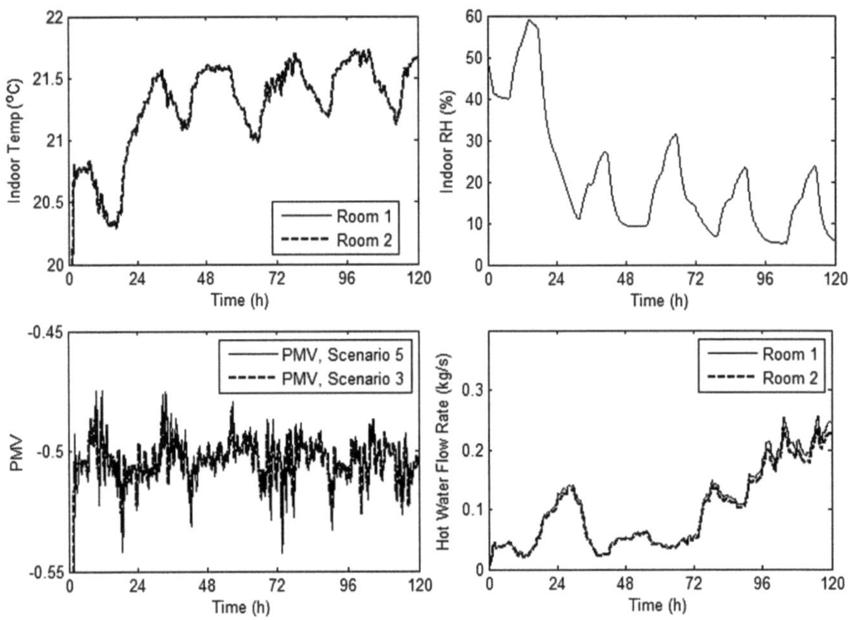

Fig. 6.64 Simulation results, SC-5

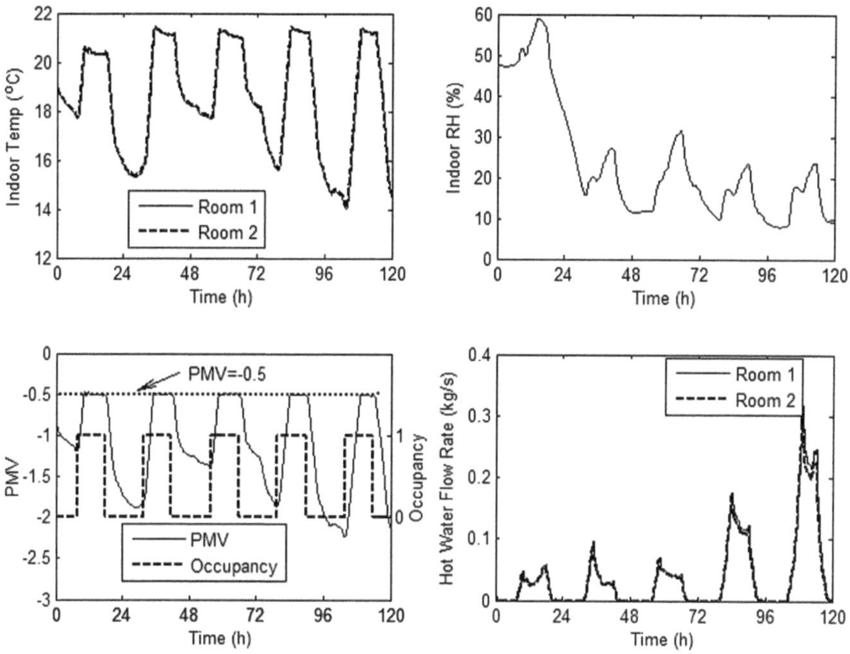

Fig. 6.65 Simulation results, SC-6-1

the occupancy switched between occupied status and unoccupied status. The simulation takes 187 s to obtain the optimums.

However, it is shown from Fig. 6.65 that the PMV value during the occupied hours can be lower than −0.5 and may cause discomfort. It is due to the various factors of the system, including heating capacity limitation, outdoor air condition, and inertia of building structure. With the energy consumption reduced by 56.7 % compared to SC-1′, the summation of squared PMV during the occupied hours increases to 39. Additional consideration should be taken to improve the performance.

In Scenario SC-6-2, the constraints are adjusted dynamically based on the outdoor air temperature prediction. The modified MPC drives the system to be within the acceptable PMV constraints with minimum energy consumption. Figure 6.66 illustrates the simulation results of SC-6-2. Compared to Scenario SC-6-1, the overall energy consumption is increased by 293 kWh, but the summation of squared PMV value is reduced to 13.44. While energy consumption is of our concern, the thermal comfort shall not be compromised. Compared to base case SC-1′ with the office schedule, the energy saving in SC-6-2 is about 41.6 %. The strategy adopted in Scenario SC-6-2 is regarded proper to improve the overall performance with both energy and thermal comfort considered.

When a space is switched from unoccupied hours to occupied hours, a smooth transition of the indoor air thermal condition is needed to ensure the space thermal

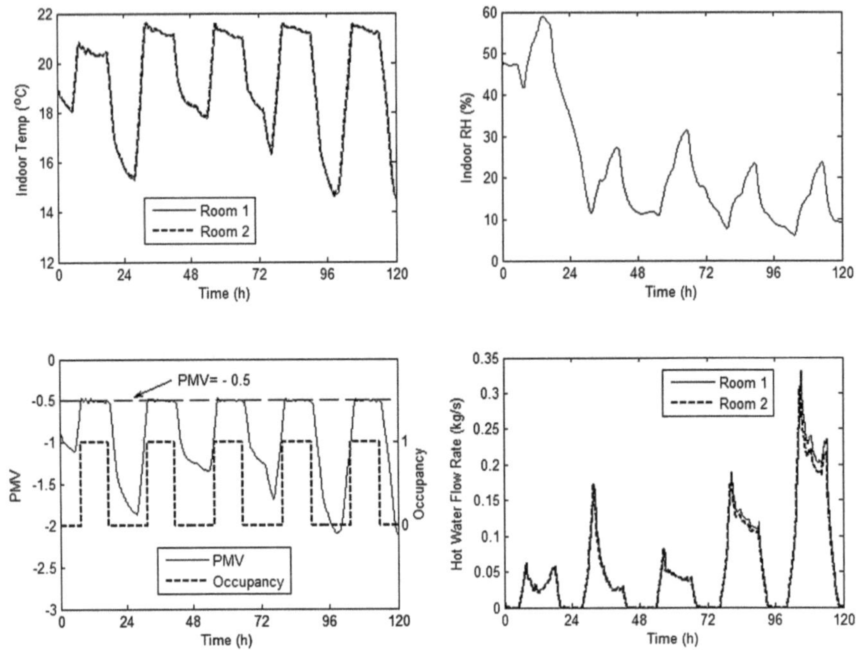

Fig. 6.66 Simulation results, SC-6-2

comfort when people come back to the building in the morning. It is termed morning warm-up in winter or cooldown in summer. The strategy is also to minimize the energy consumption by greatly relaxing the room air condition when the space is unoccupied. Figure 6.67 plots the desired set point regulation in a receding horizon manner. The set point guides the system gradually toward the comfort condition with PMV set to 0 during the day time.

Unless a heuristic and/or iterative search is conducted, we do not know what a proper profile shall be used to guarantee the smooth transition with the changing internal and external conditions. With MPC, the desire is fulfilled by penalizing the reference tracking errors and utilizing look-ahead measure on the references and constraints in this configuration. The look-ahead technique is similar to a feed-forward strategy used in conventional SISO PID control. Scenario SC-7 is designed to evaluate the performance with the techniques. The control horizon is set as 1 time interval and the prediction horizon is set as 10. Figure 6.68 plots the simulation results. As shown in the plot, the controller successfully brings the system back to the desired comfort with PMV equal 0 when the space just starts to be occupied. The summation of squared PMV during the occupied hours is reduced to 0.45. The overall energy consumption with SC-7 compared to the based condition SC-1′ decreases by about 7 %. The modification on the HW nonlinear MPC increases the computational time to about 460 s for a 5-day simulation due to the increased complexity.

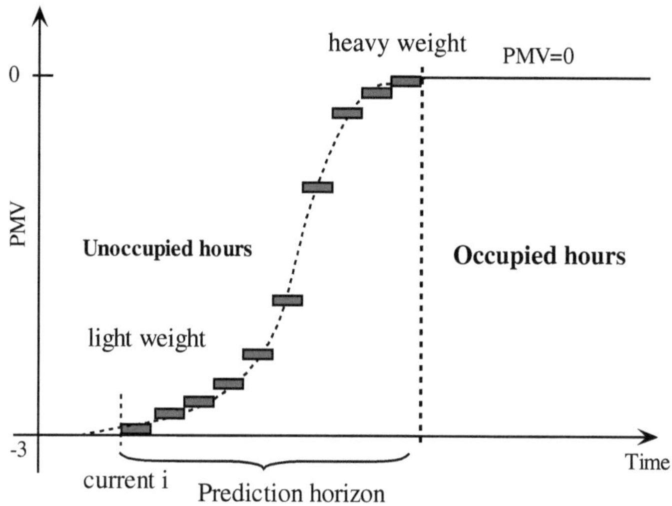

Fig. 6.67 Illustration of desired transition from unoccupied condition to occupied condition in winter

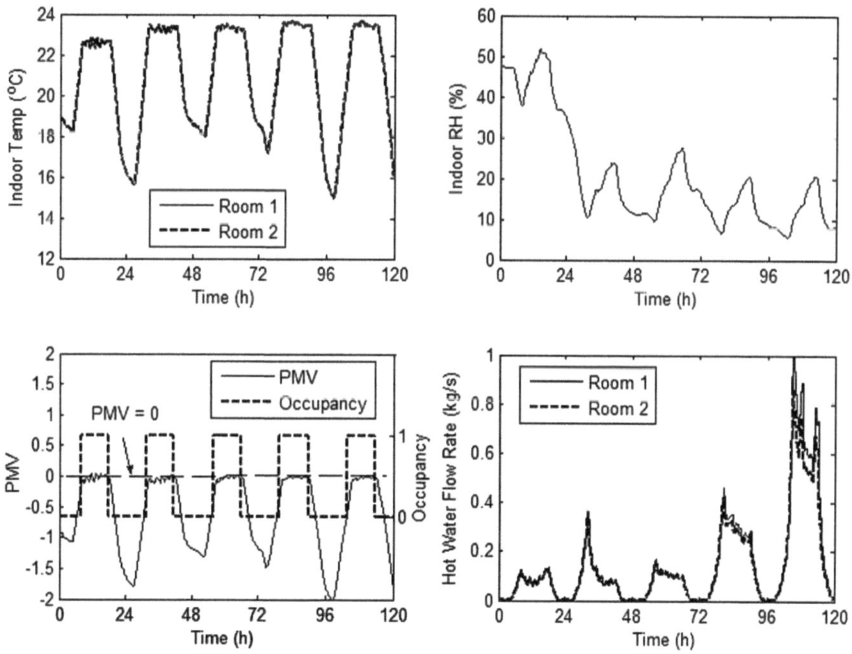

Fig. 6.68 Simulation results, SC-7

Table 6.10 Comparison of energy consumption, thermal comfort, and simulation duration

Scenario	Heating energy consumption (kWh)	Mean (PMV + 0.5) in occupied hours	Energy savings (%)	Simulation duration (s)
SC-1	3091.7	0.72	0	36.1
SC-2, 4	2895	0.56	6	76.9–78.1
SC-3, 5	1977	0.00	36.1	175.8–198.1
SC-1′	1941	0.69	0	40.6
SC-6-1	840.8	−0.25	56.7	187.4
SC-6-2	1134	0.01	41.6	205.3
SC-7	1800	0.51	7	460.0

The simulation results in terms of the energy consumption, thermal comfort, and simulation duration for the scenarios are collected in Table 6.10. In addition to the simulation duration and energy savings, an additional index based on PMV value in the occupied hours is also included. We assume −0.5 should be adopted in winter to ensure thermal comfort. The results show that the higher the mean PMV value, the lower the energy savings. With the office hour schedule, Scenario SC-6-1 stands out since cool morning hours happen which is not acceptable. Overall, with a 24/7 schedule, comfort-oriented MPC control in the test bed can bring in from 6 to 36 % energy savings; with an office schedule, comfort-oriented MPC control can offer from 7 to 41.6 % energy savings.

6.2.2.7 Small Summary

The methodology utilizes the techniques including Hammerstein, Wiener, and Hammerstein–Wiener (HW) modeling structures to convert a nonlinear system into a combination of a linear dynamic building system and nonlinear static mappings. A linear MPC can therefore be applied for online implementation to ensure much lower computational cost. Several scenarios with various MPCs are designed based on the proposed approach. The many aspects often run in predictive optimal control, including reference tracking, disturbance rejection, computational cost, energy savings, and so forth, are compared.

With the proposed approach, all the optimums for five-day operations are successfully identified in less than eight minutes. The nonlinear MIMO system is controlled either toward a minimum energy consumption condition, tracking a desired thermal comfort condition, or combined targets. The rejection capability on the unmeasured disturbances from the infiltration, internal heat gains, and so forth, is acceptable. The controller can proactively guide the system transit from unoccupied condition to occupied condition in an energy conservative manner. Furthermore, dynamic look-ahead on both the set points and the weights improves the overall performance since the future information is utilized for the preparation of a building system, where the timescale is of hours level. It can be very useful when there are energy storage capacitance, space use difference, and multiple thermal sources.

The simulation results show that thermal comfort imposed as a band constraint offers more energy savings than a fixed thermal comfort reference approach. By applying thermal comfort-oriented control, in a 24/7 schedule condition, an MPC helps acquire about 6–36 % of energy savings. In an office schedule condition, about 7–42 % of energy may be saved by using the MPC empowered thermal comfort control.

6.3 State-Space Feedback Control System Design

Compared with the transfer model (which adopts classical control theory), the state-space model (which adopts modern control theory) has outstanding advantages in the control design of MIMO system. This section mainly deals with design principle of control system based on state-space model and applications of state-space models in the control design of HVAC components. Taking the water-to-air surface heat exchanger and refrigeration system, for example, the basic method of control design based on state-space model is illustrated, and the control strategies are realized with MATLAB/Simulink tool.

6.3.1 *Basic Principle*

In order to make a system to have relatively good dynamic characteristics, system-state variables can be fed back into the input variables. Meanwhile, to ensure the system has no steady state error, we still need to introduce an output error feedback [32]. Thus, the theoretic diagram for state feedback control loop can be obtained as shown in Fig. 6.69 where R(k) is the control system desired output value, L_1 is output-feedback matrix, L_2 is state feedback matrix, $U(k)$ is control input matrix, $X(k)$ is state matrix, and $Y(k)$ is the output matrix.

From Fig. 6.69, we know that the total system control input variable is

$$U(k) = L_1[R(k) - Y(k)] - L_2X(k) \tag{6.56}$$

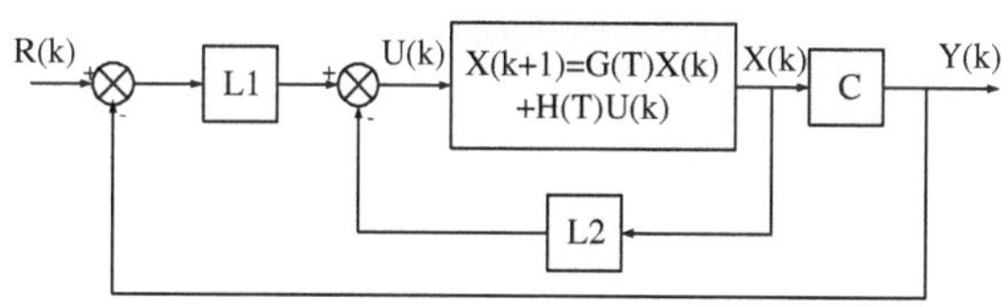

Fig. 6.69 Theoretic diagram for state-space feedback control design

Discrete form of the incremental equation of state-space model is described as below:

$$\begin{cases} \Delta X(k+1) = A\Delta X(k) + B\Delta U(k) \\ \Delta Y(k+1) = C\Delta X(k) + D\Delta U(k) \end{cases} \quad (6.57)$$

where $\Delta X(k) = X(k) - X(k-1)$; $\Delta Y(k) = Y(k) - Y(k-1)$; $\Delta U(k) = U(k) - U(k-1)$.

The following equation can be further obtained from Eq. (6.57):

$$Y(k+1) = Y(k) + C\Delta X(k+1) = Y(k) + CA(k) + CB\Delta U(k) \quad (6.58)$$

Introducing the new state vector $Z(k) = \begin{bmatrix} Y(k) \\ \Delta X(k) \end{bmatrix}$, we can obtain new state equation as below:

$$Z(k+1) = A_z Z(k) + B_z \Delta U(k) \quad (6.59)$$

where $A_z = \begin{bmatrix} I & C \cdot A \\ 0 & A \end{bmatrix}$, $B_z = \begin{bmatrix} C \cdot B \\ B \end{bmatrix}$

The design control law is as follows:

$$\Delta U(k) = L_1[R(k) - Y(k)] - L_2 \Delta X(k) \quad (6.60)$$

Let $L = [L_1 \quad L_2]$, Eq. (6.60) can be transformed into

$$\begin{aligned} \Delta U(k) &= L_1 R(k) - L_1 Y(k) - L_2 \Delta X(k) \\ &= L_1 R(k) - L \begin{bmatrix} Y(k) \\ \Delta X(k) \end{bmatrix} = L_1 R(k) - LZ(k) \end{aligned} \quad (6.61)$$

By substituting Eq. (6.61) into Eq. (6.59), we can obtain the state equation of the closed-loop system as below:

$$\begin{aligned} Z(k+1) &= A_z Z(k) + B_z[L_1 R(k) - LZ(k)] \\ &= (A_z - B_z L)Z(k) + B_z L_1 R(k) \end{aligned} \quad (6.62)$$

According to the principle of state-space feedback control, so long as we select (choose) a suitable feedback matrix L and a closed-loop system control matrix $[A_z - B_z L]$ having the desired eigenvalues, the system can have satisfactory dynamic characteristics.

6.3.2 *Control System Design for Water-to-Air Heat Exchanger*

6.3.2.1 Calculation of State-Space Feedback Matrix

The state-space model of water-to-air surface heat exchanger (i.e., air-cooling coil) is employed to carry out control system design. The controlled parameter, i.e., control of air temperature leaving the heat exchanger, is made through regulating water flow rate of the surface heat exchanger. Only the inlet air temperature of the heat exchanger is considered as the disturbance variable.

Under certain initial working conditions, we have the following state-space model of surface heat exchanger,

$$\begin{cases} \dot{x}_{\text{coil}} = A_{\text{coil}} x_{\text{coil}} + B_{\text{coil}} u_{\text{coil}} \\ y_{\text{coil}} = C_{\text{coil}} x_{\text{coil}} + D_{\text{coil}} u_{\text{coil}} \end{cases} \tag{6.63}$$

where $x_{\text{coil}} = [\Delta t_{\text{w,L}}, \Delta t_{\text{a,L}}, \Delta W_{\text{a,L}}, \Delta t_{\text{g}}]^{\text{T}}$ is system state variable; $y_{\text{coil}} = [\Delta t_{\text{w,L}}, \Delta G_{\text{w,L}}, \Delta t_{\text{a,L}}, \Delta W_{\text{a,L}}, \Delta G_{\text{a,L}}]^{\text{T}}$ is system output variables; $u_{\text{coil}} = [\Delta t_{\text{w,E}}, \Delta G_{\text{w,E}}, \Delta t_{\text{a,E}}, \Delta W_{\text{a,E}}, \Delta G_{\text{a,E}}]^{\text{T}}$ is system input variables (i.e., regulative or perturbation variables).

$$A_{\text{coil}} = \begin{bmatrix} -0.8175 & 0 & 0 & 0.6493 \\ 0 & -5.5056 & 0.0759 & 4.3120 \\ 0 & -0.0001 & -2.9362 & 0.0012 \\ 0.2004 & 0.0468 & 0.0605 & -0.5393 \end{bmatrix};$$

$$B_{\text{coil}} = \begin{bmatrix} -0.1564 & 9.5371 & 0 & 0 & 0 \\ 0 & 0 & 0.7508 & 0.0759 & 193.8230 \\ 0 & 0 & -0.0001 & 2.9336 & 0.0620 \\ 0.2004 & -160.7194 & 0.0468 & 0.0605 & -3.3213 \end{bmatrix};$$

$$C_{\text{coil}} = \begin{bmatrix} 1 & 0 & 0 & 0 \\ 0 & 0 & 0 & 0 \\ 0 & 1 & 0 & 0 \\ 0 & 0 & 1 & 0 \\ 0 & 0 & 0 & 0 \end{bmatrix};$$

$$D_{\text{coil}} = \begin{bmatrix} 0 & 0 & 0 & 0 & 0 \\ 0 & 1 & 0 & 0 & 0 \\ 0 & 0 & 0 & 0 & 0 \\ 0 & 0 & 0 & 0 & 0 \\ 0 & 0 & 0 & 0 & 1 \end{bmatrix}$$

Consider using computer for carrying out the system control design, the state-space equation is required to be discrete as follows:

$$\begin{cases} x(k+1) = A_{\text{coil,d}}(T) \cdot x(k) + B_{\text{coil,d}}(T) \cdot u(k) \\ y(k) = C_{\text{coil,d}}(T) \cdot x(k) + D_{\text{coil,d}}(T) \cdot u(k) \end{cases} \tag{6.64}$$

where T is sampling period; $A_{\text{coil,d}} = \mathrm{e}^{A_{\text{coil}}T}$; $B_{\text{coil,d}}(T) = \int_0^T e^{A_{\text{coil}} \cdot t} \mathrm{d}t \cdot B_{\text{coil}}$.

If the sampling period is set as 10 s, then the coefficient matrixes in Eq. (6.8) are as follows:

$$A_{\text{coil,d}} = \begin{bmatrix} 0.0210 & 0.0005 & 0.0013 & 0.0578 \\ 0.0147 & 0.0004 & 0.0009 & 0.0405 \\ 0 & 0 & 0 & 0 \\ 0.0178 & 0.0004 & 0.0011 & 0.0492 \end{bmatrix};$$

$$B_{\text{coil,d}} = \begin{bmatrix} 0.1521 & -320.3107 & 0.1113 & 0.2547 & -3.5213 \\ 0.3476 & -337.1684 & 0.2495 & 0.2867 & 31.6336 \\ 0.0002 & -0.1640 & 0 & 0.9992 & 0.0182 \\ 0.4452 & -431.9730 & 0.1449 & 0.3321 & -4.5749 \end{bmatrix};$$

Figure 6.70 shows the structure diagram for state-space feedback control of water-to-air surface heat exchanger. In Fig. 6.70, $R(k) = 0$ suggesting that the variations of exit air temperature of heat exchanger should be controlled as zero.

According to the principle of state-space feedback control design, in order to make the system have a relatively shorter regulation time, we need to solve $L = [L_1 \quad L_2]$ that make the eigenvalues of $[A_{\text{coil,d}} - B_{\text{coil,d}}L]$ be zero. When the controlled parameter is the exit air temperature of the surface heat exchanger, we have $C_{\text{coil,d}} = [0\ 1\ 0\ 0\ 0]$, $D_{\text{coil,d}} = [0]_{1\times 5}$. Through calculation, the system's feedback

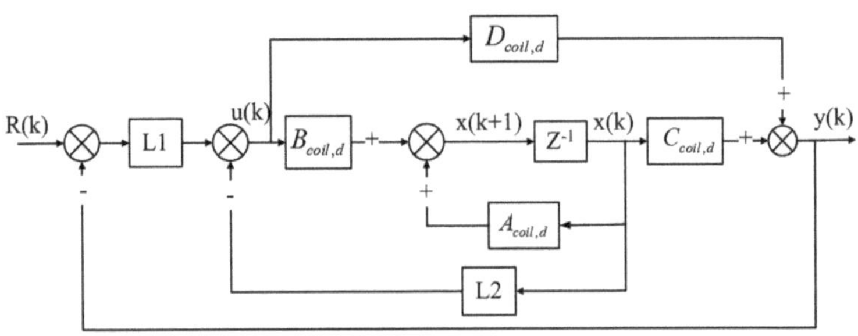

Fig. 6.70 Structural diagram for state-space feedback control of water-to-air surface heat exchanger

gain matrix can be obtained as follows: $L = [-0.0.0026\ 0\ -0.0001\ 0.0004\ 0\ 0.0002\ 0\ 0\ -0.0005]$, i.e., $L_1 = [-0.0.0026\ 0\ -0.0001\ 0.0004\ 0]$, $L_2 = [0.0002\ 0\ 0\ -0.0005]$.

6.3.3 MATLAB Simulation of the Control System

The state-space feedback control scheme of the surface heat exchanger is realized with the MATLAB/Simulink tool. Figure 6.3 shows the corresponding MATLAB model for simulation. In Fig. 6.71, $L_2\Delta x_{\text{coil}}$ and $L_1\Delta y_{\text{coil}}$ are the state-feedback and the output-feedback matrix, respectively. The $L_2\Delta x_{\text{coil}}$ and $L_1\Delta y_{\text{coil}}$ are matrix multiplication, and the example of their implementation in MATLAB is given in Fig. 6.72.

When choosing water flow rates as the regulative parameter, the simulated results of control effect under the perturbation of inlet air temperature with step increase by 5 °C and corresponding change process of the regulative parameter (i.e., water flow rate) are presented in Fig. 6.73. From the simulation results, we see that the above-designed state-space feedback matrix can obtain satisfactory control effect of the exit air temperature of surface heat exchanger. Under the perturbation of inlet air temperature with step increase by 5 °C, the maximum deviation of the exit air temperature of the surface heat exchanger from set point value is less than

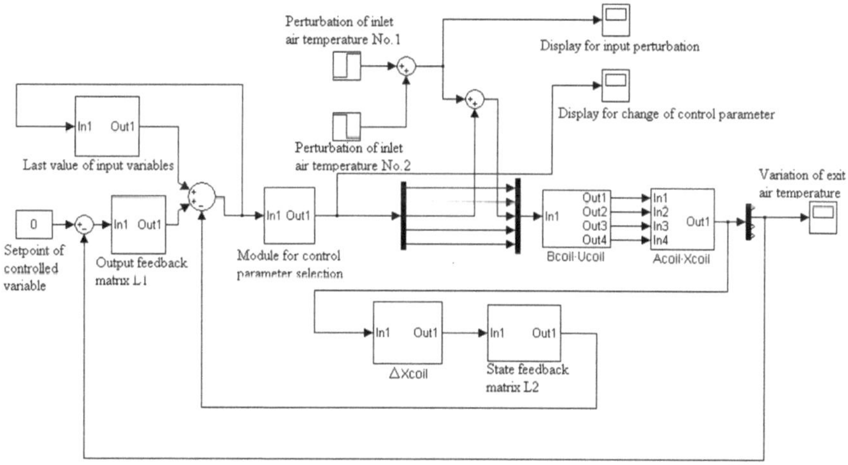

Fig. 6.71 Diagram of state-space feedback control scheme for water-to-air heat exchanger

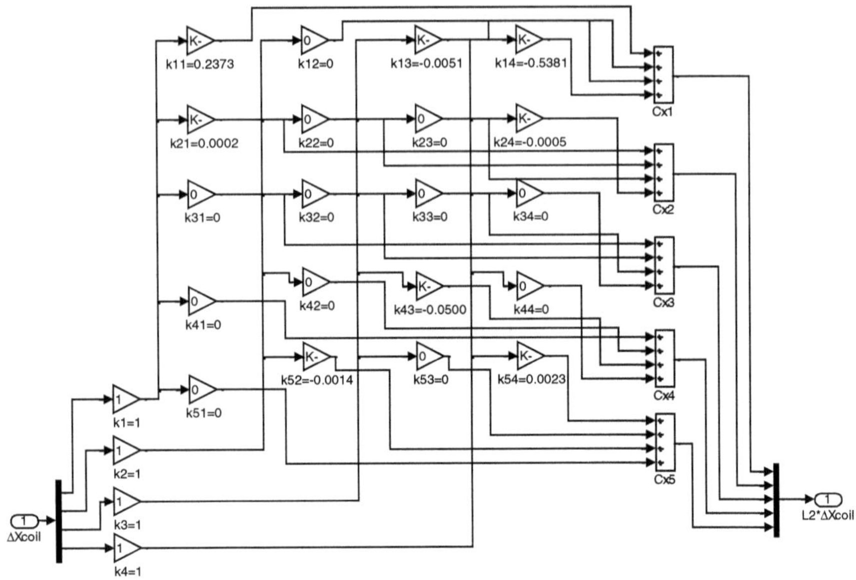

Fig. 6.72 Implementation of $L_2 \Delta x_{\mathrm{coil}}$ in MATLAB

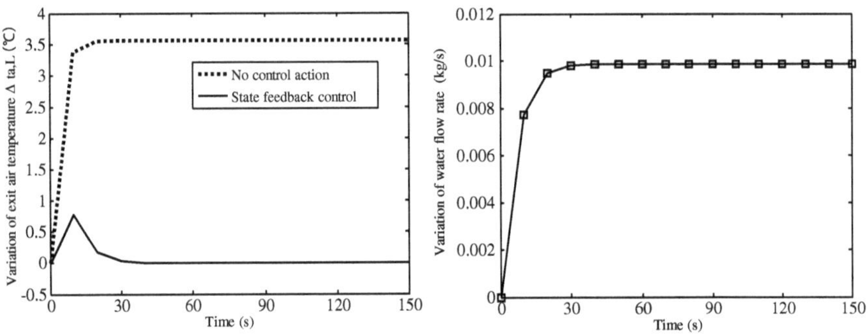

(a) Control effect (b) Change process of regulative parameter

Fig. 6.73 Results on **a** control effect of exit air temperature of surface heat exchanger under the perturbation of inlet air temperature with step increase by 5 °C, **b** change process of regulative parameter

1 °C (about 0.7 °C) when the state-space feedback control scheme is applied, otherwise (without control action) the maximum deviation of the exit air temperature will be over 3.5 °C. It takes about 30 s for the exit air temperature to attain stability after the perturbation occurs.

6.3.4 *Control System Design for Refrigeration System*

6.3.4.1 Calculation of State-Space Feedback Matrix

Consider the following state-space model of chiller under specific initial conditions:

$$\begin{cases} \dot{x}_{\text{chiller}} = A_{\text{chiller}} x_{\text{chiller}} + B_{\text{chiller}} u_{\text{chiller}} \\ y_{\text{chiller}} = C_{\text{chiller}} x_{\text{chiller}} + D_{\text{chiller}} u_{\text{chiller}} \end{cases} \tag{6.65}$$

where $x_{\text{chiller}} = [\Delta t_{\text{c}}, \Delta t_{\text{cw,L}}, \Delta t_{\text{cg}}, \Delta t_{\text{k}}, \Delta t_{\text{ew,L}}, \Delta t_{\text{eg}}]^{\text{T}}$ is system state variable; $y_{\text{chiller}} = [\Delta t_{\text{cw,L}}, \Delta G_{\text{cw,E}}, \Delta t_{\text{ew,L}}, \Delta G_{\text{ew,E}}, \Delta N_{\text{com}}, \Delta Q_{\text{c}}, \Delta \text{COP}]^{\text{T}}$ is system output variables.

$u_{\text{chiller}} = [\Delta t_{\text{cw,E}}, \Delta G_{\text{cw,E}}, \Delta t_{\text{ew,E}}, \Delta G_{\text{ew,E}}, \Delta G_{\text{rm}}]^{\text{T}}$ is system input variables (i.e., regulative or perturbation variables).

$$A_{\text{chiller}} = \begin{bmatrix} -7.3170 & 0 & 7.3080 & 0 & 0 & 0 \\ 0 & -5.4648 & 9.3621 & 0 & 0 & 0 \\ 1.4277 & 0.9104 & -3.2485 & 0 & 0 & 0 \\ 0.0989 & 0 & 0 & -22.3752 & 0 & 22.3073 \\ 0 & 0 & 0 & 0 & -3.7552 & 3.8167 \\ 0 & 0 & 0 & 3.5439 & 0.3039 & -4.1517 \end{bmatrix}$$

$$B_{\text{chiller}} = \begin{bmatrix} 0 & 0 & 0 & 0 & 28.7831 \\ -3.8973 & -23.2085 & 0 & 0 & 0 \\ 0.9104 & 3.8354 & 0 & 0 & 0 \\ 0 & 0 & 0 & 0 & -49.0332 \\ 0 & 0 & -0.0098 & 25.6544 & 0 \\ 0 & 0 & 0.3039 & -3.2699 & 0 \end{bmatrix}$$

$$C_{\text{chiller}} = \begin{bmatrix} 0 & 1.0000 & 0 & 0 & 0 & 0 \\ 0 & 0 & 0 & 0 & 0 & 0 \\ 0 & 0 & 0 & 0 & 1.0000 & 0 \\ 0 & 0 & 0 & 0 & 0 & 0 \\ 279.6932 & 0 & 0 & -239.1032 & 0 & 0 \\ -348.4032 & 0 & 0 & 239.1032 & 0 & 0 \\ -0.1063 & 0 & 0 & 0.0862 & 0 & 0 \end{bmatrix};$$

$$D_{\text{chiller}} = \begin{bmatrix} 0 & 0 & 0 & 0 & 0 \\ 0 & 1 & 0 & 0 & 0 \\ 0 & 0 & 0 & 0 & 0 \\ 0 & 0 & 0 & 1 & 0 \\ 0 & 0 & 0 & 0 & 48590 \\ 0 & 0 & 0 & 0 & 172730 \\ 0 & 0 & 0 & 0 & 0 \end{bmatrix}.$$

Likewise, the state-space model of chiller is discrete as follows:

$$\begin{cases} x(k+1) = A_{\text{chiller,d}}(T) \cdot x(k) + B_{\text{chiller,d}}(T) \cdot u(k) \\ y(k) = C_{\text{chiller,d}}(T) \cdot x(k) + D_{\text{chiller,d}}(T) \cdot u(k) \end{cases} \tag{6.66}$$

where T is sampling period; $A_{\text{chiller,d}} = \mathrm{e}^{A_{\text{chiller}}T}$; $B_{\text{chiller,d}}(T) = \int_0^T \mathrm{e}^{A_{\text{chiller}} \cdot t} \mathrm{d}t \cdot B_{\text{chiller,d}}$.

When the sampling period is set as 20 s, the coefficient matrixes in Eq. (6.66) are obtained as follows:

$$A_{\text{chiller,d}} = \begin{bmatrix} 0.0233 & 0.0201 & 0.1168 & 0 & 0 & 0 \\ 0.0404 & 0.0348 & 0.2020 & 0 & 0 & 0 \\ 0.0228 & 0.0196 & 0.1141 & 0 & 0 & 0 \\ 0.0023 & 0.0019 & 0.0109 & 0.0107 & 0.0058 & 0.0671 \\ 0.0023 & 0.0018 & 0.0109 & 0.0116 & 0.0063 & 0.0724 \\ 0.0022 & 0.0018 & 0.0104 & 0.0107 & 0.0058 & 0.0666 \end{bmatrix};$$

$$B_{\text{chiller,d}} = \begin{bmatrix} 0.8331 & -0.0118 & 0 & 0 & 21.4503 \\ 0.7135 & -4.2656 & 0 & 0 & 29.9906 \\ 0.8379 & -0.0143 & 0 & 0 & 17.6301 \\ 0.0352 & 0.0039 & 0.8939 & -3.5601 & -24.1526 \\ 0.0315 & 0.0043 & 0.9038 & 3.2193 & -22.3726 \\ 0.0317 & 0.0039 & 0.8975 & -3.5741 & -22.1437 \end{bmatrix};$$

Similarly, in order to make the system have a relatively shorter regulation time, we need to solve $L = [L_1 \quad L_2]$ that make the eigenvalues of $[A_{\text{coil,d}} - B_{\text{coil,d}}L]$ be zero. When the controlled variable is the exit coolant temperature of chiller's evaporator, we have $C_{\mathrm{coil,d}} = [0\ 0\ 0\ 0\ 1\ 0]$, $D_{\mathrm{coil,d}} = [0]1 \times 5$. Through calculation, the system's feedback gain matrix of chiller can be obtained as follows: $\mathrm{L} = [-0.0.0054\ \ 0.0005\ \ 0.0004\ \ 0.0026\ \ -0.0003\ \ -0.0001\ \ -0.0017]$, i.e., $L_1 = [-0.0.0054]$, $L_2 = [0.0005\ 0.0004\ 0.0026\ -0.0003\ -0.0001\ -0.0017]$.

6.3.4.2 MATLAB Simulation of the Control System

The above feedback gain matrix of the refrigeration system is used for the simulation with MATLAB. The regulative variable is the refrigerant flow rate (ΔGrm), and the controlled variable is the exit coolant of chiller's evaporator. Two cases of perturbation are investigated: (1). a step increase in the inlet coolant temperature of evaporator by 5 °C; (2). a step increase in the inlet coolant temperature of condenser by 10 °C

(1) inlet coolant temperature of evaporator increases by 5 °C

Figure 6.74a shows the control effect of exit coolant temperature of evaporator under the perturbation of inlet coolant temperature of evaporator with step increase by 5 °C, and Fig. 6.74b shows the corresponding change process of regulative variable (i.e., refrigerant flow rate). Simulation results show that the variation of exit coolant temperature of evaporator has a fast increase in the beginning of 10 s after the disturbance occurs, and then it decreases gradually until to a new steady state (close to its initial value) after 180 s. As shown in Fig. 6.74b, the increase rate of the regulative variable (i.e., refrigerant flow rate) decreases with the time going on, and it does not change at the time point 180 s after the perturbation happens.

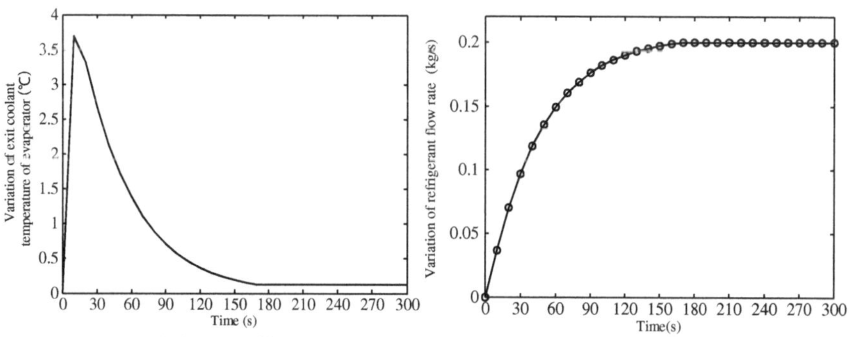

(a) Control effect

(b) Change process of regulative parameter

Fig. 6.74 Results on **a** control effect of exit coolant temperature of evaporator under the perturbation of inlet coolant temperature of evaporator with step increase by 5 °C, **b** change process of regulative parameter

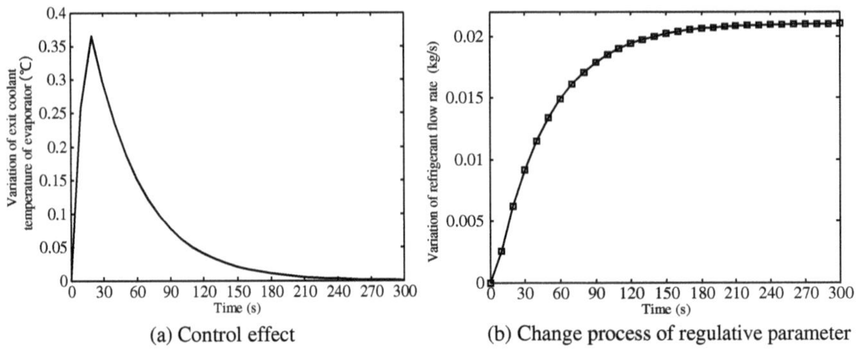

Fig. 6.75 Results on **a** control effect of exit coolant temperature of evaporator under the perturbation of inlet coolant temperature of condenser with step increase by 10 °C, **b** change process of regulative parameter

Due to the fact that the evaporator and the loaded liquid have thermal inertia, it requires some time for the increase of refrigerant flow rate to produce significant influence on the exit coolant temperature of evaporator. Since the perturbation of inlet coolant temperature of evaporator greatly impacts the exit one, it is suggested that a feed-forward control loop of the evaporator's inlet coolant temperature be added to the state feedback control system so that the control action responds quickly once the perturbations of evaporator's inlet coolant temperature occurs.

(2) Inlet coolant temperature of condenser increases by 10 °C

Figure 6.75a shows the control effect of exit coolant temperature of evaporator under the perturbation of inlet coolant temperature of condenser with step increase by 10 °C, and Fig. 6.75b presents the corresponding change process of regulative variable (i.e., refrigerant flow rate). As shown in Fig. 6.75a, the exit coolant temperature of evaporator increases rapidly in the first 20 s after the perturbation of inlet coolant temperature of condenser occurs, and then it decreases gradually until to its initial condition. Comparing Figs. 6.75a to 6.74a, it is shown that the derivation of exit coolant temperature of evaporator from initial value when subjected to the perturbation of inlet coolant temperature of condenser is much smaller than that when subjected to the perturbation of inlet coolant temperature of evaporator. This is reasonable because the evaporator's inlet coolant temperature impacts its exit one much more directly and greatly compared with the condenser's inlet coolant temperature does. So, the measurement of disturbance of inlet coolant temperature of condenser can be neglected in the real/practical closed-loop feedback control system.

References

1. Qin, S.J., Badgwell, T.A.: A survey of industrial model predictive control technology. Control Eng. Pract. **11**(7), 733–764 (2003)
2. Zhang, Y., Hanby, V.I.: Model-based control of renewable energy systems in buildings. HVAC&R Res. **12**(3a), 739–760 (2006)
3. Yuan, S., Perez, R.: Multiple-zone ventilation and temperature control of a single-duct VAV system using model predictive strategy. Energy Build. **38**(10), 1248–1261 (2006)
4. Freire, R.Z., Oliveira, G.H.C., Mendes, N.: Predictive controllers for thermal comfort optimization and energy savings. Energy Build. **40**(7), 1352–1365 (2008)
5. Kolokotsa, D., Pouliezos, A., Stavrakakis, G., Lazos, C.: Predictive control techniques for energy and indoor environmental quality management in buildings. Build. Environ. **44**(9), 1850–1863 (2009)
6. Spindler, H.C., Norfod, L.K.: Naturally ventilated and mixed-mode buildings—part II: Optimal control. Build. Environ. **44**(4), 750–761 (2009)
7. Moroşan, P.D., Bourdais, R., Dumur, D., Buisson, J.: Building temperature regulation using a distributed model predictive control. Energy Build. **42**(9), 1445–1452 (2010)
8. Ostendorp, P.M., Henze, G.P.: Model-predictive control of mixed-mode buildings with rule extraction. Build. Environ. **46**(2), 428–437 (2011)
9. Karlsson, H., Hagentoft, C.E.: Application of model based predictive control for water-based floor heating in low energy residential buildings. Build. Environ. **46**(5), 556–569 (2011)
10. Kim, S.H.: Building demand-side control using thermal energy storage under uncertainty: an adaptive multiple model-based predictive control approach. Build. Environ. **67**(9), 111–128 (2012)
11. Hazyuk, I., Ghiaus, C., Penhouet, D.: Optimal temperature control of intermittently heated buildings using model predictive control: part II—control algorithm. Build. Environ. **51**(2), 388–394 (2002)
12. Wang, S., Ma, Z.: Supervisory and optimal control of building HVAC systems: a review. HVAC&R Res. **14**(1), 3–32 (2008)
13. Findeisen, R., Allgower, F.: An introduction to nonlinear model predictive control. In: 21st Benelux Meeting on Systems and Control, Veldhoven (2002)
14. Allgower, F., Findeisen, R., Nagy, Z.K.: Nonlinear model predictive control: from theory to application. J. Chin. Inst. Chem. Eng. **35**(3), 299–315 (2004)
15. Henson, M.A.: Nonlinear model predictive control: current status and future directions. Comput. Chem. Eng. **23**, 187–202 (1998)
16. Bequette, B.W.: Nonlinear model predictive control: a personal retrospective. Can. J. Chem. Eng. **85**, 408–415 (2007)
17. Ydstie, B.E.: Extended horizon adaptive control. In: 9th IFAC World Congress, Budapest, Hungary (1984)
18. Pathwardhan, R.S., Lakshminarayanan, S., Shah, S.L.: Constrained nonlinear MPC using Hammerstein and Wiener models: PLS framework. AIChE J. **44**(7), 1611–1622 (1998)
19. Jakobsen, A.., Rasmussen, B., Skovrup, M.J., Fredsted, J.: Development of energy optimal capacity control in refrigeration systems. In: International Refrigeration and Air Conditioning Conference, pp. 329–336 (2000)
20. Leducq, D., Guilpart, J., Trystram, G.: Nonlinear predictive control of a vapor compression cycle. Int. J. Refrig. **29**, 761–772 (2006)
21. Jordan, A.J.: Linearization of nonlinear state equations. Bull. Pol. Acad. Sci.—Tech. Sci. **54** (1), 63–73 (2006)
22. Jordan, A., Nowacki, J.P.: Global linearization of non-linear state equations. Int. J. Appl. Electromagn. Mech. **19**, 637–642 (2003)
23. Friedland, B.: Control System Design: An Introduction to State-Space Methods. ISBN-10: 0486442780 (2005)
24. Ogata, K.: Modern control engineering. ISBN- **10**, 0136156738 (2009)

25. Jurinak, J.J.: Open Cycle Solid Desiccant Cooling—Component Models and system Simulations. PhD Thesis University of Wisconsin–Madison (1982)
26. Deng, K., Barooah, P., Metha, P.G., Meyn, S.P.: Building thermal model reduction via aggregation of states. In: Proceedings of ACC 2010 (2010)
27. Skartveit, A., Oleseth, J.A.: A model for the diffuse fraction of hourly global radiation. Sol. Energy **38**(4), 271–274 (1987)
28. Skartveit, A., Oleseth, J.A., Tuft, M.E.: An hourly diffuse fraction model with correction for variability and surface albedo. Sol. Energy **63**(3), 173–183 (1998)
29. Liu, B.Y.H., Jordan, R.C.: A rational procedure for predicting the long-term average performance of flat-plate solar-energy collectors. Sol. Energy **7**(2), 53–74 (1963)
30. ASHRAE.: ANSI/ASHRAE Standard 55-2004 Thermal environmental conditions for human occupancy (2004)
31. ISO.: ISO Standard 7730: ergonomics of the thermal environment—analytical determination and interpretation of thermal comfort using calculation of the PMV and PPD indices and local thermal comfort criteria (2005)
32. Li, J.: Theory and Method of Linear Control System. Xi'an Electronic and Science University press, Xi'an (2009)

Chapter 7
Combined Forecasting Models for Air-Conditioning Load Prediction

Air-conditioning load forecasting is prerequisite for the optimal control and energy-saving operation of HVAC systems. In particular for those systems that use thermal storage technology, air-conditioning load forecasting will be extremely important and indispensable. In this chapter, the principles of typical forecasting models are firstly introduced; then, two forecasting models are developed: one is 'combined forecasting model based on AHP (analytic hierarchy process)'; the other is 'RBF neural network (RBFNN) model with combined residual error correction.' The 'combined forecasting model based on AHP' can make full use of individual merits of different forecasting models and will be of good adaptability and validity. The method of analytic hierarchy process (AHP) is employed to deduce the weights of each model in the combined model. The 'RBF neural network (RBFNN) model with combined residual error correction' adopts the combined forecasting method to estimate the residual errors of forecasting results by the radial basis functions neural network (RBFNN) model and make the correction.

7.1 Typical Methods

Currently, the often used forecasting models include multiple linear regression (MLR) model, autoregressive integrated moving average (ARIMA) model, gray model (GM), and artificial neural network (ANN) model.

7.1.1 MLR Modeling

A multiple linear regression model that has n inputs $(x_1, x_2, x_3, \ldots x_n)$ and one output (y) at time t can be described by Eq (7.1) [1].

Y. Yao and Y. Yu, *Modeling and Control in Air-conditioning Systems*,
Energy and Environment Research in China, DOI 10.1007/978-3-662-53313-0_7

$$y = \beta_o + \beta_1 x_1 + \beta_2 x_2 + \ldots + \beta_k x_k + u \tag{7.1}$$

where $\beta_o, \beta_1, \cdots, \beta_k$ are constant unknown parameters, u is the random error.

For n groups of observed samples, $(y_1; x_{11}, x_{12}, \cdots, x_{1k}; u_1)_1$, $(y_2; x_{21}, x_{22}, \cdots, x_{2k}; u_2)_2, \cdots,$
$(y_n; x_{n1}, x_{n2}, \cdots, x_{nk}; u_n)_n$, the MLR model can be expressed as follows:

$$Y = XB + U \tag{7.2}$$

where

$$Y = \begin{bmatrix} y_1 \\ y_2 \\ \vdots \\ y_n \end{bmatrix}; X = \begin{bmatrix} 1 & x_{11} & x_{12} & \cdots & x_{1k} \\ 1 & x_{21} & x_{22} & \cdots & x_{2k} \\ \vdots & \vdots & \vdots & \cdots & \vdots \\ 1 & x_{n1} & x_{n2} & \cdots & x_{nk} \end{bmatrix}; B = \begin{bmatrix} \beta_o \\ \beta_1 \\ \vdots \\ \beta_k \end{bmatrix}; U = \begin{bmatrix} u_1 \\ u_2 \\ \vdots \\ u_n \end{bmatrix}$$

Assuming the estimate MLR model is:

$$\hat{Y} = X\hat{B} \tag{7.3}$$

where $\hat{B} = [\hat{\beta}_0, \hat{\beta}_1, \ldots, \hat{\beta}_k]^{\mathrm{T}}$. Then, the corresponding estimate error is:

$$Y - \hat{Y} = E = [e_1, e_2, \ldots, e_n]^{\mathrm{T}}. \tag{7.4}$$

Obviously, E is the estimate value of U.

To determine the estimate $\hat{B}$, Eq. (7.5) is given:

$$\min \sum e_i^2 = EE^{\mathrm{T}} = (Y - X\hat{B})^{\mathrm{T}}(Y - X\hat{B}). \tag{7.5}$$

Then, Eq. (7.6) exists:

$$\frac{\partial(EE^{\mathrm{T}})}{\partial \hat{B}} = -2X^{\mathrm{T}}Y + 2X^{\mathrm{T}}X\hat{B} = 0. \tag{7.6}$$

Thus, $\hat{B}$ can be solved for

$$\hat{B} = (X^{\mathrm{T}}X)^{-1}X^{\mathrm{T}}Y \tag{7.7}$$

Then, prediction of y at time m is:

$$y_{\mathrm{m}} = \hat{\beta}_o + \hat{\beta}_1 x_{1,\mathrm{m}} + \hat{\beta}_2 x_{2,\mathrm{m}} + \cdots + \hat{\beta}_{\mathrm{k}} x_{\mathrm{k,m}}. \tag{7.8}$$

MLR is a kind of forecasting model based on certain variables. In this study, the outdoor air temperature, t_{out}, the outdoor air humidity, d_{air}, and the solar radiant intensity, I_{sol} are chosen as the parameters for forecasting the residual errors of RBFNN model. Meanwhile, the time of day is also taken into account in MLR forecasting.

7.1.2 ARIMA Modeling

In the statistical approach, a statistical model is fitted to the observed data. By using an appropriate statistical model, the procedure provides a model that takes into account the characteristics of both the load profiles and the noise [2].

When y_t denotes the observation at time t and e_t is a sequence of uncorrelated variables or residual error assumed white noise, the model may be written as:

$$y_t + a_1 y_{t-1} + a_2 y_{t-2} + \ldots + a_p y_{t-p} = e_t + b_1 e_{t-1} + b_2 e_{t-2} + \ldots + b_q e_{t-q}. \quad (7.9)$$

This liner stochastic difference equation is called an autoregressive moving average model, denoted by ARMA (*p, q*). Using a time-delay operator, z^{-1} (defined by $z^{-1} y_t = y_{t-1}$), the following equation is obtained:

$$A(z^{-1}) y_t = B(z^{-1}) e_t \quad (7.10)$$

where

$$\begin{aligned} A(z^{-1}) &= 1 + a_1 z^{-1} + a_2 z^{-2} + \ldots + a_p z^{-p}; B(z^{-1}) \\ &= 1 + b_1 z^{-1} + b_2 z^{-2} + \ldots + b_q z^{-q}. \end{aligned}$$

The types of industrial time series that people wish to analyze frequently exhibit a particular kind of non-stationary behavior that can be represented by a stochastic model, which is a modified form of the ARMA model. The process is defined by the following two equations:

$$\begin{cases} A(z^{-1}) y_t = B(z^{-1}) e_t \\ y_t = \nabla^d x_t = (1 - z^{-1})^d x_t \end{cases} \quad (7.11)$$

The model corresponds to assuming that the *d*th difference of the time series $\{x_t\}$ can be represented by a stationary ARMA model. An alternative way of looking at the model for $d = 1$ results from Eq. (7.11) to give:

$$y_t = \nabla x_t = x_t - x_{t-1} \quad (7.12)$$

where ∇ is called the difference operator. In turn, ∇ has for its inverse the summation operator *S*, given by:

$$\nabla^{-1}y_t = Sy_t = y_t + y_{t-1} + y_{t-2} + \ldots = (1 + z^{-1} + z^{-2} + \ldots)y_t = (1 - z^{-1})^{-1}y_t. \tag{7.13}$$

The operator S^2y_t is similarly defined as:

$$x_t = S^2y_t = Sy_t + Sy_{t-1} + Sy_{t-2} + \ldots = \sum_{i=-\infty}^{t}\sum_{h=-\infty}^{i} y_h. \tag{7.14}$$

Also,

$$x_t = S^3y_t = \sum_{j=-\infty}^{t}\sum_{i=-\infty}^{j}\sum_{h=-\infty}^{i} y_h \tag{7.15}$$

and so on. Equation (7.11) implies that the time series $\{x_t\}$ can be obtained by integrating the stationary time series $\{y_t\}$ d times. Therefore, the model mentioned above is called autoregressive integrated moving model, denoted by ARIMA (p, d, q).

When $\{x_t\}$ contains a period component with an elementary period of s, $\nabla_s = (1 - z^{-s})$ is applied to $\{x_t\}$ d_1 times, and the ARMA (p, q) model is applied to time series $\{y_t\}$, the following model is obtained:

$$A(z^{-1})\nabla_s^{d_1}x_t = B(z^{-1})c_t. \tag{7.16}$$

Next, the periodic variation pattern is obtained from the time series $\{c_t\}$. By taking $c_{t1}, c_{t1+s}, c_{t1+2s}, \ldots$ for any time t_1 within the elementary period, the ARMA (p_1, q_1) model is applied to this time series and the following model is obtained:

$$\begin{cases} P(z^{-s}\}c_t = Q(z^{-s})e_t \\ P(z^{-s}) = 1 + \alpha_1 z^{-s} + \ldots + \alpha_{p_1} z^{-p_1 s} \\ Q(z^{-s}) = 1 + \beta_1 z^{-s} + \ldots + \beta_{q_1} z^{-q_1 s} \end{cases} \tag{7.17}$$

.

The prediction model for a time series containing an elementary period of s is obtained from Eqs. (7.16) and (7.17) as follows:

$$P(z^{-s})A(z^{-1})\nabla_s^{d_1}x_t = Q(z^{-s})B(z^{-1})e_t. \tag{7.18}$$

Furthermore, if $\{x_t\}$ has trend components and periodicity, Eq. (7.18) is rewritten as:

$$P(z^{-s})A(z^{-1})\nabla^d\nabla_s^{d_1}x_t = Q(z^{-s})B(z^{-1})e_t \tag{7.19}$$

where $\{e_t\}$ is a white-noise sequence. This model is called ARIMA $(p, d, q) \times (p_1, d_1, q_1)$. The p, d, and q are order numbers of the processes for autoregressive, integrated, and moving average components, respectively. This means that the dth deviation of the time series data is expressed by the pth-order autoregressive term and the qth-order moving average term. The p_1, d_1, *and* q_1 refer to the same orders as p, d, and q at nth previous time.

In this study, the coefficients $a_1, a_2 \ldots a_p, b_1, b_2 \ldots b_q, \alpha_1, \alpha_2 \ldots \alpha_{p_1}, \beta_1, \beta_2 \ldots \beta_{q_1}$ are estimated using historical residual errors of RBFNN model after the suitable order numbers (p, d, q, p_1, d_1, q_1) are chosen empirically. Then, the future residual errors of RBFNN model will be predicted.

7.1.3 GM Modeling

Gray forecasting model (GM) [3] has three basic operations: ① accumulated generation, ② inverse accumulated generation, and ③ gray modeling. The gray forecasting model uses the operations of accumulated generation to build differential equations. Intrinsically speaking, it has the characteristics of requiring less data. The GM (1, 1) gray model, i.e., a single variable first-order gray model, is summarized as follows:

Step 1: The initial sequence is

$$x^{(0)} = \left(x^{(0)}(1), x^{(0)}(2), \ldots, x^{(0)}(i), x^{(0)}(n)\right) \tag{7.20}$$

where $x^{(0)}(i)$ is the time series data at time i.

Step 2: Based on the initial sequence $\boldsymbol{x}^{(0)}$, a new sequence $\boldsymbol{x}^{(1)}$ is generated by the accumulated generating operation (AGO), where

$$x^{(1)} = \left(x^{(1)}(1), x^{(1)}(2), \ldots, x^{(1)}(i), x^{(1)}(n)\right) \tag{7.21}$$

$$x^{(1)}(k) = \sum_{i=1}^{k} x^{(0)}(i) \tag{7.22}$$

Step 3: The following first-order differential equation holds true:

$$\frac{\mathrm{d}x^{(1)}}{\mathrm{d}t} + ax^{(1)} = u \tag{7.23}$$

Step 4: From Step 3, we have

$$\hat{x}^{(1)}(k+1) = \left(x^{(0)}(1) - \frac{u}{a}\right)e^{-ak} + \frac{u}{a} \tag{7.24}$$

$$\hat{x}^{(0)}(k+1) = \hat{x}^{(1)}(k+1) - \hat{x}^{(1)}(k) \tag{7.25}$$

where

$$\hat{a} = \begin{bmatrix} a \\ u \end{bmatrix} = (B^T B)^{-1} B^T y_N \tag{7.26}$$

$$B = \begin{bmatrix} -0.5\left(x^{(1)}(1) + x^{(1)}(2)\right) & 1 \\ -0.5\left(x^{(1)}(2) + x^{(1)}(3)\right) & 1 \\ \vdots & \vdots \\ -0.5\left(x^{(1)}(n-1), x^{(1)}(n)\right) & 1 \end{bmatrix} \tag{7.27}$$

$$y_N = \left(x^{(0)}(2), x^{(0)}(3), \ldots, x^{(0)}(n)\right)^T \tag{7.28}$$

$\hat{x}^{(0)}(k+1)$ is the predicted value of $x^{(0)}(k+1)$ at time k+1.

7.1.4 ANN Modeling

An artificial neural network (ANN) basically consists of interconnected neurons. Each neuron or node is an independent computational unit (Fig. 7.1). The output from ANN is expressed as,

$$y = f\left[\sum (x_1 w_1 + x_2 w_2 + x_3 w_3 + \ldots) + \beta\right] \tag{7.29}$$

Fig. 7.1 Working of a neuron

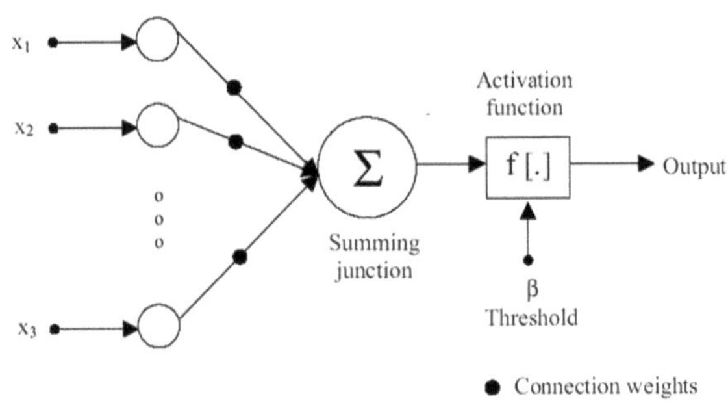

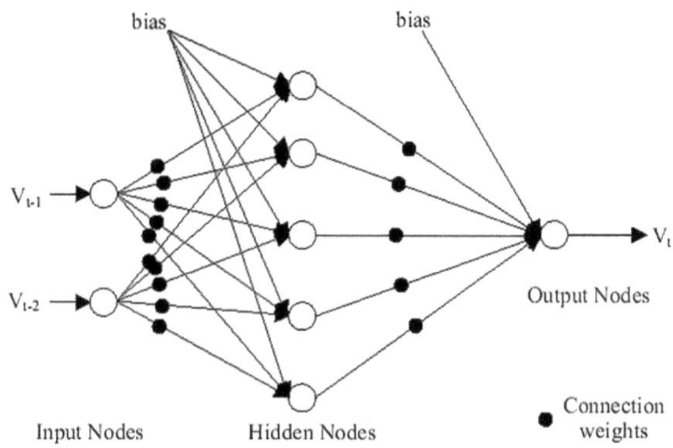

Fig. 7.2 Typical feed-forward network

where y is the output from neuron; x_1, x_2, x_3, y are the input values; w_1, w_2, w_3 are the connection weights; β is the bias value; f is the transfer function, typically sigmoidal function given by

$$f[\cdot] = \frac{1}{1 + e^{-[\cdot]}} \tag{7.30}$$

.

A typical neural network used in the current study is shown in Fig. 7.2. This is called feed forward type of network where computations proceed along the forward direction only. There are three layers of neurons, namely input, hidden, and output layer. The output obtained from the output neurons constitutes the network output [4].

The connection weights and bias values are initially chosen as random numbers and then fixed by the results of a training process. Many alternative training processes are available, out of which the present study adopted two popular schemes, namely backpropagation (BP) and cascade correlation (CC). The goal of any training algorithm is to minimize the global (mean sum squared) error E defined below:

$$E = \frac{1}{2} \sum (O_n - t_n)^2 \tag{7.31}$$

where O_n and t_n are network and target output for any nth output node. The summation has to be carried out over all output nodes for every training pattern. A pair of input and output values constitutes a training pattern.

7.2 Combined Forecasting Model Based on Analytic Hierarchy Process (AHP)

Though there are various forecasting models mentioned above, no single one has performed well enough because each model can take just several or usually only one relevant factor into consideration. In practical applications, engineers often try several kinds of models to satisfy the actual need better. The result of each forecasting model is compared, and analysis has to be done by experienced forecasters to get the best forecasting result.

To fully utilize the useful information from the models, the combined forecasting method is introduced here. It is one of the most popular subjects in the field of forecasting methods [5–7]. The theory of the combined forecasting method is based on a certain linear combination of various results from different forecast models. The fitting capacity of the combined forecasting model is greatly improved, and the forecasted result will show higher precision [8]. Formulations have been developed in the past literature [9] for the optimal combined forecasting method, whose deviation reaches the minimum and is less than that of each single forecasting method. The application of the combined forecasting method can combine separate methods and integrate merits of each model to provide a more accurate result.

7.2.1 *Principles of the Combined Forecasting Method*

For a certain forecasting problem, assume the actual value in period t is y_t (t = 1, 2, …, n) and there are m kinds of forecasting models. Let the forecasting value in period t by model i is f_{it} (i = 1, 2, …, m), then the corresponding deviation is $e_{it} = y_t - f_{it}$. Suppose the weights' vector is $W = [w_1, w_2, \ldots, w_m]^{\mathrm{T}}$, the combined forecasting model can be expressed as follows:

$$\hat{y}_t = \sum_{i=1}^{m} w_i f_{it} \quad (t = 1, 2, 3, \ldots, \mathrm{n}) \tag{7.32}$$

$$\sum_{i=1}^{m} w_i = 1 \tag{7.33}$$

.

Equation (7.32) can be rewritten as,

$$\hat{Y} = FW \tag{7.34}$$

where $\hat{Y} = [\hat{y}_1, \hat{y}_2, \ldots, \hat{y}_n]^{\mathrm{T}}$, $F = [f_{it}]_{n \times m}$.

The forecasting error of combined model can be written as:

$$e_t = y_t - f_t = \sum_{i=1}^{m} (w_i y_t) - \sum_{i=1}^{m} (w_i f_{it}) = \sum_{i=1}^{m} w_i (y_t - f_{it}) = \sum_{i=1}^{m} (w_i e_{it}). \tag{7.35}$$

Although the combined model cannot improve the forecasting accuracy essentially, it may take advantage of the 'randomness' of the errors to reduce the forecasting error. For example, when the deviations of all the models are not in the same direction, the errors can counteract partially each other in the combined forecasting.

The key of the combined forecasting method is to determine the weights of each model. There are a variety of methods available to determine the weights used in the combination of forecasts. The equal weights (EW) method that uses an arithmetic average of the individual forecasts is a very simple approach. It does not require information about the precision of the forecasts or the correlations between their errors. However, the method treats the forecasts as though they are exchangeable and indistinguishable from one another. While this may be a reasonable assumption when the models have similar error variances, it is in general not appealing. The minimum-variance (MV) method is a Bayesian approach for combining individual forecasts. The combination weights proposed by the MV method are less reliable when the data are sparse or unstable [5]. Here, the analytic hierarchy process (AHP) is employed to deduce the weights of each model.

7.2.2 *Determining Weights by Analytic Hierarchy Process (AHP)*

AHP is an intuitive method for formulating and analyzing decisions. AHP has been applied to numerous practical problems in the last few decades [10]. Because of its intuitive appeal and flexibility, many corporations and governments routinely use AHP for making major policy decisions [11]. A brief discussion of AHP is provided in this section. More detailed description of AHP and application issues can be found elsewhere [12–14]. Application of AHP to a decision problem involves four steps as below:

Step 1: **Structuring of the decision problem into a hierarchical model**

It includes decomposition of the decision problem into elements according to their common characteristics and the formation of a hierarchical model having different levels. Each level in the hierarchy corresponds to the common characteristic of the elements in that level. The topmost level is the 'focus' of the problem. The intermediate levels correspond to criteria and sub-criteria, while the lowest level contains the 'decision alternatives.'

Step 2: **Making pair-wise comparisons and obtaining the judgment matrix**

In this step, the elements of a particular level are compared pair wise with respect to a specific element in the immediate upper level. A judgment matrix is formed and used for computing the priorities of the corresponding elements. First, criteria are compared pair wise with respect to the goal. A judgment matrix, denoted as B, will be formed using the comparisons. Each entry b_{ij} of the judgment matrix is formed comparing the row element B_i with the column element B_j:

$$B = (b_{ij}) \quad (i, j = 1, 2, \ldots, \text{the number of criteria}). \tag{7.36}$$

The comparison of any two criteria C_i and C_j with respect to the goal is made using questions of the type: 'Of the two criteria C_i and C_j, which is more important and how much more?' Saaty [14] suggests the use of a 9-point scale to transform the verbal judgments into numerical quantities representing the values of b_{ij}. The scale is explained in Table 7.1. Larger number assigned to the pair-wise comparisons means larger differences between criterion levels. Thus, in comparison to the numerical mode, the verbal mode is expected to predict larger differences between criterion levels. This implies a larger range between the weights of the most preferred criterion level and the least preferred criterion level. Provided the example of a decision maker who prefers alternative 'A' slightly to alternative 'B,' the AHP interprets this verbal statement as the numerical score 3, implying that the decision maker prefers alternative A three times as much as alternative B. Given the meaning of the word 'slightly' in the regular use of language, the score 3 is probably an

Table 7.1 The semantic scale used in AHP

Intensity of importance	Definition	Description
1	Equal importance	Elements B_i and B_j are equally important
3	Weak importance of B_i over B_j	Experience and judgment slightly favor B_i over B_j
5	Essential or strong importance	Experience and judgment strongly favor B_i over B_j
7	Demonstrated importance	B_i is very strongly favored over B_j
9	Absolute importance	The evidence favoring B_i over B_j is of the highest possible order of affirmation
2,4,6,8	Intermediate	When compromise is needed, values between two adjacent judgments are used
Reciprocals of the above judgments	If B_i has one of the above judgments assigned to it when compared with B_j, then B_j has the reciprocal value when compared with B_i	A reasonable assumption

overestimation of the difference as perceived by the decision maker. The same applies to the other verbal judgments in the AHP.

The entries b_{ij} are governed by the following rules:

$$b_{ij} > 0; \quad b_{ij} = 1/b_{ji}; \quad b_{ii} = 1 \text{ for all } i. \tag{7.37}$$

Because of the above rules, the judgment matrix B is a positive reciprocal pair-wise comparison matrix.

Step 3: **Local priorities and consistency of comparisons**

Once the judgment matrix of comparisons of criteria with respect to the goal is available, the local priorities of criteria are obtained and the consistency of the judgments is determined. It has been generally agreed that priorities of criteria can be estimated by finding the principal eigenvector w of the matrix B. That is:

$$Bw = \lambda_{\max} w. \tag{7.38}$$

When the vector w is normalized, it becomes the vector of priorities of the criteria with respect to the goal. $\lambda_{\max}$ is the largest eigenvalue of the matrix B and the corresponding eigenvector w contains only positive entries. The consistency of the judgment matrix can be determined by a measure called the consistency ratio (CR), defined as:

$$\text{CR} = \frac{\text{CI}}{\text{RI}} \tag{7.39}$$

where CI is called the consistency index and RI, the random index.

CI is defined as:

$$\text{CI} = \frac{(\lambda_{\max} \quad n)}{(n-1)} \tag{7.40}$$

where n is the matrix size.

RI is the consistency index of a randomly generated reciprocal matrix from the 9-point scale with reciprocals forced. Saaty [12–14] has provided average consistencies (RI values) of randomly generated matrices (up to size 11 × 11) for a sample size of 500. The RI values for matrices of different sizes are shown in Table 7.2 [14].

If CR of the matrix is higher, it means that the input judgments are not consistent and hence are not reliable. Generally, it is acceptable only if CR < 0.10. Using a

Table 7.2 The average consistencies of random matrices (RI)

Size (n)	1	2	3	4	5	6	7	8	9
RI	0.00	0.00	0.58	0.90	1.12	1.24	1.32	1.41	1.45

very similar procedure, the local priorities of alternatives with respect to each criterion can be estimated.

Step 4 **Aggregation of local priorities**

Once the local priorities of elements of different levels are available as outlined in the previous step, they are aggregated to obtain final priorities of the alternatives. For aggregation, the following principle of hierarchic composition [14] is used: Final local priority of decision alternative= $\sum_{i=1}^{n}$ (local priority of decision alternative with respect to C_i × local priority of C_i with respect to the goal).

Note that the above is a simple weighted summation. The final priorities thus obtained represent the rating of the alternatives in achieving the focus of the problem. In the following study, local priorities of decision alternative stand for the weight of each forecasting model in the combined forecasting, respectively.

7.2.3 Combined Forecasting Model for Hourly Cooling Load Prediction Using AHP

7.2.3.1 Basic Idea

To establish the combined forecasting model using AHP, the common characteristics of cooling load prediction are ought to be known. Generally, forecasting is made on the basis of the historical data. So, the degree of fitting to the historical data is one of the elements that are under consideration during the forecasting. In addition, the adaptability and the reliability are another two important elements that are taken into account in evaluating an individual forecasting model. Adaptability refers to the ability the forecasting model has to adapt to the fickle environments, and the reliability refers to the accuracy of forecasting.

Here, we only consider the three elements (degree of fitting to the historical data, adaptability, and reliability) that impact on the effect of cooling load prediction. Thus, a hierarchical model having three levels for cooling load combined forecasting can be formed, as is shown in Fig. 7.3 [15]. In this model, cooling load

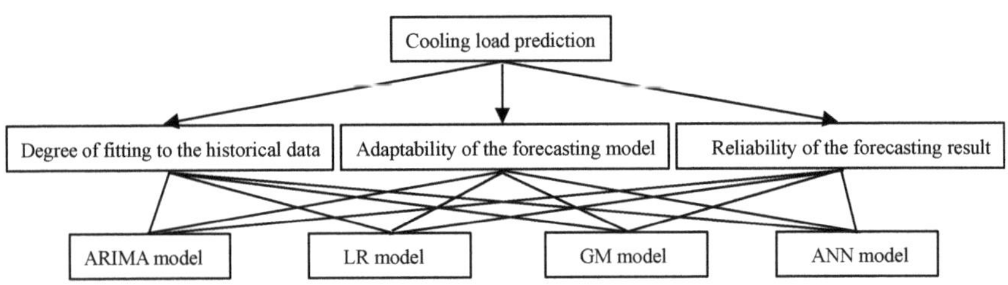

Fig. 7.3 The AHP model for cooling load combined forecasting

prediction is reckoned as the 'focus' of the problem, which is in the topmost level. The intermediate level corresponds to criteria that include the three elements mentioned above, while the lowest level contains four forecasting models that are reckoned as the 'decision alternatives'. In order to predict cooling load using the AHP combined model, the weights of each individual forecasting model must be obtained in advance by the method of AHP according to the actual situation. The following can be done manually or automatically by the AHP program:

(1) Constructing the pair-wise comparison matrices based on the experiments or expert;
(2) Calculating the priority vector for a criterion;
(3) Calculating λ_{max};
(4) Calculating the consistency index, CI;
(5) Selecting appropriate value of the random consistency ratio from Table 7.2;
(6) Calculating the consistency ratio, CR;
(7) Checking the consistency of the pair-wise comparison matrix using the value of CR to check whether the decision maker's comparisons were consistent or not;
(8) Obtaining the weights of each model and using Eq. (7.32) to make the combined forecasting.

7.2.3.2 Case Study

To further illustrate the AHP combined forecasting model, an example of hourly cooling load prediction for an office room is presented as follows.

To begin with, it is necessary to have a brief description of the room. The room is about 321.5 m^2; the exposed walls are all made of gravel concrete; the exposed windows are all double-glazing windows; the glazing rate is 15 % on the southern wall and 35 % on the northern wall; indoor heat source mainly comes from the computers and the occupants, which is about 60 $\sim$ 70 W per square meters. The room is air conditioned by fan-coil units that are equipped with thermal meters from which the cooling load can be recorded. The thermal meter uses an ultrasonic flowmeter (type TFX; measure precision is $\pm$ 1.0 %) and two Pt100 temperature sensors (measure precision is) to detect the chilled water flow rates and the inlet/outlet water temperatures, respectively, when the fan-coil unit is running. The actual cooling load of the room may be calculated by:

$$Q_o(\tau) = c_p \cdot G(\tau) \cdot [t_o(\tau) - t_i(\tau)] \tag{7.41}$$

where

$Q_o(\tau)$ the actual cooling load at the time τ, W;
$G(\tau)$ the mass flow rate of the chilled water passing through fan coil at the time τ, kg/s;
$t_o(\tau)$ the temperature of outlet water of fan coil at the time τ, °C;

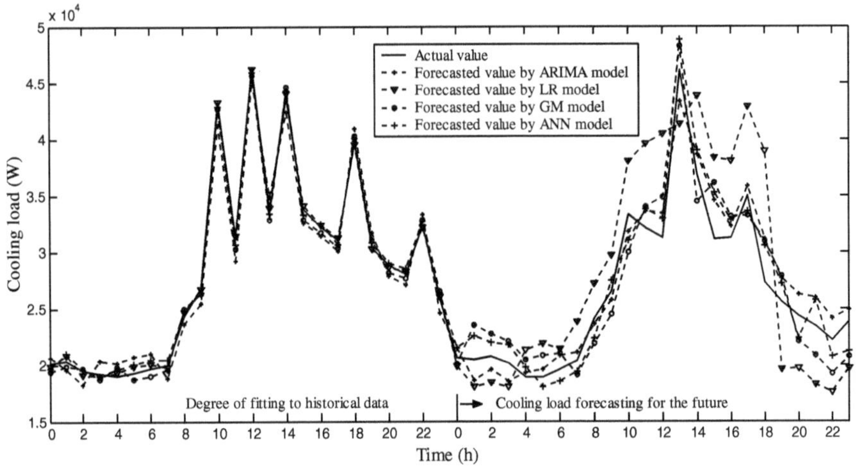

Fig. 7.4 Cooling load prediction by different forecasting models

$t_i(\tau)$ the temperature of inlet water of fan coil at the time τ, °C;
c_p the mass specific heat of the chilled water, J/(kg.°C).

To provide enough information for each forecasting model, some other thermal parameters were real timely monitored including the indoor temperatures, the outdoor temperatures, and the solar intensity. The solar intensity was measured by solar radiometer (type CE183; manufactured by CIMEL Company of France; measurement precision ± 1.0 %).

The four individual forecasting models are used to perform the prediction in advance, respectively, to make their pair-wise comparisons of the relative priority of the criteria in the intermediate level. Figure 7.4 is the result of one summer day's hourly cooling load prediction by different individual forecasting model. Figure 7.5 shows the comparisons of forecasted errors among these models. In this paper, the forecasted errors are defined as 'the actual cooling loads minus the forecasted ones.' Known from Figs. 7.4 and 7.5, LR model has the best fitting to the historical data, GM model and ANN model have a parallel better one, while ARIMA has the worst one. Thus, the pair-wise comparison matrices for the criterion I (Degree of Fitting to the Historical Data) can be obtained, as is shown in Table 7.3. The calculations for these items will be explained next for illustration purposes.

At first, the largest eigenvalue, $\lambda_{\max}$, and the corresponding principal eigenvector of the judgment matrix (B_1) can be calculated, respectively, as follows:

$$\lambda_{\max} = 4.008 \tag{7.42a}$$

$$\bar{W} = [0.1120, 0.8884, 0.3148, 0.3148]^{T}. \tag{7.42b}$$

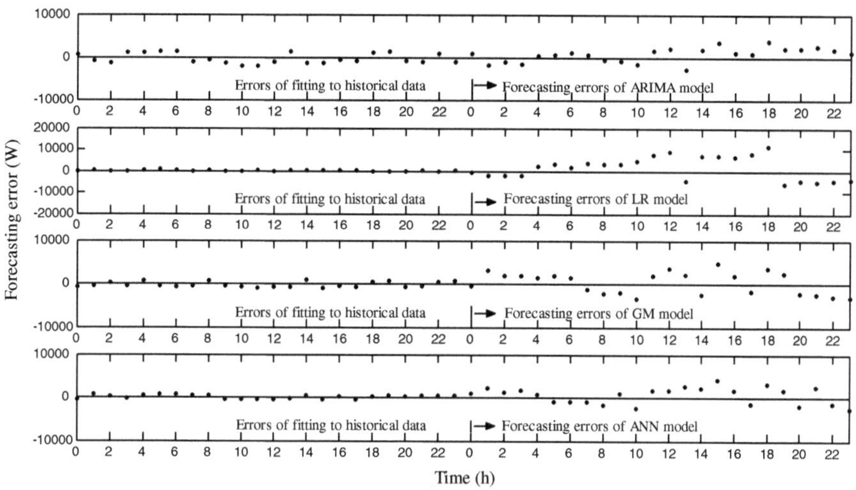

Fig. 7.5 Comparisons of errors between different forecasting models

Table 7.3 Pair-wise comparison of four forecasting models according to the criterionI (degree of fitting to the historical data)

Degree of fitting to the historical data	ARIMA model	LR model	GM model	ANN model
ARIMA model	1	1/7	1/3	1/3
LR model	7	1	3	3
GM model	3	1/3	1	1
ANN model	3	1/3	1	1

Finally, the priority vector w is obtained by normalizing $\bar{W}$:

$$w = \left[\frac{0.1120}{\sum_{i=1}^{4} \bar{W}_i}, \frac{0.8884}{\sum_{i=1}^{4} \bar{W}_i}, \frac{0.3148}{\sum_{i=1}^{4} \bar{W}_i}, \frac{0.3148}{\sum_{i=1}^{4} \bar{W}_i} \right]^{\mathrm{T}} = [0.069, 0.545, 0.193, 0.193]^{\mathrm{T}} \tag{7.43}$$

Now, estimation of the consistency ratio is as follows:
The consistency index, CI, can be calculated:

$$\mathrm{CI} = \frac{\lambda_{\max} - n}{n-1} = \frac{4.008 - 4}{4-1} = 0.00267. \tag{7.44}$$

Table 7.4 Local priority of four forecasting models according to the criterion I (degree of fitting to the historical data)

Degree of fitting to the historical data	ARIMA model	LR model	GM model	ANN model	Local priority
ARIMA model	1	1/7	1/3	1/3	0.069
LR model	7	1	3	3	0.545
GM model	3	1/3	1	1	0.193
ANN model	3	1/3	1	1	0.193

λ_{max} = 4.008; CI = 0.00267; RI = 0.90; CR = 0.00297 < 0.1 OK

Selecting appropriate value of random consistency ratio, RI, for a matrix size of four using Table 7.2, we find RI = 0.90. The consistency ratio, CR, can be calculated as follows:

$$CR = \frac{CI}{RI} = \frac{0.00267}{0.90} = 0.00297. \tag{7.45}$$

As the value of CR is less than 0.1, the judgments are acceptable. The results are shown in Table 7.4.

From the characteristics of these models, the authors think that GM model and ANN model equally have the best adaptability in the cooling load prediction, and ARIMA model is slightly worse than them, while LR model has the least adaptability. Thus, the pair-wise comparison matrices and priority vectors for adaptability can be found in Table 7.5.

Known from the forecasting results in Figs. 7.4 and 7.5, ARIMA model is the best in terms of reliability, GM model and ANN model are equally reliable. However, they are both slightly worse than ARIMA model. LR model is the worst in reliability among the four forecasting models. Thus, the pair-wise comparison matrices and priority vectors for reliability can be found in Table 7.6

In addition to the pair-wise comparison for the decision alternatives (the four forecasting models), it can be also used to set priorities for all three criteria in terms of importance of each in contributing to the overall goal (cooling load prediction). Among the three criteria, the criterion of reliability should have the topmost priority to be considered, the adaptability has the lower one, while the degree of fitting to

Table 7.5 Pair-wise comparison of four forecasting models according to the criterion II (adaptability of the forecasting model)

Adaptability of the forecasting model	ARIMA model	LR model	GM model	ANN model	Local priority
ARIMA model	1	3	1/3	1/3	0.143
LR model	1/3	1	1/5	1/5	0.064
GM model	3	5	1	1/3	0.288
ANN model	3	5	3	1	0.505

λ_{max} = 4.198; CI = 0.066; RI = 0.90; CR = 0.0733 < 0.1 OK

Table 7.6 Pair-wise comparison of four forecasting fodels according to the criterion III (reliability of the forecasting results)

Reliability of the forecasting result	ARIMA model	LR model	GM model	ANN model	Local priority
ARIMA model	1	5	3	3	0.522
LR model	1/5	1	1/3	1/3	0.078
GM model	1/3	3	1	1	0.200
ANN model	1/3	3	1	1	0.200

$\lambda_{max} = 4.044$; CI = 0.0147; RI = 0.90; CR = 0.0163 < 0.1 OK

Table 7.7 Pair-wise comparison of criteria according to the overall objective (cooling load prediction)

Cooling load prediction	Degree of fitting to the historical data	Adaptability of the forecasting model	Reliability of the forecasting result	Local priority
Degree of fitting to the historical data	1	1/5	1/7	0.078
Adaptability of the forecasting model	5	1	1	0.435
Reliability of the forecasting result	7	1	1	0.487

$\lambda_{max} = 3.013$; CI = 0.0065; RI = 0.58; CR = 0.012 < 0.1 OK

historical data is the most subordinate factor to be taken into account in the cooling load forecasting. Thus, the pair-wise comparison matrices and priority vectors for all the three criteria can be obtained in Table 7.7.

Now, the weight of each model in the combined forecasting can be found by combining the criterion priorities and the priorities of each model to each criterion, as is shown in Table 7.8. The calculations are given below for illustration purposes.

Table 7.8 Weight of each model in the combined forecasting

Forecasting model	Criterion			
	Degree of fitting to the historical data	Adaptability of the forecasting model	Reliability of the forecasting result	Final weight of each forecasting model
	0.078	0.435	0.487	
ARIMA model	0.069	0.143	0.522	0.322
LR model	0.545	0.064	0.078	0.108
GM model	0.193	0.288	0.200	0.238
ANN model	0.193	0.505	0.200	0.332

$$\text{Weight of ARIMA model} = 0.078 \times 0.069 + 0.435 \times 0.143 + 0.487 \times 0.522 = 0.322 \tag{7.46a}$$

$$\text{Weight of LR model} = 0.078 \times 0.545 + 0.435 \times 0.064 + 0.487 \times 0.078 = 0.108 \tag{7.46b}$$

$$\text{Weight of GM model} = 0.078 \times 0.193 + 0.435 \times 0.288 + 0.487 \times 0.200 = 0.238 \tag{7.46c}$$

$$\text{Weight of ANN model} = 0.078 \times 0.193 + 0.435 \times 0.505 + 0.487 \times 0.200 = 0.332 \tag{7.46d}$$

From the weights above, it is indicated that ANN model may be considered as the best model in cooling load forecasting, and ARIMA model takes second place, then GM model, and next LR model.

Using the weights of each forecasting model in Table 7.8 and the combined forecasting formula (Eq. 7.32) as well as the forecasted results by each individual forecasting model, the cooling load prediction at the same time for the office room is made once again by the combined forecasting method, as is shown in Fig. 7.6. It is easy to see from Fig. 7.6 that the forecasted results obtained by the AHP combined forecasting model have a favorable agreement with the actual ones. To further demonstrate the validity of the proposed model in this section, comparisons of forecasting errors are made between the AHP combined forecasting model and the

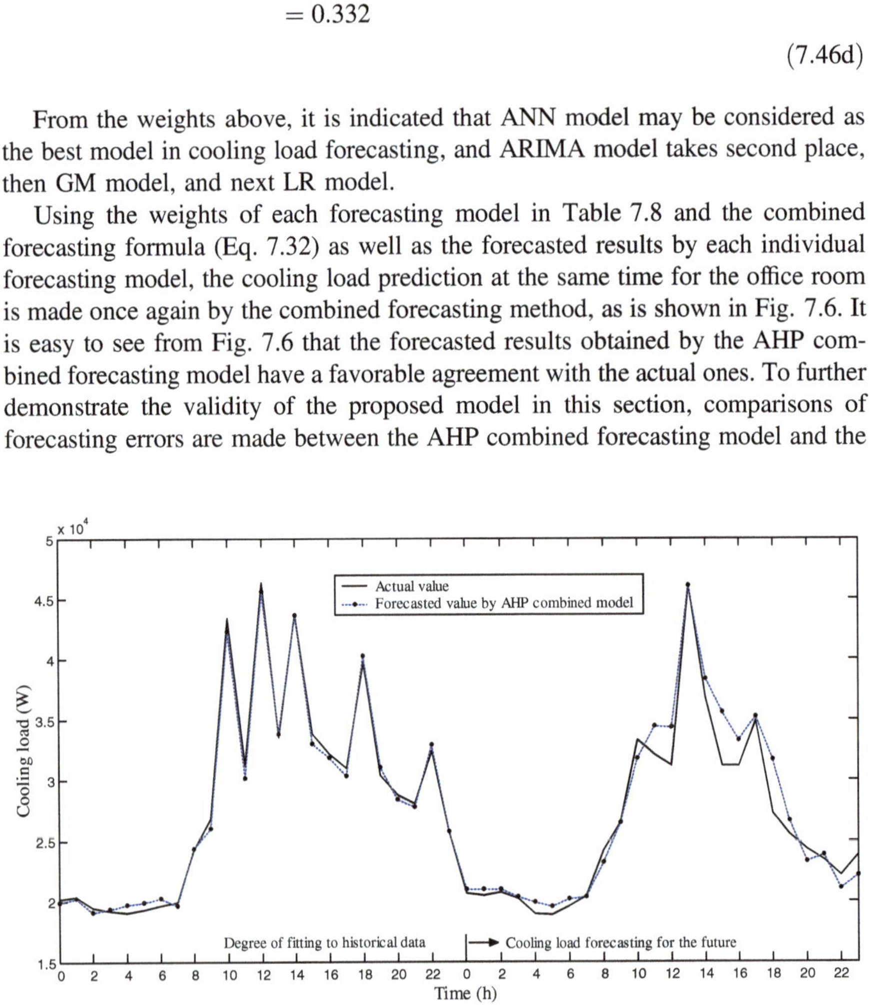

Fig. 7.6 Cooling load prediction by AHP combined forecasting model

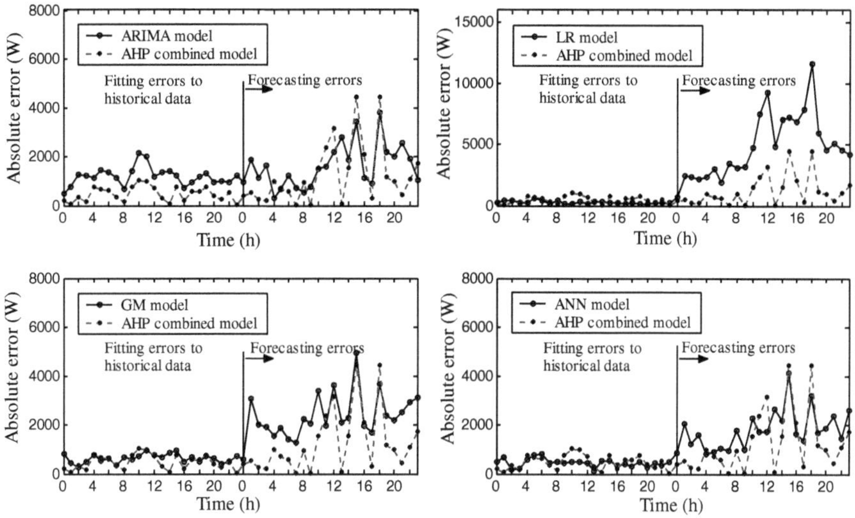

Fig. 7.7 Comparisons of forecasting errors between AHP combined model and the other ones

other forecasting ones, respectively (Please see Fig. 7.7). As is shown in Fig. 7.7, the AHP combined model has much better forecasting results than the LR model in the future 24-hour forecasting. In addition, it is easy to see from Fig. 7.7 that in the beginning of the forecasting, the AHP combined model is more accurate than the other three models (ARIMA, GM, and ANN). However, in the later forecasting hours (about after 10 h), big forecasting errors by AHP combined model occur uncertainly. Sometimes, they exceed those of ARIMA, GM, and ANN. This is because the LR model will produce increasingly large error with time going on. It indicates that the weights in Table 7.7 are suitable for the combined model to forecast the future several-hour cooling load of the building.

To know the impact of LR model on the forecasting accuracy of the combined model, only three models (ARIMA, GM, and ANN) are taken into account in the combination. Using above-mentioned AHP method, the weight of ARIMA, GM, and ANN can be obtained. They are 0.564, 0.218, and 0.219, respectively. Figure 7.8 shows the comparisons of forecasting errors between three-combined model and four-combined model. Seen from Fig. 7.8, although the forecasting errors of the three-combined model are smaller than those of the four-combined model in most cases, their gaps are very little. Therefore, it is suggested that LR model be kept in the combination for it may have useful information during the air-conditioning load forecasting.

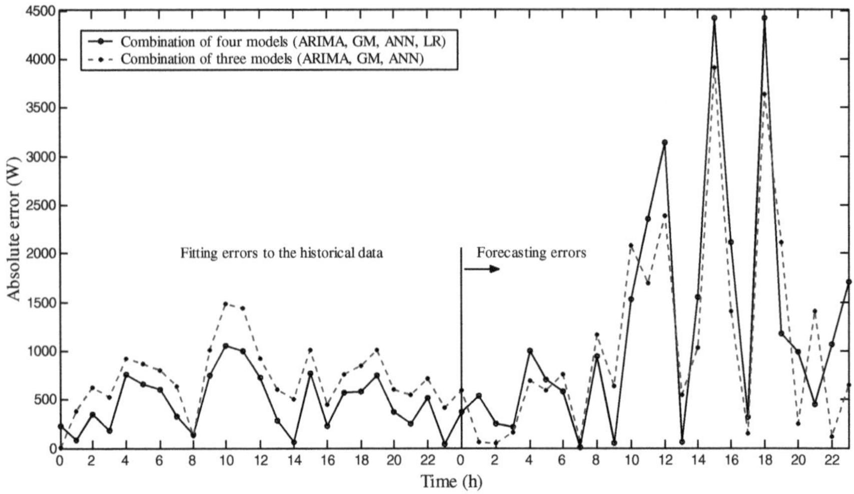

Fig. 7.8 Comparisons of forecasting errors between three-combined model and four-combined model

7.3 Forecasting Model Based on Neural Network and Combined Residual Error Correction

In this section, another approach is put forward for the AC load prediction using RBF neural network (RBFNN) with combined residual error correction. The RBFNN is chosen to forecast the air-conditioning load due to its rapid learning and generality. Meanwhile, a combined forecasting model, which is the combination of MLR, ARIMA, and GM, is established to predict the residual errors of RBFNN and make the correction.

7.3.1 Model Development

7.3.1.1 RBFNN Forecasting Model

An RBF neural network is a type of feed-forward network that learns using a supervised training technique and its output nodes form a linear combination of the radial basis functions computed by the hidden layer nodes [16]. Radial functions are a special class of functions, and their characteristic feature is that response decreases, or increases, monotonically with distance from a center point. It has been shown that RBF networks are able to approximate any reasonable continuous function mapping with a satisfactory level of accuracy [17].

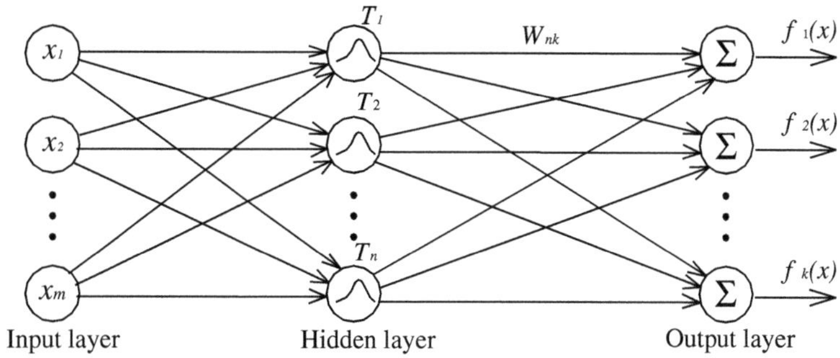

Fig. 7.9 The structure of RBF network

Normally, an RBF neural network is comprised of three different layers: an input layer, a hidden layer, and an output layer. The network structure is shown in Fig. 7.9 [18]. Here, inputs x_1, x_2, …, x_m, composing an input vector x, are applied to all neurons in the hidden layer. The hidden layer is composed of number n RBFs that are connected directly to all the elements in the output layer. A node in the hidden layer will produce a greater output when the input pattern presented is close to its center. The output of such a node will decrease as the distance from the center increases, assuming that a symmetrical basis function is used. Thus, for a given input pattern, only the neurons whose centers are close to the input pattern will produce nonzero activation values to the input stimulus. The basis function for the hidden nodes are often defined by a Gaussian exponential function shown as follows:

$$G(\|X(j) - T_i\|) = \exp\left[-\frac{1}{2}\left(\frac{\|X(j) - T_i\|}{q}\right)^2\right] \quad i = 1, 2, \cdots, n \tag{7.47}$$

where $X(j)$ is the input vector of the jth input node, $X(j) = (x_{t1}, x_{t2}, \cdots, x_{tn})$; T_i is the center of the ith RBF unit, $T_i = (t_1, t_2, \cdots, t_n)$;and q is normalization parameter.

The network output $f(x)$ is formed by a linearly weighted sum of the number of basis functions in the hidden layer. The values for the output nodes can be calculated as:

$$f_k(x) = F(X(t), W_{ik}, T_i) = \sum_{i=1}^{n} W_{ik} G(\|X(j) - T_i\|) \tag{7.48}$$

where $f_k(x)$ is the output of the kth node in the output layer, W_{ik} is the weight from the ith hidden layer neuron to the kth output layer neuron.

Known from the structure of RBF neural network described above, the transformation from the input layer to the hidden layer is nonlinear. However, the output of the network is a linear combination of the basis functions computed by the hidden layer nodes. The RBF neural network has two operating modes, namely

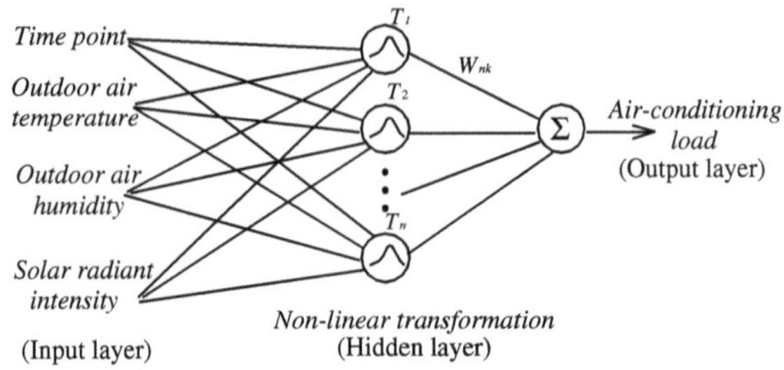

Fig. 7.10 Using RBF network to model air-conditioning load forecasting

training and testing. Training an RBF network involves determining the number of RBFs, the corresponding centers, and the output layer weight matrix. The details of algorithm and computing program can be found in literature [18, 19]. After an RBF neural network is determined and well trained, it can easily calculate the results for any new input vectors.

When RBF neural network is used to establish the air-conditioning load forecasting model, the input layer contains four neurons that receive input signals of time point ($\tau : \tau = 8 \sim 17$ for daytime; $\tau = 17 \sim 24$ and $0 \sim 8$ for nighttime), outdoor air temperature, outdoor air humidity, and solar radiant intensity. The output layer contains one neuron that produces a corresponding output, the prediction value of air-conditioning load. The RBF model for air-conditioning load forecasting is shown in Fig. 7.10. For the office buildings, different RBFNN air-conditioning load forecasting models may be set up for some special days, such as working days, double-weekend days, and festival days.

The neural network toolbox available in MATLAB [20] is used in the current work. A two-layered radial basis network is used to provide the mapping between the input parameters (independent variables) and the output parameters (dependent variables). The network in this effort is designed with the function *Solverbe* and simulated with *Simurb*, both of which are contained in the *Matlab* neural network toolbox.

7.3.1.2 Residual Error Correction Using a Combined Model

To make the RBFNN forecasting model more accurate and adaptive, it is necessary to correct the forecasting results real timely with respect to the actual situations. Assuming that the actual and the forecasted air-conditioning load at the time point τ is $Q_{\text{act}}(\tau)$ and $Q_{\text{RBF}}(\tau)$, respectively, the residual error, $\delta_Q(\tau)$, can be expressed as:

$$\delta_Q(\tau) = Q_{\text{RBF}}(\tau) - Q_{\text{act}}(\tau). \tag{7.49}$$

Obviously, if the residual error, $\delta_Q(\tau)$, is known, the final forecasting result, $Q_{\text{act},f}(\tau)$, at τ can be written as:

$$Q_{\text{act},f}(\tau) = Q_{\text{RBF}}(\tau) + \delta_Q(\tau). \tag{7.50}$$

Thus, the next key step is to establish the residual error model for the determination of the residual error, $\delta_Q(\tau)$, at the time τ, which is also a forecasting problem.

7.3.1.3 The Combined Forecasting Model for Residual Error Prediction

The theory of the combined forecasting method has been introduced in the previous section. There are also several following reasons for the application of combined forecasting model to the residual error prediction:

(1) The performance of combined model may be greatly improved because it fully utilizes the useful information from various models included.
(2) The mean deviation of combined forecasting model during a period can reach the minimum and is less than that of each single forecasting model included through optimizing the combination weights.
(3) The combined model may take advantage of the 'randomness' of the errors to reduce the forecasting error. For example, when the deviations of all the models are not in the same direction, the errors can counteract partially each other in the combined forecasting.

For a certain forecasting problem, assume the actual value in period t is y_t (t = 1, 2, …, n) and there are m kinds of forecasting models. Let the forecasting value in period t by model i is f_{it} (i − 1, 2, …, m), then the corresponding deviation is $e_{it} = y_t - f_{it}$. Suppose the weights' vector is $W = [w_1, w_2, \ldots, w_m]^{\mathrm{T}}$, the combined forecasting model and the forecasting error of combined model have been given in Eqs. (7.32) and (7.35), respectively.

Now, three typical forecasting methods, including multiple linear regression (MLR), autoregressive integrated moving average (ARIMA), and gray model (GM), are employed to establish the residual error model, respectively. The combined residual error forecasting model can be written as:

$$\delta_{e,\text{com}}(\tau) = \varpi_{\text{MLR}}\delta_{e,\text{MLR}}(\tau) + \varpi_{\text{ARIMA}}\delta_{e,\text{ARIMA}}(\tau) + \varpi_{\text{GM}}\delta_{e,\text{GM}}(\tau) \tag{7.51}$$

where $\delta_{e,\text{MLR}}(\tau), \delta_{e,\text{ARIMA}}(\tau), \delta_{e,\text{GM}}(\tau)$ are prediction values of residual error forecasted by MLR model, ARIMA model, and GM model, respectively. $\varpi_{\text{MLR}}, \varpi_{\text{ARIMA}}, \varpi_{\text{GM}}$ are the combination weights of the three forecasting models correspondingly and, at the same time, the following constraint equations exist:

$$\begin{cases} \varpi_{\text{LMR}} + \varpi_{\text{ARIMA}} + \varpi_{\text{GM}} = 1.0 \\ \varpi_{\text{MLR}} \geq 0 \\ \varpi_{\text{ARIMA}} \geq 0 \\ \varpi_{\text{GM}} \geq 0 \end{cases} \tag{7.52}$$

.

7.3.1.4 Determining the Weight Coefficients in the Combined Model

Normally, the combination weights can be dynamically optimized using the method of least error square sum (LESS) according to the historical data [21]. The equivalent equations can be written as:

$$\begin{cases} \min Z = \sum_{i=1}^{n} \left(\delta_{e,\text{com}}(\tau - i) - \delta_{Q}(\tau - i)\right)^2 \\ st. \\ \quad \varpi_{\text{LMR}} + \varpi_{\text{ARIMA}} + \varpi_{\text{GM}} = 1.0 \\ \quad \varpi_{\text{LMR}}, \varpi_{\text{ARIMA}}, \varpi_{\text{GM}} \geq 0 \end{cases} \tag{7.53}$$

.

Assuming that the forecasting deviations of single model for residual error prediction are given as follows:

$$\varepsilon_{\text{MLR}} = \delta_{e,\text{MLR}} - \delta_{Q} \tag{7.54}$$

$$\varepsilon_{\text{ARIMA}} = \delta_{e,\text{ARIMA}} - \delta_{Q} \tag{7.55}$$

$$\varepsilon_{\text{GM}} = \delta_{e,\text{GM}} - \delta_{Q} \tag{7.56}$$

Then, the deviation of combined residual error forecasting model, ε_{com}, can be deduced as:

$$\begin{aligned} \varepsilon_{\text{com}} &= \delta_{e,\text{com}} - \delta_{Q} \\ &= \varepsilon_{\text{LMR}}\varpi_{\text{LMR}} + \varepsilon_{\text{ARIMA}}\varpi_{\text{ARIMA}} + \varepsilon_{\text{GM}}\varpi_{\text{GM}} \end{aligned} \tag{7.57}$$

.

Thus, Eq. (7.53) can be expressed in the form of matrix:

$$\begin{cases} \min Z = W^{\text{T}} H W \\ st. \\ \quad e^{\text{T}} W = 1.0 \\ \quad W^{\text{T}} \geq 0 \end{cases} \tag{7.58}$$

where

$$W = \begin{bmatrix} \varpi_{\mathrm{MLR}} \\ \varpi_{\mathrm{ARIMA}} \\ \varpi_{\mathrm{GM}} \end{bmatrix}; e = \begin{bmatrix} 1 \\ 1 \\ 1 \end{bmatrix}; H = \begin{bmatrix} \varepsilon_{\mathrm{MLR}}^2 & \varepsilon_{\mathrm{MLR}}\varepsilon_{\mathrm{ARIMA}} & \varepsilon_{\mathrm{MLR}}\varepsilon_{\mathrm{GM}} \\ \varepsilon_{\mathrm{MLR}}\varepsilon_{\mathrm{ARIMA}} & \varepsilon_{\mathrm{ARIMA}}^2 & \varepsilon_{\mathrm{ARIMA}}\varepsilon_{\mathrm{GM}} \\ \varepsilon_{\mathrm{MLR}}\varepsilon_{\mathrm{GM}} & \varepsilon_{\mathrm{ARIMA}}\varepsilon_{\mathrm{GM}} & \varepsilon_{\mathrm{GM}}^2 \end{bmatrix}.$$

To solve Eq. (7.58), the following Lagrange function is introduced:

$$L = W^{\mathrm{T}}HW + \lambda(e^{\mathrm{T}}W - 1) \tag{7.59}$$

Then, by solving Eq. (6.60),

$$\begin{cases} \frac{\partial L}{\partial W} = 2HW + \lambda e = 0 \\ \frac{\partial L}{\partial \lambda} = e^{\mathrm{T}}W - 1 = 0 \end{cases} \tag{7.60}$$

The weights, $\varpi_{\mathrm{MLR}}, \varpi_{\mathrm{ARIMA}}, \varpi_{\mathrm{GM}}$, in the combined forecasting model can be finally got as:

$$W = [\varpi_{\mathrm{MLR}}, \varpi_{\mathrm{ARIMA}}, \varpi_{\mathrm{GM}}]^T = (H^{-1}e)/(e^T H^{-1}e) \tag{7.61}$$

.

7.3.1.5 RBFNN Forecasting Model with Combined Residual Error Correction

The whole model for air-conditioning load forecasting consists two parts: the RBFNN forecasting model and the combined residual error forecasting model. The forecasting process is shown in Fig. 7.11 [22]. To begin with, the RBFNN forecasting model should be established using enough historical data that include the outdoor air temperature, the outdoor air humidity, the solar radiant intensity, and the actual air-conditioning load as well as the corresponding time of day. Then, the well-trained RBFNN forecasting model is applied into the actual air-conditioning load prediction when the residual errors of RBFNN forecasting at different time are obtained using Eq. (7.35). Afterward, three residual error forecasting models including MLR, ARIMA, and GM are set up, respectively, with respect to the historical residual errors. Finally, the combined residual error forecasting model is formed and makes the corrections for RBFNN in the air-conditioning load forecasting.

The forecasting error of the model proposed in the paper can be estimated as follows:Assuming that the relative forecasting error of MLR, ARIMA, and GM are denoted as δE_{MLR}, $\delta E_{\mathrm{ARIMA}}$, and δE_{GM}, respectively, the combined relative forecasting error, δE_{Z}, can be written as:

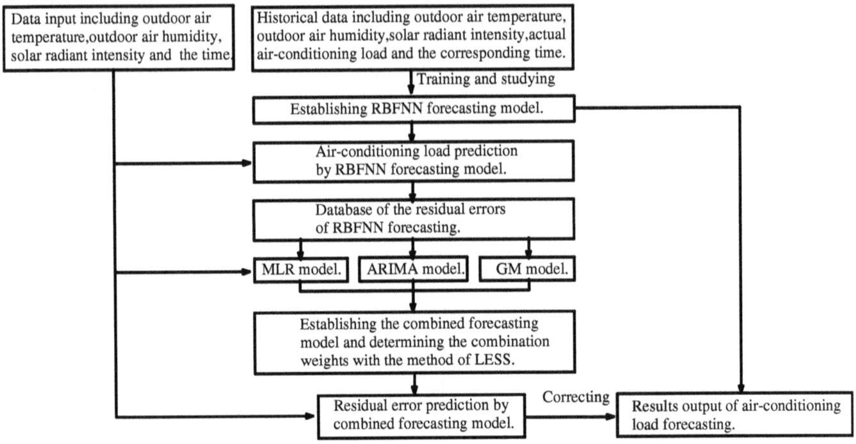

Fig. 7.11 Forecasting flowchart of RBFNN model with combined residual error correction

$$\delta E_Z = \varpi_{\mathrm{RBFNN}} \delta E_{\mathrm{MLR}} + \varpi_{\mathrm{ARIMA}} \delta E_{\mathrm{ARIMA}} + \varpi_{\mathrm{GM}} \delta E_{\mathrm{GM}} \tag{7.62}$$

Supposing the actual load and forecasted load by RBFNN at time τ are denoted as Q_{act} and $Q_{\mathrm{RBFNN},f}$, respectively, the relative forecasting error of RBFNN, $\delta E_{\mathrm{RBFNN}}$, can be expressed as:

$$\delta E_{\mathrm{RBFNN}} = \frac{\left|\Delta E_{\mathrm{RBFNN,\,act}}\right|}{Q_{\mathrm{act}}} \tag{7.63}$$

where $\Delta E_{\mathrm{RBFNN,\,act}}$ is the actual residual error of RBFNN at time τ,which is written as:

$$\Delta E_{\mathrm{RBFNN,\,act}} = Q_{\mathrm{RBFNN},f} - Q_{\mathrm{act}}. \tag{7.64}$$

When the combined forecasting model (the combination of MLR, ARIMA, and GM) is used to predict the actual residual error, $\Delta E_{\mathrm{RBFNN,\,act}}$, the relative forecasting error, δE_Z, can be expressed as:

$$\delta E_Z = \frac{\left|\Delta E_Z - \Delta E_{\mathrm{RBFNN,\,act}}\right|}{\left|\Delta E_{\mathrm{RBFNN,\,act}}\right|} \tag{7.65}$$

where ΔE_Z is the forecasted residual error of RBFNN by the combined forecasting model.

When ΔE_Z is used to correct the forecasted air-conditioning load by RBFNN, the ultimate residual error of RBFNN can be as:

$$\Delta E_{\text{cor}} = \Delta E_{\text{RBFNN, act}} - \Delta E_Z. \tag{7.66}$$

Thus, the relative forecasting error of RBFNN with combined-model correction, $\delta E^*_{\text{RBFNN}}$, can be expressed as:

$$\delta E^*_{\text{RBFNN}} = \frac{\left|\Delta E_{\text{RBFNN, act}} - \Delta E_Z\right|}{Q_{\text{act}}}. \tag{7.67}$$

Combining Eqs. (7.63), (7.65) and (7.67), the following equation can be got:

$$\delta E^*_{\text{RBFNN}} = \frac{\delta E_Z}{Q_{\text{act}}}\left|\Delta E_{\text{RBFNN, act}}\right| = \delta E_Z \delta E_{\text{RBFNN}}. \tag{7.68}$$

According to the combination theory, Eq. (7.69) exists:

$$\max(\delta E_{\text{MLR}}, \delta E_{\text{ARIMA}}, \delta E_{\text{GM}}) \geq \delta E_Z \geq \min(\delta E_{\text{MLR}}, \delta E_{\text{ARIMA}}, \delta E_{\text{GM}}). \tag{7.69}$$

Known from Eqs. (7.68) and (7.69), if the following conditions are given:

$$\delta E_{\text{MLR}} < 1 \& \delta E_{\text{ARIMA}} < 1 \& \delta E_{\text{GM}} < 1 \tag{7.70}$$

where, symbol '&' means concurrence.

Equation (7.71) comes into existence:

$$\delta E^*_{\text{RBFNN}} < \delta E_{\text{RBFNN}}. \tag{7.71}$$

Namely, the forecasting accuracy of RBFNN can be improved after the residual error correction by the combined forecasting model. However, Eq. (7.71) may not be tenable all the time because Eq. (7.70) may not be met at some time due to the complex situations.

7.3.2 *Case Study*

RBFNN model is applied to forecast the load of one certain air-conditioning system. More than 20 000 sets of data were used to build RBFNN forecasting model. Figure 7.12 shows the forecasting results of air-conditioning load in successive six days with RBFNN model. To evaluate the performance of forecasting models, absolute error and relative error are given, respectively. Absolute error is defined as 'forecasting value minus actual value,' and relative error is defined as 'the ratio of absolute error to the actual value.' Seen from the error curves in Fig. 7.12, RBFNN

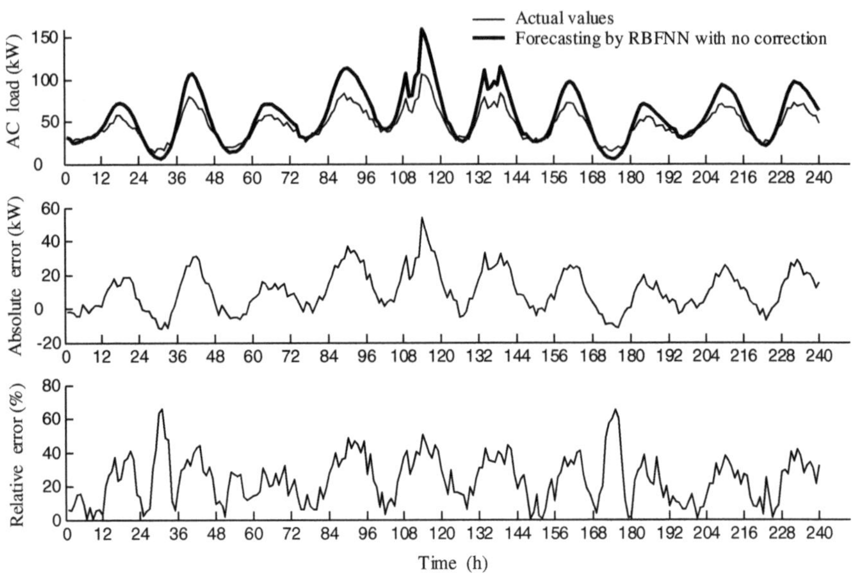

Fig. 7.12 Error analyses on RBFNN forecasting model with no correction

model has a better forecasting accuracy during the night, while in the daytime the forecasting errors are a little bigger especially in the afternoon when the relative errors are mostly above 40 %. This may be for the reason that the air-conditioning load subjects much less disturbances at night than during the day. It indicates that RBFNN model may not be enough accurate for the practical use in such case unless it should be corrected.

Figures 7.13, 7.14, and 7.15 depict the forecasting results of RBFNN model after MLR correction, ARIMA correction, and GM correction, respectively. It is clear to see that the forecasting accuracy of RBFNN has been improved visibly through residual error correction. MLR corrects the forecasting results of RBFNN with respect to the outdoor air temperature, the outdoor air humidity, and the solar radiant intensity while ARIMA and GM do it based on the time sequences. Figure 7.16 shows the forecasting results of RBFNN model after combined correction that is the combination of MLR, ARIMA, and GM. Comparing the error curves in Fig. 7.16 with those in Figs. 7.13 to 7.15, conclusion can be made that the RBFNN forecasting model with combined correction will be much more effective than that with single model correction. With the combined model correction, the relative errors of forecasting results can be within 10 % in most cases.

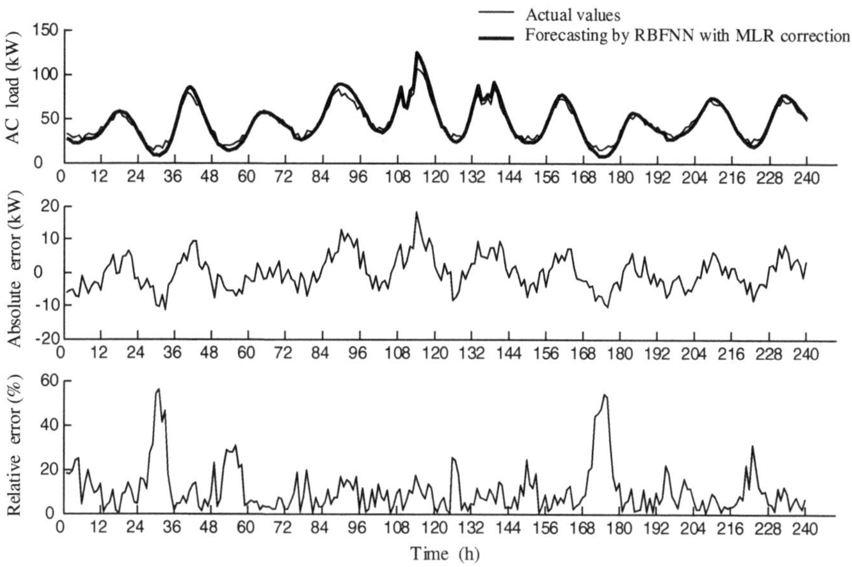

Fig. 7.13 Error analyses on RBFNN forecasting model with MLR correction

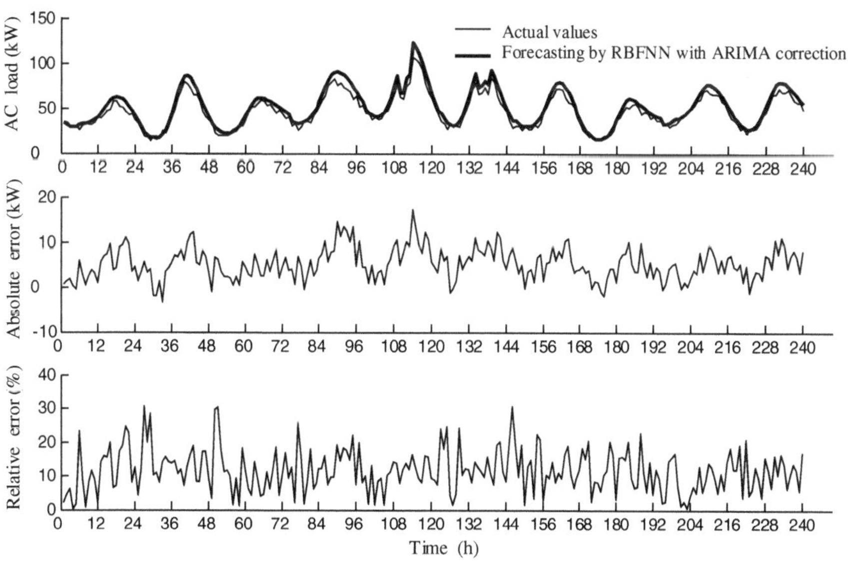

Fig. 7.14 Error analyses on RBFNN forecasting model with ARIMA correction

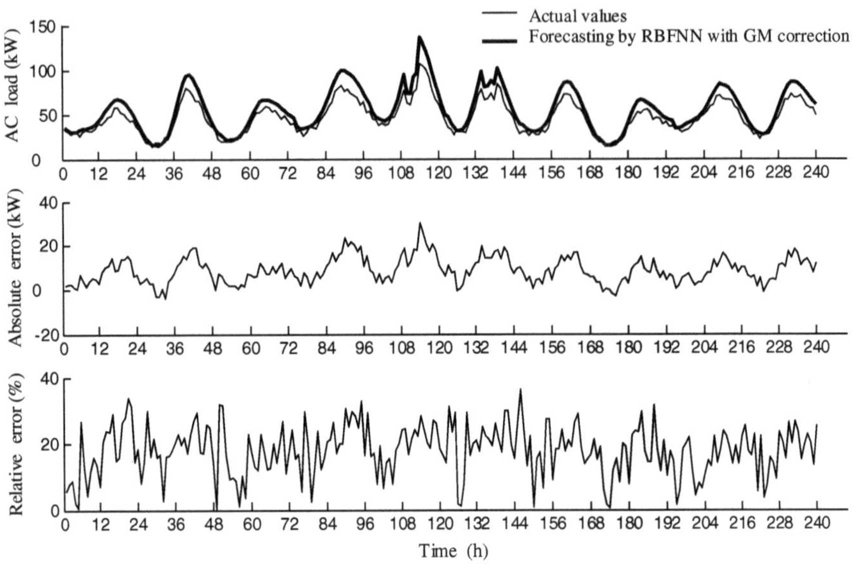

Fig. 7.15 Error analyses on RBFNN forecasting model with GM correction

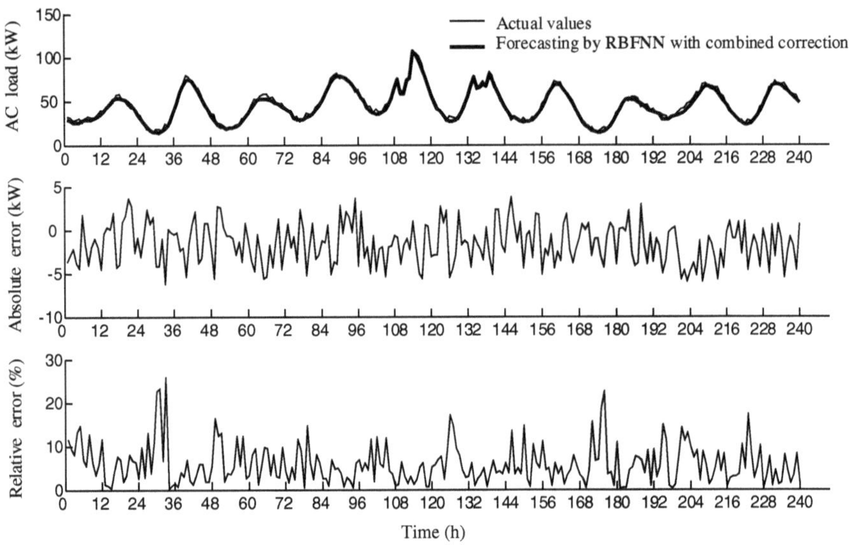

Fig. 7.16 Error analyses on RBFNN forecasting model with combined correction

References

1. Kawashima, M., Dorgan, C.E., Mitchell, J.W.: Hourly thermal load prediction for the next 24 hours by ARIMA, EWMA, LR, and an artificial neural network. ASHRAE Trans. **101**(part1), 186–200 (1995)
2. MacArthur, J.W., Mathur, A., Zhao, J.: On-line recursive estimation for load profile prediction. ASHRAE Trans. **95**(part1), 621–628 (1989)
3. Deng, J.L.: Introduction to grey system theory. J. Grey Syst. **1**(1), 1–24 (1989)
4. Kawashima, M., Dorgan, C.E.: Optimizing system control with load prediction by neural networks for an ice-storage system. ASHRAE Trans. **102**(part1), 1169–1178 (1996)
5. Kang, H.: Unstable weights in the combination of forecasts. Manage. Sci. **32**, 683–695 (1986)
6. Gupia, S., Wilton, P.: Combination of forecasts: an extension. Manage. Sci. **33**, 356–372 (1987)
7. Schmittlein, D.C.: Combining forecasts: operational adjustments to theoretically optimal rules. Manage. Sci. **36**, 1044–1056 (1990)
8. Bates, J.M., Gr, C.: The Combination of Forecast. Oper. Res. Q. **20**, 451–463 (1969)
9. Granger, C.W.J., Ramanathan, R.: Improved methods of combining forecasts. J Forecast. **3**, 197–204 (1984)
10. Shim, J.P.: Bibliographical research on the analytic hierarchy process (AHP). Socio-Economic Planning Sci. **23**, 161–167 (1989)
11. Elkarni, F., Mustafa, I.: Increasing the utilization of solar energy technologies (SET) in Jordan: analytic hierarchy process. Energy Policy **21**, 978–984 (1993)
12. Saaty, T.L.: Axiomatic foundations of the analytic hierarchy process. Manage. Sci. **32**, 841–855 (1986)
13. Saaty, T.L.: Rank generation, preservation and reversal in the analytic hierarchy process. Decis. Sci. **18**, 157–177 (1987)
14. Saaty, T.L.: Fundamentals of Decision Making and Priority Theory with The Analytic Hierarchy Process. RWS Publications, Pittsburg (2000)
15. Yao, Y., Lian, Z., Liu, S., Hou, Z.: Hourly cooling load prediction by a combined forecasting model based on analytic hierarchy process. Int. J. Therm. Sci. **43**(11), 1107–1118 (2004)
16. Hush, D.R., Horne, B.G.: Progress in supervised neural networks. IEEE Signal Process. Mag. **10**(1), 8–39 (1993)
17. Bianchini, M., Frasconi, P., Gori, M.: Learning without local minima in radial basis function networks. IEEE Trans. Neural Networks **6**(3), 749–756 (1995)
18. Wasserman, P.D.: Advanced Methods in Neural Computing. Van Nostrand Reinhold, New York (1993)
19. Haykin, S.: Neural Networks: A Comprehensive Foundation, 2nd ed. New Jersey: Prentice Hall (1999)
20. Dernuth, H., Beale, M.: Matlab Neural Network Toolbox. The Math Works Inc (1992)
21. Tang, X., Zhou, Z., Shi, Y.: The variable weighted functions of combined forecasting. Comput. Math Appl. **45**(4–5), 723–730 (2003)
22. Yao, Y., Lian, Z., Hou, Z., Liu, W.: An innovative air-conditioning load forecasting model based on RBF neural network and combined residual error correction. Int. J. Refrig **29**(4), 528–538 (2006)

Chapter 8
Energy Analysis Model for HVAC System

This chapter firstly deals with energy models of HVAC components including chiller, boiler, pump/fan, cooling tower, and water-to-air surface heat exchanger. These energy models are the basis for energy analysis on any type of HVAC system. Then, an energy evaluation program for HVAC systems is introduced, and energy savings of variable air volume (VAV) air-conditioning system compared to the other two conventional air-conditioning systems [i.e., constant-air-volume system (CAVS) and fan-coil system (FCS)] are investigated in terms of different climatic conditions. Furthermore, the year-round energy consumptions of VAV system with different air-side economizers are compared and analyzed.

8.1 Energy Models for HVAC Components

8.1.1 Chiller

The energy model of chiller is usually based on the coefficient of performance (COP) which is written as follows:

$$\mathrm{COP} = \frac{Q_{\mathrm{e}}}{N_{\mathrm{com}}} \tag{8.1}$$

where Q_{e} is heat exchange rate of chiller's evaporator (i.e., cooling capacity of chiller); kW; N_{com} is power of chiller's compressor, kW.

The COP equation can be developed according to the energy conservation law. For the steady state, the exhaust heat from the condenser of chiller is subject to the following equation:

Y. Yao and Y. Yu, *Modeling and Control in Air-conditioning Systems*,
Energy and Environment Research in China, DOI 10.1007/978-3-662-53313-0_8

$$Q_c = N_{com} + Q_e \tag{8.2}$$

where Q_c *is* heat exchange rate of chiller's condenser (i.e., condensing heat), kW.

Assuming the refrigerant in the condenser and evaporator is in two-phase state, Eq. (8.3) can be deduced from the second law of thermodynamics [1]:

$$\frac{Q_c + q_c}{T_c} - \frac{Q_e + q_e}{T_e} = 0 \tag{8.3}$$

where T_e and T_c are evaporating and condensing temperature, respectively. q_e and q_c are heat dissipation rate of chiller's evaporator and condenser, respectively, kW.

Combining Eqs. (8.1) through (8.3), the COP of chiller can be expressed as follows:

$$\text{COP} = \frac{Q_e}{\frac{T_c}{T_e}(Q_e + q_e) - q_c - Q_e} \tag{8.4}$$

Introducing partial load ratio (r):

$$r = \frac{Q_e}{Q_o} \tag{8.5}$$

Equation (8.4) can be written as follows:

$$\text{COP} = \frac{r}{\left(\frac{T_c}{T_e} - 1\right)r + \frac{q_e}{Q_o}\frac{T_c}{T_e} - \frac{q_c}{Q_o}} \tag{8.6}$$

The irreversible loss function, $f_s(T_c, T_e)$, is particularly suggested here to make Eq. (8.6) more clear.

$$f_s(T_c, T_e) = a_1 \frac{T_c}{T_e} - a_2 \tag{8.6a}$$

where $a_1 = q_e/Q_o$; $a_2 = q_c/Q_o$.

The coefficients, a_1 and a_2, can be obtained empirically from field data. As reflected by Eq. (8.6), the higher ratio of condensing temperature to evaporating one, T_c/T_e, will result in more energy dissipation and lower COP of chiller.

The evaporating temperature (T_c) and the condensing temperature (T_c) can be determined, respectively, by Eq. (8.7) and (8.8) [1]:

$$T_c = T_{w,c,E} + \frac{Q_c}{c_w G_{w,c}(1 - \exp(-UA_c/c_w G_{w,c}))} \tag{8.7}$$

$$T_e = T_{w,e,E} - \frac{Q_e}{c_w G_{w,e}(1 - \exp(-UA_e / c_w G_{w,e}))} \tag{8.8}$$

where $T_{w,c,E}$ and $T_{w,e,E}$ are inlet water temperature of condenser and evaporator, respectively, K; UA_c and UA_e are overall coefficients of heat transfer for condenser and evaporator, respectively; W/°C; c_w is specific heat of coolant (usually as water), kJ/(kg °C).

In Eq. (8.7), the total coefficient of heat transfer of condenser can be empirically written as follows:

$$UA_{con} = \left[a_3 G_{w,con}^{-0.8} + a_4 Q_{con,exh}^{-1/3} + a_5\right]^{-1} \tag{8.9}$$

where a_3, a_4, a_5 are empirical coefficients.

For the specific evaporator and condenser, the coefficients of heat transfer, UA_e and UA_c, are mainly related to the water flow rate passing through the heat exchanger. Therefore, Eq. (8.8) and (8.9) can be simplified, respectively, as follows:

$$T_c = T_{w,c,E} + \frac{Q_c}{F_c(G_{w,c})} \tag{8.10}$$

$$T_e = T_{w,e,E} - \frac{Q_e}{F_e(G_{w,e})} \tag{8.11}$$

where $G_{w,c}$ and $G_{w,e}$ are water flow rate in chiller's condenser and evaporator, respectively, kg/s; $F_c(G_{w,c})$ and $F_e(G_{w,e})$ are empirical functions dependent of $G_{w,c}$ and $G_{w,e}$, respectively.

8.1.2 Boiler

In winter, the hot water boiler is usually used for space heating. Natural gas, oil, coal, and electricity are energy sources that can be used in hot water boilers.

For fuel-consuming boilers, they can be assessed by the combustion efficiency, $\eta_{ce,boiler}$, which is the ratio of heat output from the hot water, $Q_{h,boiler}$, to the heat content rate of the fuel consumed, Q_{fuel},

$$\eta_{ce,boiler} = \frac{Q_{h,boiler}}{Q_{fuel}} \times 100\,\% \tag{8.12}$$

Usually, $\eta_{ce,boiler}$ varies from 80 to 85 % [2], while for electricity-consuming boilers, the efficiency can be considered as 100 %.

8.1.3 Pump and Fan

The energy of pump or fan with variable frequency drives can be calculated by Eq. (8.13) [3].

$$N_{\mathrm{p/f,var}} = \frac{\rho_{\mathrm{w}} H_{\mathrm{p/f}} G_{\mathrm{p/f}}}{9.8\,\eta_{\mathrm{s}} \eta_{\mathrm{m}} \eta_{\mathrm{var}}} \tag{8.13}$$

where ρ_w is density of water, kg/m^3; $H_{p/f}$ is pressure of pump or fan, m H_2O; $G_{p/f}$ is flow rate of pump or fan, m^3/s; η_s, η_m and η_{var} are transmission efficiency, motor's efficiency, and variable frequency driver's efficiency.

For convenience to apply, Eq. (8.13) can be converted into Eq. (8.14).

$$N_{\mathrm{p/f,var}} = \frac{k_{\mathrm{p/f}} \cdot X_{\mathrm{G}}^3}{\eta_{\mathrm{m}}(X_{\mathrm{G}}) \eta_{\mathrm{var}}(X_{\mathrm{G}})} \tag{8.14}$$

where

$$X_{\mathrm{G}} = \frac{G_{\mathrm{p/f}}}{G_{\mathrm{p/f,o}}} \tag{8.14a}$$

$$\eta_{\mathrm{m}}(X_{\mathrm{G}}) = B_0 + B_1 \cdot X_{\mathrm{G}} + B_2 \cdot X_{\mathrm{G}}^2 + B_3 \cdot X_{\mathrm{G}}^3 \tag{8.14b}$$

$$\eta_{\mathrm{var}}(X_{\mathrm{G}}) = C_0 + C_1 \cdot X_{\mathrm{G}} + C_2 \cdot X_{\mathrm{G}}^2 + C_3 \cdot X_{\mathrm{G}}^3 \tag{8.14c}$$

The coefficients, $k_{p/f}$, B_0, B_1, B_2, B_3, C_0, C_1, C_2, and C_3, are determined empirically by test data.

8.1.4 Cooling Tower

The energy model of cooling water tower can be established by means of tower thermal efficiency ($\varepsilon_{\mathrm{tower}}$). The heat rejected by the cooling tower can be calculated by Eq. (8.15).

$$\begin{aligned} Q_{\mathrm{tower}} &= \varepsilon_{\mathrm{tower}} G_{\mathrm{tower,a}} \left(h_{\mathrm{tower,as,E}} - h_{\mathrm{tower,a,E}}\right) \\ &= c_{\mathrm{w}} G_{\mathrm{tower,w}} \left(t_{\mathrm{tower,w,E}} - t_{\mathrm{tower,w,L}}\right) \end{aligned} \tag{8.15}$$

where G is flow rate, kg/s; h is enthalpy, J/kg; t is temperature, °C; subscripts '*a*' and 'w' denote air and water, respectively; subscripts 'E' and 'L' denote inlet and outlet, respectively; and subscript '*as*' stands for saturated moist air.

For the countercurrent cooling water tower, the thermal efficiency ($\varepsilon_{\mathrm{tower}}$) can be expressed as [4]:

$$\varepsilon_{\text{tower}} = \frac{1 - \exp[-\text{Ntu}_{\text{tower}}(1 - \text{m}^*)]}{1 - \text{m}^* \times \exp[-\text{Ntu}_{\text{tower}}(1 - \text{m}^*)]} \tag{8.16}$$

where

$$\text{m}^* = \frac{G_{\text{tower,a}}}{G_{\text{tower,w}} c_{\text{w}}} \left(\frac{h_{\text{tower,as,E}} - h_{\text{tower,as,L}}}{t_{\text{tower,w,E}} - t_{\text{tower,w,L}}} \right) \tag{8.16a}$$

$$\text{Ntu}_{\text{tower}} = A_{\text{tower}} \left(\frac{G_{\text{tower,w}}}{G_{\text{tower,a}}} \right)^{B_{\text{tower}}} \tag{8.16b}$$

The coefficients in Eq. (8.16b), A_{tower} and B_{tower}, reflect thermal performance of cooling tower [5]. They can be identified by field test data including temperature and humidity of inlet air, temperature of inlet and outlet water, and airflow rate as well as water flow rate.

Combining Eqs. (8.15) and (8.16), the outlet water temperature of the cooling tower can be obtained by Eq. (8.17).

$$t_{\text{tower,w,L}} = t_{\text{tower,w,E}} - \frac{\varepsilon_{\text{tower}} \left(h_{\text{tower,as,E}} - h_{\text{tower,a,E}} \right)}{c_{\text{w}} G_{\text{tower,w}}} \tag{8.17}$$

8.1.5 Water-to-Air Heat Exchanger

Basically, there are three types of thermal models for heat exchanger. These are as follows: (1) finite volume element model [6, 7]; (2) artificial neural network (ANN) model [8]; and (3) efficiency number of heat transfer unit (η-NTU) model [9]. The finite volume element model is pure theoretical model, which is very complicated and needs a large amount of computation time. It is suitable for the optimal structure design of heat exchanger. The ANN model is pure empirical model, and it requires large amounts of data for modeling. The η-NTU model belongs to semi-theoretical model, and it is often used for the analysis of system energy consumption or efficiency in which the cooling/heating capacity and the output variables (including exit air temperature and humidity as well as exit water temperature) of heat exchanger are required to be calculated by model under different operation conditions. The η-NTU model of the water-to-air heat exchanger mainly consists of Eqs. (8.18) through (8.24). The calculation flow is given in Fig. 8.1 [10].

$$\varepsilon_{\text{coil,2}} = 1 - \frac{t_{\text{coil,a,L}} - t_{\text{coil,s,L}}}{t_{\text{coil,a,E}} - t_{\text{coil,s,E}}} = f\left(G_{\text{coil,a}}, F_{\text{coil}}\right) \tag{8.18}$$

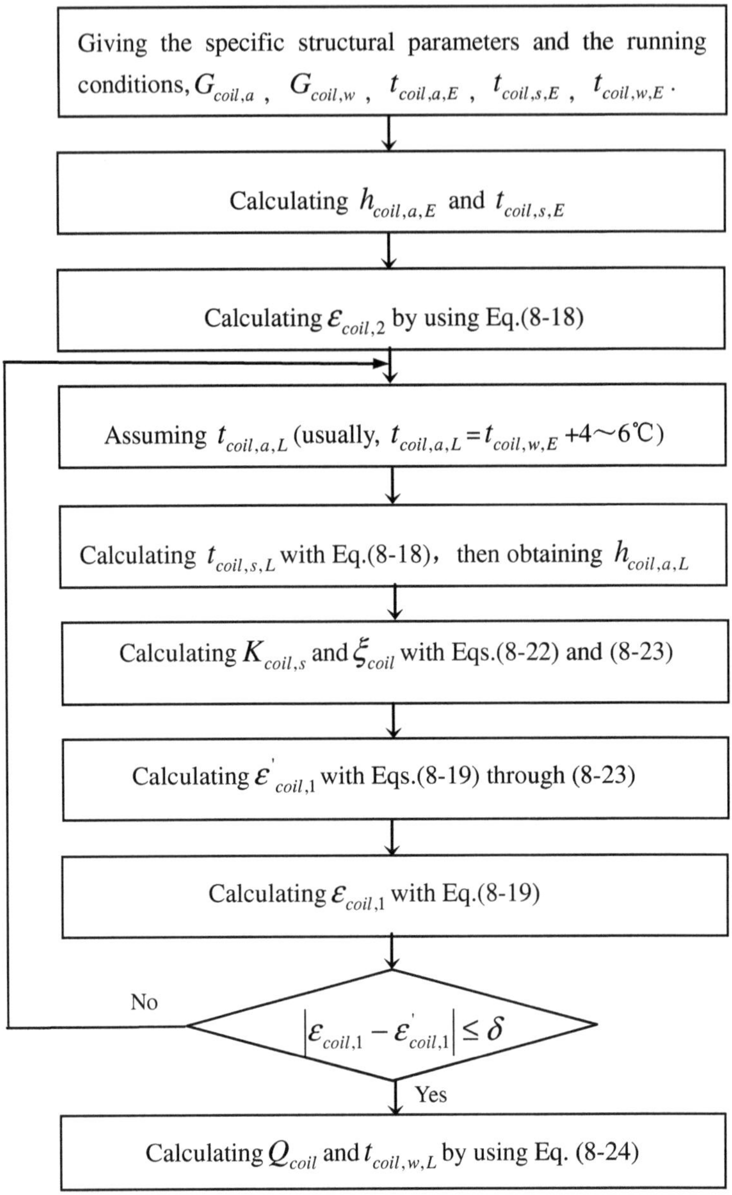

Fig. 8.1 Calculation flowchart of water-to-air surface heat exchanger model

$$\varepsilon_{\text{coil},1} = \frac{t_{\text{coil,a,E}} - t_{\text{coil,a,L}}}{t_{\text{coil,a,E}} - t_{\text{coil,w,E}}} = \frac{1 - \exp[-NTU_{\text{coil}}(1 - C_r)]}{1 - C_r \exp[-NTU_{\text{coil}}(1 - C_r)]} \quad (8.19)$$

$$C_r = \frac{\zeta_{\text{coil}} G_{\text{coil,a}} c_a}{G_{\text{coil,w}} c_w} \quad (8.20)$$

$$\text{NTU}_{\text{coil}} = \frac{F_{\text{coil}} K_{\text{coil,s}}}{\zeta_{\text{coil}} G_{\text{coil,a}} c_a} \quad (8.21)$$

$$K_{\text{coil,s}} = \left[\frac{1}{A_{\text{coil}} \cdot G^m_{\text{coil,a}} \cdot \zeta^n_{\text{coil}}} + \frac{1}{B_{\text{coil}} \cdot G^z_{\text{coil,w}}} \right]^{-1} \quad (8.22)$$

$$\xi_{\text{coil}} = \frac{h_{\text{coil,a,E}} - h_{\text{coil,a,L}}}{c_a (t_{\text{coil,a,E}} - t_{\text{coil,a,L}})} \quad (8.23)$$

$$Q_{\text{coil}} = G_{\text{coil,a}}(h_{\text{coil,a,E}} - h_{\text{coil,a,L}}) = G_{\text{coil,w}} c_w (t_{\text{coil,w,L}} - t_{\text{coil,w,E}}) \quad (8.24)$$

In Eq. (8.22), $\varepsilon_{\text{coil},1}$ is heat transfer efficiency; $\varepsilon_{\text{coil},2}$ is exposure coefficient of heat exchanger; ξ_{coil} is enhancement factor of heat exchanger due to moisture condensation; $t_{\text{coil,s,E}}$ and $t_{\text{coil,s,L}}$ are wet-bulb temperature of air entering and leaving heat exchanger, °C; and $A_{\text{coil}}, B_{\text{coil}}, m, n$ and z are empirical coefficients that are determined by experimental data.

8.2 Energy-Saving Analysis on VAV Air-Conditioning System

Variable air volume system (VAVS) is an air system that varies its supply air volume flow rate to match the reduction of space load during part-load operation to maintain a predetermined space parameter, usually air temperature, and to conserve fan power at reduced volume flow. As an important all-air HVAC system, VAVS was first introduced by Urban [11] in the late 1960s. It has considerable advantages over the CAVS, e.g., energy saving due to the variable airflow rate that varies with the rate of cooling/heating; an economy cycle can be incorporated to use outside air to provide cooling at the times when the temperature is appropriate; less size of the main ducts as the maximum demand for cooling/heating never coincides simultaneously in all spaces. Therefore, VAVS has become popular in buildings in England and other European countries since the 1980s [12]. In this section, the energy analysis of VAVS is made based on the energy models given in the previous section.

8.2.1 Evaluation Program for Energy Saving of VAV System

8.2.1.1 Assumptions

Figure 8.2 shows the schematic diagram of a general central air-conditioning system. In cooling season, the refrigerating station is running, while the boilers stop working. In such case, the valves A1 and A2 are opened, and the valves B1 and B2 are closed. In heating season, the boilers are running and the refrigerating station stops working. Accordingly, the valves B1 and B2 are opened, and the valves A1 and A2 are closed.

To make the energy simulation of various air-conditioning systems (VAVS, CAVS, and FCS), some assumptions are necessary to be made as follows:

(1) When the entering water temperature is lower than its dew-point temperature of entering air, the heat exchanger is working under the wetting condition. Meanwhile, the relative humidity of the leaving air is considered as 90 %.
(2) Neglect the thermal losses of the circulating water due to the heat transfer of water pipes.
(3) The identical equipments have the identical working performance.
(4) The chillers and boilers can adjust automatically their cooling/heating capacity to balance the thermal load demand of air-conditioned buildings. And for VAVS, the fan in the AHU (air-handling unit) can adjust the appropriate

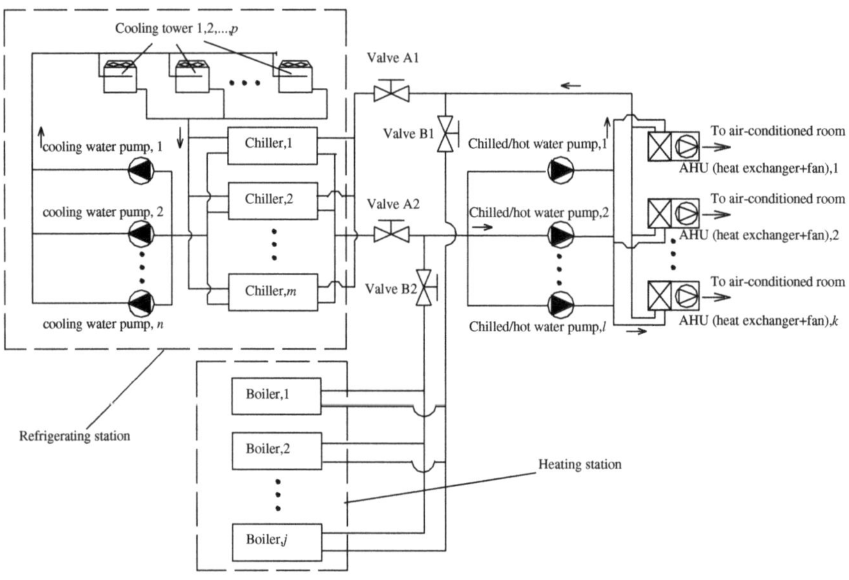

Fig. 8.2 Schematic diagram of a general central air-conditioning system

supply air volume according to the thermal load demand of air-conditioned rooms in charge.

(5) The three kinds of air-conditioning systems have the uniform operation schedule.

(6) The counting interval is set as an hour during which all the calculation conditions are considered to be unchanged.

8.2.1.2 Sub-procedures

1) *Calculation procedure for chillers' power*

The COP of chiller is the key parameter for calculating the chiller's power. As shown in Eq. (8.6), the chiller's COP is related to partial load ratio (r), evaporating temperature (T_e), and condensing temperature (T_c). The partial load ratio of chillers, r, can be got according to the cooling load demand of air-conditioned building. The evaporating temperature, T_e, is usually known in advance since it is a controlled parameter. The determination of chiller's condensing temperature, T_c, needs to couple with the entering and leaving water temperature of cooling water tower. Thus, the energy model for chiller and cooling water tower will be used for calculating chiller's power. The calculation procedure is given in Fig. 8.3.

2) *Calculation procedure for AHUs' power*

Fan is the main energy-consuming equipment in the AHU (air-handling unit). For CAV system and FC system, the power of fan is fixed as the rated input electric power during its running. For VAV system, however, the power of fans varies with the cooling/heating load changing. Therefore, the power of fan in VAV system will couple with thermal model of heat exchanger as shown in Fig. 8.1.

Assuming air-conditioning load demand is $Q_{\mathrm{h/c,room}}$, enthalpy of indoor air is $i_{\mathrm{a,room}}$, and enthalpy of supply air (i.e., outlet air of heat exchanger) is $i_{\mathrm{coil,a,L}}$; the airflow rate of fan, G_f, can be counted by:

$$G_{\mathrm{f}} = \left| \frac{1000 \cdot Q_{\mathrm{h/c,room}}}{\rho_{\mathrm{a}}\left(i_{\mathrm{coil,a,L}} - i_{\mathrm{a,room}}\right)} \right| \tag{8.25}$$

Normally, indoor air conditions (i.e., temperature and the humidity) are required for calculating air-conditioning load of room. The key step is to determine the outlet air enthalpy of heat exchanger ($i_{\mathrm{coil,a,L}}$). The calculation procedure for AHU's electric power in VAV systems is shown in Fig. 8.4.

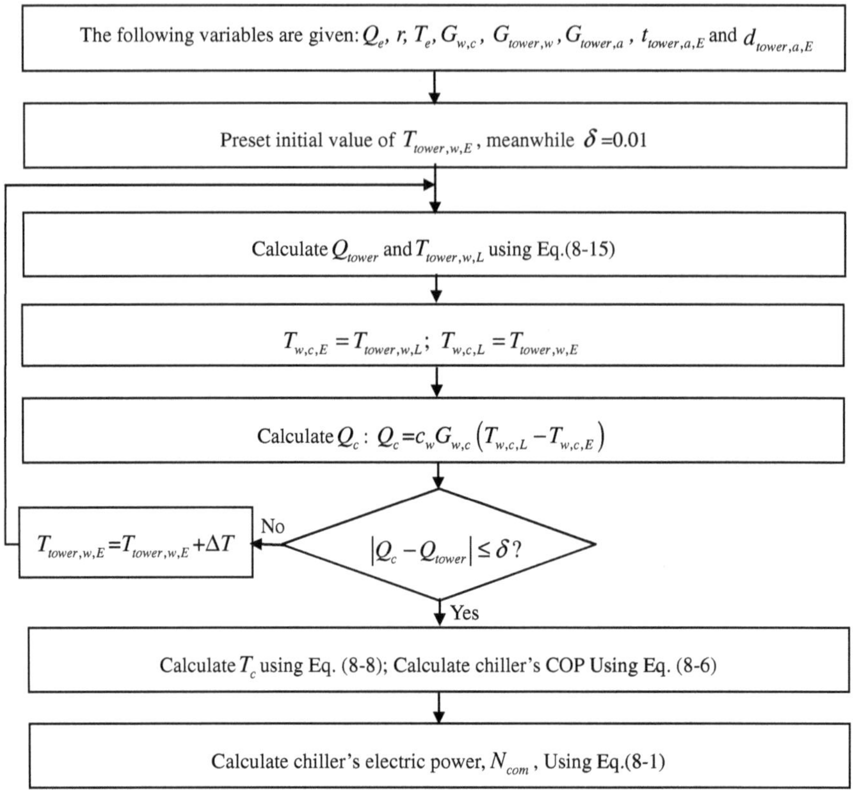

Fig. 8.3 Calculation procedure for chiller's electric power

8.2.1.3 Energy-Saving Evaluation of VAV System

The year-round energy consumption of HVAC system can be written as follows:

$$E_{\text{HVAC}} = \sum_{\tau=1}^{R_{\text{cooling}}}\left[\sum_{j=1}^{m} N_{\text{chiller},j}(\tau) + \sum_{j=1}^{n+l} N_{\text{p},j}(\tau) + \sum_{j=1}^{k} N_{\text{f},j}(\tau) + \sum_{j=1}^{p} N_{\text{t},j}(\tau)\right] + \sum_{\tau=1}^{R_{\text{heating}}}\left[\sum_{j=1}^{i} N_{\text{boiler},j}(\tau) + \sum_{j=1}^{l} N_{\text{p},j}(\tau) + \sum_{j=1}^{k} N_{\text{f},j}(\tau)\right] \tag{8.26}$$

Clearly, different operation modes of fans in terminal equipments lead to difference of year-round energy consumption among CAV system, FC system, and VAV system. For CAV system and FC system, the fans in terminal equipments run in full-load condition all the time. While for VAV system, the fans in terminal equipments run in partial load conditions in most cases.

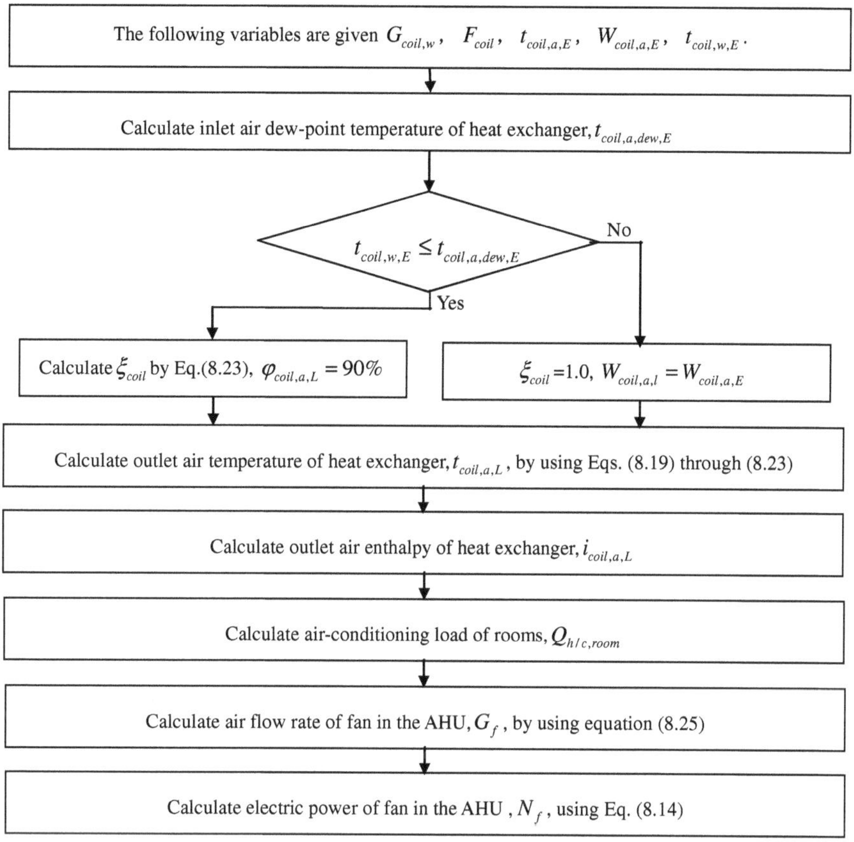

Fig. 8.4 Calculation procedure for AHU's electric power in VAV systems

The energy-saving ratio of VAV system compared to CAV system ($RE_{VAVS/CAVS}$) and to FC system ($RE_{VAVS/FCS}$) can be evaluated as follows:

$$RE_{VAVS/CAVS} = \left(1 - \frac{E_{VAVS}}{E_{CAVS}}\right) \times 100\,\% \tag{8.27}$$

$$RE_{VAVS/FCS} = \left(1 - \frac{E_{VAVS}}{E_{FCS}}\right) \times 100\,\% \tag{8.28}$$

8.2.2 Case Study

The year-round energy consumptions of VAV system, CAV system, and FC system are compared in one small air-conditioned office building in different areas of China.

8.2.2.1 Description for the Study Object

1) *The office building*

The simulation object is a small office building with 43 m in length, 17 m in width, and 26 m in height. There are six floors in total, and the standard floor height is 3.7 m. The building stands in south–north direction. The glazing rate in the north-facing wall is about 43 %, and the one in the south-facing wall is about 18 %.

Basic information about the structures of the office building is listed as follows:

(1) External wall: heavy clay mortar and porous concrete, 240 mm, with thermal resistance of 1.247 m^2 W/K;
(2) Inner wall: heavy clay mortar, 240 mm, with thermal resistance of 0.338 m^2 W/K;
(3) Roof: light roof with heat insulation by polystyrene of 100 mm, thermal resistance is 0.804 m^2 W/K; and
(4) Floor: reinforced concrete, 150 mm, with thermal resistance of 0.141 m^2 W/K.

The indoor air-conditioning is required as follows: 18 °C of temperature and more than 40–50 % of relative humidity (RH) in winter, and 24–26 °C and 50–60 %RH in summer. The amount of fresh air is required as 20 m^3/h person for both winter and summer. Inner load of the building includes the light (20 W/m^2), the equipment (20 W/m^2), and indoor occupants (53 W of heat and 0.102 g/h of moisture for one person). The occupant density in the building is one person per 5 m^2. The working schedule on weekdays is set as 8:00–18:00, and the nighttime, weekend, days, and holidays are not be taken into account.

2) *Typical cities*

Six cities standing for representative climates of China are chosen for study. They are Beijing, Xi'an, Shenyang, Guangzhou, Chengdu, and Shanghai, which stand for the north, the northwest, the northeast, the south, the southwest, and the southeast areas of China, respectively. The geographical locations of these cities are marked in Fig. 8.5. The weather data of these cities, including hourly temperature and humidity of outdoor air and hourly solar radiant intensity, can be got from the National Weather Bureau of China. Figure 8.6 gives the statistical data on the weather conditions of the six cities in a typical year.

3) *Equipments*

Usually, the chillers/boilers of HVAC system are chosen according to the maximum cooling/heating load of the building in the whole year. Total area of air-conditioned space is about 2084 m^2. The year-round cooling/heating loads of the building in different cities are calculated, respectively, using the load calculation program based on the frequency response method [13]. The year-round maximum cooling/heating loads of the building in the six cities are shown in Table 8.1. For Guangzhou city, the weather in winter is very warm; hence, the building does not require heat supply throughout the year.

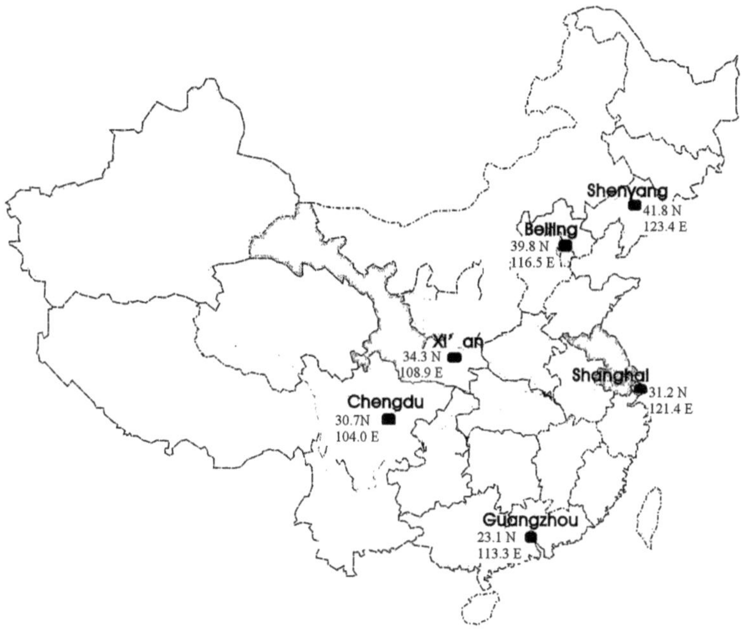

Fig. 8.5 Geographical location of the cities in China (N—north latitude; E—East longitude)

Referring to the design handbook of HVAC system [5], equipments for the three types of HVAC systems are chosen according to the climate conditions of the six cities. As listed in Table 8.2, the equipments mainly include the following:

(1) Chiller (Manufactory: TRANE): Type RTHB130 (rate cooling capacity: 250 kW; rated input electric power: 77 kW) for Beijing, Shenyang, and Chengdu; Type RTHB150 (rate cooling capacity: 313 kW; rated input electric power: 89 kW) for Xi'an, Guangzhou, and Shanghai.
(2) Boiler (Manufactory: PRECISION): Type HW16D-120B (input electric power: 120 kW) for Chengdu and Shanghai; Type HW16D-150B (input electric power: 150 kW), HW16D-180B (input electric power: 180 kW), and HW16D-210B (input electric power: 210 kW) for Xi'an, Beijing, and Shenyang, respectively.
(3) Cooling tower (Manufactory: SHANGLEN): Type XFBL-100 (cooling water volume: 100 m^3/h; electric power of motor: 3 kW) for the six cities.
(4) Cooling water pump (Manufactory: KAIQUAN): Type KQL 125/315 (rated water flow rate: 100 m^3/h; electric power of motor: 15 kW; water head: 32 m) for the six cities.
(5) Chilled water pump (Manufactory: KAIQUAN): Type KQL 125/300 (rated water flow rate: 60 m^3/h; electric power of motor: 11 kW; water head: 27 m) for the six cities.

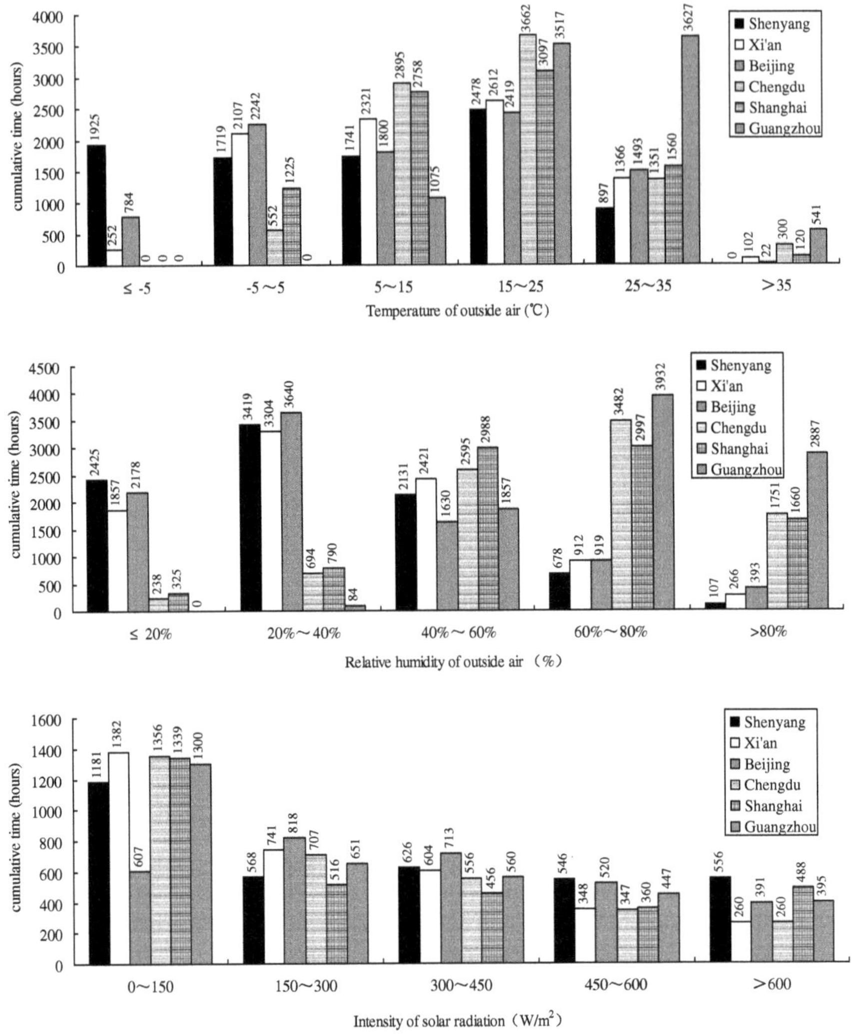

Fig. 8.6 Weather conditions of the six cities

Table 8.1 Maximum cooling/heating load of the building in the six cities (kW)

City	Beijing	Xi'an	Shenyang	Guangzhou	Chengdu	Shanghai
Maximum cooling load	−220	−307.4	−190	−257.6	−199.4	−286.3
Maximum heating load	166.7	146.7	192.8	/	80.4	97.3

Table 8.2 Main equipments for various HVAC systems

Equipments	Type	Manufactory	Number	Types of HVAC system	For which city
Chiller	RTHB130	TRANE	1	CAV, VAV, FC	Beijing, Shenyang, Chengdu
	RTHB150	TRANE	1	CAV, VAV, FC	Xi'an, Guangzhou, Shanghai
Boiler	HW16D-120B	PRECISION	1	CAV, VAV, FC	Chengdu, Shanghai
	HW16D-150B	PRECISION	1	CAV, VAV, FC	Xi'an
	HW16D-180B	PRECISION	1	CAV, VAV, FC	Beijing
	HW16D-210B	PRECISION	1	CAV, VAV, FC	Shenyang
Cooling tower	XFBL-100	SHANGLEN	1	CAV, VAV, FC	Beijing, Xi'an, Shenyang, Guangzhou, Chengdu, Shanghai
Cooling water pump	KQL 125/315	KAIQUAN	1	CAV, VAV, FC	Beijing, Xi'an, Shenyang, Guangzhou, Chengdu, Shanghai
Chilled water pump	KQL 125/300	KAIQUAN	1	CAV, VAV, FC	Beijing, Xi'an, Shenyang, Guangzhou, Chengdu, Shanghai
Terminal equipment	AHD-FS-50	DUNHAN	3	CAV	Beijing, Xi'an, Shenyang, Guangzhou, Chengdu, Shanghai
	KFPC-50	DUNHAN	3	VAV	Beijing, Xi'an, Shenyang, Guangzhou, Chengdu, Shanghai
	FP-5	DUNHAN	115	FC	Beijing, Xi'an, Shenyang, Guangzhou, Chengdu, Shanghai

(6) Terminal equipment (Manufactory: DUNHAN): Type AHD-FS-50 (rated air volume: 35,000 m^3/h; rated pressure: 1050 Pa; rated input electric power: 18.2 kW) for CAV system; Type KFPC-50 (rated air volume: 37,000 m^3/h; rated pressure: 980 Pa; rated input electric power: 16.8 kW; type of frequency convertor: VG7S) for VAV system; and Type FP-5 (rated cooling output: 2210 W; rated heating output: 1950 W; power of motor: 75 W) for FCS.

According to the data provided by the product manual, the model coefficients of equipments can be acquired, which are listed as follows:

For RTHB130 chiller model (refer to Sect. 7.1.1):

$$a_1 = 0.2033;\ a_2 = -0.0239;\ a_3 = 9.61e-5;\ a_4 = 1.74e-5;\ a_5 = 3.04e-4.$$

For RTHB150 chiller model (refer to Sect. 7.1.1):

$$a_1 = 0.1811;\ a_2 = -0.0852;\ a_3 = 8.89e-5;\ a_4 = 1.88e-5;\ a_5 = 2.96e-4.$$

For XFBL-100 cooling tower model (refer to Sect. 7.1.4):

$$A_{\text{tower}} = 0.318;\ B_{\text{tower}} = 0.149;$$

For KFPC-50 VAV fan model [refer to Eq. (7.14)]:

$$k_{\text{p/f}} = 1.205;\ B_0 = 0.023;\ B_1 = 3.83;\ B_2 = -5.23;\ B_3 = 2.39;$$
$$C_0 = 0.51;\ C_1 = 1.28; C_2 = -1.42;\ C_3 = 0.58;$$

For KFPC-50 VAV heat exchanger model [refer to Eq. (7.22)]:

$$A_{\text{coil}} = 39.7;\ B_{\text{coil}} = 332.7;\ m = 0.52;\ n = 1.03;\ z = 0.8.$$

8.2.2.2 Results and Analysis

Energy consumption of the three types of HVAC systems, i.e., CAV system, VAV system, and FC system, are investigated, respectively, in the same building and under different climates in China represented by the six cities mentioned above.

The detailed results on energy consumption of different HVAC systems in cooling and heating season are given in Table 8.3 and Table 8.4, respectively. The results manifest that the terminal equipments of VAV system are of much more energy saving compared with that of CAV system and FC system. Taking Shenyang city for example, the energy consumption of terminal equipments of VAV system is about 10.7 % of that of CAV system and 22.7 % of that of FC system in cooling season, and about 50.2 % of that of CAV system and 80.4 % of that of FC system in heating season.

Energy-saving ratio of VAV system compared to CAV system and FC system in the same building will be different from one climate to another (see the results in Table 8.5). As for the energy-saving ratio of VAV system compared to CAV system, Shenyang ranks the first place (37.6 %), then followed by Shanghai (24.2 %), Chengdu (22.9 %), Xi'an (21.8 %), Beijing (19.2 %), and Guangzhou (17.0 %). However, compared to FC system, VAV system will have the biggest energy-saving ratio in Chengdu (10.2 %) and in Shanghai (10.1 %) and have the

Table 8.3 Analysis on energy consumption of different HVAC systems in cooling season

Cities	Energy consumption of equipments (kWh)						Ratio of energy consumption of different terminal equipments	
	Chiller	Chilled and cooling water pumps	Cooling tower	Three kinds of terminal equipments			$\frac{E_{VAV,terminal}}{E_{CAV,terminal}}$	$\frac{E_{VAV,terminal}}{E_{fancoil,terminal}}$
				CAV	Fan coil	VAV		
Beijing	82,352	53,773	3102	38,127	18,768	5574	0.146	0.300
Xi'an	75,869	51,136	2950	36,748	17,848	5162	0.140	0.289
Shenyang	60,800	45,411	2619	33,753	15,850	3605	0.107	0.227
Guangzhou	122,427	71,697	5457	58,184	26,680	14,266	0.245	0.535
Chengdu	96,721	62,735	3747	43,729	20,010	6866	0.157	0.343
Shanghai	95,866	58,254	3613	39,810	19,343	6011	0.151	0.311

Table 8.4 Analysis on the energy consumption of different HVAC systems in heating season

Cities	Energy consumption of equipments (kWh)					Ratio of energy consumption of different terminal equipments	
	Boiler	Hot water pump	Three types of terminal equipments			$\frac{E_{VAV,terminal}}{E_{CAV,terminal}}$	$\frac{E_{VAV,terminal}}{E_{fancoil,terminal}}$
			CAV	FC	VAV		
Beijing	159,519	28,387	62,644	23,418	12,964	0.207	0.554
Xi'an	128,361	26,928	59,939	22,215	8307	0.139	0.374
Shenyang	244,535	28,710	63,244	39,476	31,734	0.502	0.804
Guangzhou	/	/	/	/	/	/	/
Chengdu	49,178	19,648	46,438	16,209	1924	0.041	0.119
Shanghai	72,051	22836	52,350	18,839	2883	0.055	0.153

Table 8.5 Energy-saving ratio of VAV system compared with CAV and FC system in different cities

	Beijing (%)	Xi'an (%)	Shenyang (%)	Guangzhou (%)	Chengdu (%)	Shanghai (%)
Compared to CAV system $(1-E_{VAV}/E_{CAV}) \times 100\ \%$	19.2	21.8	37.6	17.0	22.9	24.2
Compared to FC system $(1-E_{fancoil}/E_{CAV}) \times 100\ \%$	6.4	8.2	4.6	5.5	10.2	10.1

smallest energy-saving ratio in Shenyang (4.6 %). The energy-saving ratio of VAV system is one of the most important factors that influence the payback period of the system. Generally, the higher energy-saving ratio the VAV system is, the shorter payback period of the system will have. In this case, compared with the FC system, the payback period of VAV system in Chengdu and Shanghai will be shorter than that in other cities, which indicates that for the small office building, VAV system may be more suitable for such climate as Chengdu and Shanghai than the other kind of climates.

8.3 Energy Analysis on VAV Air-Conditioning System with Different Air-Side Economizers

Although the variable air volume (VAV) system has an overwhelming energy-saving characteristic compared with the other types of HVAC systems, its energy performance can be further improved through some new technologies, e.g., adopting a new energy management system to control the VAV AC system via the occupants' requests [14], optimizing the supply air temperature [15], and developing the innovative low-temperature differential VAV system [16]. An economizer cycle is considered as one of the popular energy-saving measures in air-conditioning systems [17, 18] because it eliminates or reduces refrigeration cooling by utilizing the free cooling capacity of ambient environment. Some studies have been made on the energy performance of economizer in different HVAC applications, such as constant-air-volume (CAV) systems [19], variable air volume (VAV) systems [20], and single-duct or dual-duct air-handling units [21, 22]. The control algorithms and programs [20, 21, 23] were developed as well for economizer implementation in practice. Since economizer cycle needs additional components and more sophisticated control programs than non-economizer one, it is necessary to evaluate the benefits brought about by the economize cycle. So, having a good knowledge of energy savings of economizer strategy is very important to decide whether or not the economizer be worthy of applying in an actual situation. As indicated by some studies, the local climatic condition is one of the critical factors that impact the energy savings of economizer cycle. Budaiwi [24] investigated the potential energy conservation when economizer cycles were employed under three climatic conditions in Saudi Arabia, i.e., hot-humid summer and mild winter, moderate summer and winter, as well as hot-dry summer and cold winter. The study results show that the greatest energy saving by an economizer cycle is achieved in the hot-dry summer and cold winter climate with about 13 % reduction in cooling requirement. The least energy saving (about 3 %) corresponds to the hot-humid summer and mild winter climate. Yiu et al. [25] made a comparative study on energy savings of economizers in urban, suburban, and rural areas of Hong Kong. They concluded that the potential for use of economizers in suburban and rural areas would be significantly higher than that in urban areas. The results may be explained by the fact that the climates in suburban and rural areas are cooler than those in urban area, and hence, more free cooling capacity can be available from outside air.

The economizer cycle is usually realized by two kinds of economizer control methods. One is air-side economizer control, which utilizes free cooling capacity of outdoor air directly to offset the space cooling load. The other is water-side economizer control, which uses the outdoor air to cool the condenser water in a cooling tower and then cool the mixture of fresh and return air to provide a part of space cooling load. The air-side method may be more advantageous than the water-side one in many aspects, e.g., better indoor air quality, lower investment cost, and reduction of maintenance [26]. Hence, the air economizer is discussed here. Basically, air economizer can be enthalpy-based or temperature-based. The operation of an enthalpy-based economizer control system is based on the comparison of air enthalpies between outdoor and indoor, while a temperature-based one is based on the comparison of air dry-bulb temperatures.

To assist HVAC designers and operators have a general knowledge of air economizers in terms of energy savings under different climates, a year-round energy simulation program for VAV system with different types of air-side economizers (i.e., the enthalpy-based and the dry-bulb temperature-based) is developed in this section.

8.3.1 Scheme for Air Economizer Cycle [27]

8.3.1.1 Study Object—Single-Zone VAV System

Single-zone VAV system, which is widely used in assembly office buildings, shopping malls, and factories, is studied here. The schematic diagram for a typical single-zone VAV air-conditioning system is shown in Fig. 8.7. The mixture of fresh air and recirculating air is handled by the filter and the surface heat exchanger in turn and then passes through the supply fan and the supply duct before it is discharged to the single-space air-conditioned zone. Meantime, certain amount of space air enters into the return air system. One part of return air is recirculated for

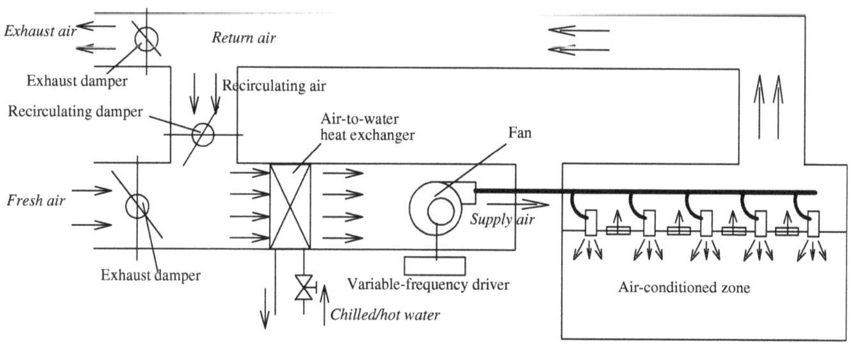

Fig. 8.7 Schematic diagram for single-zone VAV air-conditioning system [27]

the sake of energy saving, and the other is exhausted. In summer cooling season, the space cooling load is burdened by chilled water that comes from cooling plant. When air economizer cycle is used during transition season (i.e., spring and fall), the proportion of fresh air in total supply air varies from minimum to 100 %. During winter, if the heat gain is greater than the space transmission and infiltration loss, a cooling mode operation is still needed, and hence, the cold outside air can be utilized for free cooling. If the heat gain is smaller than the space transmission and infiltration loss, the space requires heating load, and the mixture of outdoor and recirculating air is needed to be heated by the surface heat exchanger through which the hot water flows.

8.3.1.2 Energy Models of Units

Energy models of units are the basis for the year-round energy simulation of VAV system, and great effort has been made for such work (refer to Sect. 8.1 for detail). When unit models are integrated into a system, they will interact with each other through coupling variables. Taking the cooling mode of VAV system for example, the block diagram for the system simulation is given in Fig. 8.8. The coupling variables, marked by bracket, are the temperatures of cooling/chilled water leaving from or entering into equipment. If the thermal loss during water transportation is neglected, the coupling variables are subject to the following equations:

$$\left.\begin{aligned} t_{\mathrm{tower,w,L}} &= t_{\mathrm{w,c,E}} \\ t_{\mathrm{w,e,L}} &= t_{\mathrm{w,coil,E}} \\ t_{\mathrm{w,coil,L}} &= t_{\mathrm{w,e,E}} \\ t_{\mathrm{w,c,L}} &= t_{\mathrm{tower,w,E}} \end{aligned}\right\} \tag{8.29}$$

Normally, the general iterative method can be used to solve these coupling variables based on which the energy simulations of the system are performed.

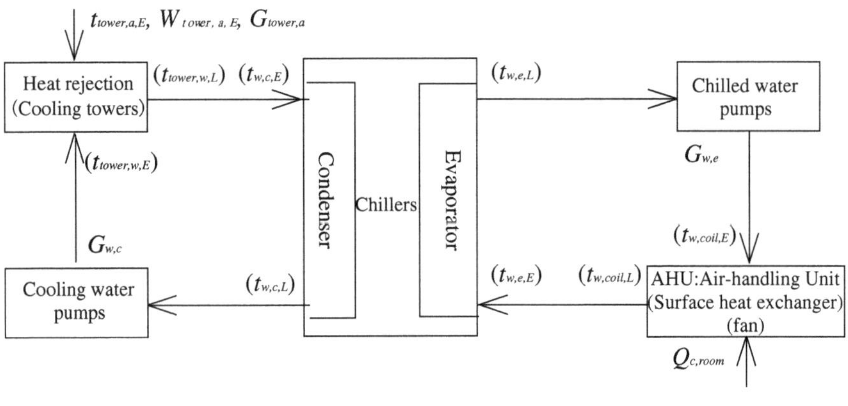

Fig. 8.8 Block diagram of coupling variables among equipments in cooling mode

8.3.1.3 Air-Conditioning Load of Building

In the course of year-round energy simulation of VAV system, it is assumed that the hourly air-conditioning load of the building be exactly known, based on which different mock operation schemes are implemented. Many models for hourly air-conditioning load calculation have been developed in the past decades [28, 29]. In this study, the frequency response method, which was first put forward by Hittle et al. [30] and applied to dynamic thermal metering in central air-conditioning system [13], is employed to calculate the building's hourly air-conditioning load throughout the year.

8.3.1.4 Year-Round Running Scheme with Air-Side Economizers

The operation of the enthalpy-based economizer scheme is based on the comparisons of outdoor air enthalpy (kJ/kg in unit) with the enthalpy of the recirculating air. As the recirculating air enthalpy is usually fixed according to the thermal comfort requirement of occupants, the outdoor, exhaust, and recirculating air dampers are modulated with the change of the outdoor enthalpy. Generally speaking, the air economizer cycle occurs in cooling operation season when the outdoor enthalpy is smaller than the recirculating air enthalpy. Similarly, the temperature-based economizer scheme is based on the comparisons of dry-bulb temperature (°C in unit) of outdoor air with that of the recirculating air. The differential temperature economizer is on when the outdoor temperature (dry-bulb) is lower than the recirculating air temperature (dry-bulb) in cooling mode operation. Basically, the year-round operation scheme can be divided into five main sub-schemes in terms of ambient conditions, and sequence of an air economizer operation scheme can be described in Fig. 8.9. The character, *x*, in *x*-coordinate stands for outdoor air temperature as the temperature-based economizer is employed and for outdoor air enthalpy as the enthalpy-based economizer is employed. The on-and-off air economizer cycle

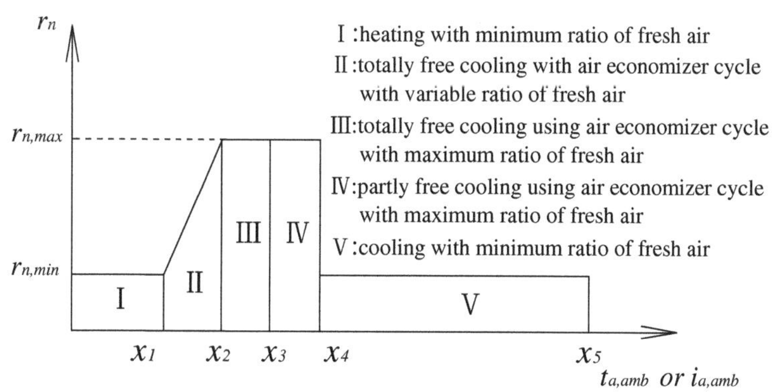

Fig. 8.9 Schematic for year-round operation scheme of VAV system

mainly depends on ambient air enthalpy or temperature. The operation of an enthalpy-based economizer control system is based on the comparisons of the enthalpy of the outdoor air, $i_{a,amb}$, with x_1, x_2, x_3 and x_4 in enthalpy unit (kJ/kg), and a temperature-based one is based on the comparisons of outdoor air temperature, $t_{a,amb}$, with x_1, x_2, x_3 and x_4 in temperature unit (°C).

In mode I, the air-conditioning system is operated for heating the building when the outdoor air dampers are modulated to the minimum opening degree. When the outdoor air is in the range of x_1–x_2 (mode II) or x_2–x_3 (mode III), the air-side economizers bring in up to variable ratio or maximum outdoor air to utilize its free cooling capacity, rather than using refrigeration cooling, and to offset the space cooling load. When the outdoor air temperature (or enthalpy) is higher than x_3, the refrigeration cooling begins to be required. However, free cooling capacity can be still obtained from outdoor air when the outdoor air temperature (or enthalpy) is lower than x_4 (mode IV). And when the outdoor air temperature (or enthalpy) is higher than x_4, no free cooling capacity of outdoor air can be utilized any more. In such case, the outdoor damper is adjusted to the minimum opening degree for a minimum requirement of fresh air.

To realize the year-round energy-saving strategy in VAV system using the air-side economizers, these critical values, i.e., x_1, x_2, x_3 and x_4, are determined by using Eqs. (8.30) through (8.37), respectively.

For differential enthalpy economizer (HEC),

$$x_1 = i_{\text{rec,winter}} - Q_{\text{room,h}} / \left(\rho_{\text{a}} \cdot G_{\text{f,min}}\right) \tag{8.30}$$

$$x_2 = i_{\text{rec,summer}} - Q_{\text{room,c}} / \left(\rho_{\text{a}} \cdot G_{\text{o,f}}\right) \tag{8.31}$$

$$x_3 = i_{\text{rec,summer}} - Q_{\text{room,c}} / \left(\rho_{\text{a}} \cdot G_{\text{f,min}}\right) \tag{8.32}$$

$$x_4 = i_{\text{rec,summer}} \tag{8.33}$$

For differential dry-bulb economizer (TEC),

$$x_1 = t_{\text{rec,winter}} - Q_{\text{room,h}} / \left(\rho_{\text{a}} \cdot G_{\text{f,min}}\right) \tag{8.34}$$

$$x_2 = t_{\text{rec,summer}} - Q_{\text{room,c}} / \left(c_{\text{a}} \cdot \rho_{\text{a}} \cdot G_{\text{o,f}}\right) \tag{8.35}$$

$$x_3 = t_{\text{rec,summer}} - Q_{\text{room,c}} / \left(c_{\text{a}} \cdot \rho_{\text{a}} \cdot G_{\text{f,min}}\right) \tag{8.36}$$

$$x_4 = t_{\text{rec,summer}} \tag{8.37}$$

In Eqs. (8.30) through (8.37), $G_{\text{f,min}}$ is the minimum fresh air volume required for IAQ (indoor air quality) demand; $t_{\text{rec}}(i_{\text{rec}})$ is recirculating air temperature (enthalpy), e.g., indoor air temperature (enthalpy).

Energy consumption of system for each running mode can be calculated, respectively, by:

$$E(I) = \sum_{\lambda=1}^{\kappa=\frac{\tau_1}{\Delta\tau}} \left[\left(Q_{\text{fuel}}^{\lambda\cdot\Delta\tau} + N_{\text{p}}^{\lambda\cdot\Delta\tau} + N_{\text{f}}^{\lambda\cdot\Delta\tau}\right)\cdot\Delta\tau\right] \tag{8.38}$$

$$E(II) = \sum_{\lambda=1}^{\kappa=\frac{\tau_2}{\Delta\tau}} \left[\left(N_{\text{f}}^{\lambda\cdot\Delta\tau}\right)\cdot\Delta\tau\right] \tag{8.39}$$

$$E(III) = \sum_{\lambda=1}^{\kappa=\frac{\tau_3}{\Delta\tau}} \left[\left(N_{\text{f}}^{\lambda\cdot\Delta\tau}\right)\cdot\Delta\tau\right] \tag{8.40}$$

$$E(IV) = \sum_{\lambda=1}^{\kappa=\frac{\tau_4}{\Delta\tau}} \left[\left(N_{\text{chiller}}^{\lambda\Delta\tau} + N_{\text{p}}^{\lambda\Delta\tau} + N_{\text{f}}^{\lambda\Delta\tau}\right)\cdot\Delta\tau\right] \tag{8.41}$$

$$E(V) = \sum_{\lambda=1}^{\kappa=\frac{\tau_3}{\Delta\tau}} \left[\left(N_{\text{chiller}}^{\lambda\Delta\tau} + N_{\text{p}}^{\lambda\Delta\tau} + N_{\text{f}}^{\lambda\Delta\tau}\right)\cdot\Delta\tau\right] \tag{8.42}$$

where, τ_1, τ_2, τ_3, τ_4 and τ_5 denote the operation time of system in modeI, II, III, IV and V, respectively. The superscript, $\lambda\Delta\tau$, stands for certain time point.

In order to make the comparisons of energy saving performance of the two economizers, the ratio of energy saving, ESR, is calculated by Eq. (8.43).

$$\text{ESR}_{(\text{TEC/HEC})} = \frac{E_{\text{NEC}} - E_{\text{TEC/HEC}}}{E_{\text{NEC}}} \times 100\% \tag{8.43}$$

where $E_{\text{TEC/HEC}}$ denotes energy consumption of VAV system combined with temperature-based economizer cycle or enthalpy-based one. E_{NEC} denotes energy consumption of VAV system without any economizers.

8.3.2 *Case Study*

The small office building with VAV air-conditioning system studied in previous section (Sect. 8.2) is reused for the air-side economizer cycle study. The detailed information about the building, VAV system's equipments, as well as climatic conditions of the six cities has been illustrated in the Sect. 8.2.2.1. In the simulation, the supply water temperature for air-handing units in cooling and heating mode is kept as 7 and 60 °C, respectively. The air-conditioning system operates from 7:00 a.m. to 20:00 p.m. for every working day. Thermal loss of the circulating water and air during transportation is neglected.

By using the model and program developed, year-round energy consumptions of the VAV system with different air economizers are obtained under different climatic conditions of cities in China. The results shown in Fig. 8.10 correspond to the indoor air temperature of 26 °C in summer and 21 °C in winter as well as the indoor air relative humidity of 60 % throughout the year. It manifests that the energy-saving performance of VAV system would be further improved if the air economizer cycle was applied to the system.

The climatic conditions may be one of the important factors that influence the energy-saving effect of air economizers. It seems that the air economizers will bring about more energy savings (about 10–20 % in ESR) under the climatic condition of hot-humidity in summer and not very cold in winter (typically like Chengdu, Shanghai and Guangzhou) than that (about 5–10 % in ESR) under the cool-dry summer and cold winter climatic condition (typically like Shenyang, Beijing, and Xi'an). This is because the energy savings caused by economizer cycle mainly occur in summer when cooling is needed, and the cooling season in southern cities of China (commonly characterized as hot-humidity summer) is longer than that in northern ones (commonly characterized as cool-dry summer). Therefore, the longer economizer cycle time will be available in the southern cities (see Fig. 8.11), which provides more chances of energy saving for the air economizers. It is shown from Fig. 8.10 and 8.11 that although the temperature-based economizer (TEC) has a shorter economizer cycle time than the enthalpy-based one (HEC) as applied to the northern cities, the former still achieves a higher ESR than the latter. And in the southern cities, TEC has a longer economizer cycle time, but is of less energy saving than HEC. The results can be explained with the help of enthalpy–humidity

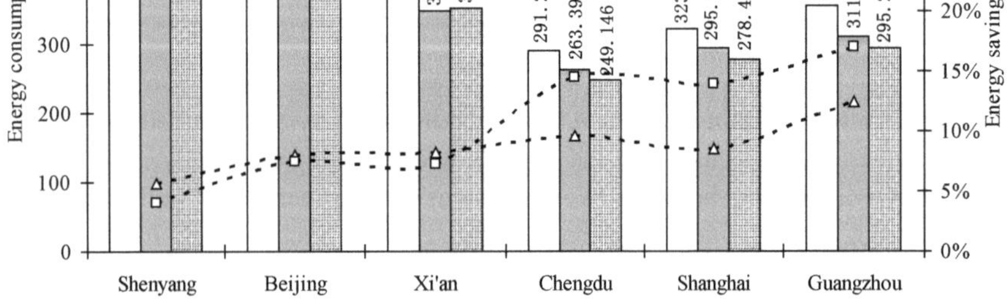

Fig. 8.10 Comparisons of year-round energy consumption of VAV system between with and without economizer scheme

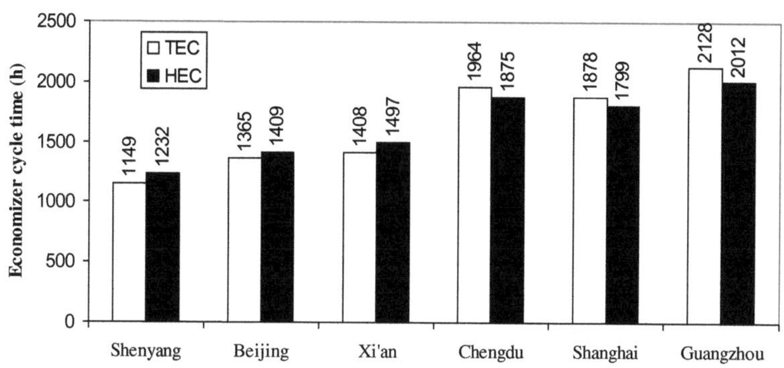

Fig. 8.11 Economizer cycle time of different air economizers in different cities

diagram of air-conditioning process as shown in Fig. 8.12. The enthalpy line and the dry-bulb temperature line in Fig. 8.12 are used for HEC and TEC scheme, respectively. When the state of outside air is within the dotted area A, the enthalpy-based economizer (HEC) cycle is turned off (i.e., in the mode of minimum fresh air recirculation), while the temperature-based economizer (TEC) is turned on (i.e., in the mode of full fresh air recirculation). For those cities characterized by hot-humidity summer (e.g., Chengdu, Shanghai, and Guangzhou), there will be numerous outdoor air states that fall inside the dotted area A in summer; while for those northern cities (e.g., Shenyang, Beijing, and Xi'an), many outdoor air states in summer will be more likely to fall inside the shaded area B. Therefore, the HEC

Fig. 8.12 Enthalpy–humidity diagram for air-conditioning process

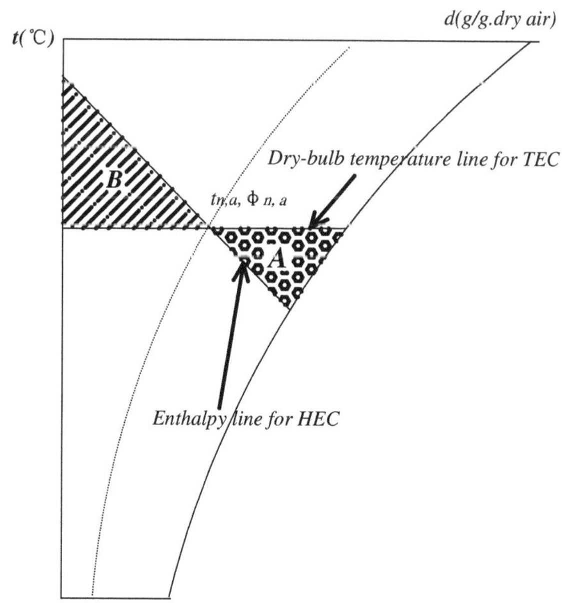

will result in the longer time of economizer cycle in the northern cities and the shorter time in the southern cities than the TEC. However, it is not always the case that the economizer cycling mode will bring about energy savings. As shown in Fig. 8.12, economizer mode with full outdoor air would consume more energy than non-economizer mode with minimum outdoor ventilation when the weather conditions fall inside the dotted area A or the shaded area B. So, we can safely conclude HEC will be certain to have a better energy-saving performance than TEC in the southern areas of China that have a long period of hot and humid climate, but no better than TEC when applied in northern China where the dry climate is dominated throughout the year.

Since economizer cycle only occurs in cooling season, the use of air economizers will not affect the energy consumption of boilers in VAV system, but will produce great influence on the energy consumption of the other equipments, i.e., chillers, pumps, and fans. As shown in Fig. 8.13, the year-round energy consumption by chillers and pumps will reduce greatly (about 20 % in ESR) when the air economizers are adopted in VAV system. This is because these equipments are often postponed to operate due to the utilization of free outdoor cooling capacity. However, more energy of fans will be needed in the case of air economizer cycle. The reason is that the air supply temperature or enthalpy in the economizer case is usually higher than that under refrigerant cooling mode, and a larger volume of air (which requires higher fan power) is required for cooling demand of the building under economizer mode. It can be known as well from Fig. 8.13 that the chillers and the pumps will consume more energy under HEC scheme than under TEC scheme in the northern cities, while case is contrary in the southern cities. Since chillers and pumps occupy a large proportion of energy of the whole air-conditioning system, the results of ESR in chillers and pumps in Fig. 8.13 can help to understand why TEC will have a better energy saving performance in the northern cities and a poorer one in the southern cities when compared with HEC.

The indoor air temperature is often considered as the most important control variable in HVAC system because it produces great influence on both indoor thermal comfort and energy consumption of buildings [31]. Figure 8.14 shows the energy consumptions of the VAV system with different air economizers under different indoor air temperatures in cooling season. It shows that the ESR due to the air economizers will increase with the rising of the indoor air temperature. It is reasonable because a longer economizer time can be available if the indoor air temperature (which is directly or indirectly used for HEC and TEC control) is set as a higher value. Hence, more free cooling capacity of outdoor air can be utilized, and more energy savings can be achieved by the air economizers under a higher indoor air temperature in cooling season. From the slope of the ESR curves, it seems that the effect of indoor air temperature on the energy saving performance of HEC is greater than that of TEC in the southern cities where the climate is typically hot and humid.

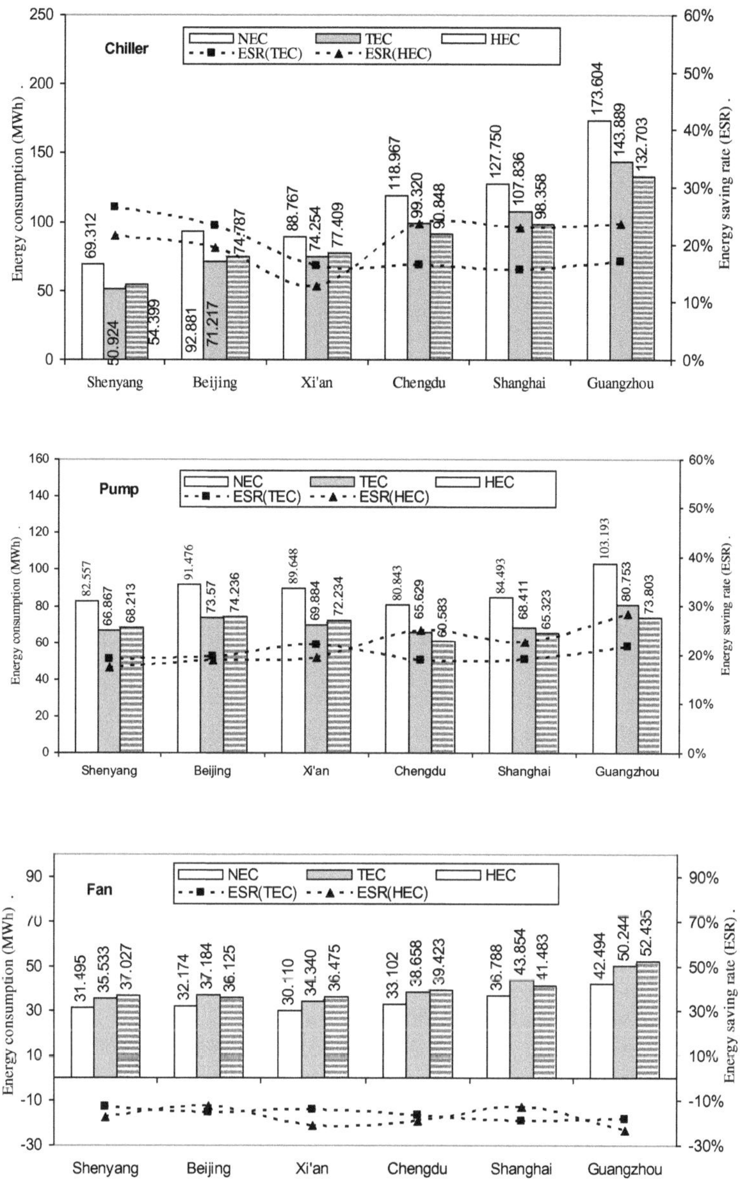

Fig. 8.13 Comparisons of year-round energy consumption of main equipments between non-economizer and economizer scheme

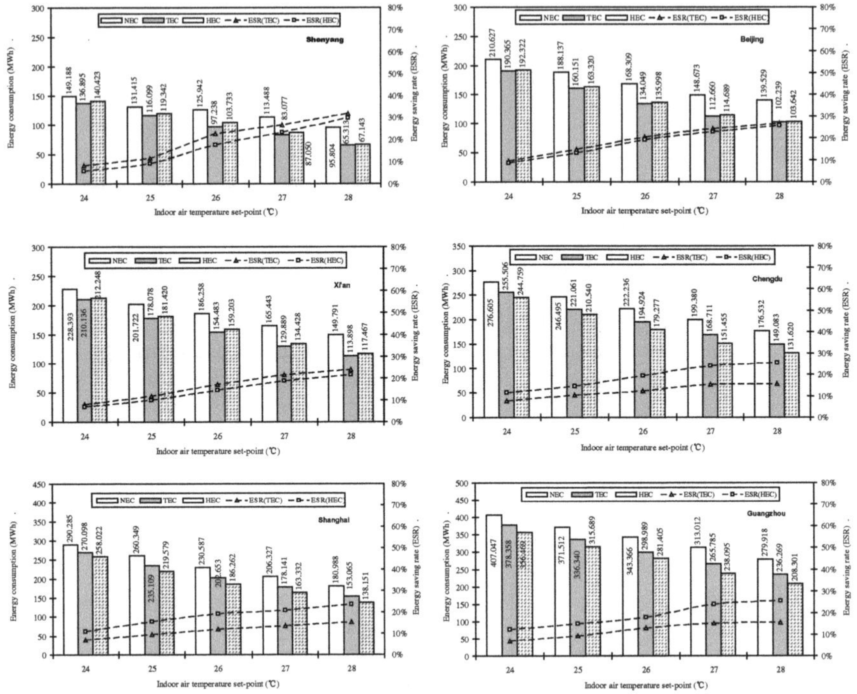

Fig. 8.14 Effect of indoor air temperature set point on the energy savings of the two kinds of economizers in cooling operation season

References

1. Browne, M.W., Bansal, P.K.: Steady-state model of centrifugal liquid chillers. Int. J. Refrig. **21**(5), 343–358 (1998)
2. Wang, S.K.: Heating systems furnaces and boilers. In: Handbook of Air Conditioning and Refrigeration, second edn. McGraw-Hill, New York (2000)
3. Michel, A.B., Bernard, B.: Pumping energy and variable frequency drives. ASHRAE J. **12**, 37–39 (1999)
4. Jorge, F., Armando, C.O.: Thermal behavior of closed wet cooling towers for use with chilled ceilings. Appl. Therm. Eng. **20**(13), 1225–1236 (2000)
5. ASHRAE: ASHRAE Handbook—HVAC Systems and Equipment. ASHRAE Inc, Atlanta (2008)
6. Mirth, D.R., Ramadhyani, S., Hittle, D.C.: Thermal performance of chilled-water cooling coils operating at low water velocities. ASHRAE Trans. **99**(1), 43–53 (1993)
7. Khan, A.Y.: Heat and mass transfer performance analysis of cooling coils at part-load operating conditions. ASHRAE Trans. **100**(1), 54–62 (1994)
8. Istamoglu, Y.: A new approach for the prediction of the heat transfer rate of the wire-on-tube type heat exchanger—use of an artificial neural network model. Appl. Therm. Eng. **23**(2), 243–249 (2003)
9. Xue, D.: Air Conditioning. Tsinghua University Press, Beijing (2006) (in Chinese)

10. Yao, Y., Lian, Z., Hu, Y., Hou, Z.: The method of thermal metering on the air-side for the fan-coil units. ASHRAE Trans. **110**(part 2), 325–334 (2004)
11. Urban, R.A.: Design considerations and operating characteristics of variable volume systems. ASHRAE J. **12**(2), 77–84 (1969)
12. Amold, D.: Air conditioning in office buildings after World War Two. ASHRAE J. **41**(7), 33–41 (1999)
13. Yao, Y., Lian, Z., Hu, Y., Hou, Z.: Energy-cost allocation based on the theory of frequency response. Appl. Energy **79**(4), 371–383 (2004)
14. Murakami, Y., Terano, M., Mizutani, K., Harada, M., Kuno, S.: Field experiments on energy consumption and thermal comfort in the office environment controlled by occupants' requirements from PC terminal. Build. Environ. **42**(12), 4022–4027 (2007)
15. Engdahl, F., Johansson, D.: Optimal supply air temperature with respect to energy use in a variable air volume system. Energy Build. **36**(3), 205–218 (2004)
16. Ke, M.T., Weng, K.L., Chiang, C.M.: Performance evaluation of an innovative fan-coil unit: low-temperature differential variable air volume FCU. Energy Build. **39**(6), 702–708 (2007)
17. Taras, M.F.: Is economizer cycle justified for AC applications? ASHRAE J. **47**(7), 38–44 (2005)
18. Fisk, W.J., Seppanen, O., Faulkner, D., Huang, J.: Economic benefits of an economizer system: energy savings and reduced sick leave. ASHRAE Trans. **111**(2), 673–679 (2005)
19. Brandemuehl, M.J., Braun, J.E.: Impact of demand-controlled and economizer ventilation strategies on energy use in buildings. ASHRAE Trans. **105**(2), 80–88 (1999)
20. Nassif, N., Moujaes, S.: A new operating strategy for economizer dampers of VAV system. Energy Build. **40**(1), 289–299 (2008)
21. Liu, M., Claridge, D.E., Park, B.Y.: An advanced economizer controller for dual-duct air-handling systems—with a case application. ASHRAE Trans. **103**(2), 156–163 (1997)
22. Joo, I.S., Liu, M.: Economizer application in dual-duct air-handling units. ASCE **9**(4), 126–133 (2003)
23. Wang, S., Xu, X.: A robust control strategy for combining DCV control with economizer control. Energy Convers. Manag. **43**(18), 2569–2588 (2002)
24. Budaiwi, I.M.: Energy performance of the economizer cycle under three climatic conditions in Saudi Arabia. Int. J. Ambient Energy **22**(2), 83–94 (2001)
25. Yiu, J.C.M., Wang, S.W., Yik, F.W.H.: Assessment of practical applications of outdoor air economizer in Hong Kong. Build. Serv. Eng. Res. Technol. **21**(3), 187–198 (2000)
26. Wang., S.K.: Air systems: variable-air-volume systems. In: Handbook of Air Conditioning and Refrigeration, 2nd edn. McGraw-Hill, New York (2000) (Chapter 21)
27. Yao, Y., Wang, L.: Energy analysis on VAV system with different air-side economizers in China. Energy Build. **42**(8), 1220–1230 (2010)
28. Iu, I., Fisher, D.E.: Application of conduction transfer functions and periodic response factors in cooling load calculation procedures. ASHRAE Trans. **110**(2), 829–841 (2004)
29. Chantrasrisalai, C., Iu, I., Eldridge, D.S.: Experimental validation of design cooling load procedures: the heat balance method. ASHRAE Trans. **109**(2), 160–173 (2003)
30. Hittle, D.C., Pedersen, C.O.: Calculating building heating loads using the frequency response of multi-layered slabs. ASHRAE Trans. **87**(2), 545–557 (1981)
31. Yao, Y., Lian, Z., Hou, Z., Liu, W.: Optimal indoor air temperature considering energy savings and thermal comfort in the Shanghai area. In: The 6th International Conference for Enhanced Building Operation, Shenzhen, China, Maximize Comfort: Temperature, Humidity, and IAQ: vol. 1-1-3 (2006)

Chapter 9
Optimal Control of HVAC System Aiming at Energy Conservation

Although the efficiency of individual components has increased considerably in HVAC systems, there still exits much energy waste as much as 30 % of total energy consumption due to various non-optimal operations [1]. Improving the energy efficiency with better operations of HVAC systems is thus critical for energy and environmental sustainability. This chapter presents two optimal control models for HVAC system aiming at energy conservation, one is air-side synergetic control, and the other is global optimization control.

9.1 Air-Side Synergic Control

9.1.1 Background and Basic Idea

In an HVAC system, operation of the air-side secondary system has a significant influence on the overall performance of a building energy system. Due to the high dynamics and correlations among the AHU components and the terminals, it is prone to energy inefficiency with improper operations [2]. Many practical control and operation approaches are proposed to improve the air-side system performance, exemplary, resetting the supply air pressure [3], supply air temperature [4, 5], ventilation air intake [6], or terminal minimum/maximum airflow rates [7]. The common point of these approaches is to identify the inferior operations, which could lead to unintended simultaneous heating and cooling, and introduce optimal measures to minimize energy waste. However, there are two challenges not well addressed with AHU system operations in office buildings:

(1) Utilization of internal heat gain in winter is limited.

Compared to those of earlier times, modern office buildings have higher internal heat gains since more electricity-powered equipment is employed, such as computers,

Y. Yao and Y. Yu, *Modeling and Control in Air-conditioning Systems*,
Energy and Environment Research in China, DOI 10.1007/978-3-662-53313-0_9

copy machines, printers, and even small data servers. By analyzing a hypothetical multi-story office building, it is found that, if the internal heat gain can be fully transferred and utilized, no external heat source is needed for that virtual building during the occupied periods when the outside air temperature is at or above −5 °C [8].

(2) The sensible and latent coil load adversely coupled in some partial load conditions.

Moisture load in office buildings mainly comes from the outside air intake, either the mechanical ventilation or unorganized infiltration [9]. To maintain indoor thermal comfort, the working condition of the cooling coil needs to satisfy the bigger one of the sensible load and the latent load. In humid mild weather, when the space has low sensible cooling load, the conventional cooling and dehumidification operation could cause huge terminal reheat and waste energy.

The existing approaches to tackling these issues mainly rely on adding more thermal components in the air-loop, for example, heat recovery wheel [10], desiccant wheel [11], and so on. However, high initial cost and maintenance difficulty limit the applications of these approaches. Here, synergic control model is proposed to synergize the AHUs with enhanced control in open space office buildings as illustrated in Fig. 9.1. Open environment exists in many office buildings. To

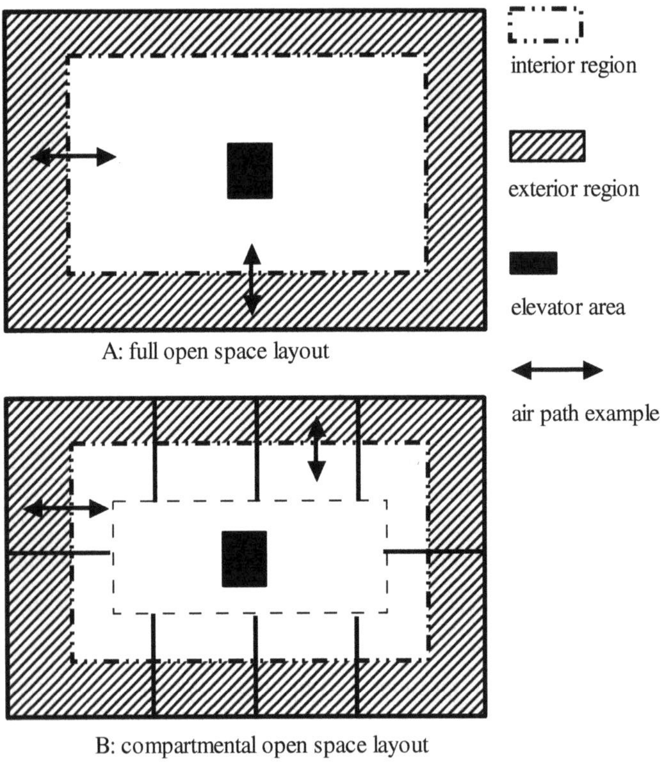

Fig. 9.1 Illustration of office building space layout

minimize energy waste due to simultaneous heating and cooling, the space is typically divided into two regions, with one or more zones in one region. Each zone is served by a dedicated AHU. However, it is not always practical to have a rigid boundary between exterior zones and interior zones. The air terminals of the systems serving different zones might even be put in the same physical room. The ventilation air from different systems in the occupant space mixes and networks spontaneously. The continuous to open space feature provides a path for the indoor air recirculation across the space. The indoor air is credited for recirculation with, or without if the air is adequately fresh, mixing with the outside air intake. Besides the purpose of maintaining the acceptable IAQ (indoor air quality), outside air introduced into a building is also used to keep a positive building pressure and provide free cooling to the building in favorable weather conditions. Nevertheless, in conventional building systems, the operation of outside air intake is designed and performed within the individual AHUs. According to ASHRAE 60.1-2004, performance-based approach, in addition, is stated for IAQ control to allow innovations that can result in indoor air contaminant concentration equal to or lower than the prescriptive approach [12].

The proposed synergic control model here adopts the performance-based approach and utilizes the heat gain discrepancy and heating/cooling mode differences across AHU systems to boost up the energy efficiency. The outside air quantity of AHUs maintains the IAQ in both regions as good as or even higher than the minimum requirement. Modern direct digital control makes it feasible for a building level communication and coordination across the AHUs for the collaboration. The innovation is generated based on the facts that: (1). indoor air flows and mixes well with occupants' activities, air convection, and pressure gradients; (2). the AHUs can have two operation patterns: one for the interior region and the other for the exterior region as shown in Fig. 9.1.

9.1.2 Mathematic Deduction of Synergic Control Model

9.1.2.1 Assumptions

The synergic control model for AHUs of open space office buildings is based on the following assumptions:

(1) The air transferred from a region to another is well mixed with the air within the space of the recipient region;
(2) Maintaining IAQ in a space is generally to keep the steady-state average contaminants' concentration in the air within an acceptable range;
(3) Indoor air conditioning control has good set-point tracking and does not fluctuate within the time intervals of the analysis;
(4) Operation of a system in the steady-state condition is the main part that determines the magnitude of energy uses. While the dynamics in the transient

state is also important or could even be dominant in some circumstances, the analysis of IAHU (Integrated air-handling units) mainly focuses on the steady-state.

9.1.2.2 Model Development

A core region plus perimeter region zoned AHU system layout is widely accepted in office buildings to lower the heating and cooling mode compromising. Depending on the space factors, such as area and orientation, two or more AHUs might be deployed within each region. With the identical operation and outside air control strategy of AHUs in the same region, the AHUs are synergized and consolidated into two virtual AHUs (VAHUs) for the two regions in a building. Figure 9.2 illustrates the consolidated two VAHUs with key parameters denoted. The airflow rates and ratios of the system are calculated based on the two VAHUs. Exemplary, the total supply airflow rate of each VAHU is the summation of all the AHUs in the corresponding region of the building. The notations of the variables in the figure are in accordance with the nomenclature and equations throughout the paper. In real implementation, the synergic control variables are processed and distributed to individual AHU if there are more than one AHU existing in a region.

When the AHU system is in operation, the involved AHUs in the supply region returns air at a lower rate than that of the supplied air and naturally increase the pressure in the region. The reception region is manipulated accordingly to a lower positive pressure so that the air can flow into it from the supply region. The air distribution components, including the supply fans, return fans, and dampers, are modulated in following sections.

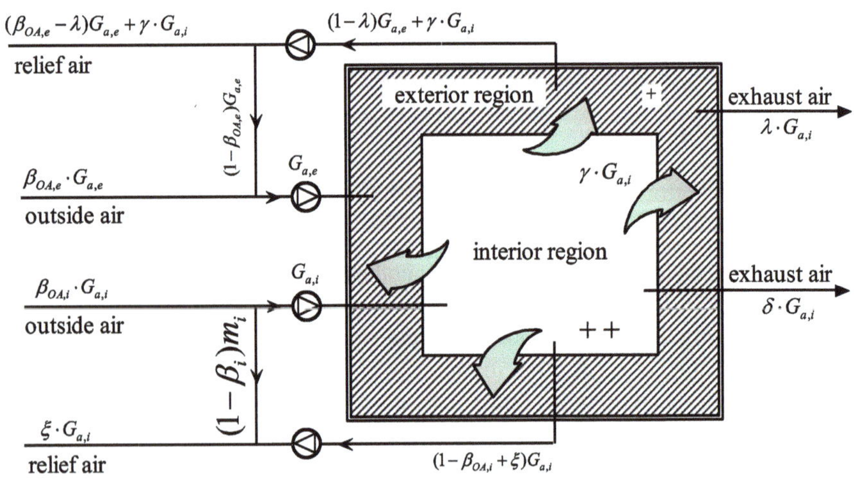

Fig. 9.2 Illustration and notation of consolidated open space AHU system

(1) Ventilation constraints

To achieve an acceptable IAQ indoors, the OA intakes of the two VAHUs are mutually correlated and determined in sequence. Concentration of indoor contaminants (e.g., carbon dioxide) is generally regarded as a performance index of IAQ and should be maintained within an acceptable low level. In a control volume, the net-in and net-out mass difference and the mass of contaminants generated within the volume are equal to the change rate of contaminants within the control volume.

For the interior region, the mass flow rate of contaminants coming from the OA is a function of the supply airflow rate, OA ratio, and the OA contaminant concentration:

$$G_{\text{in},i} = \beta_{\text{OA},i} \cdot G_{\text{a},i} \cdot \varsigma_{\text{o}} \tag{9.1}$$

where $G_{\text{in},i}$ = mass flow rate of contaminants entering the interior region (kg/s), $\beta_{\text{OA},i}$ = outside air flow ratio, $G_{\text{a},i}$ = air mass flow rate supplied to the interior region (kg/s), ς_{o} = contaminants concentration in the outside air (kg/kg).

The mass of contaminants leaving the zone is:

$$G_{\text{out},i} = -\beta_{\text{OA},i} \cdot G_{\text{a},i} \cdot \varsigma_i \tag{9.2}$$

where $G_{\text{out},i}$ = mass flow rate of contaminants leaving the interior region (kg/s), ς_i = contaminants concentration in the air of interior region (kg/kg).

The change rate of the contaminants within the region is:

$$G_{\text{cv},i} = \frac{\text{d}\varsigma_i}{\text{d}\tau} \tag{9.3}$$

where $G_{\text{cv},i}$ = contaminants concentration change rate (kg/(kg s)), τ = time (s).

Under the mass conservation law, the balance equation in the control volume of the interior region is therefore obtained as:

$$\frac{-\beta_{\text{OA},i} G_{\text{a},i}\varsigma_i + \beta_{\text{OA},i} G_{\text{a},i}\varsigma_{\text{o}} + G_{\text{contam},i}}{M_{\text{a,v},i}} = \frac{\text{d}\varsigma_i}{\text{d}\tau} \tag{9.4}$$

where $G_{\text{contam},i}$ = generating rate of contaminants in the interior region (kg/s), $M_{\text{a,v,i}}$ = air mass in the control volume of interior region (kg).

A similar mass balance deduction holds for the exterior region:

$$\frac{-\beta_{\text{OA,e}} G_{\text{a,e}}\varsigma_{\text{e}} + \beta_{\text{OA,e}} G_{\text{a,e}}\varsigma_{\text{o}} + G_{\text{contam,e}}}{M_{\text{a,v,e}}} = \frac{d\varsigma_{\text{e}}}{d\tau} \tag{9.5}$$

where $\beta_{\text{OA,e}}$ = outside air flow rate ratio, $G_{\text{a,e}}$ = air mass flow rate supplied to the exterior region (kg/s), ς_{e} = contaminants concentration in the air of exterior region (kg/kg), $M_{\text{a,v,i}}$ = air mass in the control volume of exterior region (kg), $G_{\text{contam,e}}$ = generating rate of contaminants in the exterior region (kg/s).

In the previous Eqs. (9.1)–(9.5), the interior region and the exterior region are treated as separate ones. There is, as in conventional operation, no controlled interaction of air or contaminants between the two regions. In addition, collaboration between the two regions for contaminant concentration control is considered.

To better present the analysis, we adopt subscript (IAQ,*i*) to denote the relevant variables of the interior region in normal AHU operation and (IAQ,dsg,*e*) to describe a similar circumstance for the exterior region when an independent OA intake is performed.

To satisfy the IAQ requirement, the OA ratios for the two regions should be ensured derived from Eqs. (9.4) and (9.5):

$$\beta_{\mathrm{IAQ},i} = \frac{\frac{d\varsigma_i}{d\tau} M_{\mathrm{a,v},i} + G_{\mathrm{contam},i}}{G_{\mathrm{a},i}(\varsigma_{\mathrm{IAQ},i} - \varsigma_{\mathrm{o}})} \tag{9.6}$$

$$\beta_{\mathrm{IAQ,dsg,e}} = \frac{\frac{d\varsigma_{\mathrm{e}}}{d\tau} M_{\mathrm{a,v,e}} + G_{\mathrm{contam,e}}}{G_{\mathrm{a,e}}(\varsigma_{\mathrm{IAQ,dsg,e}} - \varsigma_{\mathrm{o}})} \tag{9.7}$$

where $\beta_{\mathrm{IAQ},i}$ = minimum outside air intake ratio to maintain the IAQ in the interior region, $\varsigma_{\mathrm{IAQ},i}$ = acceptable contaminants concentration ratio for IAQ consideration in the interior region (kg/kg), $\beta_{\mathrm{IAQ,dsg,e}}$ = minimum outside air intake ratio to maintain the IAQ in the exterior region, $\varsigma_{\mathrm{IAQ,dsg,e}}$ = acceptable contaminants ratio for IAQ consideration in the exterior region (kg/kg).

In a steady state, there is no concentration change in either region, which means $\frac{d\varsigma_i}{d\tau} = \frac{d\varsigma_{\mathrm{e}}}{d\tau} = 0$. Therefore, Eqs. (9.6) and (9.7) evolve into:

$$\beta_{\mathrm{IAQ},i} = \frac{G_{\mathrm{contam},i}}{G_{\mathrm{a},i}(\varsigma_{\mathrm{IAQ},i} - \varsigma_{\mathrm{o}})} \tag{9.8}$$

$$\beta_{\mathrm{IAQ,dsg,e}} = \frac{G_{\mathrm{contam,e}}}{G_{\mathrm{a,e}}(\varsigma_{\mathrm{IAQ,dsg,e}} - \varsigma_{\mathrm{o}})} \tag{9.9}$$

Meanwhile, with any arbitrary OA intake ratio β_{OA}, which maintains an acceptable contaminant concentration no worse than the minimum requirement, the mass balance of the contaminants and the air exist. For the interior region with such a $\beta_{\mathrm{OA},i,}$ the following two equations can be established for the steady-state condition:

$$\beta_{\mathrm{OA},i} G_{\mathrm{a},i} \varsigma_{\mathrm{o}} + G_{\mathrm{contam},i} \quad \beta_{\mathrm{OA},i} G_{\mathrm{a},i} \varsigma_i = 0 \tag{9.10}$$

$$\beta_{\mathrm{OA},i} = \gamma + \delta + \xi \tag{9.11}$$

where γ = air flow rate ratio recirculated from the interior region into the exterior region, δ = air flow rate ratio exhausted from the interior region, ξ = air flow rate ratio relived from the interior region.

To fulfill a synergic control operation for the energy benefits, the interior region is designated as the primary OA intake region with controlled $\varsigma_i \leq \varsigma_e$. As shown in Fig. 9.2, the amount of still-fresh room air transferred from the interior region to the exterior region is defined as $\gamma \cdot G_{a,i}$. The supplementary OA intake from the exterior AHU itself is $\beta_{IAQ,e} \cdot G_{a,e}$. This additional amount of OA is to ensure that the IAQ requirement is met if the air transfer across the regions alone does not satisfy the IAQ in the exterior region. When there is no additional direct OA intake needed from the exterior region VAHU, $\beta_{IAQ,e} \cdot G_{a,e}$ becomes negative or zero.

Similarly, based on the mass conservation of contaminants, the following equation can be established for the exterior region:

$$\beta_{IAQ,e} G_{a,e} \varsigma_o + G_{contam,e} + \gamma G_{a,i} \varsigma_i - (\beta_{IAQ,e} G_{a,e} + \gamma G_{a,i}) \varsigma_e = 0 \tag{9.12}$$

where $\beta_{IAQ,e}$ = supplementary outside air intake ratio in AHUs operation mode.

We normalize the supply airflow rates as ratios and define $\phi = G_{a,i}/G_a$. Equation (9.12) can be written as:

$$\beta_{IAQ,e} G_{a,e} \varsigma_o - \beta_{IAQ,e} G_{a,e} \varsigma_e + \gamma \varphi G_a \varsigma_i - \gamma \varphi G_a \varsigma_e + G_{contam,e} = 0 \tag{9.13}$$

where $G_a = \left(G_{a,i} + G_{a,e}\right)$ is the total air flow rate supplied into the building (kg/s).

From Eq. (9.10), it is known that:

$$\varsigma_i = \varsigma_o + \frac{G_{contam,i}}{\beta_{OA,i} G_{a,i}} \tag{9.14}$$

Replacing ς_i in Eq. (9.13) with Eq. (9.14), we acquire the explicit expression for $\beta_{IAQ,e}$:

$$\beta_{IAQ,e} = \frac{1}{G_{a,e}(\varsigma_e - \varsigma_o)} \left[G_{contam,e} + \gamma \varphi G_a (\varsigma_o - \varsigma_e) + \gamma \frac{G_{contam,i}}{\beta_{OA,i}}\right] \tag{9.15}$$

Meanwhile, Eqs. (9.8) and (9.9) can be further generalized as:

$$G_{contam,i} = (\varsigma_i - \varsigma_o) \beta_{OA,i} G_{a,i} = (\varsigma_{IAQ,i} - \varsigma_o) \beta_{IAQ,i} G_{a,i} \tag{9.16}$$

$$G_{contam,e} = (\varsigma_{IAQ,dsg,e} - \varsigma_o) \beta_{IAQ,dsg,e} G_{a,e} \tag{9.17}$$

Substituting the two equations into Eq. (9.15) and noticing that the contaminants concentration ς_e in the exterior region should be a value that can ensure an acceptable IAQ, which means $\varsigma_e \leq \left(\varsigma_{IAQ,dsg,e} = \varsigma_{IAQ,i}\right)$ (the acceptable IAQ standard holds the same for both regions with $\varsigma_{IAQ,dsg,e} = \varsigma_{IAQ,i}$), we can obtain:

$$\beta_{IAQ,e} = \beta_{IAQ,dsg,e} - \frac{\gamma \varphi}{1 - \varphi} \left(1 - \frac{\beta_{IAQ,i}}{\beta_{OA,i}}\right) \tag{9.18}$$

To this end, the relationship of the mutually correlated constraints, $\beta_{\mathrm{IAQ,e}}$ and $\beta_{\mathrm{OA},i,}$ for IAQ in AHU system operation is obtained. The following section is dedicated to finding the AHUs operation schema of the OA intakes in both regions to achieve a synergization in various scenarios.

(2) Decision-making schema

The ultimate goal of operating AHUs in office buildings is to reduce the thermal energy consumption on the air side and ensure an acceptable or even improved IAQ. The AHUs are coordinated in terms of the OA intakes and system operation. The optimums are obtained by evaluating the cost function in different conditions.

The total thermal energy consumption for cooling and heating the building with two consolidated AHUs in a time period is the integration of the power given as below:

$$Q_{\mathrm{thm}} = Q_{\mathrm{hc,i}} + Q_{\mathrm{cc,i}} + Q_{\mathrm{rh,i}} + Q_{\mathrm{hc,e}} + Q_{\mathrm{cc,e}} + Q_{\mathrm{rh,e}} \tag{9.19}$$

where Q_{thm} = total thermal power of VAHUs in the building (W), $Q_{\mathrm{hc,i}}\left(Q_{\mathrm{hc,e}}\right)$ = thermal power of heating coils in interior region and exterior region, respectively, (W), $Q_{\mathrm{cc,i}}\left(Q_{\mathrm{cc,e}}\right)$ = thermal power of cooling coils in interior region and exterior region, respectively, (W), $Q_{\mathrm{rh,i}}\left(Q_{\mathrm{rh,e}}\right)$ = thermal power of reheats in interior region and exterior region, respectively, (W).

This is a constraint-varying optimization problem since the operation scenario changes under the dynamic indoor and outdoor conditions. The costs of individual thermal components for both regions are:

$$Q_{\mathrm{hc}} = \max(0, G_{\mathrm{a}}(h_{\mathrm{c}} - h_{\mathrm{mix}})) \tag{9.20}$$

$$Q_{\mathrm{cc}} = \max(0, G_{\mathrm{a}}(h_{\mathrm{mix}} - h_{\mathrm{c}})) \tag{9.21}$$

$$Q_{\mathrm{rh}} = \max(0, G_{\mathrm{a}}(h_{\mathrm{s}} - h_{\mathrm{c}})) \tag{9.22}$$

where h_{c} = specific enthalpy of supply air (J/kg), h_{mix} = specific enthalpy of mixed air (J/kg), h_{s} = specific enthalpy of discharge air (J/kg).

To facilitate the interpretation, Fig. 9.3 illustrates the configuration of a simplified AHU system. The essential thermal components, including a heating coil, cooling coil, and reheat coil, and system states of the air at different location are denoted.

In addition to the IAQ constraint, there are dynamic constraints inherited from the original system (Eqs. (9.23)–(9.26) below) for this problem. First of all, the OA ratios should be no less than the minimum requirements regardless of the operation mode.

$$\beta_{\mathrm{OA},i} \in [\beta_{\mathrm{IAQ},i}, 1] \tag{9.23}$$

$$\beta_{\mathrm{OA,e}} \in [\max(\beta_{\mathrm{IAQ,e}}, 0), 1] \tag{9.24}$$

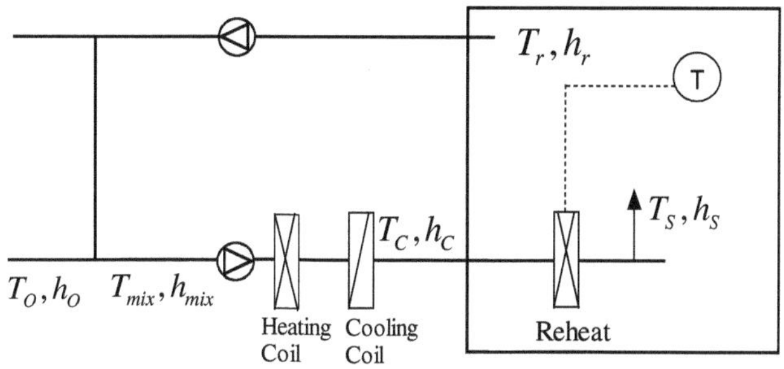

Fig. 9.3 Thermal components of a typical AHU system

In transition seasons or winter, if a region is just in cooling mode, (termed as economizer/free-cooling mode), outside air becomes favorable to replace the room air for less thermal energy consumption. It is likely that more than IAQ required amount of OA should be introduced into the space. The OA intake constraints for the regions in economizer/free-cooling operation are:

$$\beta_{\mathrm{OA},i} \in [\beta_{\mathrm{eco},i}, 1] \tag{9.25}$$

$$\beta_{\mathrm{OA,e}} \in [\beta_{\mathrm{eco,e}}, 1] \tag{9.26}$$

where $\beta_{\mathrm{eco}} = \frac{T_{\mathrm{r}} - T_{\mathrm{c}}}{T_{\mathrm{r}} - T_{\mathrm{o}}}$, T_{r} = room air temperature (°C), T_{o} = outside air temperature (°C), T_{c} = supply air temperature (°C).

Outside air condition typically determines the thermal process mode (heating/cooling, outside air intake amount, etc.) of an air conditioning system via the heat transfer through the envelope and load in the direct OA intake. The system control strategy for optimal thermal energy consumption is therefore highly related to the OA condition. The different scenarios illustrated in Fig. 9.4 are analyzed to retain the operation strategy for AHUs:

The approach to find the optimum of AHUs for each scenario is based on the engineering transformation and mathematic analysis. The variable thermal energy consumption in a system can always be transformed into the summation of an unchangeable component and a changeable component for a given condition. The meaning of unchangeable and changeable is relative from the operation point of view. The unchangeable component is the thermal load in the corresponding region, which is mainly determined by the intrinsic factors of the space (e.g., occupants, insulation level). The changeable component is the additional thermal energy needed for the space conditioning derived from a specific operation. It varies when a different operation algorithm of AHU is applied.

The non-negative cost functions are defined in the sub-domain of $0 \leq \gamma \leq \beta_{\mathrm{OA},i} \leq 1$ on $(\beta_{\mathrm{OA},i}, \gamma)$ plane. The equivalent cost functions in synergic control model for AHUs are summarized in Table 9.1 for different outside air conditions. The detailed deduction can be found in the appendix later in this

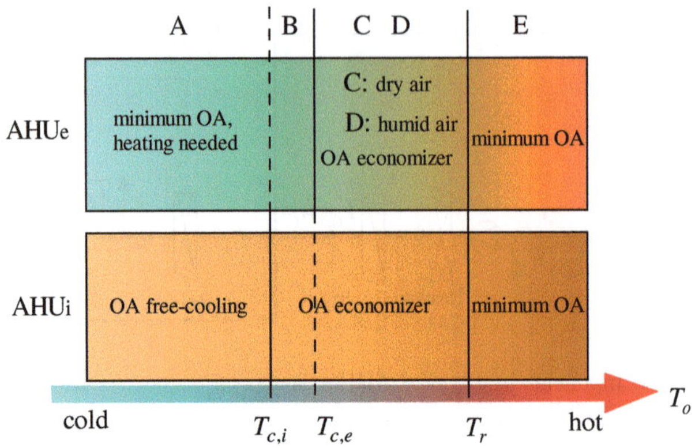

Fig. 9.4 AHU operation scenarios in office buildings

Table 9.1 Scenarios and cost functions for synergic control operation

No.	Equivalent cost function
A	$f = \beta_{\mathrm{IAQ,dsg,e}}(1-\varphi) - \gamma\varphi(1-\frac{\beta_{\mathrm{IAQ},i}}{\beta_{\mathrm{OA},i}}) + \varphi\beta_{\mathrm{OA},i}$
B	$f = \beta_{\mathrm{IAQ,dsg,e}}(1-\varphi) - \gamma\varphi(1-\frac{\beta_{\mathrm{IAQ},i}}{\beta_{\mathrm{OA},i}}) - \varphi\beta_{\mathrm{OA},i}$
C	$f = -\beta_{\mathrm{IAQ,dsg,e}}(1-\varphi) + \gamma\varphi(1-\frac{\beta_{\mathrm{IAQ},i}}{\beta_{\mathrm{OA},i}}) - \varphi\beta_{\mathrm{OA},i}$
D/E	$f = f_1 - f_2$ $f_1 = 2(h_{\mathrm{s,e}} - h_{\mathrm{c,e}})(1-\varphi)$ $f_2 = (h_{\mathrm{o}} - h_{\mathrm{r}})(\varphi\beta_{\mathrm{OA},i} - \varphi\beta_{\mathrm{IAQ},i} - (1-\varphi)\beta_{\mathrm{IAQ,dsg,e}})$

Table 9.2 AHUs OA intake schema

No.	T_{o} condition	$\beta_{\mathrm{OA},i}$	$\beta_{\mathrm{OA},e}$
A	$T_{\mathrm{o}} < T_{\mathrm{c},i}$	$\max(\beta_{\mathrm{IAQ},i}, \beta_{\mathrm{eco},i})$	$\max(\beta_{\mathrm{IAQ,e}}, 0)$
B	$(T_{\mathrm{c},i}, T_{\mathrm{c,e}}]$	1	$\max(\beta_{\mathrm{IAQ,e}}, 0, \beta_{\mathrm{eco,e}})$
C	$(T_{\mathrm{c,e}}, T_{\mathrm{r}}], h_{\mathrm{o}} < h_{\mathrm{r}}$	1	1
D: $(T_{\mathrm{c,e}}, T_{\mathrm{r}}], h_{\mathrm{o}} \geq h_{\mathrm{r}}$			
D1	mild weather	$[\frac{\gamma\varphi\beta_{\mathrm{IAQ},i}}{\gamma\varphi - \beta_{\mathrm{IAQ,dsg,e}}(1-\varphi)}]$, $[\beta_{\mathrm{IAQ},i}]$	[0], $[\beta_{\mathrm{IAQ,dsg,e}}]$
D2	normal	$[\beta_{\mathrm{IAQ},i}]$	$[\beta_{\mathrm{IAQ,dsg,e}}]$
E: $T_{\mathrm{r}} \leq T_{\mathrm{o}}$			
E1	mild weather	$[\frac{\gamma\varphi\beta_{\mathrm{IAQ},i}}{\gamma\varphi - \beta_{\mathrm{IAQ,dsg,e}}(1-\varphi)}]$, $[\beta_{\mathrm{IAQ},i}]$	[0], $[\beta_{\mathrm{IAQ,dsg,e}}]$
E2	normal	$[\beta_{\mathrm{IAQ},i}]$	$[\beta_{\mathrm{IAQ,dsg,e}}]$

chapter. The optimum OA intake schema in synergic control operation is provided in Table 9.2.

(3) Parameter and feasibility analysis

Five airflow rate ratios, $\beta_{IAQ,dsg,e}$, $\beta_{IAQ,i}$, φ, $\beta_{OA,i}$, γ, are involved in the analytical expression of $\beta_{IAQ,e}$. These parameters are determined sequentially when the AHU system is operated. In an economizer/free-cooling mode, two more OA ratios, $\beta_{eco,i}$ and $\beta_{eco,e}$, should be considered as well.

$\beta_{IAQ,dsg,e}$ and $\beta_{IAQ,i}$ are the initial minimum OA intake ratios in the exterior region and interior region, respectively. The two variables are defined for ventilation purpose when the AHUs operate individually. They are generally regarded as design constants with a fixed value. These two variables have a positive correlation with $\beta_{IAQ,e}$, as indicated in their derivatives. In other words, $\beta_{IAQ,e}$ increases if either of the two variables increases. In synergic control decoupling operation (D1 and E1 in Table 9.2), the two variables, along with the airflow ratios (φ and $(1-\varphi)$), jointly determine the feasibility of such a conversion from normal AHU operation into synergic control operation.

$\beta_{eco,i}$ and $\beta_{eco,e}$ come into effect in an economizer/free-cooling mode when the outside air is favorable with a temperature lower than the room air temperature. For most time of a free-cooling season, $\beta_{eco,i}$ is greater than $\beta_{IAQ,i}$. Figure 9.5 shows that, even when the OA temperature drops to −17 °C, the OA ratio is above 0.25 with the supply air temperature at 12.8 °C. This provides extra credit to the

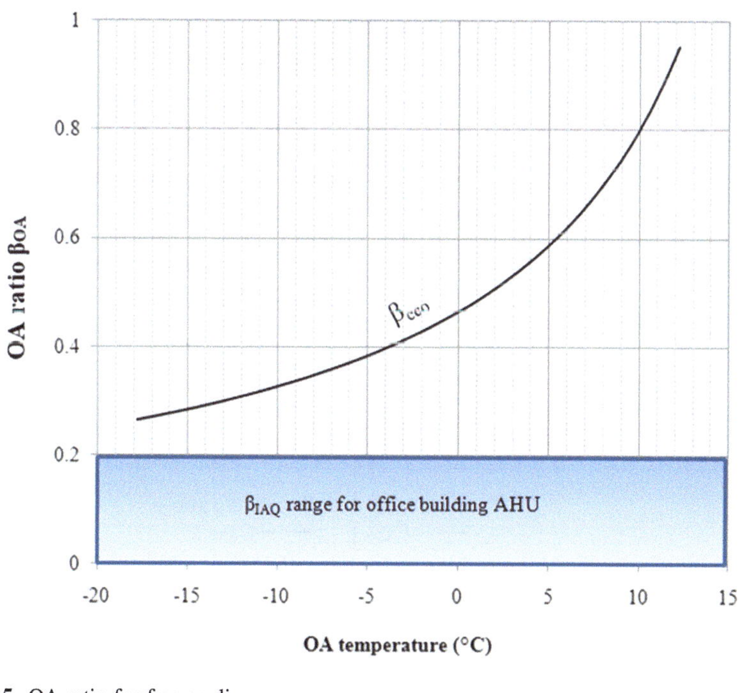

Fig. 9.5 OA ratio for free cooling

freshness of the interior region air when it is in economizer/free-cooling mode. The room air in the interior region can be circulated into the exterior region for the purpose of ventilation.

As a supply air flow rate ratio φ is influenced by two airflow rates: that of the interior region and that of the exterior region. It has the trend to decrease when the OA temperature increases in summer. In winter, the airflow ratio decreases slowly when the OA temperature drops. The derivative of Eq. (9.18) over φ is given as:

$$\frac{\partial \beta_{\mathrm{IAQ,e}}}{\partial \varphi} = \left(\frac{\beta_{\mathrm{IAQ},i}}{\beta_{\mathrm{OA},i}} - 1\right)\frac{\gamma}{(1-\varphi)^2} \tag{9.27}$$

From Eq. (9.27), it can be seen that the value is always negative as $\beta_{\mathrm{OA},i} > \beta_{\mathrm{IAQ},i}$ in AHUs. Therefore, φ has an inverse correlation with $\beta_{\mathrm{IAQ,e}}$. In other words, with given values of the other variables, a greater φ results in a smaller $\beta_{\mathrm{IAQ,e}}$. Figure 9.6 illustrates the change pattern of $\beta_{\mathrm{IAQ,e}}$ and φ with arbitrary variables in two cases (Case 1: $\beta_{\mathrm{IAQ,dsg,e}} = 0.1$, $\gamma = 0.25$, $\beta_{\mathrm{IAQ},i} = 0.1$, $\beta_{\mathrm{OA},i} = 0.3$; Case 2: $\beta_{\mathrm{IAQ,dsg,e}} = 0.1$, $\gamma = 0.4$, $\beta_{\mathrm{IAQ},i} = 0.1$, $\beta_{\mathrm{OA},i} = 0.5$). In Case 1, when the supply airflow rate ratio is 0.4 or higher, there is no direct OA intake requirement as supplement on the exterior

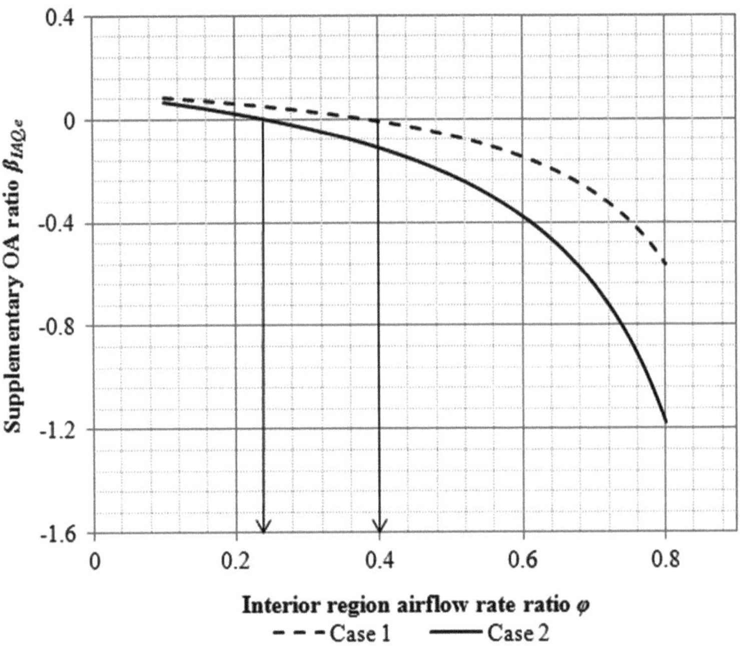

Fig. 9.6 Exterior zone OA ratio for two cases

region to satisfy the IAQ. Case 2 shows, if the direct OA intake to the interior region is higher than 0.5, the minimum supply airflow rate ratio can be less than 0.25.

$\beta_{OA,i}$ is inversely correlated to $\beta_{IAQ,e}$. the higher the OA intake from the interior region, the lower the OA needed from the exterior region. In most mechanical heating/cooling seasons, less OA is generally desired for both regions if the IAQ can be maintained with that amount of OA intake. In a load decoupling mode, $\beta_{OA,i}$ is obtained by setting $\beta_{IAQ,e}$ in Eq. (9.18) to zero. In winter, because the interior region has cooling load in office buildings, $\beta_{OA,i}$ evolves to $\beta_{eco,i}$ as a constraint determined by the internal thermal load. $\beta_{IAQ,e}$ is consequently obtained thereafter. $\beta_{OA,i}$ also has a mutual dependency on the airflow ratio (γ) recirculated from the interior region to the exterior region with the relation $0 \leq \gamma \leq \beta_{OA,i} \leq 1$.

The derivative of $\beta_{IAQ,e}$ on γ has the following expression:

$$\frac{\partial \beta_{IAQ,e}}{\partial \gamma} = \left(\frac{\beta_{IAQ,i}}{\beta_{OA,i}} - 1 \right) \frac{\varphi}{(1 - \varphi)} \tag{9.28}$$

It can be seen that the derivative is always negative since $\beta_{OA,i}$ is greater than $\beta_{IAQ,i}$ in AHUs. This means that $\beta_{IAQ,e}$ decreases when γ increases. It is easy to understand from physics that the recirculated fresh air from the interior region naturally reduces the direct OA demand through the exterior region AHUs.

Assuming $\beta_{IAQ,e}$ is zero in Eq. (9.18), we acquire the following relationship between γ and $\beta_{OA,i}$:

$$\gamma = \frac{(1 - \varphi)\beta_{IAQ,dsg,e}}{\left(1 - \frac{\beta_{IAQ,i}}{\beta_{OA,i}}\right)\varphi} \tag{9.29}$$

Figure 9.7 depicts γ and $\beta_{OA,i}$ based on Eq. (9.29) with $\beta_{IAQ,dsg,e} = \beta_{IAQ,i} = 0.15$. With a given airflow rate ratio φ, when the OA intake ratio of the interior region is small (for example, $\beta_{OA,i} \leq 0.4$), the recirculation air ratio γ increases quickly if $\beta_{OA,i}$ decreases. If γ increases into the infeasible area ($\gamma \geq \beta_{OA,i}$), an AHU system has to take at least part of the OA directly from the exterior region AHUs to satisfy the IAQ. If this situation is not desired, the AHU system cannot be operated.

In the deduction of synergetic control operation for cooling decoupling mode, it has been stated that shifting OA intake might not save energy in conditions D1 and E1. The operation could be complex and needs an investigation.

Introducing a partial load ratio r $\left(r = \frac{h_r - h_{s,e}}{h_r - h_{c,e}}\right)$ and referring to the section 'Appendix,' the equivalent cost function of AHUs can be obtained as below:

$$f = (h_o - h_r)(\varphi\beta_{OA,i} - \varphi\beta_{IAQ,i} - (1 - \varphi)\beta_{IAQ,dsg,e}) - 2(h_r - h_{c,e})(1 - r)(1 - \varphi) \tag{9.30}$$

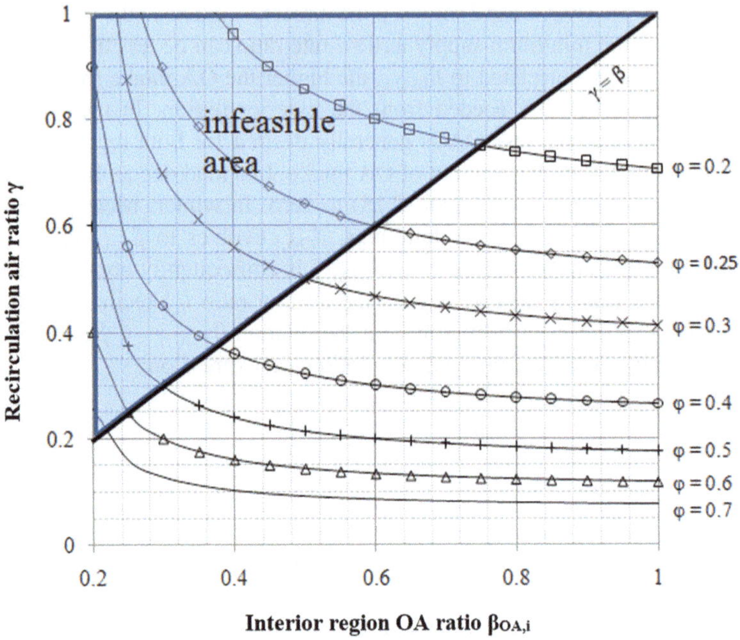

Fig. 9.7 Minimum recirculation air ratio γ with different $\beta_{OA,i}$

Correspondingly, with all other variables defined, β_i should also satisfy the following equation to avoid a penalty when the cost is greater than the savings:

$$\beta_{OA,i} \leq \frac{2(h_r - h_{c,e})(1-r)(1-\varphi)}{(h_o - h_r)\varphi} + \frac{(\varphi\beta_{IAQ,i} + (1-\varphi)\beta_{IAQ,dsg,e})}{\varphi} \tag{9.31}$$

Evaluation of this equation is difficult in a real application. From Fig. 9.7, it can be seen that the minimum OA intake β_i in decoupling mode lies on the line where $\beta_{OA,i} = \gamma$. It makes:

$$\beta_{OA,i} = \frac{\beta_{IAQ,dsg,e}(1-\varphi) + \beta_{IAQ,i}\varphi}{\varphi} \tag{9.32}$$

This is an extreme which can guarantee the satisfaction of inequality of Eq. (9.32) in conditions D1 and E1.

The feasibility of AHUs thermal decoupling in mild weather also depends on the real-time interior region supply airflow rate ratio and recirculation capability (γ_{max} is the maximum allowable value of γ) between the two regions. The recirculation capability can be considered as the maximum recirculation ratio that can be accepted by the exterior region return fan capacity and the space layout. Therefore, the OA intake $\beta_{OA,i}$ in the interior region is also subject to the following inequality:

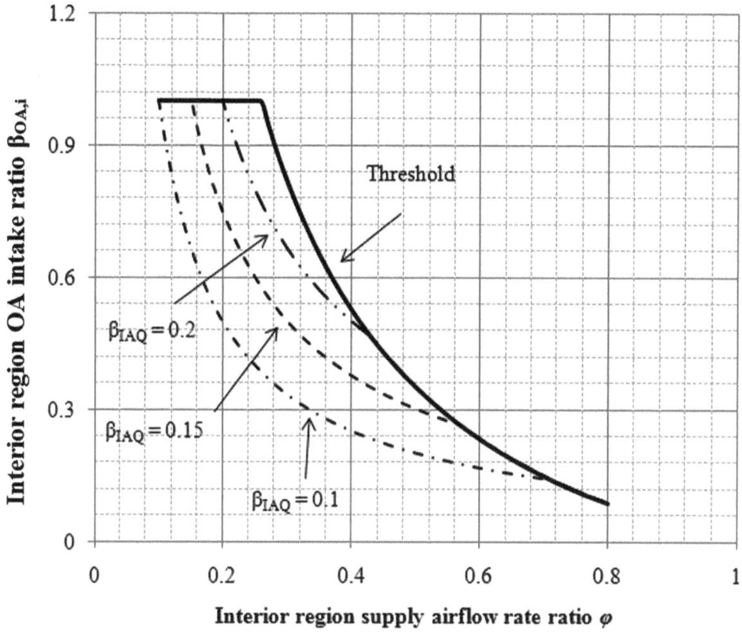

Fig. 9.8 Feasible $\beta_{OA,i}$ with different φ ($x = 0.30$)

$$\beta_{OA,i} \leq \frac{\gamma_{max} \cdot (1 - \varphi)}{\varphi} \tag{9.33}$$

The inequality relationship can be plot as Fig. 9.8. For example, in a given condition, when the interior region airflow ratio is 0.6 and the minimum OA intake for both regions is equal to or above 0.15, the synergetic control is not suitable for thermal decoupling operation. Figure 9.8 also gives the information of minimum supply airflow rate ratios in interior region where the corresponding $\beta_{OA,i}$ for IAQ in synergetic control equals to 1. If the interior region air ratio drops to be lower than the corresponding critical points, thermal decoupling operation is not feasible.

9.1.3 Control Logic Details

The general synergetic control logic includes the following five steps:

Step 1: Compute the proper supply fan speed and supply air temperature for the selected AHUs. The supply airflow rate should be kept higher than the minimum as in normal AHU operation to ensure indoor air circulation.

Step 2: Determine the scenario of synergetic control based on the outside air temperature, outside air enthalpy, and VAHU operation mode. If multiple

AHUs exist in one region, consolidate and update the airflow rate ratios for $VAHU_i$ and $VAHU_e$.

Step 3: Calculate the new outside air intake and airflow allocations based on the decision schema. Determine the feasibility of synergetic control with the new outside air intakes and airflow reallocation among VAHUs. If it is feasible, go to Step 4. Otherwise, preserve original AHU operation and return to Step 1.

Step 4: Realize IAHU optimization supervisory level control. Distribute the steady-state set points of outside air intake and return airflow rate to the VAHUs. Modulate the outside air dampers to maintain the outside air intake set points. Control the airflow modulating components, including exhaust fans, return fan VFDs, and relief dampers, for the new airflow allocations.

Step 5: Switch back to normal AHUs operation of Step 1 if any of the conditions do not support the chosen scenario: the outside air temperature, the outside air humidity, the airflow ratio, or the outside air ratio.

The control flow charts are shown in Figs. 9.9, 9.10 and 9.11 as the decision-making procedure. Figure 9.9 illustrates mainly the actions from Step 1 to

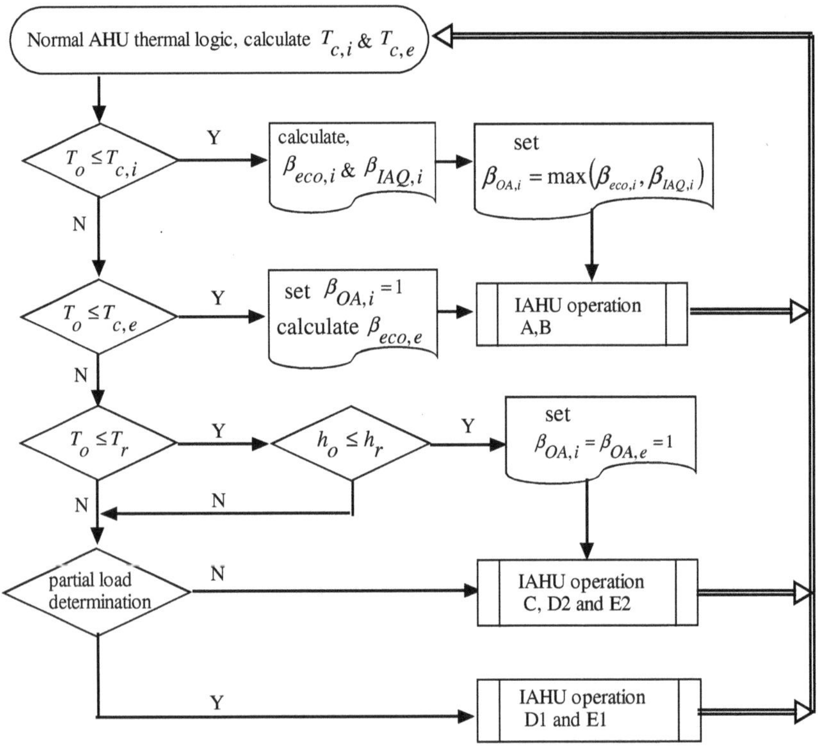

Fig. 9.9 Control algorithm determining the operation scenario

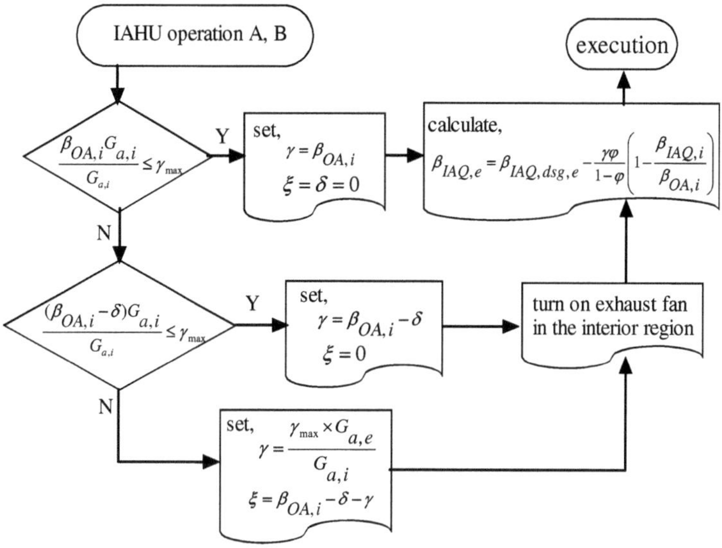

Fig. 9.10 Control algorithm of operation scenario A and B

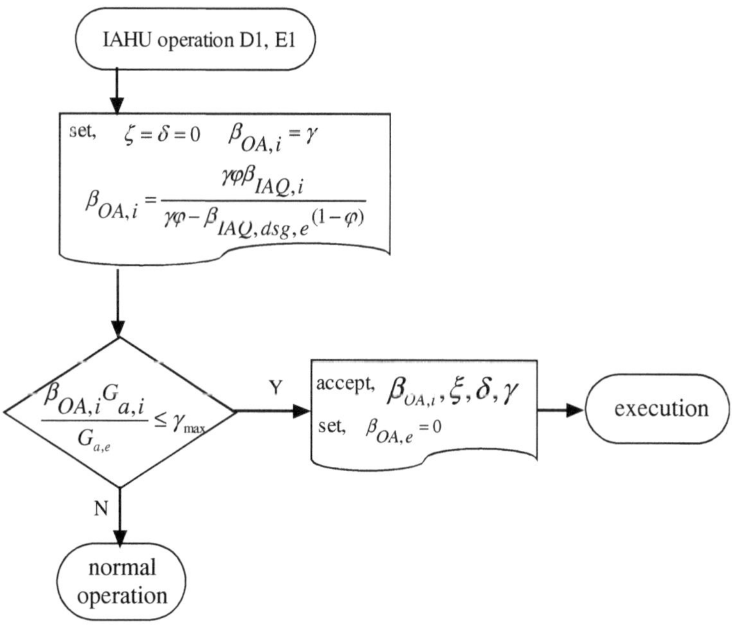

Fig. 9.11 Control algorithm of operation scenario D1 and E1

Step 3. Figure 9.10 provides the supervisory control algorithm for scenario A and scenario B. This operation is to utilize the internal heat gain from the interior region for warming up the outside air intake when VAHUi and VAHUe are in opposite operation modes. Figure 9.11 shows the process of IAHU operation in scenario D1 and scenario E1. Coil load decoupling is performed in the two conditions to avoid unnecessary energy waste of using reheat in mild weather. The outside air (OA) intake is directed to $VAHU_i$ for the whole building ventilation in synergetic operation. Operation scenario C, D2, and E2 are identical to normal zoned-AHU operation without airflow reallocation among the AHUs.

The synergetic control usually does not change the original thermal conditioning logics in the VAHUs. However, there is only one exception, when the $VAHU_e$ is in a partial load mode in mild weather. With the synergetic control, the supply air temperature limitation on $VAHU_e$ can be freed up to accommodate sensible load only. The partial load determination for thermal decoupling operation in the synergetic control might vary for different AHU and terminal configurations. The switching logic based on a combination of timing and thresholds should be developed onsite for implementation when a building system characteristic is clearly understood.

For the same purpose of airflow allocation in AHUs, with different system layouts, the execution measures might differ. Exhaust fans, variable frequency drivers (VFDs), and relief air dampers are the components that could be utilized to fulfill the purpose in the implementation. If a system is originally constant air volume, typically serving as the primary air with terminal induction units for the perimeter region in office buildings, a VFD is recommended on the return air fan for AHUs and energy conservation consideration.

9.1.4 *Case Study*

9.1.4.1 Consolidation and Decomposition

To reduce the thermal compromising due to the heating and cooling mode difference across the spaces, a zoned-AHU system layout with interior and exterior region virtually separated is widely adopted in office buildings. The synergetic control model incorporates the practice of conventional AHU operation and supervises the outside air intake allocation and airflow recirculation. The indoor air can be circulated via the air paths across the interior and exterior regions when AHUs are in operation.

Depending on the building area and space factors, e.g., orientation and function, one or more AHUs might be deployed within the core region and the perimeter region, respectively. The AHUs are dynamically selected and categorized into two types, termed supplier and receiver, based on the predefined operation rules. Consolidation and distribution are the essential and dynamic preprocess and postprocess in synergetic control implementation for buildings with multiple AHUs.

Consolidation is the process that sorts and groups the AHUs in the same region before determining a proper synergetic operation during the decision interval. Distribution is the opposite process, which takes the system supervisory level decisions and executes them among the selected AHUs.

The AHUs in the same region feature a similar thermal load and operation pattern. Therefore, the AHUs in an office building can be consolidated into two AHUs for the whole conditioned space: one as a provider and the other as a receiver of ventilation air. In an ideal synergization condition, all the AHUs in the interior region can be consolidated into one supplier while all the AHUs in the exterior region into one receiver. Under some extreme circumstances, if an AHU in a given region is very unique from others in terms of its thermal operation, it will be excluded from the synergetic control. An example would be that if an AHU in the exterior region were not in heating mode in winter while the others are, the AHUs suitable for synergetic operation are then virtually consolidated as either the supply or the reception AHU. To facilitate the description, VAHU is used to represent the consolidated virtual AHU in synergetic operation. $\mathrm{VAHU_i}$ denotes the supplier and usually the interior one, and $\mathrm{VAHU_e}$ denotes the receiver and usually the exterior one. The AHUs not attending the synergetic operation remain the original control logics.

The consolidated airflow rate of a VAHU is the summation of that of the grouped AHUs. The total supply airflow rates for the two VAHUs are given by:

$$G_{\mathrm{a},i} = \sum_{j=1}^{l} G_{\mathrm{a},i,j} \tag{9.34}$$

$$G_{\mathrm{a,e}} = \sum_{k=1}^{n} G_{\mathrm{a,e},k} \tag{9.35}$$

where $G_{\mathrm{a},i}$ = consolidated air mass flow rate supplied to the interior region (kg/s), $G_{\mathrm{a},i,j}$ = air mass flow rate supplied to the jth zone in the interior region (kg/s), $G_{\mathrm{a,e}}$ = consolidated air mass flow rate supplied to the exterior region (kg/s), and $G_{\mathrm{a,e},k}$ = air mass flow rate supplied to the kth zone in the exterior region (kg/s). l is the total number of the selected AHUs for $\mathrm{VAHU_i}$ and n is the total number of the selected AHUs for $\mathrm{VAHU_e}$. l and n are variable.

Normalized airflow rates are adopted in the synergetic operation for convenience. The flow rate ratio of supply air from $\mathrm{VAHU_i}$ is given by:

$$\varphi = \frac{G_{\mathrm{a},i}}{G_{\mathrm{a},i} + G_{\mathrm{a,e}}} \tag{9.36}$$

The outside air intakes in normal AHU operation, e.g., economizer outside air intakes and designed minimum outside air intakes, are similarly normalized into ratios.

$$\beta_{\text{eco},i} = \frac{\sum_{j=1}^{l} (\beta_{\text{eco},i,j} G_{\text{a},i,j})}{G_{\text{a},i}} \tag{9.37}$$

$$\beta_{\text{IAQ},i} = \frac{\sum_{j=1}^{l} (\beta_{\text{IAQ},i,j} G_{\text{a},i,j})}{G_{\text{a},i}} \tag{9.38}$$

$$\beta_{\text{IAQ,dsg,e}} = \frac{\sum_{k=1}^{n} (\beta_{\text{IAQ,dsg,e,k}} G_{\text{a,e}}, k)}{G_{\text{a,e}}} \tag{9.39}$$

where $\beta_{\text{eco},i}$ = consolidated outside air ratio to the interior region in economizer/free-cooling mode, $\beta_{\text{eco},i,j}$ = outside air ratio of the *j*th AHU in the interior region in economizer/free-cooling mode, $\beta_{\text{IAQ},i}$ = consolidated minimum outside air ratio to the interior region for IAQ consideration, $\beta_{\text{IAQ},i,j}$ = minimum outside air ratio of the *j*th AHU in the interior region for IAQ consideration, $\beta_{\text{IAQ,dsg,e}}$ = consolidated minimum outside air ratio to the exterior region for IAQ consideration, and $\beta_{\text{IAQ,dsg,e},k}$ = minimum outside ratio of the *k*th AHU in the exterior region for IAQ consideration.

9.1.4.2 Building and System Model

Many researchers have pointed out that a large commercial building can be modeled as two zones: an internal one and a perimeter one, for energy consumption analysis. Knebel [13] presented the idea to simplify energy analysis using a modified BIN method. Katipamula and Claridge [14] modeled a commercial building as two zones and considered it adequate for commercial building load simulation. Very accurate results were found by using the simplified zoning in case studies [15, 16]. Liu et al. [15] applied the simplified hourly modeling for the purpose of automated calibration and fault detection. Their studies support that a two-region simplified hourly model works well with properly defined interior and exterior zones for commercial buildings. The detail of the building load model is collected in the appendix for reference.

For air-based conditioning systems in commercial office buildings, the main thermal components are primary heating coils, cooling coils, and reheat coils. The model is provided below to describe the system states (flow rate, temperature, humidity ratio, etc.) of the air and energy consumption rate of the system.

(1) Supply airflow rates

With different system configurations, the supply airflow rate is determined differently. In a constant air volume system, the airflow rate does not change. In a variable air volume system, the supply airflow rate is determined by the sensible load of the space and the supply air temperature set point. It is also constrained by a minimum airflow rate to ensure the ventilation and air distribution.

For a variable air volume system, the supply airflow rate is given by:

$$G_a = \max\left(G_{a,\min}, \frac{Q}{c_p \times (T_r - T_c)}\right) \tag{9.40}$$

where Q = space thermal load in W, c_p = specific heat of air in J/(kg K), G_a = supply air mass flow rate in kg/s, and T = air temperature in °C. Subscript r means room air, c means cold deck.

Thermal energy:

The thermal energy consumptions are mainly from the heating coils, cooling coils, and reheat coils:

$$Q_{hc} = G_a \times c_p \times (T_r - T_{mix}) \tag{9.41}$$

$$Q_{cc} = G_a \times (h_{mix} - h_c) \tag{9.42}$$

$$Q_{rh} = G_a \times c_p \times (T_{dis} - T_c) \tag{9.43}$$

where Q = thermal power in W and h = specific enthalpy of moist air in J/(kg). Subscript *hc* means heating coil, *cc* means cooling coil, *mix* means mixed air, and *dis* means discharge air.

In some conventional systems, there are induction units deployed along the external walls. The energy consumption of the induction coils is simulated as:

$$Q_{id} = n_r \times G_a \times (h_{id} - h_r) \tag{9.44}$$

where n_r = induction ratio. Subscript *id* means induction unit.

(2) Air states

The status of mixed air can be simulated by using:

$$T_{mix} = T_o\beta_{OA} + (1 - \beta_{OA})T_r \tag{9.45}$$

$$h_{mix} = h_o\beta_{OA} + (1 - \beta_{OA})h_r \tag{9.46}$$

$$W_{mix} = W_o\beta_{OA} + (1 - \beta_{OA})W_r \tag{9.47}$$

where β_{OA} = outside air intake ratio, W is air humidity ratio in g/kg.

Discharge air temperature is determined by the space sensible load and should not be lower than T_c, which is specified in the control algorithm:

$$T_{dis} = \frac{Q}{G_a \times c_p} + T_c \tag{9.48}$$

The room air enthalpy is acquired based on the room air temperature and the room air humidity:

$$h_r = T_r \times c_p + W_r \times h_{fg} \tag{9.49}$$

where h_{fg} = specific enthalpy of water vapor in J/kg.

The room air humidity is determined by:

$$W_r = \frac{W + \frac{G_a}{\rho} \times W_c + ACH_{inf} \times V \times W_o}{\frac{G_a}{\rho} + ACH_{inf} \times V} \tag{9.50}$$

where ACH_{inf} = air change rate due to infiltration (/hr), V = air volume of the space m^3, ρ = air density kg/m^3.

The return air humidity w_r and the supply air humidity w_c are coupled when the coil is dry. Under this circumstance, the supply air humidity w_c equals the mix air humidity w_{mix}. Iteration with Eqs. (9.47) and (9.50) is needed until the two converge. Or, the following equation can be used directly, to solve for the room air humidity.

$$W_r = \frac{W_o \times \frac{G_a}{\rho} \times \beta_{OA} + W + ACH_{inf} \times V \times W_o}{\frac{G_a}{\rho} \times \beta_{OA} + ACH_{inf} \times V} \tag{9.51}$$

With a secondary coil in induction units, the induction air states are needed.

$$T_{id} = ((n_r + 1) \times T_{dis} - T_c)/n_r \tag{9.52}$$

$$h_{id} = T_{id} \times c_p + W_{id} \times h_w \tag{9.53}$$

In winter, both the primary coil and the secondary induction coil are in dry mode, the humidity ratio does not change across the coils, and Eq. (9.51) is applicable.

In summer, the primary cooling coil and the secondary coil have four possible statuses: both dry, both wet, and one dry plus the other wet. The return air humidity W_r, the supply air humidity W_c, and the induction air humidity W_{id} are therefore coupled. If both coils are dry, Eq. (9.51) is utilized to evaluate the room air humidity.

With a wet primary cooling coil, the room air humidity is expressed as:

$$W_r = \frac{\frac{G_a}{\rho} \times (W_c + W_{id} \times n_r) + W + ACH_{inf} \times V \times W_o}{\frac{G_a}{\rho} \times (n_r + 1) + ACH_{inf} \times V} \tag{9.54}$$

If the primary coil is dry, it needs to be re-evaluated as:

$$W_r = \frac{\frac{G_a}{\rho} \times (\beta_{OA} + W_{id} \times n_r) + W + ACH_{inf} \times V \times W_o}{\frac{G_a}{\rho} \times (n_r + \beta_{OA}) + ACH_{inf} \times V} \tag{9.55}$$

The humidity of the induction air in Eqs. (9.54) and (9.55) is W_c if the secondary coil is wet, and W_r if the secondary coil is dry. After obtaining the room air humidity, Eq. (9.47) should be re-evaluated to solve for the mixed air humidity.

To this end, an hourly steady-state secondary system modeling structure is presented. It can be used to simulate constant air volume and variable air volume system, with or without terminal heating/cooling coils. The main air states and the thermal energy consumptions of the components can be predicted with this model.

9.1.4.3 Case Building Simulation and Results

A case building located in Omaha, Nebraska, USA, is selected to demonstrate the IAHU operation and its performance. The weather data, building information, and occupancy schedule are fed into the simulation engine to predict the building load. It is assumed that the room air for each controlled zone is well mixed and the dynamic condition is satisfied through the system's local control loops.

(1) Case building information

The case building has 12 floors above and one floor below the ground. The first two floors are mixed-use space, with lobby, courts and offices, and air-conditioned by one AHU. From the 3rd floor to the 12th floor is the open space office tower. The longitude of the building runs from north to south. Double-layer tinted glazes with aluminum frames are installed on the external walls. The shading is provided with manually adjustable internal blinds. The tower office area is suitable for an IAHU operation because of the typical system and space layout. The area is about 1756 m^2 (82.3 m × 21.3 m) per floor, with 95 % air conditioned. Figure 9.12 shows the outside view and inside view of the office building. The office tower is occupied by about 620 people and operated in average 60 h per week. A building automation system is deployed in the building and can be programmed for a desired advanced control.

Outside view from south east　　Inside view of an office

Fig. 9.12 Pictures of the case building

The tower area is conditioned by three AHUs (AHU-1, 2, and 3). The cooling and heating thermal sources, i.e., chilled water and steam, are provided and metered by a local company. AHU-1 system, a single duct VAV system with 60 electrical reheat equipped terminal boxes, serves the interior region. Two constant air systems (AHU-2 and AHU-3), with 495 induction terminals each, condition the perimeter space along the external wall. All supply and return fans in the three AHUs are equipped with VFDs. In the interior region, the conditioned air is evenly distributed and returned via the slot diffusers on the lighting panels. In the exterior region, the air is supplied into the space through the induction units. The room air is returned via the slot openings on the lighting panels.

(2) Operation information

The operation algorithm of the existing AHUs is conventional, as described below.

(1) *System ON/OFF:*

The system starts at 4:00 am and shuts down at 10:00 pm from Monday to Friday in normal operation. During hot summer and cold winter, the system is kept on constantly 24/7.

(2) *Supply air temperature:*

AHU-1 is a year-around cooling-only system serving the interior region. The discharge air temperature is set at 12.7 °C (55 °F) (adjustable). For AHU-2/3, the discharge air temperature is maintained at 12.7 °C (55 °F) (adjustable) when the OAT is greater than 21.1 °C (70 °F), and at 23.9 °C (75 °F) when the OAT is less than 12.7 °C (55 °F). A linear reset as depicted in Fig. 9.13 is applied when the OAT is in between.

(3) *Outside air intake:*

The systems maintain an OA intake no less than the minimum via the individual AHU to ensure the ventilation.

(4) *Economizer:*

A fixed dry bulb air temperature-based economizer is used for the systems. It is enabled when the OA temperature is below 21.1 °C (70 °F). The return air dampers and the OA dampers are controlled to maintain the mixed air temperature set point 12.7 °C (55 °F) (adjustable), provided the minimum OA intake is satisfied. Table 9.3 summarizes the original OA intake schema of the AHUs.

(3) Other inputs

The building and system information are collected from the building drawings, onsite observation, and control documentations. The designed indoor air temperature is 23.9 °C (75 °F) in summer and 22.2 °C (72 °F) in winter. The system design airflow rate for AHU-1 is 42.8 m^3/s (90696cfm) and 25.5 m^3/s (54000cfm) for

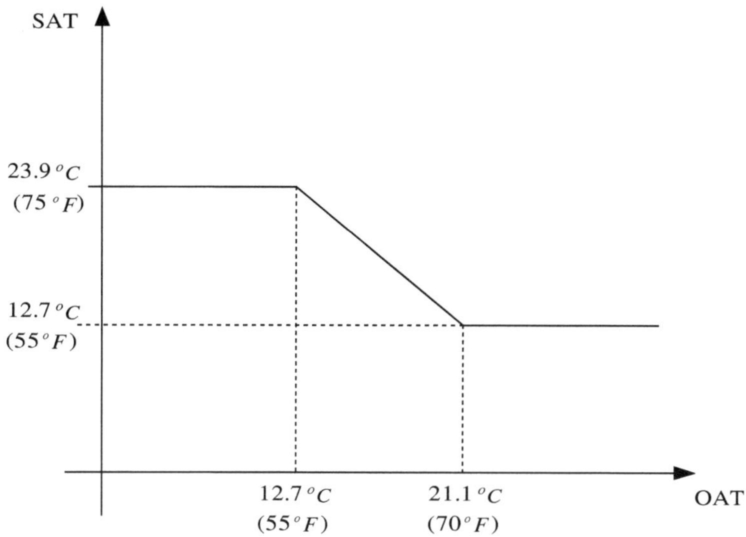

Fig. 9.13 Supply air temperature for AHU-2/3

Table 9.3 OA intake schedule in the AHUs

No.	OA condition	Interior OA ratio ($\beta_{OA,i}$)	Exterior OA ratio ($\beta_{OA,e}$)
A	$T_o \leq T_{c,i}$	$\max(\beta_{IAQ,e}, \beta_{eco,i})$	$\beta_{IAQ,e}$
B	$(T_{c,i}, T_{c,e}]$	1	$\max(\beta_{IAQ,dsg,e}, \beta_{eco,e})$
C	$(T_{e,c}, T_{eco}]$	1	1
D/E	$T_{eco} \leq T_o$	$\beta_{IAQ,i}$	$\beta_{IAQ,e}$

AHU-2/3, respectively. The minimum OA intake is 3.13 m^3/s (6636CFM) in AHU-1 and 3.43 m^3/s (7263CFM) in AHU-2/3. The minimum airflow rate ratio of AHU-1 is 30 % and AHU-2/3 100 %.

The other main simulation inputs of the building features are collected below in Table 9.4:

(4) Results and analysis

The hourly simulation is performed for the normal AHU and the improved IAHU operation with the control algorithms and parameters provided above. The partial load determination in condition D and E for decoupling operation of the two exterior region AHUs is based on the principle that no terminal reheat should be on for a given simulation hour. The weather data is TMY3 for EnergyPlus provided by the US Department of Energy [17]. Omaha locates in climate zone 5A and has cold winters and hot summers. The hottest temperature reaches above 37.7 °C (100 °F) while the lowest drops to −23.9 °C (−11 °F). In the occupied hours, it has the

Table 9.4 Main inputs of the building model

Building and system Info	
Conditioned floor area: 16,722.5 m^2 (180,000 ft^2). Interior region: 8318.5 m^2 (85,940 ft^2), exterior region: 8738.4 m^2 (94,060 ft^2)	
Walls *U* value: 0.842 W/m^2 K (0.148 BTU/h °F ft^2)	
East	1703.6 m^2 (18,337 ft^2)
West	2269 m^2 (24,423 ft^2)
North	2787 m^2 (30,000 ft^2)
South	2787 m^2 (30,000 ft^2)
Roof: 1672.3 m^2 (18,000 ft^2), U value: 0.559 W/m^2 K (0.099 BTU/h °F ft^2)	
Windows *U* value: 3.6 W/m^2 K (0.633 BTU/h °F ft^2)	
East	1630 m^2 (17,550 ft^2)
West	1160 m^2 (12,480 ft^2)
North	242 m^2 (2600 ft^2)
South	242 m^2 (2600 ft^2)
Maximum lighting power density: 15.61 W/m^2 (1.45 W/ft^2) Maximum equipment power density: 10.78 W/m^2 (1 W/ft^2) Maximum occupant's density: 0.0366/m^2 (0.0034/ft^2) Infiltration air change rate: 0 ACH	

Table 9.5 Hours of different OA conditions during occupied time [T_r set as 23.9 °C (75 °F)]

No.	OA condition	Hours	
A	$T_o \leq T_{c,i}$	1508	
B	$(T_{c,i}, T_{c,e}]$	224	
C	$(T_{c,e}, T_r], h_o < h_r$	169	
D	$(T_{c,e}, T_r], h_o \geq h_r$	488	448, mild weather
E	$T_o > T_r$	896	

longest duration for condition A and condition D/E. The distribution of the hours is grouped into A, B, C, and D/E for comparison in Table 9.5.

In Fig. 9.14, the thermal energy consumption of normal AHU and IAHU and the energy savings are plot against the OA temperature for this given building. As shown in the figure, when the OA temperature is below 10 °C (50 °F), the IAHU energy-saving capability gradually increases first and then decreases as the temperature drops. This is because the OA intake gradually decreases when the interior region slides in free cooling mode. The amount of OA intake from VAHU$_i$ satisfies both the interior region thermal load and the building's overall ventilation requirement at high and free cooling temperature. When the OA temperature keeps dropping until below a turning point, i.e., roughly −5 °C (23 °F), a supplementary

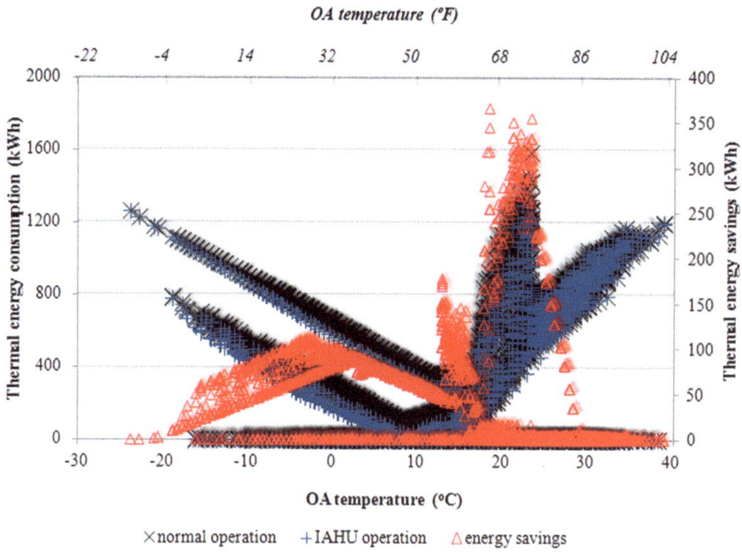

Fig. 9.14 Thermal energy performance comparison vs OA temperature

OA intake from the exterior region $VAHU_e$ is needed, and gradually increases in quantity to satisfy the overall building IAQ. The energy savings still exist but decrease along with the OA temperature thereafter.

Another temperature band with great energy savings is between 12.7 °C (55 °F) and 25 °C (77 °F) when the partial load is most likely to happen. The savings are mainly from the coil load decoupling operation. The stream of dehumidified cold air from the AHUs in normal operation has to be reheated up before it is distributed into the space when the partial load mode drops below the threshold of the system. A great amount of energy is saved by converting the normal operation AHUs into IAHU.

As collected in Table 9.5, condition A and D/E have the longest duration in the total occupied hours. An IAHU in this climate as Omaha might generate the most accumulative savings in these two scenarios. The simulation results in Fig. 9.15 show that, while the operations consume most energy in scenario A and D/E, the opportunity of improving energy efficiency for the scenarios is also obvious. The normal operation in scenario A consumes about 25 % more thermal energy than that of IAHU. About 110,000kWh (375MMBTU) thermal heating could be saved in this scenario, as plot in Fig. 9.16. While the accumulative energy savings in scenario B is small due to the short duration, it has high hourly energy savings (92 kWh/hr). The energy savings of applying IAHU can be much higher in a climate where B has longer duration. The savings in scenario D/E come from the coil latent and sensible load decoupling operation. IAHU consumes about 50,000 kWh

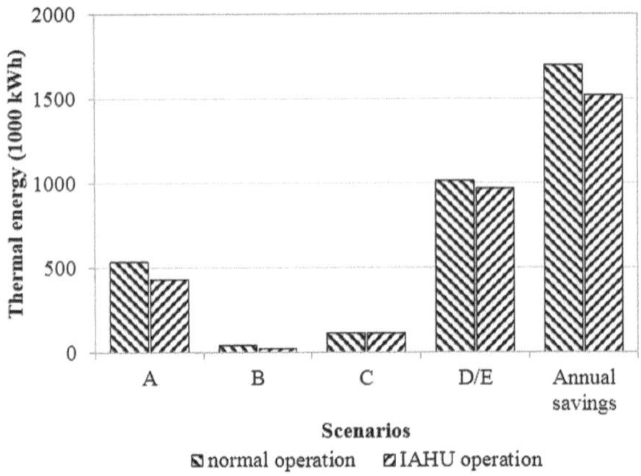

Fig. 9.15 Thermal energy performance comparison in scenarios

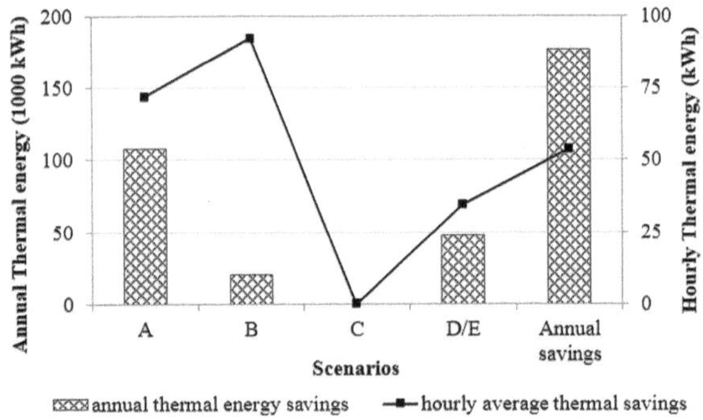

Fig. 9.16 Thermal energy savings in scenarios

(171MMBTU) less thermal energy than the normal operation by decoupling the sensible and latent load.

In the four scenarios shown in Fig. 9.16, condition B has the highest hourly saving potential, followed by condition A and then D/E. Since condition C has identical operation as normal AHUs, it does not yield any energy savings. By utilizing IAHU, the total thermal energy savings in this case building is 176,000 kWh (600MMBTU), accounting for about 10 % of the total thermal energy consumption. The hourly thermal savings reach 55kWh (187MBTU). The savings equal to around 11 kWh/m^2 year (3.5MBTU/ft^2 year) on the annual energy use intensity. Among the total annual savings in all the scenarios, roughly 80 % are saved as thermal energy for heating (or reheating) purpose. While heating in many office buildings is directly provided by steam, gas combustion, or electricity, the energy-saving effect of using IAHU on heating is conspicuous.

A set of parametric runs are also conducted with this model to evaluate the impact of different maximum circulation ratio x which is jointly determined by the space layout and system capacity ratios. Figure 9.17 gives the result comparison for

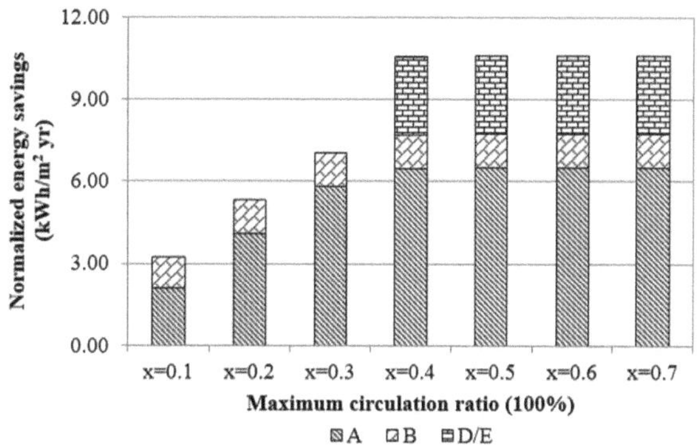

Fig. 9.17 Normalized energy saving of using IAHU for different maximum circulation ratio

an x changing between 0.1 and 0.7. In the simulations, the other parameters and inputs remain the same. If the system allows only up to 10 % supply air recirculation, the savings can be around 3 kWh/m^2 year (950BTU/ft^2 year). It is about one-third of the full potential when x is allowed up to 40 %. The savings increase when the limit is freed up from 0.1 to 0.4. No decoupling operation is conducted with x less than 0.4 since, in these conditions, the OA intake from the interior region of IAHU is much higher than that of the normal operation in the case building. The simulation also shows that, once the limit is allowed to be equal or above 0.4, no more energy savings can be achieved by loosening the circulation constraint.

9.2 Global Optimization Control

The global optimization model for the overall energy-saving control of the central air-conditioning system is to be developed in this section based on the local energy models of equipments. The method of decomposition–coordination, an effective way to settle the high-dimensional optimization problems, is introduced to solve the global model possibly concerned to a large-scale air-conditioning system in which a large number of decision variables are to be optimized. The global optimization model is then used in a central air-conditioning system, and the energy-saving potentials are analyzed based on the simulation results.

9.2.1 Model Development

The optimization model normally consists of objective function(s) and a series of constraint equations. The tasks of the optimization model mainly include two aspects as below: (1) Determining how many equipments (such as chiller, cooling tower, and

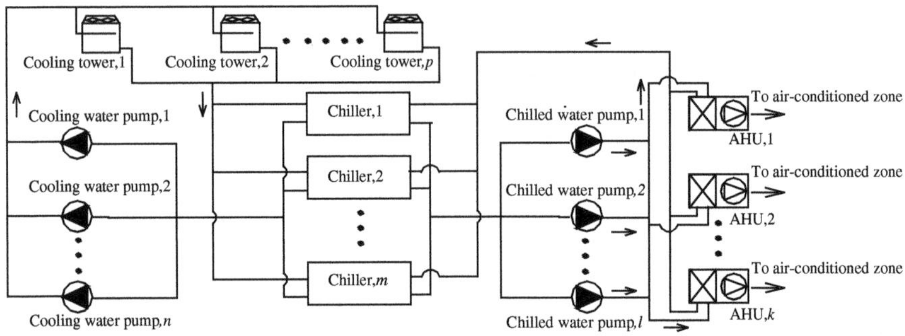

Fig. 9.18 Schematic for the central air-conditioning system

cooling/chilled water pump.) should be turned on to meet the cooling load requirements; (2) Acquiring optimal running conditions of the equipments aiming at minimum energy consumption. The first task can be regarded as an integer-programming problem with the control variables being the specific combination of running equipments, and the second is a nonlinear programming problem [18]. The central air-conditioning system is a large-scale complex system that consists of numerous coupled components. In order to make the problem more explicit, the local optimization models in terms of the type of equipment are presented separately, based on which the global one for the overall system is integrated.

As shown in Fig. 9.18, a conventional central air-conditioning system mainly includes chillers (*m* sets), cooling water pumps (*n* sets), chilled water pumps (*l* sets), cooling water towers (*p* sets), and AHUs (Air-handling Units, *k* sets). Thus, it can be divided into five subsystems in terms of the equipment categories. As shown in Fig. 9.19, the subsystems interact with each other mainly in the form of energy transfer through the working fluid (water).

The local optimization models of the subsystems are presented as follows.

(1) The local optimization model for chillers:

$$\begin{cases} N_{\text{chiller}}\left(C_{\text{chiller}}, G_{\text{w,c}}, t_{\text{w,c,E}}\right) = \min\left\{\sum\limits_{i=1}^{m} N_{\text{com},i}\left(t_{\text{e},i}, t_{\text{c},i}, r_i\right)\right\} \\ \text{s.t.} \quad C_{\text{chiller}} \geq [1, 1, \cdots, 1]_{m\times 1}^{\text{T}}[t_{\min}] \end{cases} \tag{9.56}$$

where

$$C_{\text{chiller}} = \left[t_{\text{e},1}, t_{\text{e},2}, \cdots t_{\text{e,m}}\right]^{\text{T}} \tag{9.56a}$$

$$N_{\text{com},i}\left(t_{\text{e},i}, t_{\text{c},i}, r_i\right) = \frac{Q_{\text{e},i}}{COP_i\left(t_{\text{e},i}, t_{\text{c},i}, r_i\right)} \tag{9.56b}$$

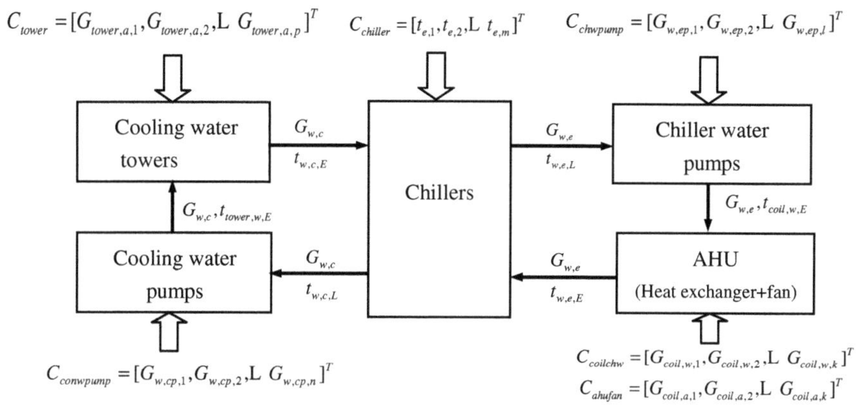

Fig. 9.19 Relationship among subsystems in the central air-conditioning system

$$t_{c,i} = f(t_{w,c,E}, G_{w,c,i}) \; i \in [1, m] \tag{9.56c}$$

COP_i denotes the coefficient of performance of the ith chiller; t_{min} is the lower limit of the evaporating temperature of chillers for safety consideration. The entering water temperature of condenser, $t_{w,c,E}$, is related to the outlet water temperature of cooling water tower. Assuming all the cooling water towers burden equal amount of cooling water, $t_{w,c,E}$ can be calculated by Eqs. (9.57)–(9.60).

$$t_{w,c,E} = \sum_{i=1}^{p} (t_{tower,w,L})_i \bigg/ p + \Delta t_{conw,loss,1} \tag{9.57}$$

$$(t_{tower,w,L})_i = f\left(G_{tower,a,i}, \frac{1}{p} \sum_{i=1}^{m} (G_{w,c,i}), h_{tower,as,E}, h_{tower,a,E}, t_{tower,w,E} \right) \quad i \in [1, p] \tag{9.58}$$

$$t_{tower,w,E} = \sum_{i=1}^{m} \left[G_{w,c,i} \cdot (t_{w,c,L})_i \right] \bigg/ \sum_{i=1}^{m} (G_{w,c,i}) + \Delta t_{conw,loss,2} \tag{9.59}$$

$$(t_{w,c,L})_i = \frac{Q_{c,i}}{c_w G_{w,c,i}} + t_{w,c,E} \quad i \in [1, m] \tag{9.60}$$

$\Delta t_{conw,loss,1}$ and $\Delta t_{conw,loss,2}$ is the change of cooling water temperature due to heat loss of water pipes between the towers and the chillers.

(2) The local optimization model for chilled water pumps:

$$\begin{cases} N_{\mathrm{ep}}\left(C_{\mathrm{chwpump}}\right) = \min\left\{\sum_{i=1}^{l} N_{\mathrm{ep},i}\left(G_{\mathrm{w,ep},i}\right)\right\} \\ \text{s.t.} \quad C_{\mathrm{chwpump}} \le \begin{bmatrix} 1 & 0 & 0 & \cdots & 0 \\ 0 & 1 & 0 & \cdots & 0 \\ \vdots & \vdots & \vdots & \vdots & \vdots \\ 0 & 0 & 0 & 0 & 1 \end{bmatrix}_{l\times l} \left[G_{\mathrm{w,ep,max},1}, G_{\mathrm{w,ep,max},2}, \cdots G_{\mathrm{w,ep,max},l}\right]_{l\times 1}^{\mathrm{T}} \\ C_{\mathrm{chwpump}} \ge \begin{bmatrix} 1 & 0 & 0 & \cdots & 0 \\ 0 & 1 & 0 & \cdots & 0 \\ \vdots & \vdots & \vdots & \vdots & \vdots \\ 0 & 0 & 0 & 0 & 1 \end{bmatrix}_{l\times l} \left[G_{\mathrm{w,ep,min},1}, G_{\mathrm{w,ep,min},2}, \cdots G_{\mathrm{w,ep,min},l}\right]_{l\times 1}^{\mathrm{T}} \\ G_{\mathrm{w,e}} = [1, 1, \cdots 1]_{1\times l} C_{\mathrm{chwpump}} \end{cases} \tag{9.61}$$

where $C_{\mathrm{chwpump}} = \left[G_{\mathrm{w,ep},1}, G_{\mathrm{w,ep},2}, \cdots, G_{\mathrm{w,ep,l}}\right]^{\mathrm{T}}$; $G_{\mathrm{w,ep},i}$, $G_{\mathrm{w,ep,max},i}$ and $G_{\mathrm{w,ep,min},i}$ denote, respectively, the actual, the maximum, and the minimum water flow rate of the ith ($i \in [1, l]$) chilled water pump.

(3) The local optimization model for cooling water pumps

$$\begin{cases} N_{\mathrm{cp}}\left(C_{\mathrm{conwpump}}\right) = \min\left\{\sum_{i=1}^{n} N_{\mathrm{cp},i}\left(G_{\mathrm{w,cp},i}\right)\right\} \\ \text{s.t.} \quad C_{\mathrm{conwpump}} \le \begin{bmatrix} 1 & 0 & 0 & \cdots & 0 \\ 0 & 1 & 0 & \cdots & 0 \\ \vdots & \vdots & \vdots & \vdots & \vdots \\ 0 & 0 & 0 & 0 & 1 \end{bmatrix}_{n\times n} \left[G_{\mathrm{w,cp,max},1}, G_{\mathrm{w,cp,max},2}, \cdots G_{\mathrm{w,cp,max},l}\right]_{n\times 1}^{\mathrm{T}} \\ C_{\mathrm{conwpump}} \ge \begin{bmatrix} 1 & 0 & 0 & \cdots & 0 \\ 0 & 1 & 0 & \cdots & 0 \\ \vdots & \vdots & \vdots & \vdots & \vdots \\ 0 & 0 & 0 & 0 & 1 \end{bmatrix}_{n\times n} \left[G_{\mathrm{w,cp,min},1}, G_{\mathrm{w,cp,min},2}, \cdots G_{\mathrm{w,cp,min},l}\right]_{n\times 1}^{\mathrm{T}} \\ G_{\mathrm{w,c}} = [1, 1, \cdots 1]_{1\times n} C_{\mathrm{conwpump}} \end{cases} \tag{9.62}$$

where $C_{\mathrm{conwpump}} = \left[G_{\mathrm{w,cp},1}, G_{\mathrm{w,cp},2}, \cdots, G_{\mathrm{w,cp},n}\right]^{\mathrm{T}}$; $G_{\mathrm{w,cp},i}$, $G_{\mathrm{w,cp,max},i}$, and $G_{\mathrm{w,cp,min},i}$ denote, respectively, the actual, the maximum, and the minimum water flow rate of the ith ($i \in [1, n]$) cooling water pump.

(4) The local optimization model for cooling water towers

$$\begin{cases} N_{\text{towerfan}}(C_{\text{tower}}) = \min\left\{\sum_{i=1}^{p} N_{\text{towerfan},i}(G_{\text{tower},a,i})\right\} \\ \text{s.t.} \\ C_{\text{tower}} \le \begin{bmatrix} 1 & 0 & 0 & \cdots & 0 \\ 0 & 1 & 0 & \cdots & 0 \\ \vdots & \vdots & \vdots & \vdots & \vdots \\ 0 & 0 & 0 & 0 & 1 \end{bmatrix}_{p\times p} [G_{\text{tower},a,\max,1}, G_{\text{tower,a,max},2}, \cdots G_{\text{tower,a,max},p}]_{p\times 1}^{\mathrm{T}} \\ C_{\text{tower}} \ge \begin{bmatrix} 1 & 0 & 0 & \cdots & 0 \\ 0 & 1 & 0 & \cdots & 0 \\ \vdots & \vdots & \vdots & \vdots & \vdots \\ 0 & 0 & 0 & 0 & 1 \end{bmatrix}_{p\times p} [G_{\text{tower,a,min},1}, G_{\text{tower,a,min},2}, \cdots G_{\text{tower,a,min},p}]_{p\times 1}^{\mathrm{T}} \end{cases} \tag{9.63}$$

where $C_{\text{tower}} = [G_{\text{tower,a},1}, G_{\text{tower,a},2}, \cdots, G_{\text{tower,a,p}}]^{\mathrm{T}}$; $G_{\text{tower,a},i}$, $G_{\text{tower,a,max},i}$, and $G_{\text{tower,a,min},i}$.

denote, respectively, the actual, the maximum, and the minimum air flow rate of the ith ($i \in [1, p]$) cooling water tower.

(5) The local optimization model for AHUs

$$\begin{cases} N_{\text{ahufan}}(C_{\text{ahufan}}) = \min\left\{\sum_{i=1}^{k} N_{\text{ahufan},i}(G_{\text{coil,a},i})\right\} \\ \text{s.t.} \quad C_{\text{ahufan}} \le \begin{bmatrix} 1 & 0 & 0 & \cdots & 0 \\ 0 & 1 & 0 & \cdots & 0 \\ \vdots & \vdots & \vdots & \vdots & \vdots \\ 0 & 0 & 0 & 0 & 1 \end{bmatrix}_{k\times k} [G_{\text{coil,a,max},1}, G_{\text{coil,a,max},2}, \cdots G_{\text{coil,a,max},k}]_{k\times 1}^{\mathrm{T}} \\ C_{\text{ahufan}} \ge \begin{bmatrix} 1 & 0 & 0 & \cdots & 0 \\ 0 & 1 & 0 & \cdots & 0 \\ \vdots & \vdots & \vdots & \vdots & \vdots \\ 0 & 0 & 0 & 0 & 1 \end{bmatrix}_{k\times k} [G_{\text{coil,a,min},1}, G_{\text{coil,a,min},2}, \cdots G_{\text{coil,a,min},k}]_{k\times 1}^{\mathrm{T}} \\ M_{\Delta t}^{\mathrm{T}} \le [1, 1, \cdots, 1]_{k\times 1}^{\mathrm{T}} \Delta t_{\text{air,max}} \\ C_{\text{ahufan}} M_{\Delta t}[c_{\text{a}}] = \varsigma \cdot [1, 1, \cdots 1]_{1\times k} \cdot Q_{\text{room}} \\ G_{\text{w,e}} = [1, 1, \cdots 1]_{1\times k} C_{\text{ahuchw}} \end{cases} \tag{9.64}$$

where

$$C_{\text{ahufan}} = [G_{\text{coil,a},1}, G_{\text{coil,a},2}, \ldots, G_{\text{coil,a},k}]^{\mathrm{T}} \tag{9.64a}$$

$$C_{\text{ahuchw}} = [G_{\text{coil,w},1}, G_{\text{coil,w},2}, \ldots, G_{\text{coil,w},k}]^{\mathrm{T}} \tag{9.64b}$$

$$M_{\Delta t} = \left[(t_{\text{air,room}} - t_{\text{coil,a,L}})_1, (t_{\text{air,room}} - t_{\text{coil,a,L}})_2, \ldots, (t_{\text{air,room}} - t_{\text{coil,a,L}})_k\right]_{1\times k} \tag{9.64c}$$

$$\varsigma = [\varsigma_1, \varsigma_2, \ldots \varsigma_k]_{k\times 1}^{\mathrm{T}} \tag{9.64d}$$

$$Q_{\text{room}} = \left[Q_{\text{room},1}, Q_{\text{room},2}, \cdots Q_{\text{room,k}}\right]^{\text{T}}_{k\times 1} \tag{9.64e}$$

$G_{\text{coil,a,max},k}$ is the maximum air flow rate of fan available in the kth AHU; $G_{\text{coil,a,min},k}$ is the minimum air flow rate that is determined by the IAQ (Indoor Air Quality) demand; $\Delta t_{\text{air,max}}$ is the maximum temperature difference between the supply air and the indoor air.

ς_i is the ratio of sensible cooling load of the ith $(i \in [1,k])$ AHU, which is defined as:

$$\varsigma_i = \frac{Q_{\text{tload},i}}{Q_{\text{room},i}} \tag{9.65}$$

where $Q_{\text{tload},i}$ and $Q_{\text{room},i}$ denote, respectively, sensible and total cooling load burdened by the AHU.

The supply air temperature of the *i*th $(i \in [1,k])$ AHU, $(t_{\text{coil,a,L}})_i$, can expressed as:

$$(t_{\text{coil,a,L}})_i = f(G_{\text{coil,a},i}, G_{\text{coil,w},i}, t_{\text{coil,a,E}}, t_{\text{coil,w,E}}) \tag{9.66}$$

It can be calculated by the heat exchanger model. The critical step is the determination of the entering water temperature, $t_{\text{coil,w,E}}$, of the cooling coil in the AHU. Presuming the chilled water is allocated evenly among all the cooling coils, $t_{\text{coil,w,E}}$ can be obtained by using Eqs. (9.67)–(9.69):

$$t_{\text{coil,w,E}} = \frac{\sum_{i=1}^{m} (t_{\text{w,e,L}})_i}{m} + \Delta t_{\text{chw,loss},1} \tag{9.67}$$

$$(t_{\text{w,e,L}})_i = t_{\text{w,e,E}} - \frac{Q_{\text{e},i}}{c_{\text{w}} G_{\text{w,e},i}} \quad i \in [1,m] \tag{9.68}$$

$$t_{\text{w,e,E}} = \frac{\sum_{i=1}^{k} \left[G_{\text{coil,w},i} \cdot (t_{\text{w,coil,L}})_i\right]}{\sum_{i=1}^{m} (G_{\text{w,e},i})} + \Delta t_{\text{chw,loss},2} \tag{9.69}$$

In Eqs. (9.67) and (9.69), $\Delta t_{\text{chw,loss},1}$ and $\Delta t_{\text{chw,loss},2}$ are temperature change of chilled water due to heat loss of water transmission pipes. The constrain equation, $C_{\text{ahufan}} M_{\Delta t}[c_{\text{a}}] = \varsigma \cdot [1, 1, \ldots, 1]_{1\times k} \cdot Q_{\text{room}}$, means the cooling capacity of AHU should be adjusted to meet the cooling demand of air-conditioned room (Q_{room}). For the model solving, the cooling load of air-conditioned room, Q_{room}, must be given in advance. Taking into account the lagging effect of control system, the predicted air-conditioning load by forecasting model is used here.

Combining the above local models, the global optimization model of central air-conditioning system can be summarized as below:

$$\begin{cases} N_{\mathrm{sys}} = \min \left\{ \begin{array}{l} \sum_{i=1}^{m} \left[N_{\mathrm{com},i}\left(t_{\mathrm{e},i}, t_{\mathrm{c},i}\left(G_{\mathrm{w,cp}}, C_{\mathrm{tower}}\right), r_i\right)\right] + \sum_{i=1}^{l} N_{\mathrm{ep},i}\left(G_{\mathrm{w,ep},i}\right) + \\ \sum_{i=1}^{n} N_{\mathrm{cp},i}\left(G_{\mathrm{w,cp},i}\right) + \sum_{i=1}^{p} N_{\mathrm{towerfan,i}}\left(G_{\mathrm{tower,a},i}\right) + \sum_{i=1}^{k} N_{\mathrm{ahufan},i}\left(G_{\mathrm{coil,a},i}\right) \end{array} \right\} \\ \mathrm{s.t.} \quad C_{\mathrm{chiller}} \geq [1, 1, \cdots, 1]_{m\times 1}^{\mathrm{T}} [t_{\min}] \\ \begin{bmatrix} C_{\mathrm{chwpump}} \\ C_{\mathrm{conwpump}} \\ C_{\mathrm{tower}} \\ C_{\mathrm{ahufan}} \end{bmatrix} \leq \begin{bmatrix} [\Lambda]_{l\times l} & 0 & 0 & 0 \\ 0 & [\Lambda]_{n\times n} & 0 & 0 \\ 0 & 0 & [\Lambda]_{p\times p} & 0 \\ 0 & 0 & 0 & [\Lambda]_{k\times k} \end{bmatrix} \begin{bmatrix} G_{w,ep,\mathrm{max}} \\ G_{w,cp,\mathrm{max}} \\ G_{\mathrm{tower,a,max}} \\ G_{\mathrm{coil,a,max}} \end{bmatrix} \\ \begin{bmatrix} C_{\mathrm{chwpump}} \\ C_{\mathrm{conwpump}} \\ C_{\mathrm{tower}} \\ C_{\mathrm{ahufan}} \end{bmatrix} \geq \begin{bmatrix} [\Lambda]_{l\times l} & 0 & 0 & 0 \\ 0 & [\Lambda]_{n\times n} & 0 & 0 \\ 0 & 0 & [\Lambda]_{p\times p} & 0 \\ 0 & 0 & 0 & [\Lambda]_{k\times k} \end{bmatrix} \begin{bmatrix} G_{\mathrm{w,ep,min}} \\ G_{\mathrm{w,cp,min}} \\ G_{\mathrm{tower,a,min}} \\ G_{\mathrm{coil,a,min}} \end{bmatrix} \\ M_{\Delta t}^{\mathrm{T}} \leq [1, 1, \ldots, 1]_{k\times 1}^{\mathrm{T}} \Delta t_{\mathrm{air,max}} \\ C_{\mathrm{ahufan}} M_{\Delta t} [c_a] = \varsigma \cdot [1, 1, \cdots 1]_{1\times k} \cdot Q_{\mathrm{room}} \\ \sum_{i=1}^{m} \left(G_{\mathrm{w,e},i}\right) = [1, 1, \cdots 1]_{1\times l} C_{\mathrm{chwpump}} = [1, 1, \cdots 1]_{1\times k} C_{\mathrm{ahuchw}} \\ \sum_{i=1}^{m} \left(G_{\mathrm{w,c},i}\right) = [1, 1, \cdots 1]_{1\times n} C_{\mathrm{conwpump}} \end{cases} \tag{9.70}$$

where

$$[\Lambda] = \begin{bmatrix} 1 & 0 & 0 & \cdots & 0 \\ 0 & 1 & 0 & \cdots & 0 \\ \vdots & \vdots & \vdots & \vdots & \vdots \\ 0 & 0 & 0 & 0 & 1 \end{bmatrix} \tag{9.70a}$$

9.2.2 *Decomposition–Coordination Algorithm for Model Solution*

It may be challengeable to solve a large-scale as well as highly nonlinear optimization problem using the conventional methods (e.g., the direct search simplex method and the gradient method). Even significant increases in computational power are not sufficient to offset the ever-increasing complexity of energy system such as the central air-conditioning system. Therefore, it is necessary to employ a more efficient and reliable method to solve Eq. (9.70) in which a larger number of variables are to be determined. The decomposition techniques have been widely studied in the mathematical programming [19] and have been proved to be an effective way to globally solve the highly dimensional optimization problems [20–22].

The main idea of the decomposition method is to split the large-scale system into many weakly coupled subsystems, partition variable or constraint sets to isolate easier sub-problems and decentralize global optimal decisions among local decision levels [23]. The decomposition of the collocation matrix is now performed element by element that can reduce the dimension of the original problem and allow us to solve the sub-problems with a much smaller size than the overall problem in parallel, thus reducing the storage requirements and the computational effort.

The critical step of the decomposition method is the choice of coordination variables. As for the central air-conditioning system, the evaporating temperatures of chillers are chosen as the coordination variables because of their extremely important influence on the overall energy consumption of the whole system. In addition, the whole system can be easily split into two weakly coupled subsystems as the evaporating temperatures of chillers are temporarily fixed. Therefore, Eq. (9.70) can be divided into the two sub-problems, MOP1 (Eq. 9.71) and MOP2 (Eq. 9.72), if the evaporating temperatures of chillers, $[t_{e,1}, t_{e,2}, \ldots, t_{e,m}]^{T}$, are given in advance.

$$\mathrm{MOP1} = \begin{cases} \min\left\{ \sum_{i=1}^{l} N_{ep,i}\left(G_{w,ep,i}\right) + \sum_{i=1}^{k} N_{ahufan,i}\left(G_{coil,a,i}\right) \right\} \\ \text{s.t.} \quad \begin{bmatrix} C_{chwpump} \\ C_{ahufan} \end{bmatrix} \leq \begin{bmatrix} [\Lambda]_{l\times l} & 0 \\ 0 & [\Lambda]_{k\times k} \end{bmatrix} \begin{bmatrix} G_{w,ep,max} \\ G_{coil,a,max} \end{bmatrix} \\ \begin{bmatrix} C_{chwpump} \\ C_{ahufan} \end{bmatrix} \geq \begin{bmatrix} [\Lambda]_{l\times l} & 0 \\ 0 & [\Lambda]_{k\times k} \end{bmatrix} \begin{bmatrix} G_{w,ep,min} \\ G_{coil,a,min} \end{bmatrix} \\ M_{\Delta t}^{T} \leq [1, 1, \cdots, 1]_{k\times 1}^{T} \Delta t_{air,max} \\ C_{ahufan} M_{\Delta t} [c_{a}] = \varsigma \cdot [1, 1, \cdots 1]_{1\times k} \cdot Q_{room} \\ \sum_{i=1}^{m} \left(G_{w,e,i}\right) = [1, 1, \cdots 1]_{1\times l} C_{chwpump} = [1, 1, \cdots 1]_{1\times k} C_{ahuchw} \end{cases} \tag{9.71}$$

$$\mathrm{MOP2} = \begin{cases} \min\left\{ \begin{array}{l} \sum_{i=1}^{m} \left[N_{com,i}\left(t_{e,i}, t_{c,i}\left(G_{w,cp}, C_{tower}\right), r_{i}\right)\right] + \\ \sum_{i=1}^{n} N_{cp,i}\left(G_{w,cp,i}\right) + \sum_{i=1}^{p} N_{towerfan,i}\left(G_{tower,a,i}\right) \end{array} \right\} \\ \text{s.t.} \quad \begin{bmatrix} C_{conwpump} \\ C_{tower} \end{bmatrix} \leq \begin{bmatrix} [\Lambda]_{n\times n} & 0 \\ 0 & [\Lambda]_{p\times p} \end{bmatrix} \begin{bmatrix} G_{w,cp,max} \\ G_{tower,a,max} \end{bmatrix} \\ \begin{bmatrix} C_{conwpump} \\ C_{tower} \end{bmatrix} \geq \begin{bmatrix} [\Lambda]_{n\times n} & 0 \\ 0 & [\Lambda]_{p\times p} \end{bmatrix} \begin{bmatrix} G_{w,cp,min} \\ G_{tower,a,min} \end{bmatrix} \\ \sum_{i=1}^{m} \left(G_{w,c,i}\right) = [1, 1, \cdots 1]_{1\times n} C_{conwpump} \end{cases} \tag{9.72}$$

The sub-problem MOP1 is to search the optimal running states of the chilled water pumps and the optimal flow rates of water and air passing through the cooling coils aiming at the minimum energy consumptions of the chilled water pumps and the fans in AHUs. The sub-problem MOP2 is to find the optimal running states of the cooling water pumps and towers to achieve the minimum energy consumptions of these equipments.

Usually, the sub-problem MOP2 concerns about a small number of decision variables. So, the conventional method is used for the MOP2 solution. When the total cooling water flow rate, $G_{w,c}$, is chosen as the coordination variable, the sub-problem MOP2 can be further decomposed as MOP2(1) and MOP2(2), two sub-problems with much smaller size that are easy to be solved by using the conventional methods [e.g., the penalty function method for solving the MOP2(1) and the direct search method for the MOP2(2)]. The calculation flowchart for the sub-problem MOP2 is shown in Fig. 9.20.

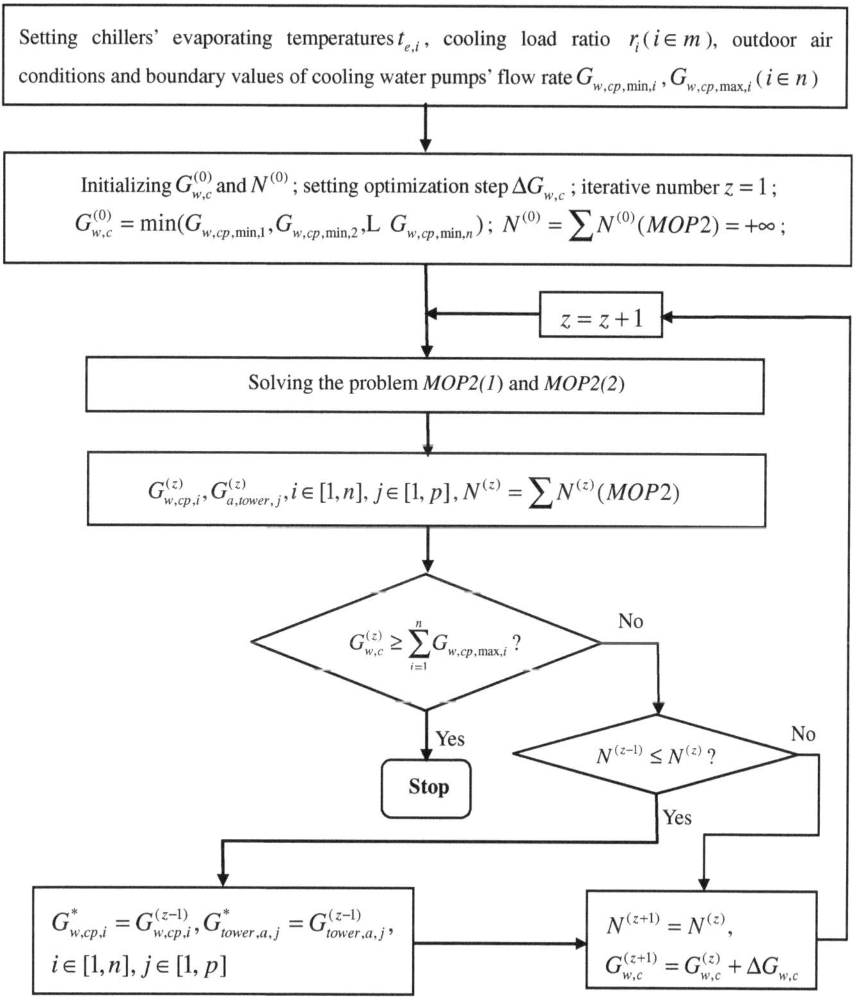

Fig. 9.20 Calculation flowchart of the direct search method for problem MOP2

$$\mathrm{MOP2(1)} = \begin{cases} \min\left\{\sum_{i=1}^{n} N_{\mathrm{cp},i}\left(G_{\mathrm{w,cp},i}\right)\right\} \\ \text{s.t.} \quad [\Lambda]_{n\times n} G_{\mathrm{w,cp,min}} \le C_{\mathrm{conwpump}} \le [\Lambda]_{n\times n} G_{\mathrm{w,cp,max}} \\ \sum_{i=1}^{m}\left(G_{\mathrm{w,c},i}\right) = [1, 1, \cdots 1]_{1\times n} C_{\mathrm{conwpump}} \end{cases} \tag{9.73}$$

$$\mathrm{MOP2(2)} = \begin{cases} \min\left\{\sum_{i=1}^{m}\left[N_{\mathrm{com},i}\left(t_{\mathrm{e},i}, t_{\mathrm{c},i}\left(G_{\mathrm{w,cp}}, C_{\mathrm{tower}}\right), r_i\right)\right] + \sum_{i=1}^{p} N_{\mathrm{towerfan},i}\left(G_{\mathrm{tower,a},i}\right)\right\} \\ \text{s.t.} \quad [\Lambda]_{p\times p} G_{\mathrm{tower,a,min}} \le [C_{tower}] \le [\Lambda]_{p\times p} G_{tower,a,\mathrm{max}} \end{cases} \tag{9.74}$$

The sub-problem MOP1, however, may arise the high-dimensional problem due to the possibly large number of AHUs (each AHU has three decision variables to be optimized) included in the system. The main calculation procedures are clarified as below.

To begin with, the sub-problem MOP1 must be converted into a convex problem equivalent to the original one (the sub-problem MOP1) using the penalty function method [23]. Introducing the penalty factor, ρ ($\rho \gg 0$), and the relaxation vectors for all the decision variables.

($\sigma_{\Delta t,\mathrm{max}} = [\sigma_{\Delta t}, \sigma_{\Delta t}, \cdots, \sigma_{\Delta t}]^{\mathrm{T}}_{k\times 1}$, $\sigma_{\mathrm{w,ep,max}} = [\sigma_{\mathrm{w,ep,max},1}, \sigma_{\mathrm{w,ep,max},2}, \cdots, \sigma_{\mathrm{w,ep,max},l}]^{\mathrm{T}}$, $\sigma_{\mathrm{w,ep,min}} = [\sigma_{\mathrm{w,ep,min},1}, \sigma_{\mathrm{w,ep,min},2}, \cdots, \sigma_{\mathrm{w,ep,min},l}]^{\mathrm{T}}$, $\sigma_{\mathrm{coil,a,max}} = [\sigma_{\mathrm{coil,a,max},1}, \sigma_{\mathrm{coil,a,max},2}, \cdots, \sigma_{\mathrm{coil,a,max},k}]^{\mathrm{T}}$, $\sigma_{\mathrm{coil,a,min}} = [\sigma_{\mathrm{coil,a,min},1}, \sigma_{\mathrm{coil,a,min},2}, \cdots, \sigma_{\mathrm{coil,a,min},k}]^{\mathrm{T}}$), the sub-problem MOP1with inequality constraints is equivalent to the locally convex problem MOP1(E) with equality constraints.

$$\mathrm{MOP1(E)} = \begin{cases} \min\begin{cases} \sum_{i=1}^{l} N_{\mathrm{ep},i}\left(G_{\mathrm{w,ep},i}\right) + \sum_{i=1}^{k} N_{\mathrm{ahufan},i}\left(G_{\mathrm{coil,a},i}\right) + \rho\left\|C_{\mathrm{chwpump}} + \sigma_{\mathrm{w,ep,max}} - G_{\mathrm{w,ep,max}}\right\|^2 \\ + \rho\left\|C_{\mathrm{chwpump}} - \sigma_{\mathrm{w,ep,min}} - G_{\mathrm{w,ep,min}}\right\|^2 + \rho\left\|C_{\mathrm{ahufan}} + \sigma_{\mathrm{coil,a,max}} - G_{\mathrm{coil,w,max}}\right\|^2 + \\ \rho\left\|C_{\mathrm{ahufan}} - \sigma_{\mathrm{coil,a,min}} - G_{\mathrm{coil,w,min}}\right\|^2 + \rho\left\|M_{\Delta t}^{T} + \sigma_{\Delta t,\mathrm{max}} - [1, 1, \ldots, 1]^{\mathrm{T}}_{k\times 1} \Delta t_{\mathrm{air,max}}\right\|^2 \\ + \rho\left\|C_{\mathrm{ahufan}} M_{\Delta t}[c_{\mathrm{a}}] - \varsigma \cdot [1, 1, \ldots, 1]_{1\times k} \cdot Q_{\mathrm{room}}\right\|^2 \end{cases} \\ \text{s.t.} \quad \begin{bmatrix} C_{\mathrm{chwpump}} \\ C_{\mathrm{ahufan}} \end{bmatrix} + \begin{bmatrix} \sigma_{\mathrm{w,ep,max}} \\ \sigma_{\mathrm{coil,a,max}} \end{bmatrix} - \begin{bmatrix} [\Lambda]_{l\times l} & 0 \\ 0 & [\Lambda]_{k\times k} \end{bmatrix} \begin{bmatrix} G_{\mathrm{w,ep,max}} \\ G_{\mathrm{coil,a,max}} \end{bmatrix} = 0 \\ \begin{bmatrix} C_{\mathrm{chwpump}} \\ C_{\mathrm{ahufan}} \end{bmatrix} - \begin{bmatrix} \sigma_{\mathrm{w,ep,min}} \\ \sigma_{\mathrm{coil,a,min}} \end{bmatrix} - \begin{bmatrix} [\Lambda]_{l\times l} & 0 \\ 0 & [\Lambda]_{k\times k} \end{bmatrix} \begin{bmatrix} G_{\mathrm{w,ep,min}} \\ G_{\mathrm{coil,a,min}} \end{bmatrix} = 0 \\ M_{\Delta t}^{T} + \sigma_{\Delta t,\mathrm{max}} - [1, 1, \ldots, 1]^{\mathrm{T}}_{k\times 1} \Delta t_{\mathrm{air,max}} = 0 \\ C_{ahufan} M_{\Delta t}[c_a] - \varsigma \cdot [1, 1, \ldots, 1]_{1\times k} \cdot Q_{\mathrm{room}} = 0 \\ [1, 1, \ldots, 1]_{1\times l} C_{\mathrm{chwpump}} - [1, 1, \ldots, 1]_{1\times k} C_{\mathrm{ahuchw}} = 0 \end{cases} \tag{9.75}$$

The expression of the augmented Lagrangian function associated to the problem *MOP1(E)* can be written as:

$$
\begin{aligned}
& L_{\mathrm{MOP1(E)}}\left(\begin{array}{l} C_{\mathrm{chwpump}}, C_{\mathrm{ahufan}}, C_{\mathrm{ahuchw}}, \lambda_{\mathrm{ahu}}, \mu_{\mathrm{chwpump,max}}, \mu_{\mathrm{chwpump,min}}, \mu_{\mathrm{ahufan,max}}, \\ \mu_{\mathrm{ahufan,min}}, \mu_{\Delta t,\mathrm{max}}, \sigma_{\mathrm{w,ep,max}}, \sigma_{\mathrm{w,ep,min}}, \sigma_{\mathrm{coil,a,max}}, \sigma_{\mathrm{coil,a,min}}, \sigma_{\Delta t,\mathrm{max}}, \eta, \rho \end{array}\right) \\
& = \sum_{i=1}^{l} N_{\mathrm{ep},i}\left(G_{\mathrm{w,ep},i}\right) + \sum_{i=1}^{k} N_{\mathrm{ahufan},i}\left(G_{\mathrm{a,coil},i}\right) + \rho\left\|C_{\mathrm{chwpump}} + \sigma_{\mathrm{w,ep,max}} - G_{\mathrm{w,ep,max}}\right\|^2 \\
& \quad + \rho\left\|C_{\mathrm{chwpump}} - \sigma_{\mathrm{w,ep,min}} - G_{\mathrm{w,ep,min}}\right\|^2 + \rho\left\|C_{\mathrm{ahufan}} + \sigma_{\mathrm{coil,a,max}} - G_{\mathrm{coil,a,max}}\right\|^2 \\
& \quad + \rho\left\|C_{\mathrm{ahufan}} - \sigma_{\mathrm{coil,a,min}} - G_{\mathrm{coil,a,min}}\right\|^2 + \rho\left\|M_{\Delta t}^T + \sigma_{\Delta t,\mathrm{max}} - [1,1,\ldots,1]_{k\times 1}^{\mathrm{T}} \Delta t_{\mathrm{air,max}}\right\|^2 \\
& \quad + \rho\left\|C_{\mathrm{ahufan}} M_{\Delta t}[c_{\mathrm{a}}] - \varsigma\cdot[1,1,\ldots,1]_{1\times k}\cdot Q_{\mathrm{room}}\right\|^2 + \lambda_{\mathrm{ahu}}\left[C_{\mathrm{ahufan}} M_{\Delta t}[c_a] - \varsigma\cdot[1,1,\ldots,1]_{1\times k}\cdot Q_{\mathrm{room}}\right] \\
& \quad + \mu_{\mathrm{chwpump,max}}\left[C_{\mathrm{chwpump}} + \sigma_{\mathrm{w,ep,max}} - G_{\mathrm{w,ep,max}}\right] + \mu_{\mathrm{chwpump,min}}\left[C_{\mathrm{chwpump}} - \sigma_{\mathrm{w,ep,min}} - G_{\mathrm{w,ep,min}}\right] \\
& \quad + \mu_{\mathrm{ahufan,max}}\left[C_{\mathrm{ahufan}} + \sigma_{\mathrm{coil,a,max}} - G_{\mathrm{coil,a,max}}\right] + \mu_{\mathrm{ahufan,min}}\left[C_{\mathrm{ahufan}} - \sigma_{\mathrm{coil,a,min}} - G_{\mathrm{coil,a,min}}\right] \\
& \quad + \mu_{\Delta t,\mathrm{max}}\left[M_{\Delta t}^T + \sigma_{\Delta t,\mathrm{max}} - [1,1,\cdots 1]_{k\times 1}^T \Delta t_{\mathrm{air,max}}\right] + \eta\left([1,1,\cdots 1]_{1\times l} C_{\mathrm{chwpump}} - [1,1,\ldots,1]_{1\times k} C_{\mathrm{ahuchw}}\right)
\end{aligned} \tag{9.76}
$$

where

$$\lambda_{\mathrm{ahu}} = \left[\lambda_{\mathrm{ahu},1}, \lambda_{\mathrm{ahu},2}, \ldots, \lambda_{\mathrm{ahu},k}\right] \tag{9.76a}$$

$$\mu_{\mathrm{chwpump,max}} = \left[\mu_{\mathrm{chwpump,max},1}, \mu_{\mathrm{chwpump,max},2}, \ldots, \mu_{\mathrm{chwpump,max},l}\right] \tag{9.76b}$$

$$\mu_{\mathrm{chwpump,min}} = \left[\mu_{\mathrm{chwpump,min},1}, \mu_{\mathrm{chwpump,min},2}, \ldots, \mu_{\mathrm{chwpump,min},l}\right] \tag{9.76c}$$

$$\mu_{\mathrm{ahufan,max}} = \left[\mu_{\mathrm{ahufan,max},1}, \mu_{\mathrm{ahufan,max},2}, \ldots, \mu_{\mathrm{ahufan,max},k}\right] \tag{9.76d}$$

$$\mu_{\mathrm{ahufan,min}} = \left[\mu_{\mathrm{ahufan,min},1}, \mu_{\mathrm{ahufan,min},2}, \ldots, \mu_{\mathrm{ahufan,min},k}\right] \tag{9.76e}$$

$$\mu_{\Delta t,\mathrm{max}} = \left[\mu_{\Delta t,\mathrm{max},1}, \mu_{\Delta t,\mathrm{max},2}, \ldots, \mu_{\Delta t,\mathrm{max},k}\right] \tag{9.76f}$$

Then, the function, $L_{\mathrm{MOP1(E)}}$, can be decomposed into multiple sub-problems, $L_{\mathrm{MOP1(E)},i}^{(a)}$ $(i \in [1,l])$ and $L_{\mathrm{MOP1(E)},j}^{(b)}$ $(j \in [1,k])$, as below:

$$
\begin{aligned}
& L_{\mathrm{MOP1(E)},i}^{(a)}\left(G_{\mathrm{w,ep},i}, \mu_{\mathrm{chwpump,max},i}, \mu_{\mathrm{chwpump,min},i}, \sigma_{\mathrm{w,ep,max},i}, \sigma_{\mathrm{w,ep,min},i}, \eta, \rho\right) \\
& = N_{\mathrm{ep},i}\left(G_{\mathrm{w,ep},i}\right) + \rho\left(G_{\mathrm{w,ep},i} + \sigma_{\mathrm{w,ep,max},i} - G_{\mathrm{w,ep,max},i}\right)^2 \\
& \quad + \rho\left(G_{\mathrm{w,ep,min},i} - \sigma_{\mathrm{w,ep,min},i} - G_{\mathrm{w,ep,min},i}\right)^2 + \mu_{\mathrm{chwpump,max},i}\left[G_{\mathrm{w,ep},i} + \sigma_{\mathrm{w,ep},i} - G_{\mathrm{w,ep,max},i}\right] \\
& \quad + \mu_{\mathrm{chwpump,min},i}\left[G_{\mathrm{w,ep},i} - \sigma_{\mathrm{w,ep,min},i} - G_{\mathrm{w,ep,min},i}\right] + \eta G_{\mathrm{w,ep},i}
\end{aligned} \tag{9.77}
$$

$$
\begin{aligned}
& L_{\mathrm{MOP1(E)},j}^{(b)}\left(G_{\mathrm{coil,a},j}, G_{\mathrm{coil,w},j}, \sigma_{\mathrm{coil,w,max},j}, \sigma_{\mathrm{coil,a,min},j}, \sigma_{\Delta t,\mathrm{max},j}, \lambda_{\mathrm{ahu},j}, \mu_{\mathrm{ahufan,max},j}, \mu_{\mathrm{ahufan,min},j}, \mu_{\Delta t,\mathrm{max},j}, \eta, \rho\right) \\
& = N_{\mathrm{ahufan},i}\left(G_{\mathrm{a,coil},i}\right) - \eta G_{\mathrm{coil,w},j} + \rho\left(G_{\mathrm{coil,a},i} + \sigma_{\mathrm{coil,a,max},j} - G_{\mathrm{coil,a,max},j}\right)^2 \\
& \quad + \rho\left(G_{\mathrm{coil,a},i} - \sigma_{\mathrm{coil,a,min},j} - G_{\mathrm{coil,a,min},j}\right)^2 + \rho\left(\left(t_{\mathrm{air,room}} - t_{\mathrm{coil,a,L}}\right)_j + \sigma_{\Delta t,\mathrm{max},j} - \Delta t_{\mathrm{air,max}}\right)^2 \\
& \quad + \rho\left(c_{\mathrm{a}} G_{\mathrm{coil,a},j}\cdot\left(t_{\mathrm{air,room}} - t_{\mathrm{coil,a,L}}\right)_j - Q_{\mathrm{room},j}\right)^2 + \lambda_{\mathrm{ahu},j}\left[c_{\mathrm{a}} G_{\mathrm{coil,a},j}\cdot\left(t_{\mathrm{air,room}} - t_{\mathrm{coil,a,L}}\right)_j - Q_{\mathrm{room},j}\right] \\
& \quad + \mu_{\mathrm{ahufan,max},j}\left[G_{\mathrm{coil,a},j} + \sigma_{\mathrm{coil,a,max},j} - G_{\mathrm{coil,a,max},j}\right] + \mu_{\mathrm{ahufan,min},j}\left[G_{\mathrm{coil,a},j} - \sigma_{\mathrm{coil,a,min},j} - G_{\mathrm{coil,a,min},j}\right] \\
& \quad + \mu_{\Delta t,\mathrm{max},j}\left[\left(t_{\mathrm{air,room}} - t_{\mathrm{coil,a,L}}\right)_j + \sigma_{\Delta t,\mathrm{max},j} - \Delta t_{\mathrm{air,max}}\right]
\end{aligned} \tag{9.78}
$$

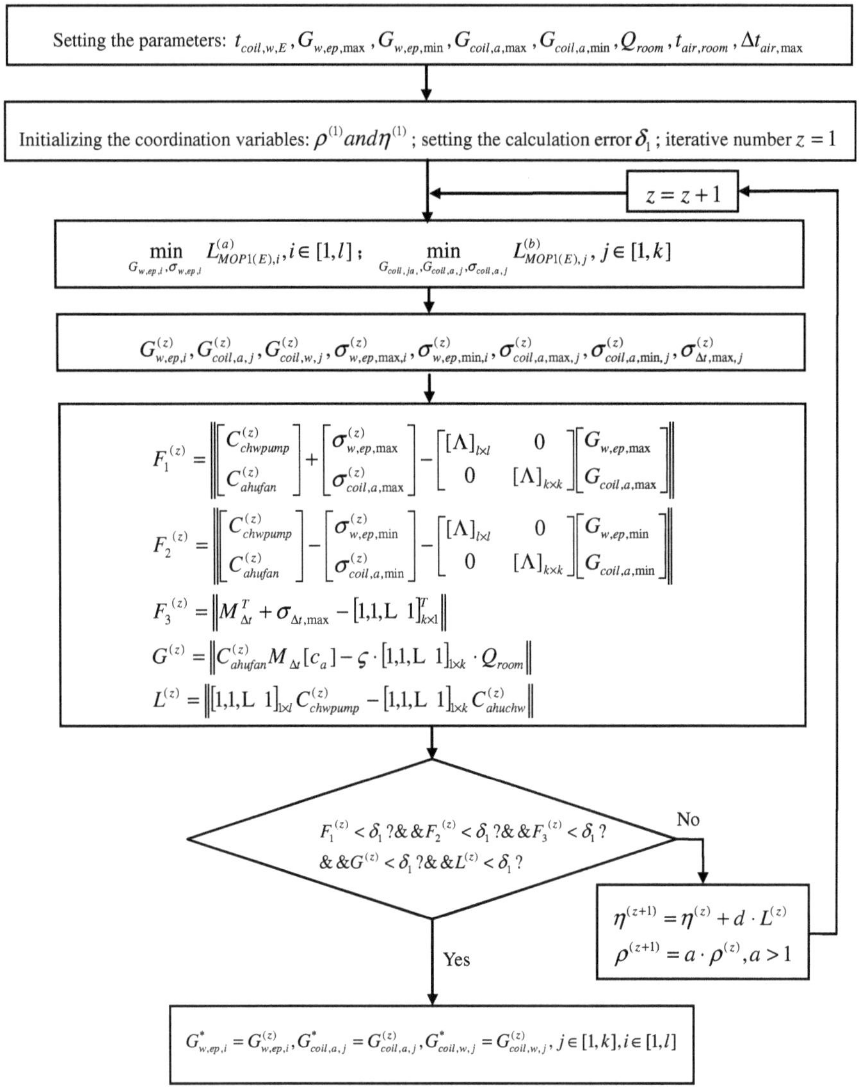

Fig. 9.21 Calculation flowchart of decomposition–coordination method for problem MOP1

Clearly, the common variables, η and ρ, in Eqs. (9.77) and (9.78), are usually chosen as the coordination parameters in solving the multiple sub-problems, $L^{(a)}_{\mathrm{MOP1(E)},i}$ $(i \in [1,l])$ and $L^{(b)}_{\mathrm{MOP1(E)},j}$ $(j \in [1,k])$, when the decomposition–coordination method is applied. The calculation flowchart for the problem *MOP1*solving is shown in Fig. 9.21.

As mentioned above, the global optimization model (Eq. 9.70) can be solved by using the decomposition–coordination method when a great number of decision variables are to be optimized. The decomposition–coordination method decomposes a large-scale optimization problem into many sub-problems with small size, which potentially improves the calculation efficiency and makes it possible for the global optimization scheme to be implemented in a large system. The global optimization model for the optimal running states of central air-conditioning system (Eq. 9.70) is solved based on a series of specific conditions (e.g., the dry and wet-bulb temperature of outdoor air, the cooling load demand, and the indoor air temperature.), and its calculation flow is given in Fig. 9.22.

9.2.3 Case Study

9.2.3.1 System Description

A central air-conditioning system mainly consists of two centrifugal chillers (Type: 19XL-4343455CQ; Rated cooling capacity:1580 kW; Rated power: 292 kW), two chilled water pumps (type: KQW200-400(I); Rated water flow: 280 m^3/h; Rated power: 75 kW; Water head: 54mH_2O), two cooling water pumps (type: KQW200-400(II); Rated water flow: 374 m^3/h; Rated power: 75 kW; Water head: 44mH_2O), two cooling tower (Type: DBHZ-400; Rated water flow: 400 m^3/h; Rated air flow: 165,000 m^3/h), and some AHUs (air-handling units) with different air-handling capacities. These AHUs include four large-capacity ones typing AHUJW40 (Surface heat-exchanger: JW40; Rated cooling capacity: 350 kW; Fan: KT4-38-22-NO7.4E; Rated air flow rate: 32,000 m^3/h; Rated fan power: 15 kW), six medium capacity ones typing AHUJW30 (Surface heat-exchanger: JW30; Rated cooling capacity: 250 kW; Fan: KT4-38-22-NO6.3E; Rated air flow rate: 25,000 m^3/h; Rated fan power: 11 kW), and four small-capacity ones typing AHUJW20 (Surface heat-exchanger: JW20; Rated cooling capacity: 145 kW; Fan: KT4-68-22-NO5E; Rated air flow rate: 16,000 m^3/h; Rated fan power: 7.5 kW). Although the central air-conditioning system has been equipped with an EMCS (Energy Management Control System), the control schemes are only limited to the on–off operations of individual equipments. For example, two chillers together with all cooling/chilled water pumps are to be put into operation when the cooling load of building attains peak, while only one chiller together with one cooling water pump and one chilled water pump is needed if the cooling load is relatively lower. It is still believed that much energy will be wasted in most cases when the equipments (the chillers, the pumps, and the fans) are in full-load operations that exceeds actual requirements a lot. Therefore, further energy saving may be achieved through a global optimization scheme.

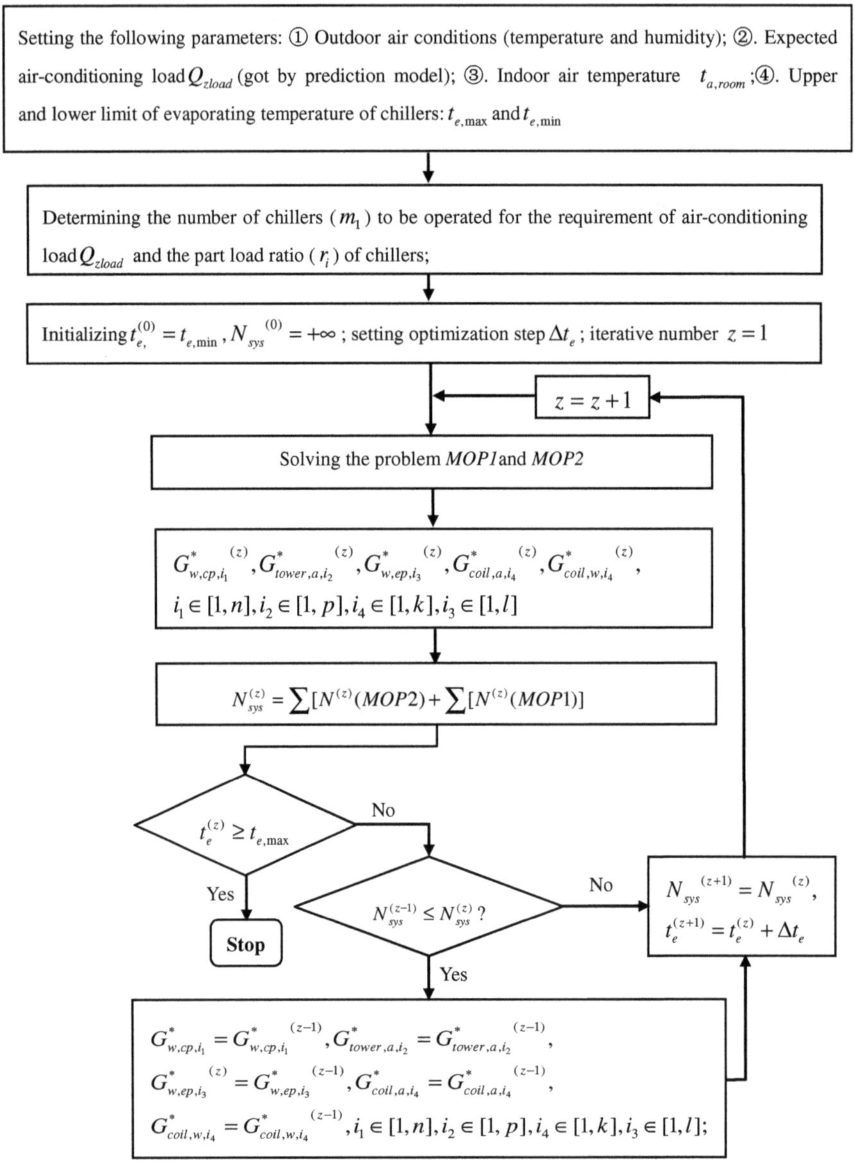

Fig. 9.22 Calculation flowchart of the global optimization for central air-conditioning system

9.2.3.2 Field Test and Empirical Equations

To establish the global optimization model of the air-conditioning system, the energy models of all equipments in this system ought to be firstly determined based on a large number of field test data. The test parameters include: outdoor air temperature and humidity, cooling water (entering and leaving) temperature of towers, chillers' condenser water (inlet and outlet) temperature and flow rate, supply and return chilled water temperature and flow rate, chillers' condensing and evaporating temperature, water/air temperature (entering and leaving), and flow rate of water-to-air heat exchanger in AHUs as well as electrical parameters (current and voltage) of all power equipments including chillers, water pumps, and fans.

Most of the above parameters can be automatically collected by the EMCS (Energy Management Control System) except for the air flow rates of AHUs and cooling water towers that are tested with a digital anemometer (Type: TESTO 405; precision: ± 3 % of reading). The test apparatus in the EMCS (Energy Management Control System) mainly includes electromagnetic water flow meters (Type: LDQ-98A; Precision: ±0.5 % of reading) for measuring cooling/chilled water flow rates, STWB-Pt100 temperature transducers (Precision: ±0.1 %) for measuring water and air temperature, HMI500 humidity transducers (Precision: ±2 %) for measuring ambient air humidity, and TBC500US Hall Closed-loop current/voltage transducers (Precision: ±0.5 %) for observing equipments' energy consumption. The water flow rates of heat exchangers (in AHUs) are obtained by flow meter installed on the water inlet of the heat exchangers. The other parameters in chiller, e.g., evaporating and condensing temperature, inlet and exit water temperature of evaporator, and condenser and input electric power, are obtained by its self-monitoring system that are connected to the EMCS through the RS-485 communication protocol.

(1) Empirical equations for chillers' model

The key of chiller's COP model (Eq. 8.6) is the determination of the irreversible loss function $f_s(T_c, T_e)$ (Eq. 8.6a), which is established based on the test parameters including the evaporating and condensing temperature, the water flow rate, and the entering and leaving water temperature of the chiller's condenser as well as the chiller's input electric power. Table 9.6 gives part of test data on chiller's COP under different running conditions. By using the test data, the irreversible loss functions, $f_s(T_c, T_e)$, are recognized as follows:

$$r = 0.5\text{:}\, f_s(T_c, T_e) = 0.1879T_c/T_e - 0.1520 \tag{9.79a}$$

$$r = 0.6\text{:}\, f_s(T_c, T_e) = 0.1490T_c/T_e - 0.1080 \tag{9.79b}$$

$$r = 0.7\text{:}\, f_s(T_c, T_e) = 0.1330T_c/T_e - 0.0868 \tag{9.79c}$$

$$r = 0.8\text{:}\, f_s(T_c, T_e) = 0.1558T_c/T_e - 0.1029 \tag{9.79d}$$

$$r = 0.9\text{:}\, f_s(T_c, T_e) = 0.2034T_c/T_e - 0.1562 \tag{9.79e}$$

$$r = 1.0\text{:}\, f_s(T_c, T_e) = 0.1734T_c/T_e - 0.1192 \tag{9.79f}$$

To validate the chiller's COP model, numerous tests (totally 245) have been performed under different evaporating temperatures ranging from 3 to 6 °C,

Table 9.6 Chiller's COP under different running conditions (test data)

t_c	r											
	0.5				0.6				0.7			
	t_e				t_e				t_e			
	4.0	5.0	6.0	7.0	4.0	5.0	6.0	7.0	4.0	5.0	6.0	7.0
32.0	4.66	4.82	4.92	5.01	5.10	5.27	5.38	5.45	5.33	5.38	5.44	5.50
33.0	4.56	4.72	4.82	4.90	4.99	5.15	5.27	5.33	5.21	5.32	5.40	5.41
34.0	4.46	4.61	4.71	4.81	4.88	5.04	5.15	5.24	5.09	5.27	5.33	5.38
35.0	4.37	4.51	4.61	4.70	4.77	4.93	5.03	5.15	4.99	5.15	5.26	5.31
36.0	4.28	4.41	4.51	4.61	4.67	4.82	4.92	5.04	4.88	5.04	5.15	5.26
37.0	4.12	4.30	4.41	4.52	4.57	4.72	4.82	4.92	4.77	4.93	5.04	5.11
38.0	4.10	4.23	4.32	4.41	4.47	4.62	4.72	4.81	4.67	4.82	4.94	5.02
39.0	3.99	4.17	4.21	4.33	4.35	4.51	4.62	4.72	4.53	4.69	4.82	4.93
40.0	3.92	4.12	4.18	4.24	4.26	4.34	4.48	4.57	4.45	4.55	4.63	4.77
41.0	3.85	4.09	4.11	4.18	4.15	4.21	4.30	4.48	4.32	4.41	4.50	4.59
t_c	r											
	0.8				0.9				1.0			
	t_e				t_e				t_e			
	4.0	5.0	6.0	7.0	4.0	5.0	6.0	7.0	4.0	5.0	6.0	7.0
32.0	5.31	5.48	5.50	5.59	5.61	5.84	5.96	6.09	5.89	5.96	6.01	6.06
33.0	5.19	5.36	5.47	5.50	5.53	5.78	5.83	5.98	5.73	5.84	5.95	6.04
34.0	5.08	5.25	5.36	5.41	5.40	5.59	5.71	5.86	5.57	5.71	5.84	5.95
35.0	4.9	5.13	5.24	5.31	5.28	5.47	5.58	5.74	5.46	5.55	5.69	5.87
36.0	4.86	5.02	5.13	5.19	5.15	5.35	5.46	5.65	5.37	5.46	5.54	5.68
37.0	4.75	4.91	5.01	5.11	5.09	5.23	5.35	5.47	5.23	5.37	5.48	5.55
38.0	4.65	4.81	4.91	4.97	4.98	5.12	5.23	5.35	5.15	5.21	5.36	5.41
39.0	4.51	4.67	4.76	4.84	4.90	5.04	5.12	5.21	5.03	5.12	5.25	5.36
40.0	4.39	4.46	4.60	4.72	4.85	4.94	5.01	5.12	4.92	5.01	5.16	5.25
41.0	4.26	4.39	4.47	4.58	4.69	4.81	4.95	5.01	4.85	4.94	5.04	5.12

different condensing temperatures ranging from 32 to 38 °C, and different cooling load ratio ranging from 0.5 to 0.9. The statistical results on the errors of chiller's COP model are listed in Table 9.7. It can be seen that the maximum relative error of COP model does not exceed 12 %, and most of them (over 90 %) are within 10 %.

(2) Empirical equations for condensing temperature

Condensing temperature is not a controlled variable in a chiller. As known from Eq. (8.7), chiller's condensing temperature (t_c) is related to inlet cooling water temperature ($T_{w,c,E}$) and cooling water flow rate ($G_{w,c}$) as well as condensing heat (Q_c) of condenser. To estimate the condensing temperature, the function $F_c(G_{w,c})$ in Eq. (8.7) should be determined in advance by test parameters including condensing temperature and inlet and exit cooling water temperature as well as cooling water flow rate of condenser. The detailed test data for obtaining $F_c(G_{w,c})$ is presented in Fig, 9.23, based on which $F_c(G_{w,c})$ can be expressed as

Table 9.7 Error analysis on the empirical COP model

Range of relative error[a] (%)	Number of test	Percentage of total[b] (%)	Range of relative error[a] (%)	Number of test	Percentage of tota[b] (%)
0–1	29	11.8	6–7	21	8.6
1–2	42	17.1	7–8	19	7.8
2–3	30	12.2	8–9	17	6.9
3–4	27	11.0	9–11	4	1.6
4–5	24	9.8	11–12	6	2.4
5–6	26	10.6	12–	0	0

[a]Relative error of empirical COP model = $|\text{model } result - test\ data|/test\ data \times 100\,\%$
[b]Total number of test is 245

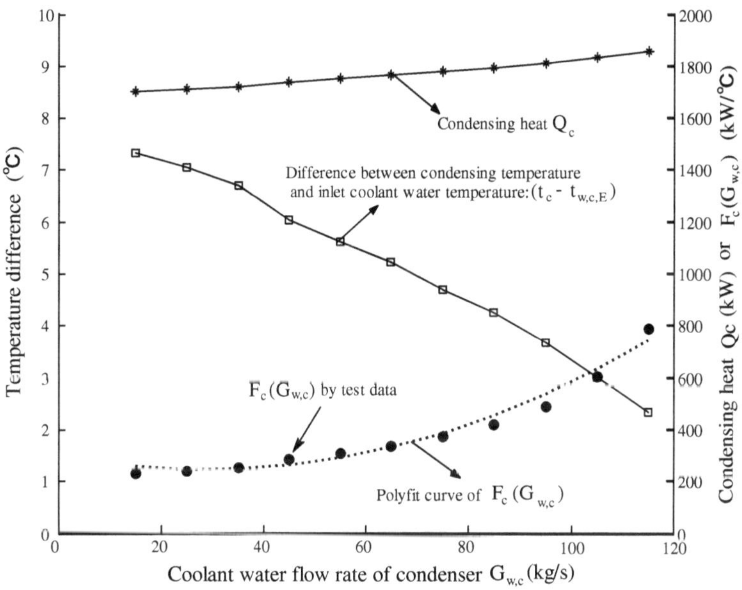

Fig. 9.23 Experimental data for obtaining the function $F_c(G_{w,c})$

$$F_c(G_{w,c}) = 0.068 \times G_{w,c}^2 - 4.07G_{w,c} + 303.67 \tag{9.80}$$

(3) Empirical equations for exit chilled water temperature

In the global optimization operation model of central air-conditioning system, the evaporating temperatures of chillers are controlled variables. Although the higher evaporating temperature will result in a higher COP of chiller, it will inevitably increase chilled water flow rate (which means more energy consumption of chilled pump) to meet the requirement of air-conditioning load. So, there should be an optimal evaporating temperature of chiller under which the whole system's energy efficiency is the highest. During the global optimization, the exit chilled water temperature of chiller is another important intermediate variable which is often considered to be the inlet chilled water temperature of AHUs, and hence, its calculation model depending on evaporating temperature (t_e), chilled water flow rate ($G_{w,e}$), and cooling capacity of chiller (Q_e) must be established.

The exit chilled water temperature of evaporator ($t_{w,e,L}$) is usually calculated by

$$t_{w,e,L} = t_{w,e,E} - \frac{Q_e}{c_w G_{w,e}} \tag{9.81}$$

Combining Eq. (9.81) and (8.11), the exit chilled water temperature of evaporator ($t_{w,e,L}$) can be expressed by Eq. (9.82)

$$t_{w,e,L} = t_e + \left[\frac{Q_e}{F_e(G_{e,w})} - \frac{Q_e}{c_w G_{w,e}}\right] \tag{9.82}$$

To obtain the expression of $F_e(G_{e,w})$, the test data on evaporating temperature and inlet and exit chilled water temperature as well as chilled water flow rate of evaporator are processed and presented in Fig. 9.24. According to the test data, $F_e(G_{e,w})$ can be modeled as:

$$F_e(G_{e,w}) = 0.0086 \times G_{e,w}^2 + 1.65 \times G_{e,w} + 120.68 \tag{9.83}$$

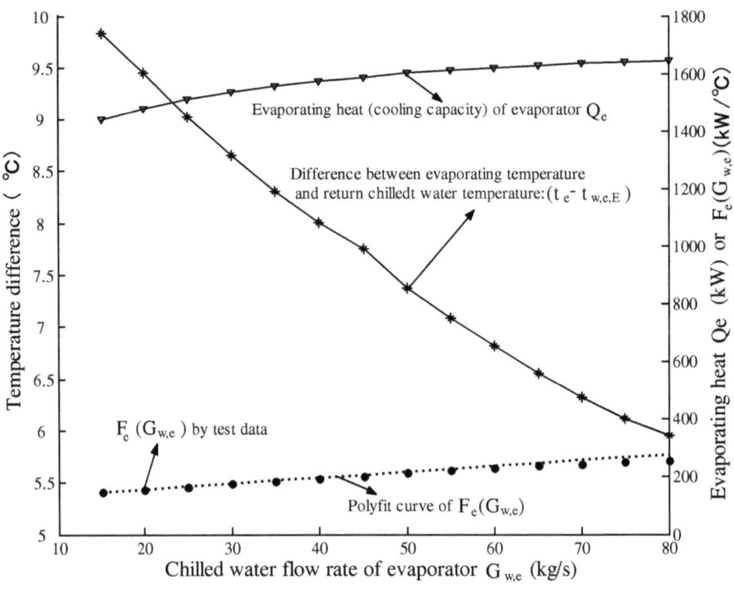

Fig. 9.24 Experimental data for obtaining the function $F_e(G_{e,w})$

(4) Empirical equations for cooing water tower's model

$\mathrm{Ntu}_{\mathrm{tower}}$ is the key parameter to calculate the outlet water temperature of cooling water towers, known from Eqs. (8.15) and (8.16), the $\mathrm{Ntu}_{\mathrm{tower}}$ can be modeled by the following variables: inlet air temperature and humidity, inlet and outlet water temperature, air flow rate, and water flow rate. Through nonlinear fitting of the test data, the $\mathrm{Ntu}_{\mathrm{tower}}$ of the two towers in this system can be empirically written as below:

$$\mathrm{Ntu}_{\mathrm{tower},1} = 1.853 \cdot \left(\frac{G_{\mathrm{tower,w}}}{G_{\mathrm{tower,a}}}\right)^{-0.585} \tag{9.84}$$

$$\mathrm{Ntu}_{\mathrm{tower},2} = 1.764 \cdot \left(\frac{G_{\mathrm{tower,w}}}{G_{\mathrm{tower,a}}}\right)^{-0.628} \tag{9.85}$$

(5) Empirical equations for the water-to-air heat exchangers' model

In heat exchanger model, the equations to be determined are the exposure coefficient $\varepsilon_{\mathrm{coil},2}$ (Eq. 8.18) and overall heat transfer coefficient $K_{\mathrm{coil},s}$ (Eq. 8.22). They can be empirically obtained through relevant tested variables including inlet and exit air conditions (temperature and humidity), and air flow rate as well as water flow rate passing through the heat exchanger. The empirical equations for $\varepsilon_{\mathrm{coil},2}$ and $K_{\mathrm{coil},s}$ of heat exchanger are given as below in terms of different types in this air-conditioning system:

For the water-to-air heat exchangers (JW20) in the small-capacity AHUs (AHUJW20):

$$\varepsilon_{\mathrm{coil},2} = -0.0664 + 0.501 \cdot G_{\mathrm{coil},a} \tag{9.86}$$

$$K_{\mathrm{coil},s} = \left[\frac{1}{27.9 G_{\mathrm{coil},a}{}^{0.53} \xi_{\mathrm{coil}}^{1.03}} + \frac{1}{108.0 G_{\mathrm{coil},w}{}^{0.8}}\right]^{-1} \tag{9.87}$$

For the water-to-air heat exchangers (JW30) in the medium-capacity AHUs (AHUJW30):

$$\varepsilon_{\mathrm{coil},2} = -0.0454 + 0.391 \cdot G_{\mathrm{coil},a} \tag{9.88}$$

$$K_{\mathrm{coil},s} = \left[\frac{1}{25.2 G_{\mathrm{coil},a}{}^{0.53} \xi_{\mathrm{coil}}{}^{1.04}} + \frac{1}{82.9 G_{\mathrm{coil},w}{}^{0.8}}\right]^{-1} \tag{9.89}$$

For the water-to-air heat exchangers (JW40) in the large-capacity AHUs (AHUJW30):

$$\varepsilon_{\mathrm{coil},2} = -0.0212 + 0.294 \cdot G_{\mathrm{coil},a} \tag{9.90}$$

$$K_{\text{coil},s} = \left[\frac{1}{17.6 G_{\text{coil},a}{}^{0.58} \xi_{\text{coil}}{}^{102}} + \frac{1}{40.4 G_{\text{coil},w}{}^{0.8}} \right]^{-1} \tag{9.91}$$

(6) Empirical equations for the pumps/fans' model

All the power models of the pumps and the fans as well as the efficiency equations of inverters are established mainly based on the performance data provided by the relevant product manufacture and be corrected by the field test data. They are listed as follows:

For the chilled water pumps:

$$N_{\text{chwpump,var}}(X_{\text{G}}) = \frac{70.69 k_{\text{chwpump}} X_{\text{G}}{}^{3}}{(1 - \mathrm{e}^{-9.04 X_{\text{G}}}) \cdot \eta_{\text{f}}(X_{\text{G}})} \tag{9.92}$$

where the correct coefficient, k_{chwpump}, is 0.92 and 0.95, respectively, for the two chilled water pumps.

For the cooling water pumps:

$$N_{\text{conwpump,var}}(X_{\text{G}}) = \frac{76.62 k_{\text{conwpump}} X_{\text{G}}^{3}}{(1 - \mathrm{e}^{-8.54 X_{\text{G}}}) \cdot \eta_{\text{f}}(X_{\text{G}})} \tag{9.93}$$

where the correct coefficient, k_{conwpump}, is 0.96 and 0.93, respectively, for the two cooling water pumps.

For the fans (KT4-68-22-NO5E) in the small-capacity AHUs (AHUJW20):

$$N_{\text{ahufanNO5E,var}}(X_{\text{G}}) = \frac{7.8 k_{\text{ahufanA}} X_{\text{G}}^{3}}{(1 - \mathrm{e}^{-7.81 X_{\text{G}}}) \cdot \eta_{\text{f}}(X_{\text{G}})} \tag{9.94}$$

where the correct coefficient, k_{ahufanA}, is 1.05, 0.97, 0.94, and 1.03, respectively, for the four fans of KT4-68-22-NO5E.

For the fans (KT4-38-22-NO6.3E) in the medium-capacity AHUs (AHUJW30):

$$N_{\text{ahufanNO6.3E,var}}(X_{\text{G}}) = \frac{11.3 k_{\text{ahufanB}} X_{\text{G}}^{3}}{(1 - \mathrm{e}^{-7.23 X_{\text{G}}}) \cdot \eta_{f}(X_{\text{G}})} \tag{9.95}$$

where the correct coefficient, k_{ahufanB}, is 1.07, 0.92, 0.96, 1.04, 1.06, and 1.03, respectively, for the six fans of KT4-38-22-NO6.3E.

For the fans (KT4-38-22-NO7.4E) in the large-capacity AHUs (AHUJW40):

$$N_{\text{ahufanNO7.4E,var}}(X_{\text{G}}) = \frac{15.5 k_{\text{ahufanC}} X_{\text{G}}^{3}}{(1 - \mathrm{e}^{-6.46 X_{\text{G}}}) \cdot \eta_{\text{f}}(X_{\text{G}})} \tag{9.96}$$

where the correct coefficient, k_{ahufanC}, is 0.95, 0.94, 1.06, and 1.05, respectively, for the four fans of KT4-38-22-NO7.4E.

For the fans of the cooling water towers:

$$N_{\mathrm{towerfan,var}}(X_{\mathrm{G}}) = \frac{10.1 k_{\mathrm{towerfan}} X_{\mathrm{G}}^{3}}{(1 - e^{-8.88 X_{\mathrm{G}}}) \cdot \eta_{\mathrm{f}}(X_{\mathrm{G}})} \tag{9.97}$$

where the correct coefficient, k_{towerfan}, is 0.95 and 0.93, respectively, for the two towers' fans.

In Eqs. (9.92) through (9.97), the inverters' efficiency, $\eta_{\mathrm{f}}(X_{\mathrm{G}})$, is modeled as:

$$\eta_{\mathrm{f}}(X_{\mathrm{G}}) = 0.507 + 1.28 X_{\mathrm{G}} - 1.42 X_{\mathrm{G}}^{2} + 0.584 X_{\mathrm{G}}^{3} \tag{9.98}$$

9.2.3.3 Results and Discussion

The global optimization model, based on the above empirical equations, was used to produce hourly the optimal running conditions of the whole system in an operation day of August. The optimization parameters include: the chillers' evaporating temperature, the (cooling/chilled water) pumps' flow rate, the flow rate of water, and air passing through the cooling coils in AHUs as well as the cooling towers' air flow rate.

The system's operation time was from 8:00 a.m. to 21:00 p.m. The controller log on the day showed that only a half number of equipments in the cooling plant were operated before 11:00 a.m., while in the following time, all the equipments were put into operation due to the increasing cooling load of the building. The running pumps and fans in the system were actually in full-load operation because no inverters were installed to adjust them. The actual evaporating temperature of chiller was set as 5.5 °C.

The hourly dry-bulb and wet-bulb temperatures of outdoor air on the day were shown in Fig. 9.25. The hourly cooling loads on the AHUs, as plotted in Fig. 9.26, were obtained from the testing data of entering and leaving water temperatures and

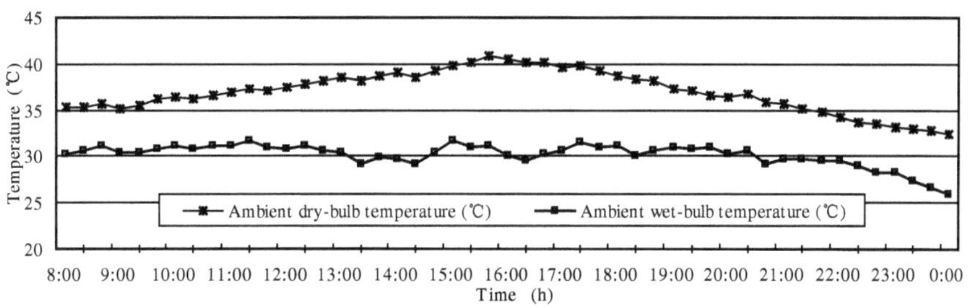

Fig. 9.25 The dry-bulb and wet-bulb temperatures of outdoor air on the focused day

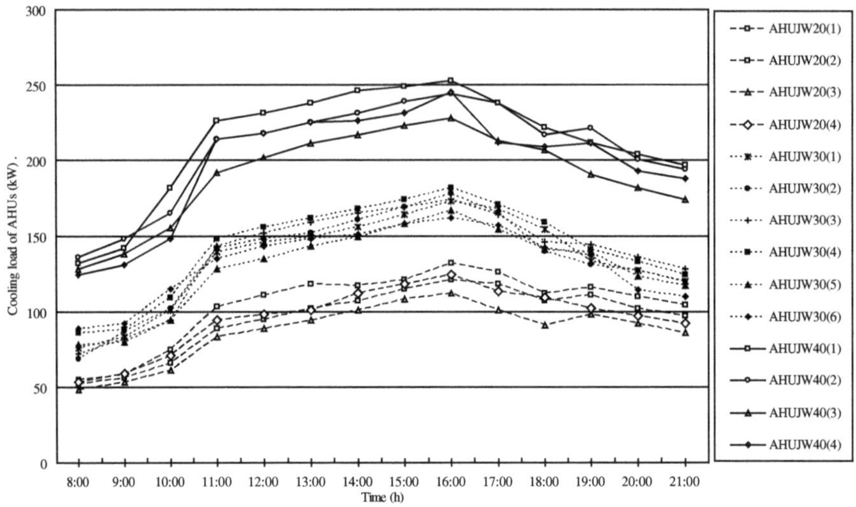

Fig. 9.26 Hourly cooling load burdened by the AHUs on the experimental day

water flow rates of cooling coils in AHUs. The upper and lower limits on the equipments' operation were given as below for the calculation:

(1) Chillers' evaporating temperature was not allowed to be lower than 3.0 °C;
(2) Minimum flow rate of the cooling water pumps and the chilled water pumps was 20 and 25 kg/s, respectively;
(3) Lower limit of air flow rate of the cooling tower fans was set as 5 kg/s, and the fans in AHUs allowed the minimum air flow rates that were not lower than 20 % of the rated capacity.
(4) Indoor air temperature was kept as 25 °C and the ratio of fresh air intake of AHUs was set as 25 %.

To verify the valid of the decomposition–coordination algorithm, the results obtained by the direct search method was referred to. The optimization results of running conditions of the cooling plant are numerically listed in Table 9.8 for this verification. It indicates that the decomposition–coordination method will lead to almost the same results as the direct search method. However, the former method had the much higher calculation efficiency than the latter one. Under the same computer hardware conditions (Intel Pentium CPU frequency: 1000 MHz; RAM: 512 MB), it needed about 380 s for the direct search method to complete one-cycle calculation, while the computation time would drop to about 50 s when the decomposition–coordination method was employed. It is believed that the advantage of high-efficiency calculation of the decomposition–coordination method will be more highlighted in solving the optimization problem for the over-large-scale systems.

As seen from Table 9.8, the evaporating temperature (of chillers in this system) on the focused operation day should be kept as 3.0 °C (the lower limit value); the

Table 9.8 Optimal running conditions for the cooling plant on the experimental day

Optimization variables	Time												
	8:00	9:00	10:00	11:00	12:00	13:00	14:00	15:00	16:00	17:00	18:00	19:00	20:00
$t_{e,1}$ (°C)	3.0 3.0[a]	3.0 3.0[a]	3.0 3.0[a]	3.0 3.0[a]	3.0 3.0[a]	3.0 3.0[a]	3.0 3.0[a]	3.0 3.0[a]	3.0 3.0[a]	3.0 3.0[a]	3.0 3.0[a]	3.0 3.0[a]	3.0 3.0[a]
$t_{e,2}$ (°C)	/	/	/	3.0 3.0[a]	3.0 3.0[a]	3.0 3.0[a]	3.0 3.0[a]	3.0 3.0[a]	3.0 3.0[a]	3.0 3.0[a]	3.0 3.0[a]	3.0 3.0[a]	3.0 3.0[a]
$G_{w,ep,1}$ (kg/s)	0.0	0.0	0.0	0.0	0.0	0.0	0.0	0.0	0.0	0.0	0.0	0.0	0.0
$G_{w,ep,2}$ (kg/s)	34.0 38.8[a]	34.0 38.8[a]	34.0 38.8[a]	43.0 42.9[a]	45.0 45.1[a]	49.0 48.8[a]	50.0 50.3[a]	53.0 52.8[a]	59.0 59.3[a]	52.0 51.8[a]	45.0 44.7[a]	45.0 44.7[a]	41.0 40.8[a]
$G_{w,cp,1}$ (kg/s)	0.0	0.0	0.0	0.0	0.0	0.0	0.0	0.0	0.0	0.0	0.0	0.0	0.0
$G_{w,cp,2}$ (kg/s)	28.0 27.8[a]	28.0 27.8[a]	28.0 27.8[a]	28.0 27.8[a]	28.0 27.8[a]	28.0 27.8[a]	28.0 27.8[a]	28.0 27.8[a]	28.0 27.8[a]	28.0 27.8[a]	28.0 27.8[a]	28.0 27.8[a]	28.0 27.8[a]
$G_{tower,a,1}$ (kg/s)	19.5 19.3[a]	19.5 19.3[a]	19.5 19.3[a]	18.0 17.8[a]	18.0 17.8[a]	18.0 17.8[a]	18.0 17.8[a]	18.0 17.8[a]	18.0 17.8[a]	18.0 17.8[a]	18.0 17.8[a]	18.0 17.8[a]	17.8 17.9[a]
$G_{tower,a,2}$ (kg/s)	18.5 18.4[a]	18.5 18.4[a]	18.5 18.4[a]	17.0 16.8[a]	17.0 16.8[a]	17.0 16.8[a]	17.0 16.8[a]	17.0 16.8[a]	17.0 16.8[a]	17.0 16.8[a]	17.0 16.8[a]	17.0 16.8[a]	17.0 16.8[a]

The optimal values marked by ‘a’ were obtained by using the direct search method

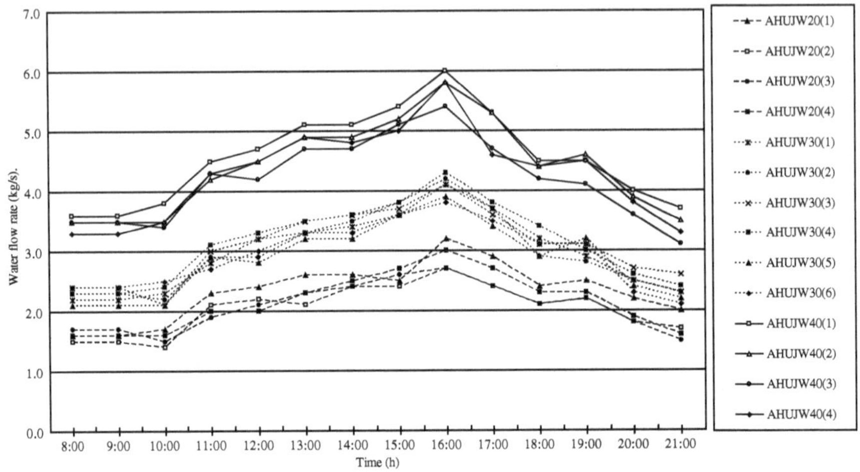

Fig. 9.27 Hourly optimization results on chilled water flow rate in each AHU

total cooling water flow rate should be kept as 28 kg/s (which is much smaller than the rated capacity (78 kg/s) of the cooling water pumps in this system), and the total chilled water flow rate increases with the cooling load demand increasing, but the biggest value (about 59.0 kg/s) is still much lower than the rated capacity of the chilled water pumps (103 kg/s); the air flow rate passing through the cooling towers hardly changes throughout the experimental day. The hourly optimization results hint that the lower evaporating temperature of chillers will be in favor of energy saving of the whole system, and all the pumps and fans in the cooling plant operate in a stably optimal state except for the chilled water pumps that are required to be adjusted according to the variations of cooling load demand.

The hourly optimization results on water and air flow rate in every AHU are shown in Figs. 9.27 and 9.28, respectively. It can be seen that the water-side adjustment takes place only after 10:00 a.m. It is reasonable that whether or not the local water flow rates have a change mainly be restricted by the variations of the total water flow rate. Actually, the water-side adjustment in the AHUs is just the rational allocation of the total chilled water flow rate provided by the pumps. Before 10:00 a.m., the total chilled water flow rate (optimized) does not vary, and this will not cause any changes in the local water flow rates of the AHUs. So, the AHUs' cooling capacity is adjusted through regulating the air flow rate. From 10:00 a.m. to 16:00 p.m., the total chilled water increases with the cooling load demand increasing, and the local chilled water flow rates in the AHUs will increase accordingly. The air flow rates in the AHUs have a big increase till to the near maximum point at 11:00 a.m., but hardly increase afterwards even the local cooling loads continue rising. In such case, the water-side adjustments become dominant to meet the increasing cooling load demand. After 16:00 p.m., both the air flow rate and the water flow rate tend to decrease with the decline of the cooling load.

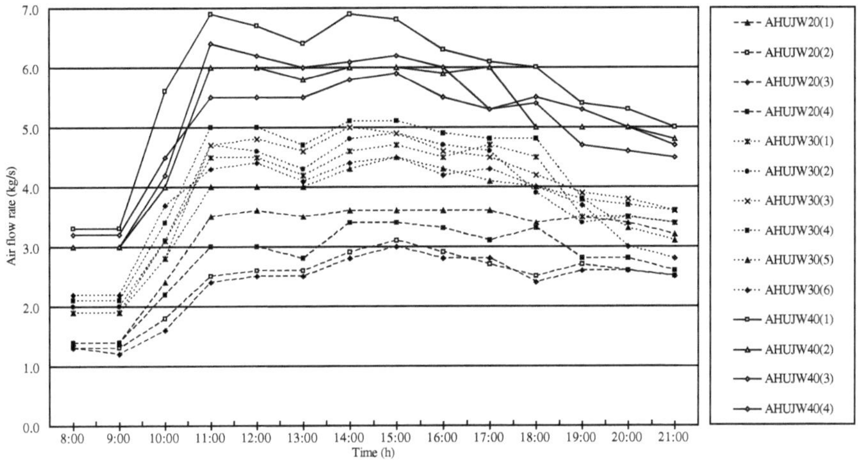

Fig. 9.28 Hourly optimization results on air flow rate in each AHU

The hourly electric powers of various equipments under optimal and non-optimal operation are plotted, respectively, in Figs. 9.29, 9.30 and 9.31. The results indicate that the powers of pumps and fans in this system would drop much if the global optimization control scheme were applied to the system, but it is the contrary for the chillers' power. It means that the global optimization does not necessarily lead to local optimization. The *SCOP* (*System Coefficient Of Performance*), which is expressed by Eq. (9.99), is suggested to analyze the energy efficiency of the whole system.

$$\text{SCOP} = \frac{\sum_{i=1}^{m1} Q_{\text{e},n}}{\sum_{i=1}^{m1} N_{\text{chiller},i} + \sum_{j=1}^{m2} N_{\text{pump},j} + \sum_{k=1}^{m3} N_{\text{fan},k}} \tag{9.99}$$

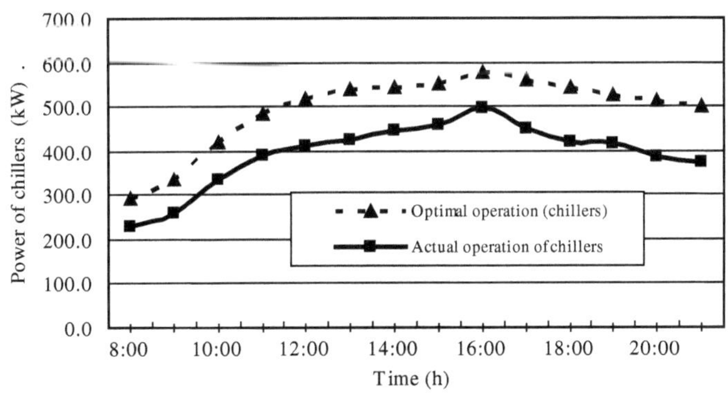

Fig. 9.29 Comparisons of chillers' electric power between optimal and actual operation

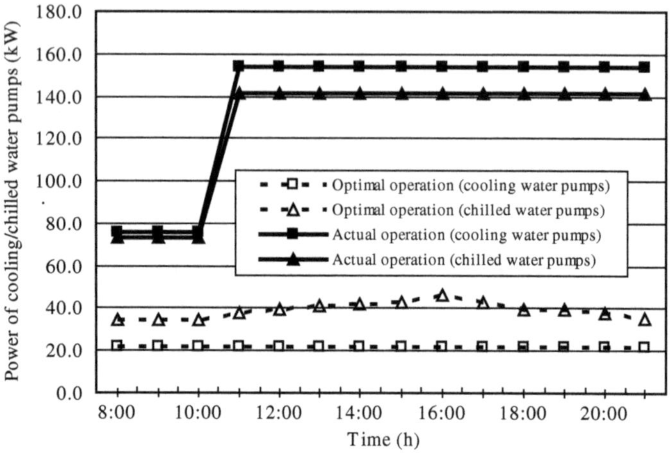

Fig. 9.30 Comparisons of power of cooling/chilled water pumps between optimal and actual operation

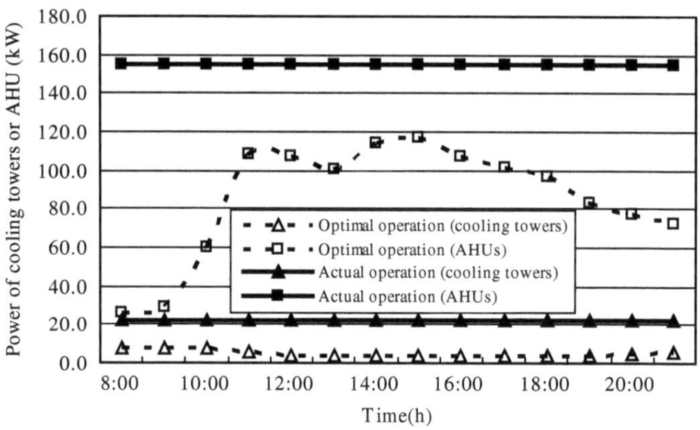

Fig. 9.31 Comparisons of power of cooling towers and AHUs between optimal and actual operation

where $m1$, $m2$ and $m3$ denote, respectively, the number of chillers, pumps, and fans included in the central air-conditioning system.

The hourly power and *SCOP* of the system on the focused operation day is shown in Fig. 9.32. The *SCOP* under optimal operation is obviously higher than that under non-optimal operation. Particularly, the gap between them before 10:00 a.m is much bigger than that of the other time of the day (the narrowest gap occurs at about 16:00 p.m.). The trend may have some relation with the variations of cooling load of the day. It seems that the lower the cooling load is, the bigger the

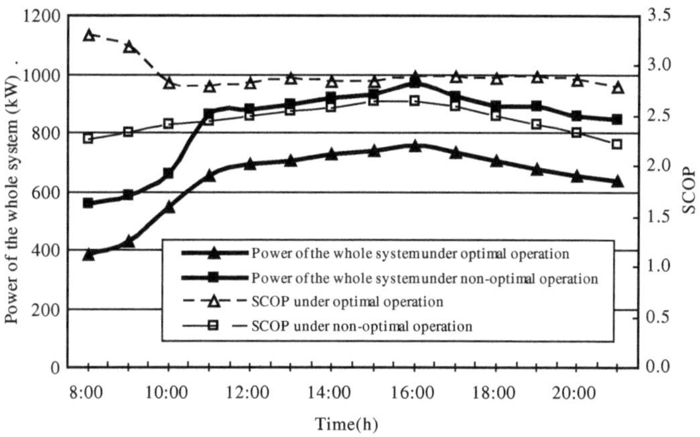

Fig. 9.32 Electric power and SCOP of the whole system under optimal and actual operation

gap between the two SCOP curves (one is under the optimal operation, the other under non-optimal operation) will be. It indicates that the energy saving brought by the globally optimal operation will be more significant when the system runs under the lower load conditions.

Energy analysis on the system under the optimal and actual (non-optimal) operation is made here, as shown in Fig. 9.33. The results in Fig. 9.33a show that the global optimization scheme will save the system about 2635 kWh of electrical energy on the experimental day. The energy saving mainly comes from the pumps and the fans that should be emphasized in practical engineering. Big changes will take place in the percentage of energy consumption of individual equipment in the total energy consumption of the system after the optimization scheme is applied. As shown in Fig. 9.33b, the energy consumption percentage of chillers in the system on the experimental day would rise from 47 % under the non-optimal operation to 76 % under the optimal operation, while that of the cooling water pumps, the chilled water pumps, and the AHUs as well as the cooling towers would drop from 16, 15, 19, and 3 % to 3, 6, 13, and 1 %, respectively, (Fig. 9.33).

Appendix

The deduction of the optimums of outside air intake by using analytical approach and equivalent cost function for the different outside air scenarios is briefly given below.

A: $T_{\mathrm{o}} \leq T_{\mathrm{c},i}$

It is winter, and free cooling with outside air is available for VAHU in the interior region. The possible components of thermal energy consumption of IAHU include heating and reheat. The cost function can be rewritten as:

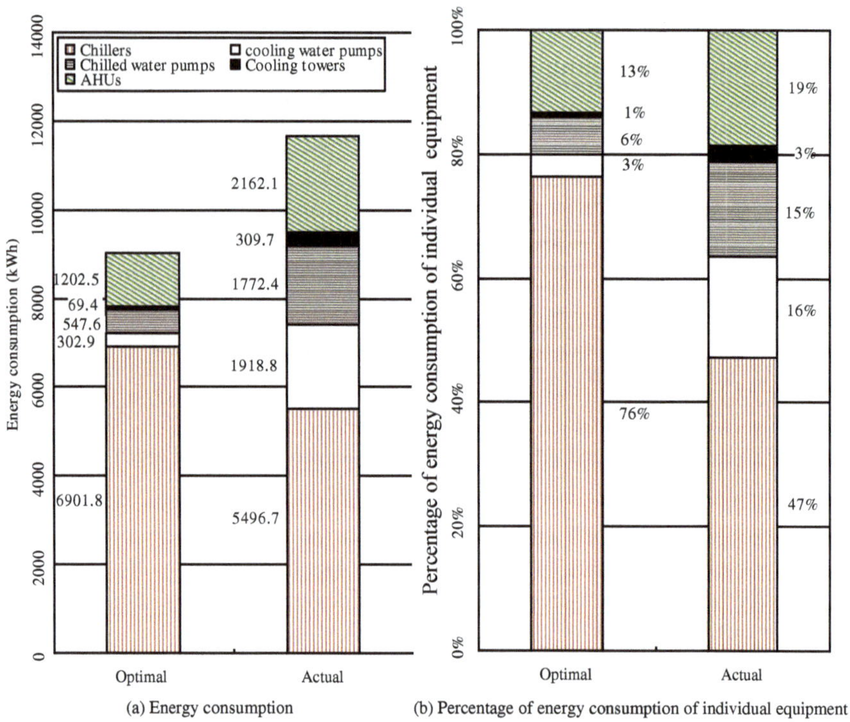

Fig. 9.33 Analysis on **a** energy consumption and **b** percentage of energy consumption of individual equipment under optimal and actual operation on the experimental day

$$Q_{\mathrm{thm}} = Q_{\mathrm{hc},i} + Q_{\mathrm{rh},i} + Q_{\mathrm{hc},e} + Q_{\mathrm{rh},e} \tag{9.100}$$

The heating energy for the interior region is:

$$Q_{\mathrm{hc},i} = G_{\mathrm{a},i} c_{\mathrm{p}} \left(T_{\mathrm{c},i} - T_{\mathrm{mix},i} \right) \tag{9.101}$$

where c_{p} = air specific heat (J/(kg K)).

The reheat, if there is any, is:

$$Q_{\mathrm{rh},i} = G_{\mathrm{a},i} c_{\mathrm{p}} \left(T_{\mathrm{s}} - T_{\mathrm{c},i} \right) \tag{9.102}$$

The mixed air temperature is a function of outside air intake and outside air and room air temperature:

$$T_{\mathrm{mix}} = T_{\mathrm{o}} \beta_{\mathrm{OA}} + (1 - \beta_{\mathrm{OA}}) T_{\mathrm{r}} \tag{9.103}$$

Combining Eqs. (9.101)–(9.103), we obtain:

$$Q_{hc,i} + Q_{rh,i} = G_{a,i}c_p(T_{c,i} - T_r) + G_{a,i}c_p\beta_{OA,i}(T_r - T_o) + G_{a,i}c_p(T_s - T_r) + G_{a,i}c_p(T_r - T_{c,i}) \quad (9.104)$$

It is known that the third term on the right side of Eq. (9.104) is the thermal load in the interior region, and the first term cancels out the last term. Therefore, the final expression of heating cost in the interior region is:

$$Q_{hc,i} + Q_{rh,i} = G_{a,i}c_p\beta_{OA,i}(T_r - T_o) - \mathrm{Load_i} \quad (9.105)$$

where $\mathrm{Load_i}$ is an absolute value of the thermal load in interior region (W).

The same deduction holds true for the perimeter region. Thus, the final thermal energy consumption of the entire system is:

$$Q_{thm} = G_{a,i}c_p\beta_{OA,i}(T_r - T_o) - \mathrm{Load_i} + G_{a,e}c_p\beta_{OA,e}(T_r - T_o) - \mathrm{Load_e} \quad (9.106)$$

where $\mathrm{Load_e}$ is an absolute value of the thermal load in exterior region (W).

Since outside air is the cooling source for the interior region in this condition, $\beta_{OA,i}$ is a dependent variable and constrained by the maximum of Eqs. (9.23) and (9.25). To minimize the thermal consumption, we want the two terms with the outside air ratio β to be minimized. The total amount of outside air intake for the building is given as:

$$f = \beta_{IAQ,dsg,e}(1 - \varphi) - \gamma\varphi\left(1 - \frac{\beta_{IAQ,i}}{\beta_{OA,i}}\right) + \varphi\beta_{OA,i} \quad (9.107)$$

This is the equivalent cost function of this outside air condition. It is easy to find that this function does not have extremum on the definition domain, since the discriminant is always zero for all points within the domain:

$$\Delta = AC - B^2 \quad (9.108)$$

where $A = f''_{\gamma\gamma}$, $C = f''_{\gamma\beta_{OA,i}}$, and $B = f''_{\beta_{OA,i}\beta_{OA,i}}$

Therefore, the extremum exists only at the boundary. Meanwhile, the partial derivative on $\beta_{OA,i}$ has a form as:

$$\frac{\partial f}{\partial \beta_{OA,i}} = \varphi\left(1 - \frac{\gamma\varphi\beta_{IAQ,i}}{\beta_{OA,i}^2}\right) \quad (9.109)$$

Since $\beta_{OA,i} \geq \beta_{IAQ,i}$ and $\gamma \leq \beta_{OA,i}$, the derivative is constantly positive and the extremum of the function is the lower boundary. The partial derivative on γ is negative and has a form as:

$$\frac{\partial f}{\partial \gamma} = -\varphi\left(1 - \frac{\beta_{\text{IAQ},i}}{\beta_{\text{OA},i}}\right) \tag{9.110}$$

In summary, the minimum locates where $\beta_{\text{OA,i}}$ takes the low boundary and γ takes the upper boundary.

$$\beta_{\text{OA},i} = \max(\beta_{\text{IAQ},i}, \beta_{\text{eco},i}) \tag{9.111}$$

$$\beta_{\text{OA,e}} = \max(\beta_{\text{IAQ,e}}, 0) \tag{9.112}$$

B: $T_{\text{c},i} \leq T_{\text{o}} \leq T_{\text{c,e}}$

In this scenario, the interior region is in a cooling mode while the exterior region in either a cooling or a heating mode. The cost function is different from scenario A with:

$$Q_{\text{thm}} = Q_{\text{cc},i} + Q_{\text{rh},i} + Q_{\text{hc,e}} + Q_{\text{rh,e}} \tag{9.113}$$

$$Q_{\text{cc},i} + Q_{\text{rh},i} = G_{\text{a},i}c_{\text{p}}\left(T_{\text{mix},i} - T_{\text{c},i}\right) + G_{\text{a},i}c_{\text{p}}\left(T_{\text{s},i} - T_{\text{c},i}\right) \tag{9.114}$$

$$Q_{\text{hc,e}} + Q_{\text{rh,e}} = G_{\text{a,e}}c_{\text{p}}\left(T_{\text{c,e}} - T_{\text{mix,e}}\right) + G_{\text{a,e}}c_{\text{p}}\left(T_{\text{s,e}} - T_{\text{c,e}}\right) \tag{9.115}$$

T_{mix} is given by Eq. (9.103). After the substitution and rearrangement, the cost function can be put as:

$$Q_{\text{thm}} = (T_{\text{r}} - T_{\text{o}})\left(G_{\text{a,e}}c_{\text{p}}\beta_{\text{OA,e}} - G_{\text{a},i}c_{\text{p}}\beta_{\text{OA},i}\right) + 2G_{\text{a},i}c_{\text{p}}\left(T_{\text{r}} - T_{\text{c},i}\right) + \text{Load}_{\text{e}} - \text{Load}_{\text{i}} \tag{9.116}$$

To minimize the thermal energy consumption, it is desirable to decrease the value of the first term on the right side, which means to increase $\beta_{\text{OA,i}}$ and decrease $\beta_{\text{OA,e}}$. This is beneficial to the entire system, since the corresponding outside air ratios for IAQ consideration also have an inverse correlation in IAHU. Thus, we have the equivalent cost function from the first term of Eq. (9.116):

$$f = \beta_{\text{IAQ,dsg,e}}(1 - \varphi) - \gamma\varphi\left(1 - \frac{\beta_{\text{IAQ},i}}{\beta_{\text{OA},i}}\right) - \varphi\beta_{\text{OA},i} \tag{9.117}$$

Similar to the analysis of condition A, the extremum lies at the boundary instead of on the inside. The partial derivatives on both variables are negative. Therefore, the upper boundary value on both $\beta_{\text{OA},i}$ and γ minimizes the energy consumption.

$$\beta_{\text{OA},i} = 1 \tag{9.118}$$

$$\beta_{\text{OA,e}} = \max(\beta_{\text{IAQ,e}}, 0, \beta_{\text{eco,e}}) \tag{9.119}$$

C* and *D: $T_{c,e} \leq T_o \leq T_r$

In this condition, both regions need mechanical cooling. It is possible that the enthalpy of outside air is higher than that of indoor air. If so, a dehumidification might be involved in the air conditioning process. The thermal energy consumption is therefore defined in terms of air enthalpy.

The thermal energy cost in each region is:

$$Q_{thm,i} = G_{a,i}\left(h_{mix,i} - h_{c,i}\right) + G_{a,i}\left(h_{s,i} - h_{c,i}\right) \tag{9.120}$$

$$Q_{thm,e} = G_{a,e}\left(h_{mix,e} - h_{c,e}\right) + G_{a,e}\left(h_{s,e} - h_{c,e}\right) \tag{9.121}$$

The mixed air enthalpy for both VAHUs can be put as:

$$h_{mix} = \beta_{OA} h_o + (1 - \beta_{OA}) h_r \tag{9.122}$$

Taking the interior region for a deduction, we have:

$$Q_{thm,i} = G_{a,i}(\beta_{OA} h_o + (1 - \beta_{OA}) h_r - h_{c,i}) + G_{a,i}(h_{s,i} - h_{c,i}) \tag{9.123}$$

Since $h_{s,i} - h_{c,i} = (h_{s,i} - h_{r,i}) + (h_{r,i} - h_{c,i})$, replacing the last term in the previous equation:

$$\begin{aligned} Q_{thm,i} &= G_{a,i}(\beta_{OA} h_o + (1 - \beta_{OA}) h_r - h_{c,i}) + G_{a,i}(h_{s,i} - h_r) + G_{a,r}(h_r - h_{c,i}) \\ &= 2G_{a,i}(h_r - h_{c,i}) + G_{a,i}\beta_{OA,i}(h_o - h_r) - \mathrm{Load_i} \end{aligned} \tag{9.124}$$

A similar expression can be obtained for the thermal energy consumption in exterior region. The total building's thermal energy consumption, after combination and rearrangement, under this condition is:

$$\begin{aligned} Q_{thm} = {} & 2G_{a,i}(h_r - h_{c,i}) + 2G_{a,e}(h_r - h_{c,e}) - \mathrm{Load_i} - \mathrm{Load_e} + (h_o \\ & - h_r)(G_{a,i}\beta_{OA,i} + G_{a,e}\beta_{OA,e}) \end{aligned} \tag{9.125}$$

For *C*: dry air condition ($h_o \leq h_r$) the last term in Eq. (9.125) is negative. This means the higher the OA intake, the lesser the energy consumption. The equivalent cost function is as follows:

$$f = -\beta_{IAQ,dsg,e}(1 - \varphi) + \gamma\varphi\left(1 - \frac{\beta_{IAQ,i}}{\beta_{OA,i}}\right) - \varphi\beta_{OA,i} \tag{9.126}$$

This is an opposite function to that of condition A. The upper boundary of $\beta_{OA,i}$ and lower boundary γ gives the minimum. No recirculation is needed between the two regions.

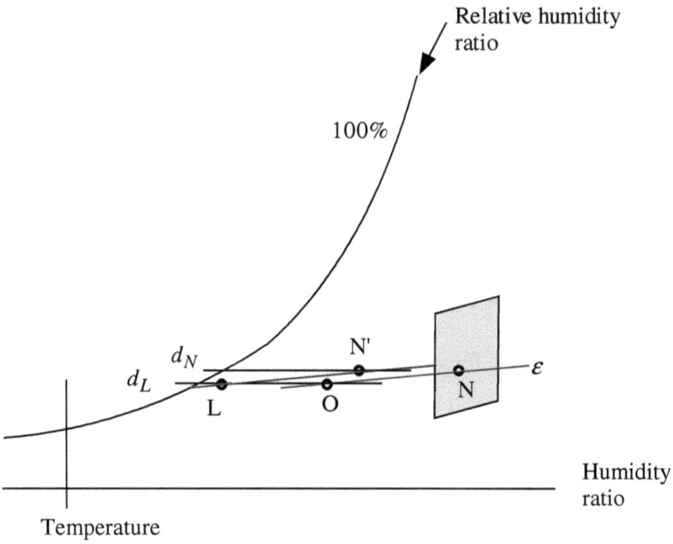

Fig. 9.34 Thermal process of partial load condition

$$\beta_{\mathrm{OA},i} = \beta_{\mathrm{OA,e}} = 1 \tag{9.127}$$

For ***D***: humid air condition $(h_{\mathrm{o}} \geq h_{\mathrm{r}})$

With conventional cooling dehumidification approach, the mixed air is processed to the corresponding dew point (e.g., 12.8 °C) of room air before being distributed into the space.

A complication could occur in humid mild weather when the sensible load is very low. As illustrated in Fig. 9.34, the conditioned supply air needs to be reheated from point L to O in order to attain an acceptable indoor air status *N* in the diamond area. Or the space will be overcooled to *N'*. Under this circumstance, a decoupled sensible load and latent load processing is desired. The analysis for an optimum system operation is conducted briefly.

Equation (9.125) holds true for the humid condition. Dropping out the unchangeable terms, the changeable parts are recollected as below:

$$f = 2G_{\mathrm{a,e}}(h_{\mathrm{r}} - h_{\mathrm{c,e}}) + (h_{\mathrm{o}} - h_{\mathrm{r}})(G_{\mathrm{a},i}\beta_{\mathrm{OA},i} + G_{\mathrm{a,e}}\beta_{\mathrm{OA,e}}) \tag{9.128}$$

During a normal operation, the first term on the right side is regarded as a fixed value with a given supply airflow rate since $h_{\mathrm{c,e}}$ is fixed by the room air dew point for the purpose of dehumidification. The total airflow rate is modulated to meet the building load. It is easy to find that the minimum points exist at the low boundary of the two OA ratios, with $\beta_{\mathrm{OA},i} = \beta_{\mathrm{IAQ},i}$ and correspondingly $\beta_{\mathrm{OA,e}} = \beta_{\mathrm{IAQ,dsg,e}}$ There is no controlled air transfer between the two regions.

When a partial load happens in humid mild weather, the following inequity exists:

$$\left(h_{s,e} = h_r - \frac{\text{Load}_e}{G_{a,e}}\right) \geq h_{c,e} \tag{9.129}$$

In this condition, the airflow rate is first decreased until the minimum threshold. If the sensible load keeps dropping across this point, a potential reheat is needed in the system which wastes energy to balance the load discrepancy. The additional cost is likely to be eliminated if $h_{c,e}$ can be readjusted to only cover the sensible load. This is achieved by reducing $\beta_{OA,e}$ to zero and introducing the outside air from the interior VAHU. The OA intake from the interior region β_i is reversely obtained with Eq. (9.13). With this operation, the zone sensible load and latent load are decoupled.

During such a decoupling process in IAHU, the first term on the right side in Eq. (9.128) decreases and the second term increases. The evaluation of the question becomes a magnitude comparison of the savings and the cost of the two terms.

The normalized energy saving from the decoupled first term is:

$$f_1 = 2(h_{s,e} - h_{c,e})(1 - \varphi) \tag{9.130}$$

And the extra energy cost of the second term, $(h_o - h_r)((G_{a,i}\beta_{OA,i} + G_{a,e}\beta_{OA,e}) - (G_{a,i}\beta_{IAQ,i} + G_{a,e}\beta_{IAQ,dsg,e}))$, can be normalized as:

$$f_2 = (h_o - h_r)(\varphi\beta_{OA,i} - \varphi\beta_{IAQ,i} - (1 - \varphi)\beta_{IAQ,dsg,e}) \tag{9.131}$$

In this operation, $h_{s,e}$ is determined by the real-time sensible load in the conditioned exterior region. The optimum can be found by comparing f_1 and f_2 with real-time measurements.

When $h_{s,e} >> h_{c,e}$, $f_1 \geq f_2$ is generally, if not always, true in humid mild weather, because the ratio term $2 \cdot (1 - \varphi)$ in f_1 is much greater than the ratio term $(\varphi\beta_{OA,i} - \varphi\beta_{IAQ,i} - (1 - \varphi)\beta_{IAQ,dsg,c})$ in f_2. Therefore, in decoupling mode, the minimum OA intake can be found by setting $\beta_{OA,e}$ to zero in Eq. (9.13):

$$\beta_{OA,i} = \frac{\gamma\varphi\beta_{IAQ,i}}{\gamma\varphi - \beta_{IAQ,dsg,e}(1 - \varphi)} \tag{9.132}$$

$$\beta_{OA,e} = 0 \tag{9.133}$$

Otherwise, a normal operation should be retained.

E: $T_r \leq T_o$

Since the OA temperature is greater than the room air temperature, both regions are in a mechanical cooling mode. The deduction is similar to condition D with identical results obtained.

References

1. DOE: Energy Efficiency Trends in Residential and Commercial Buildings. The Office of Energy Efficiency and Renewable Energy (2008)
2. Liu, M., Claridge, E.D.: Improving building energy system performance by continuous commissioning. In: Proceedings of the 3rd National Commissioning Conference, San Diego, CA (1997)
3. Liu, M., Feng, J.: Impacts of static pressure reset on VAV system air leakage, fan power, and thermal energy. ASHRAE Transactions. **116**(1), 428–430 (2010)
4. Ke, Y., Mumma, A.S.: Optimized supply-air temperature (SAT) in variable-air-volume (VAV) systems. Energy **22**(6), 601–614 (1997)
5. Fredrik, E., Dennis, J.: Optimal supply air temperature with respect to energy use in a variable air volume system. Energy Build. **36**(3), 205–218 (2004)
6. Reddy, T.A., Liu, M., Claridge, D.E.: A study of energy use and satisfactory zone ventilation of different outdoor air ventilation strategies for terminal reheat variable air volume systems. Energy Build. **29**(1), 65–75 (1998)
7. Yu, Y., Xu, K.: A smart logic for conference room terminal box of single duct VAV system. In: Proceedings of 7th International Conference for Enhanced Building Operations 2007. San Francisco, California (2007)
8. ASHRAE: ASHRAE Handbook- HVAC Systems and Equipment. American Society of Heating, Refrigeration and Air-conditioning Engineers, Inc., Atlanta, GA (2008)
9. Wang, G., Liu, M.: Optimal outside air control for air handling units with humidity control. In: Proceedings of the 6th International Conference for Enhanced Building Operations, Shenzhen, China (2006)
10. Dhital, P., Besant, R., Schoenau, G.J.: Integrating run-around heat exchanger systems into the design of large office buildings. ASHRAE Transactions. **101**(2), 979–999 (1995)
11. Harriman, L.G., Judge, J.: Dehumidification equipment advances. ASHRAE J. **44**(8), 22–29 (2002)
12. ASHRAE: Ventilation for acceptable indoor air quality. ASHRAE/IES Standard 62.1-2004 (2004)
13. Knebel, D.E.: Simplified energy analysis using the modified BIN method. ASHRAE (1983). ISBN 0-910110-39-5
14. Katipamula, S., Claridge, D.E.: Use of simplified system models to measure retrofit energy savings. J. Sol.Energy Eng. **115**, 57–68 (1993)
15. Liu, M., Claridge, D.E.: Application of calibrated HVAC system models to identify component malfunctions and optimize the operation and control schedules. In: Proceedings of Solar Energy Engineering, ASME, Maui, Hawaii (1995)
16. Liu, M., Claridge, D.E.: Use of calibrated HVAC system models to optimize system operation. J. Sol. Energy Eng. **120**, 131–138 (1998)
17. EnergyPlus: Engineering Document, Version 6.0, U.S. Department of Energy (2010)
18. Sun, J., Reddy, A.: Optimal control of building HVAC and R systems using complete simulation-based sequential quadratic programming (CSB-SQP). Build. Environ. **40**(5), 657–669 (2005)
19. Wang, Y.J., Ying, L.: Global optimization for special reverse convex programming. Comput. Math Appl. **55**(6), 1154–1163 (2008)
20. Arturo, M.C., Lorenz, T.B.: A stable elemental decomposition for dynamic process optimization. J. Comput. Appl. Math. **120**(1), 41–57 (2000)
21. An, L.T.H., Tao, P.D., Thoai, N.V.: Combination between global and local methods for solving an optimization problem over the efficient set. Eur. J. Oper. Res. **142**(2), 258–270 (2002)

22. Huang, Y.J., Reklaitis, G.V., Venkatasubramanian, V.: Model decomposition based method for solving general dynamic optimization problems. Comput. Chem. Eng. **26**(6), 863–873 (2002)
23. Abdelouahed, H.: Two-level primal–dual proximal decomposition technique to solve large scale optimization problems. Appl. Math. Comput. **160**(3), 921–938 (2005)

Chapter 10
Modeling and Control Strategies for VAV Systems

10.1 Background and Research Status

In a heating, ventilating, and air-conditioning (HVAC) system, operation of the air-side system has a significant influence on the overall performance of a building energy system. For example, in a worst-case scenario instability in the air-side economizer could trigger instability at the central chilled water plant or vice versa. In addition, because the economizer introduces extra outdoor air into the building, it could impact building pressure across the interior and exterior of the building and between the zones, which are served by multiple systems. Due to the high dynamics and correlations among the air-handling unit (AHU) components and the terminals, air systems—including variable air volume (VAV)—are prone to energy inefficiency with improper operations.

The term VAV first came into existence in the mid- to late 1960s after studies by Urban [1]. After the world energy crisis of 1970s, these systems gained popularity in the UK and other countries across Europe in the early 1980s as part of the efforts by engineers to come up with energy-efficient air-conditioning systems [2]. This was necessary due to the cost of energy that was increasing at that time and has continued to increase up to date. VAV systems by definition are simply air-conditioning (AC) systems that are designed to promote existence of constant temperature in a conditioned space by varying the volume of air supplied to the conditioned space instead of varying the temperature of supplied air [3]. Therefore, these systems vary supply air volume at a constant temperature in order to meet the demand caused by the changing heat load in the conditioned space [4]. Generally, VAV systems can be broadly classified into two categories as chilled water VAV air-conditioning systems and direct expansion (DX) cooling coil VAV air-conditioning systems.

The basic components of a VAV air-conditioning system include a central AHU with a variable speed supply fan (can vary volume of air), coils used for heating or cooling, controls, filters, mixing box, return or relief fan, air-supply duct, VAV

Y. Yao and Y. Yu, *Modeling and Control in Air-conditioning Systems*,
Energy and Environment Research in China, DOI 10.1007/978-3-662-53313-0_10

terminal unit (device) connected to thermostats and supply diffusers, and return duct or plenum. VAV systems work on the principle of opening or closing mechanical dampers or by modulating the airflow through mixing boxes powered by VAV fans as loads in various conditioned spaces of a building. For instance, if a given conditioned space requires more cooling, the damper to that space is opened wider to increase the inflow of cold air until the required temperature is achieved. During the opening of the damper, there is pressure drop in the supply duct which signals the supply fan to increase air delivery. On the other hand, if an area is too cool and requires temperature rise, the damper is gradually closed so as to reduce the inflow volume of cold air. This is usually applied in combination with variable speed drives (VSDs). The result is decrease in airflow which results in cutting down fan power needed thus saving energy [5]. In a further effort to reduce the energy requirements, most VAV systems utilize the return air in order to cut down the power requirement and energy use when the outdoor temperature is higher than exhaust air temperature [6].

Figure 10.1 depicts the AHU used in the VAV system. It maintains the supply air temperature to the terminal boxes for air-conditioning. There are four controllers regulating the supply air temperature, dampers, supply air static pressure, and return airflow rate. The supply air temperature is controlled by TC-1 which modulates the openness of valves on the cooling and heating coils. DC-1 controls the three dampers to ensure the minimum outdoor air intake for ventilation and also utilize

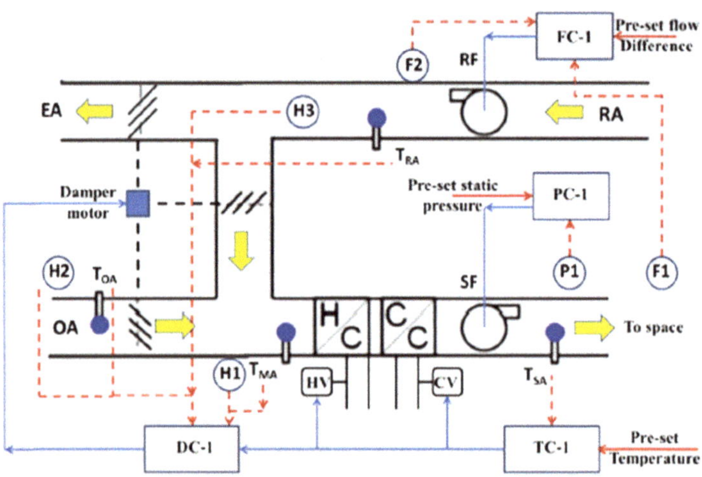

TC-1: Supply air temperature controller — HV: Heating coil valve
DC-1: Damper controller — CV: Cooling coil valve
PC-1: Supply air static pressure controller — SF: Supply fan
FC-1: Return air flow rate controller — RF: Return fan
T: Temperature sensor; H: Humidity sensor; P: Pressure sensor; F: Flow sensor.

Fig. 10.1 A typical VAV air-handling unit system

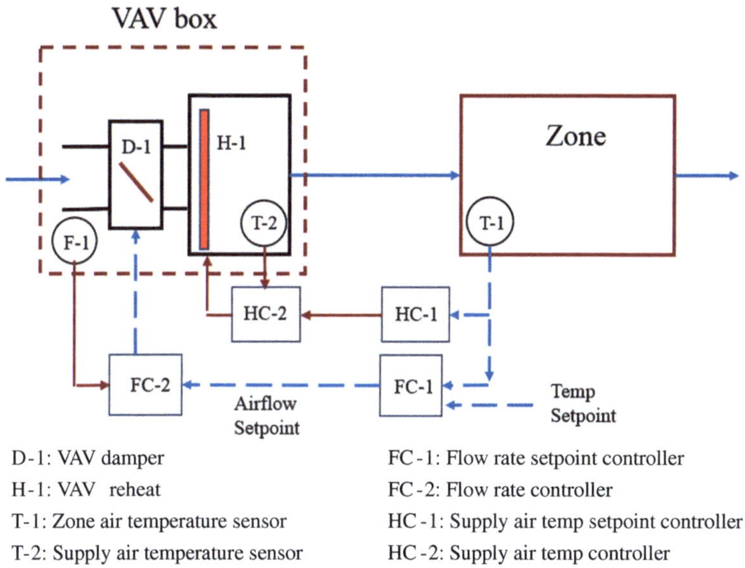

Fig. 10.2 Illustration of VAV terminal box with dual maximum control

free cooling of the outdoor air when it is mild. TC-1 is linked to DC-1 for appropriately mixing air temperature. With a fixed supply air temperature, the AHU responses to the changing demand from the downstream terminals by adjusting the supply airflow rate. The variation is reflected by the static pressure in the supply air duct. The supply air fan is regulated by PC-1 to track the set point. The return air fan is controlled by FC-1 to circulate the room air back to the AHU. It is also used to balance the airflow rate in and out of the building and thus the building pressure.

The air downstream in each individual zone is regulated by terminal boxes connecting to the supply air duct. The pressure-independent VAV boxes with reheat are employed as terminal units. The box is depicted as in Fig. 10.2. This unit regulates the temperature and the amount of air into the space to maintain the zone temperature. FC-1 is used to compute the setpoint of supply airflow rate according to the variation of cooling load, which is reflected by the room air temperature sensor. If the supply flow varies due to the change of the static pressure, the measured flow and the setpoint of FC-1 are used as inputs to FC-2 for regulating the damper position. Controller HC-1 gives the supply air temperature setpoint for HC-2 to enable dual maximum control as in ASHRAE Standard 90.1. The reheat is regulated by HC-2 in the heating mode to maintain the supply.

VAV systems are based on the indoor load or other parameters. The indoor air parameters are made to achieve the design requirements of the air-conditioning system through the adjustment of the quantity of supply air in the conditioning system in order to supply air at constant temperature into the conditioned space [7]. Since most of the time the air-conditioning system operates at part-load conditions, decreasing the quantity of supply air makes the air-conditioning cooling capacity to match the indoor heat load. This can reduce fan energy consumption. The system has been widely used in the Mainland China since the energy efficiency of the VAV air-conditioning system is significant. This technology is maturing, but is still at its infancy stage when it comes to application in the marine industry.

These systems have several advantages over other HVAC systems [8–12]. These include less fan capacity compared to constant volume systems since in VAV systems only the needed air is used; greater flexibility with respect to varying loads; improved indoor environment; the system can incorporate an economizer to utilize the outside air to provide cooling at times when temperature is appropriate; reduced size of the main ducts since there is no simultaneous coincidence of the maximum cooling/heating load demand in all spaces. The above features not only make VAV air-conditioning systems to have high design requirements, but also make the control of the VAV systems more complex such that the traditional control methods become ineffective. To develop more effective methods of control, simulation models for VAV system are needed to be established.

Numerous studies have been accomplished in regard to modeling VAV systems. Wang and Burnett [13] used component flow resistance coefficients derived from field measurements together with fan curve based on design data to establish a steady-state system model. Modulation of static pressure setpoint was realized via an ideal controller. The control enables the critical zone damper to remain at 100 % open while providing the required airflow demands. In their study, the authors investigated the effect of load distribution on performance of static pressure reset by carrying out simulations on two load scenarios. One scenario had uniform load demands across all the VAV terminal boxes while the other scenario had maximum zone load that was higher than average load. The authors concluded that uniformly distributed loads could increase energy savings when comparing static pressure reset and constant pressure control.

Khoo et al. [14] developed steady-state nonlinear models for three VAV terminal units. This study concluded that VAV terminal units could not be accurately represented by approximations that depend on data from dampers only, and damper models could underestimate the volume flow rate by more than 50 %.

Parameshwaran et al. [15] developed steady-state models of a VAV air-conditioning system. They used the models to formulate and solve constrained optimal control problem. A two-objective genetic algorithm was applied to the modeled system with setpoints of control inputs such as supply air temperature, duct static pressure, and conditioned space air temperature were directed into a Fuzzy Logic Controller for tuning. The optimization process led to substantial amount of fan and compressor energy saving.

Demand controlled ventilation (DCV) is defined as automatic ventilation that uses occupancy to make adjustments. This strategy reduces outdoor air intake rates below design rates when the actual occupancy of zones served by the system goes below the designed value [16]. The approach comprises hardware, software, and strategy for control and is an important part of building's ventilation design. The main shortcoming of DCV is that its implementation may be too complicated and require proper hardware and software to operate. Also, location and number of CO_2 sensors or real-time data pose a challenge to the efficiency of the control algorithm if the system's outdoor air (OA) intake is dynamic. As part of their efforts to improve DCV, Yu et al., designed new integrated demand controlled ventilation (IDCV) for single-duct VAV system with conference rooms [17]. This logic resets both the minimum and maximum airflow rates of the terminal boxes based on the occupancy. The authors used a model for one office building to demonstrate energy savings and show how the indoor air ventilation can be satisfied under different conditions. This methodology can ensure acceptable Indoor Air Quality (IAQ) and energy savings with lower AO intake ratio. Further, Yu et al. [18] designed a smart logic for conference room VAV terminal unit of single-duct system. Evaluation of their control algorithm was via simulation. They used the simulated results of thermal performance and energy consumption to investigate conventional and improved control logic sequences. Results showed that the proposed strategy can improve both IAQ and energy saving.

Cho [19] developed an algorithm for terminal unit control with variable minimum rate of airflow and used it in conventional single-duct VAV terminal box control sequences. Validation of simulation results was performed through evaluation of the actual building for IAQ, comfort, and energy usage. The energy consumption and thermal performance of terminal units operating under two control algorithms were compared. The results showed that the ratio of constant minimum rate of airflow caused significant concurrent heating and cooling cycles; the terminal box could maintain room thermal comfort conditions to meet various load changes in addition to reducing fan power and saving reheat energy; and the energy usage of the variable minimum rate of airflow ratio was smaller compared to that of the conventional constant minimum rate of airflow ratio.

Hartman [20] developed a concept known as Terminal Regulated Air Volume (TRAV). This concept uses real-time terminal unit airflow demand to directly control supply fan speed instead of satisfying duct static pressure setpoint value [20, 21]. In utilizing TRAV control strategy, adequate flow is signified by nearly full or completely closed dampers while inadequate flow is signified by dampers which are fully or partially opened. In case of inadequate flow, the actual airflow of every damper is compared to its airflow setpoint value. Later on, Hartman [22] developed a fan speed control algorithm for his previous studies. Englander [23] carried out research on VAV system fan speed control and VAV terminal devices control. His control strategy was based on constant static pressure control strategy of minimizing the fan speed. Adjustments have been made to explore the VAV air-conditioning system energy consumption and fan performance before and after improvement; based on the response of VAV terminal unit inlet pressure variations, he used direct digital control (DDC) of VAV terminal unit control strategy to optimize the experimental study. Later on, Englander and Norford [24] suggested two control strategies. They modulated fan speed by utilizing the error signal of primary airflow from one or more zones. They compute the error signal as the maximum or average deviation of the setpoint from measured airflows. Rather than determining the difference between the setpoint and the measured airflows, Warren and Norford [25] proposed a control strategy in which terminal box controllers trigger an alarm whenever the system could not meet required airflow. The number of alarms determines setpoint value for adjusting static pressure. When the flow sensors function properly, the system can save up to 42 % of fan power energy consumption.

Wei et al. [26] designed improved reset algorithm for controlling air volume in which the air volume control logic was connected parallel to the reset schedule [26]. In this strategy, the position of the damper and the static pressure of the highest zone formed the basis for comparing the output. The smaller output signal was then selected and sent to the variable frequency drive (VFD) controller. The shortcoming of this method is that it is not applicable for buildings with pneumatic terminal unit controllers.

Nassif and Moujaes [27] proposed split-signal damper control strategy. In this control strategy, the outdoor air was controlled only by one damper while other dampers were kept fully opened. They found that since the two modulating devices (dampers) were usually fully opened under occupied conditions, the strategy could facilitate realization of minimum static pressure drop in inlet air dampers and this resulted in minimum energy consumption in both supply and return fans.

Haasl et al. [28] proposed the polling of terminal units at an interval of five minutes and that the adjustment of the setpoint value of static pressure be carried out between a minimum and maximum value to ensure that at least one damper is always maintained at 95 % open position. This type of static pressure reset control was modified by Song et al. [29]. The authors carried out the modification process by using VSD speed as the basis for varying the minimum setpoint value. The speed range was from 30 to 70 %. Pang et al. [30] polled each terminal box at intervals of twenty minutes and used the maximum damper position using a dead band from 85 to 95 % as the basis for adjusting the setpoint value.

Liu developed an airflow control for VAV AC systems [31]. The author created a fan airflow station (FAS) whose role was to calculate airflow using measured fan speed and fan head. This was done in order to avoid inaccurate airflow measurements. Liu and Liu [32] as well as Wu et al. [33] modified the previous strategy developed by Liu et al. [34] in their efforts to address system stability. Their modifications accounted for the system load distribution profile using a load factor that increased with zone load ratio.

Zheng et al. [35] suggested that an effective control strategy for envelop-dominated buildings was linearly resetting the static pressure setpoint with reference to outdoor air temperature. In their strategy, energy savings were heavily dependent on the minimum airflow setting and the OA temperature range. After studying variable speed fan control in air-conditioning systems, Murphy [36] claimed that fan control could significantly reduce system energy use but with challenges of ensuring proper ventilation that was not common in constant fan speed control [36].

Meng et al. [37] developed steady-state and dynamic models of a multi-zone VAV air-conditioning system as well as variable water volume air-conditioning system. They introduced optimal control for energy saving in large-scale system. The study achieved substantial energy saving through integration of iterative learning and sequential quadratic programming based on steady-state hierarchical optimization control scheme.

Recently, Koulani [38] investigated response and energy-saving potential of a VAV system working under static pressure reset control strategy. The author concluded that static pressure reset (SPR) control was able to maintain low pressure setpoints throughout system operation without compromising comfort compared to constant static pressure control.

In this chapter, modular models of key components in VAV system are firstly to be established based on Simulink tool. The modular approach helps model developers add new components to a complex system without big changes to existing code, update documentation, and maintain code much more effectively. Then, the component models are integrated into a complete VAV system, and several control schemes are to be investigated based on the system model. Finally, the basic control logic and functional test scripts development for VAV system are introduced.

10.2 Modular Modeling with Simulink Tool

Simulink tool is add-on software in MATLAB software developed by MathWorks [39]. Simulink tool has proved to be a versatile tool to control design since any regulatory process of a system can be established by utilizing standard blocks found in Simulink library as shown in Fig. 10.3 [40]. New blocks can be created from existing ones and integrated into sub-systems to represent well-defined parts of an entire VAV system. The Simulink tool has a wide adaptation, clear structures, and processes as well as fine simulation. It is close to reality and has high efficiency, flexible, etc. It is widely used in the modeling and simulation of linear and non-linear systems.

The basic methods to use the Simulink tool are summarized as below:

Through left clicking on file-open, we can open an existing model as shown in Fig. 10.4. This will allow us to choose the folder in which the previous projects are stored.

Through clicking on file-new-model, we can create a new blank model (see Fig. 10.5). Then, select the desired block and drag it into the blank model as shown in Fig. 10.6.

Modules can be freely copied and moved. Double clicking on the module allows us to enter and modify module parameters. For instance, clicking on 'Constant'

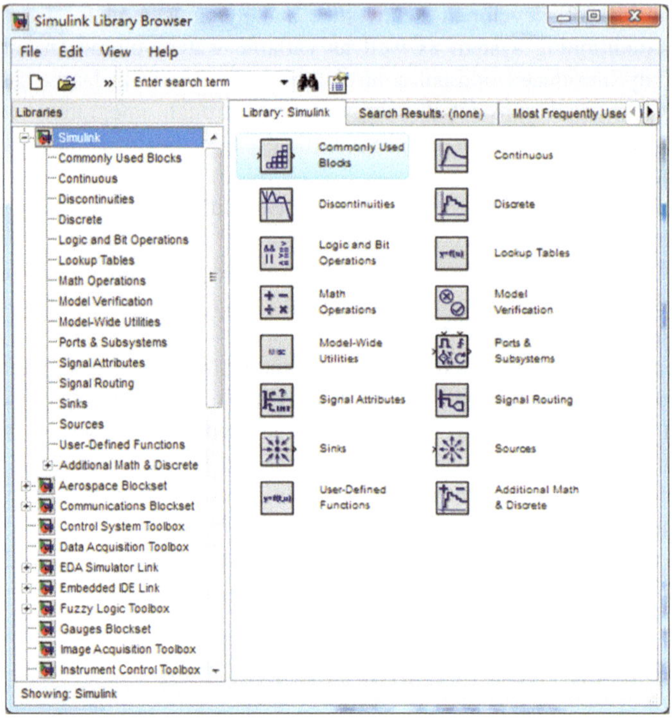

Fig. 10.3 Simulink library

module, a window pops up as shown in Fig. 10.7, which allows the 'Constant Value' to be modified.

After modifying module parameters, desired features will be connected to the simulation modules in accordance with requirements. Click the button to start the simulation as shown in Fig. 10.8.

Figure 10.8 can be modeled into a sub-system by selecting all blocks except the scope block. To achieve this, left click the mouse button at a point outside the blocks to be selected in the model. This will generate rectangle around all of the required blocks. Then, with the left mouse button still held, drag the mouse away from the original location. The highlighted blocks will be as shown below. The selected blocks are indicated by small squares at the block corners. With the required blocks already selected, they can be encapsulated in a sub-system by following the path, edit-create sub-system as shown in Fig. 10.9.

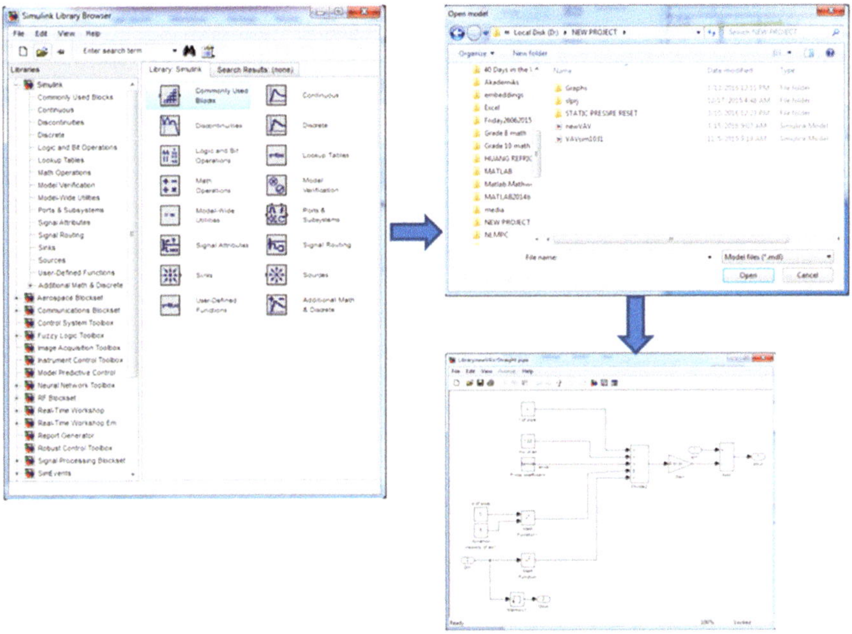

Fig. 10.4 Opening a stored project in Simulink tool

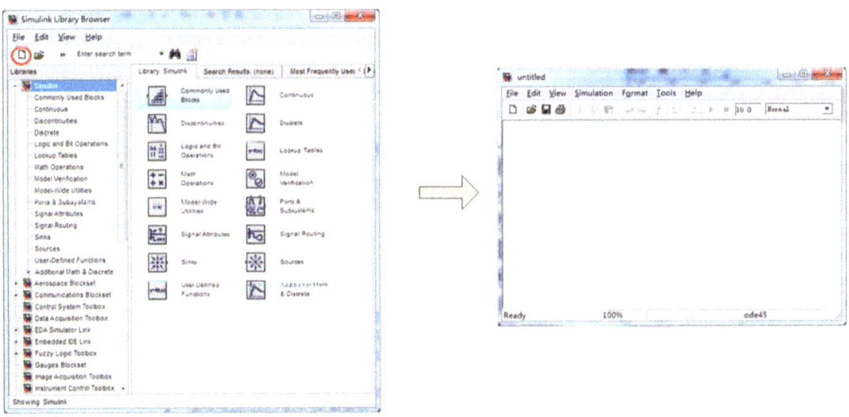

Fig. 10.5 Creating a new project in Simulink tool

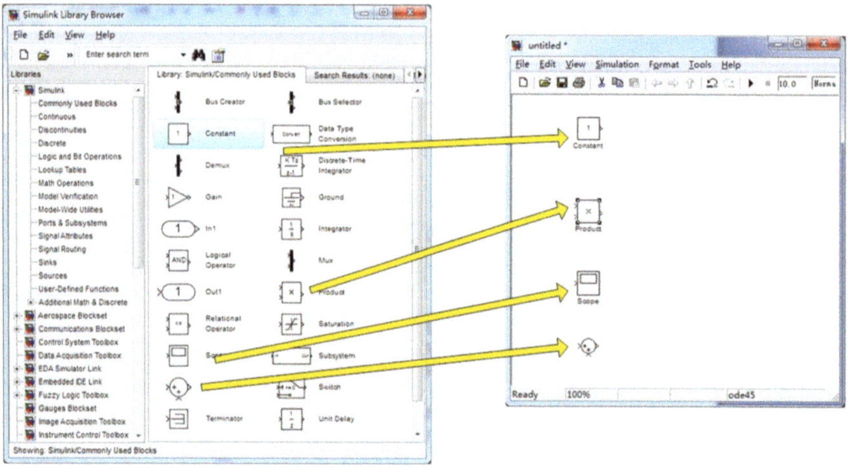

Fig. 10.6 Dragging blocks into a new project in Simulink tool

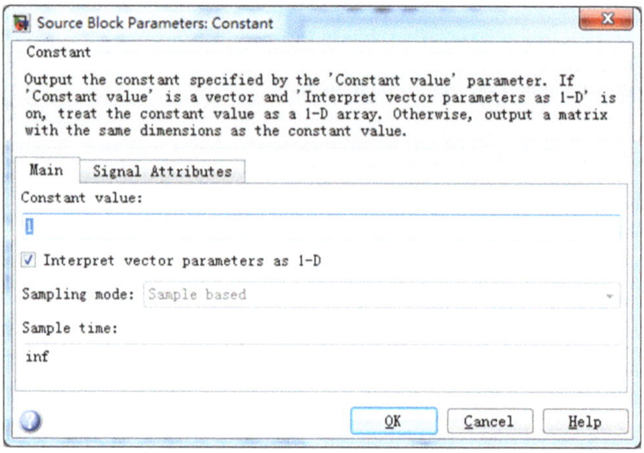

Fig. 10.7 Modifying module parameters in Simulink tool

10.3 Model Library for Components of VAV System

10.3.1 VAV Terminal Unit

This is the most important among all the VAV system components. Therefore, it is crucial to establish an accurate model for VAV terminal units. In modeling a VAV terminal unit, we do not consider its control loop rather we only focus on the air transport loop. Here, we will model single-duct VAV terminal unit. The principle of

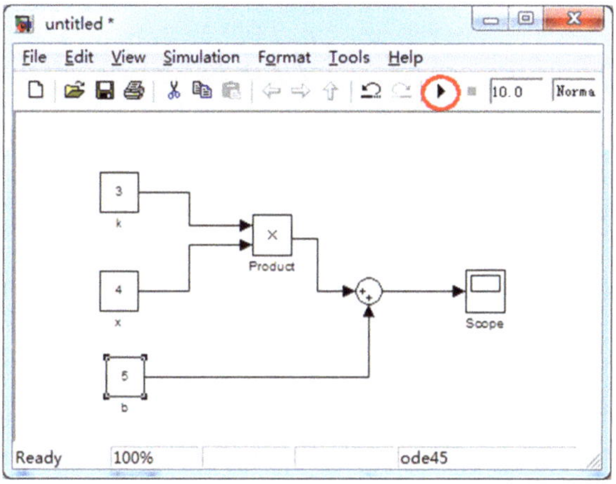

Fig. 10.8 Running simulation in Simulink tool

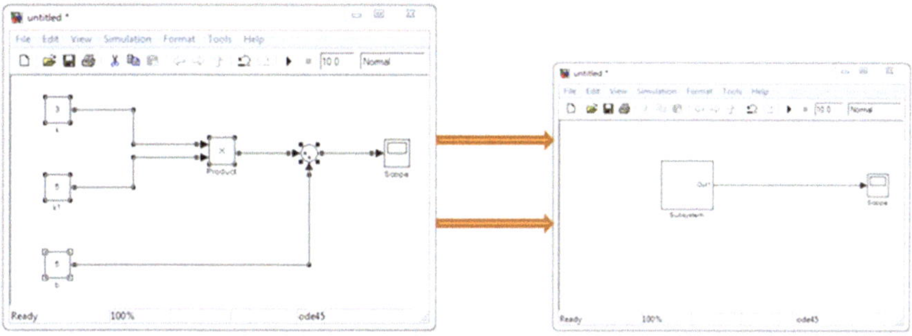

Fig. 10.9 Creating sub-system in Simulink

operation of the single-duct VAV terminal devices also known as restrictive-type terminal devices is based on the deviation between room temperature and setpoint temperature. The objective of supply air adjustment is achieved by changing the cross-sectional area of airflow. This is executed through either opening or closing of the terminal unit damper (valve). The main function of a damper is to control airflow rate in ducts of a VAV system. It accomplishes control of flow rate by varying resistance to airflow in the supply duct leading to a given conditioned space. When the damper rotates in an air stream, it regulates air passage to the space. A common VAV terminal unit mainly contains controller, air speed sensor, room temperature sensor, regulating valve, and other components. VAV terminal device can be regarded as a variable resistance element and the resistance coefficients can be seen as a function of the opening degree of the damper. The relationship between airflow pressure loss and the damper opening in a VAV terminal device can be expressed by Darcy–Weisbach equation as below [41, 42]:

$$\Delta P = k\frac{\rho_{\mathrm{a}}}{2}\left(\frac{G_{\mathrm{a}}}{A}\right)^2 \tag{10.1}$$

In which ΔP—pressure loss through the terminal device (Pa); k—damper drag coefficient (friction factor); ρ_{a}—air density (kg/m^3); A—damper flow cross-sectional area when the terminal device is fully opened (m^2); G_{a}—volume flow rate at the terminal device (m^3/s);

Damper drag coefficient (k) can be calculated by the following formula as [43, 44];

$$\mathrm{k} = \exp(c_1 + c_2\theta^{c_3}) \tag{10.2}$$

where θ is the damper angle in radians, $\theta = 0$ when the damper is fully opened and $\theta = \frac{\pi}{2}$ when the damper is fully closed. The constant coefficients, c_1, c_2 and c_3, are determined by damper resistance characteristics. They can be obtained through experimental tests.

However, in some cases (e.g., the case study later in this chapter), pressure loss model of a VAV terminal device may be better to be expressed by Eq. (10.3).

$$\Delta P = a_1(1 + 100 \times C)^{a_2}G_{\mathrm{a}}^2 \tag{10.3}$$

where C is the percentage of opening degree, $C = \frac{\pi/2-\theta}{\pi/2} \times 100\,\%$; a_1 and a_2 are undetermined coefficients. The unit of (G_{a}) in Eq. (10.3) is m^3/h for the convenience of data fitting. The Simulink block diagrams for Eq. (10.3) are presented in Fig. 10.10a.

There are two ways for the usage of Eq. (10.3): One is to determine pressure loss of a VAV terminal device according to its percentage of opening degree (C) and airflow rate (G_{a}), and the corresponding sub-system package is presented in Fig. 10.10b; the other is to determine the percentage of opening degree (C) of a VAV terminal device when the inlet static pressure (P_{in}) and airflow rate (G_{a}) is given, and the corresponding sub-system package is presented in Fig. 10.10c.

10.3.2 Variable Speed Fan

For a variable speed fan, design operation performance conditions (including the fan total pressure and power) and the volumetric flow can be expressed by [45]:

$$P_{\mathrm{t,o}} = b_4G_{\mathrm{o}}^4 + b_3G_{\mathrm{o}}^3 + b_2G_{\mathrm{o}}^2 + b_1G_{\mathrm{o}} + b_0 \tag{10.4}$$

where $P_{\mathrm{t,o}}$ is the total pressure at design conditions (Pa); G_{o} is fan volumetric flow rate at design conditions (m^3/s); b_0, b_1, b_2, b_3 and b_4, are coefficients to be determined.

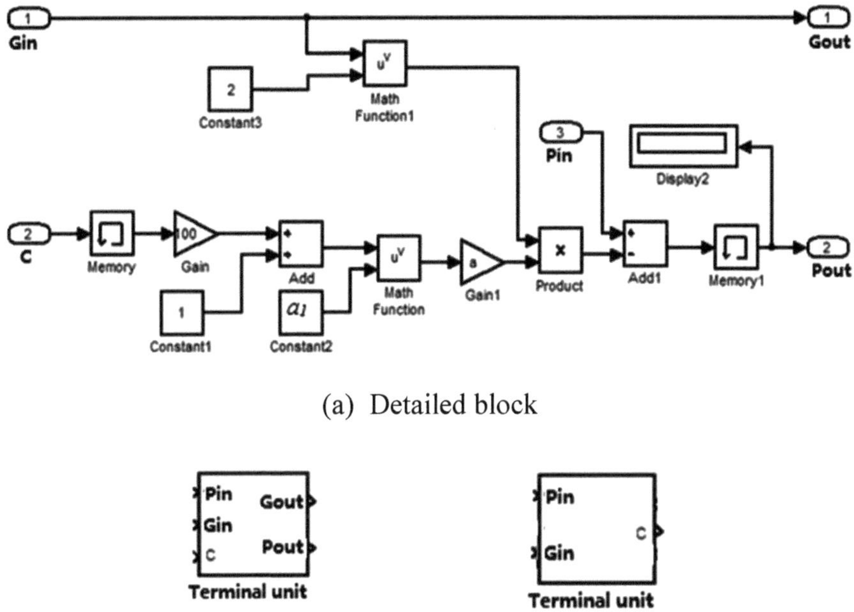

(a) Detailed block

(b) Subsystem package (c) Subsystem package

Fig. 10.10 Simulink block for terminal unit

For two sets of operation conditions or geometric similar fans and same working fluid based on fan affinity laws, the following equations come into existence:

$$\frac{G}{G_o} = \frac{n}{n_o} \tag{10.5}$$

$$\frac{P_t}{P_{t,o}} = \left(\frac{n}{n_o}\right)^2 \tag{10.6}$$

where n_o is fan speed at design conditions (rpm); n is fan speed at any conditions (rpm); G (m^3/s) and P (Pa) are volumetric flow rate and total pressure of the fan under the fan speed of n.

Therefore, at any fan speed, the total pressure (P_t) and power (N) of the fan can be written as:

$$P_t = a_4\left(\frac{n_o}{n}\right)^2 G^4 + a_3\frac{n_o}{n}G^3 + a_2G^2 + a_1\frac{n}{n_o}G + a_0\left(\frac{n}{n_o}\right)^2 \tag{10.7}$$

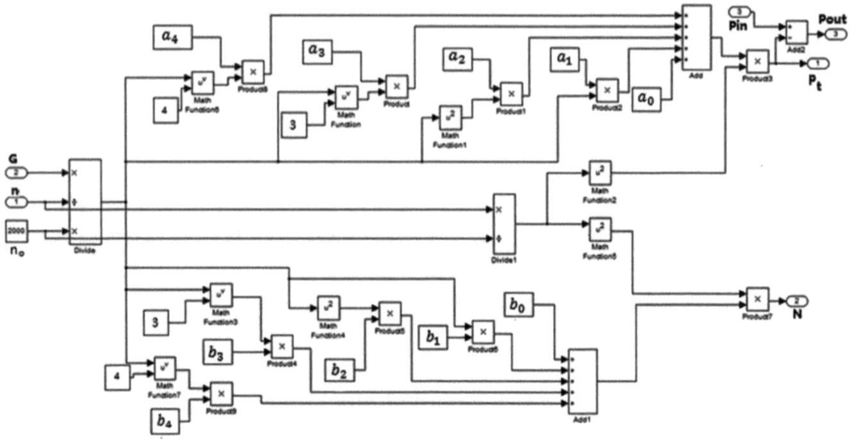

(a) Detailed block

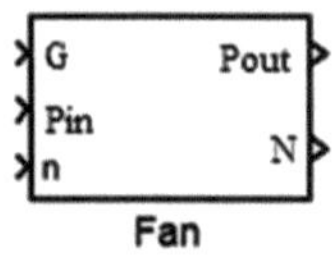

(b) Subsystem package

Fig. 10.11 Simulink block for fan model

$$N = b_4\left(\frac{n_o}{n}\right)^2 G^4 + b_3\left(\frac{n_o}{n}\right)G^3 + b_2G^2 + b_1\left(\frac{n}{n_o}\right)G + b_0\left(\frac{n}{n_o}\right)^2 \tag{10.8}$$

where $a_4, a_3, a_2, a_1, a_0, b_4, b_3, b_2, b_1$ and b_0 are undetermined coefficients.

In the fan model, the input variables are fan speed (n) and the volumetric flow needed (G). The output variables are fan power (N) and outlet pressure (P_{out}) The simulink block corresponding to Eqs. (7) and (8) for fan model is presented in Fig. 10.11.

10.3.3 Air Ducts

The air duct sub-systems mainly consist of straight ducts, elbow ducts, reducers, and three-way (tee) ducts. The main concern of modeling is the pressure drop generated by airflow inside the duct. The tee-pipeline requires the consideration of flow distribution problems of the main and the branch ducts.

The analysis of pressure loss due to flow patterns in rectangular ducts is more complex than that of circular ducts. The hydraulic diameter is often employed to study the flow resistance in non-circular ducts, which is defined as below:

$$d_e = \frac{4A}{U} \tag{10.9}$$

where d_e is hydraulic diameter or equivalent diameter (m); A is the cross-sectional area (m^2); and U is perimeter (m).

The two components of airflow resistance in the pipe are frictional resistance ΔP_f and local resistance ΔP_l. The frictional resistance is the resistance generated along the way due to the existence of friction between the air and pipeline, while the local resistance is due to changes in size of flow cross-sectional area, changes in the direction of flow of the fluid, or flowing through the resistance component to obstruct flow formation.

Resistance along the pipeline can be calculated by:

$$\Delta P_f = \lambda \frac{l_d}{d_e} \times \frac{\rho_a u_a^2}{2} \tag{10.10}$$

where l_d is length of duct, m; ρ_a is air density, kg/m^3; u_a is airflow velocity in the duct, m/s; λ is coefficient of resistance along the way, or friction coefficient, and it could be calculated by Eq. (10.11) [46]:

$$\frac{1}{\sqrt{\lambda}} = -2\log\left[\frac{\mathrm{k_s}}{3.7d_e} + \frac{2.51}{(Re)\sqrt{\lambda}}\right]. \tag{10.11}$$

where $\mathrm{k_s}$ is absolute roughness; Re is Reynolds number, $Re = \frac{\rho_a u_a d_e}{\mu}$; μ is the dynamic viscosity of fluid (kg/m s).

For a smooth pipe, the friction factor can be given by using Blasius equation as below [47];

$$\lambda = \frac{0.3164}{Re^{0.25}} \tag{10.12}$$

The local resistance (ΔP_l) is calculated by the following formula [48]:

$$\Delta P_l = \xi \frac{\rho_a u_a^2}{2} \tag{10.13}$$

where ξ is coefficient of local resistance.

10.3.3.1 Straight Duct

For the straight duct model, the relationship between the input and output parameters can be expressed as:

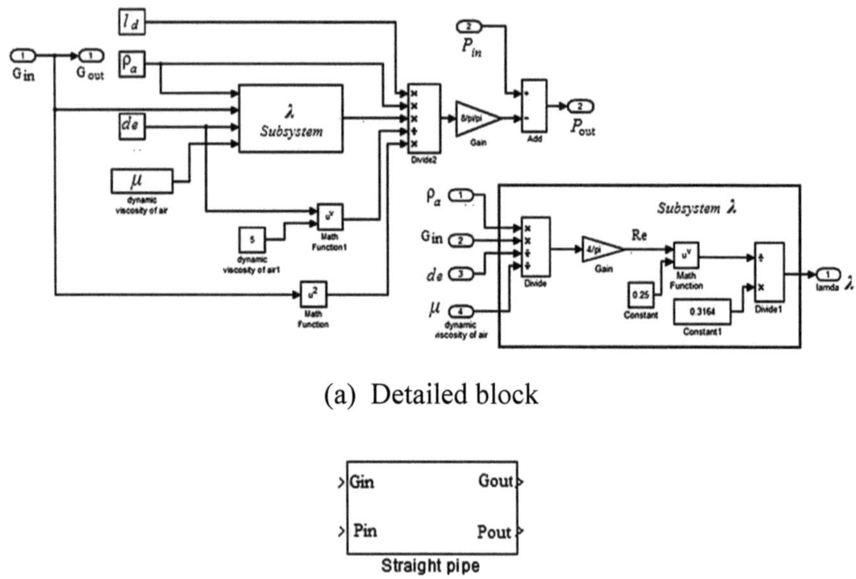

(a) Detailed block

(b) Subsystem package

Fig. 10.12 Simulink block for straight duct model. **a** Detailed block. **b** Sub-system package

$$G_{\text{in}} = G_{\text{out}} \tag{10.14}$$

$$P_{\text{out}} = P_{\text{in}} - \lambda \frac{l_{\text{d}}}{d_{\text{e}}} \times \frac{\rho_{\text{a}} u_{\text{a}}^2}{2} \tag{10.15}$$

Equation (10.15) can be expressed in terms of volumetric flow quantity (G_{in}) as below:

$$P_{\text{out}} = P_{\text{in}} - \frac{8}{\pi^2} \lambda \frac{l_{\text{d}}}{d_{\text{e}}^5} \times \rho_{\text{a}} G_{\text{in}}^2 \tag{10.16}$$

The Simulink block diagram for straight duct is shown in Fig. 10.12.

10.3.3.2 Elbow Duct

An elbow is a kind of duct fitting where the direction of airflow changes. The change of airflow direction results into creation of centrifugal force such that the force acts toward the outer wall of the duct. Pressure increases while velocity decreases at the outer wall as air flows from the straight duct to the curve duct. At the same time, there is pressure decrease and velocity increase at the inner wall.

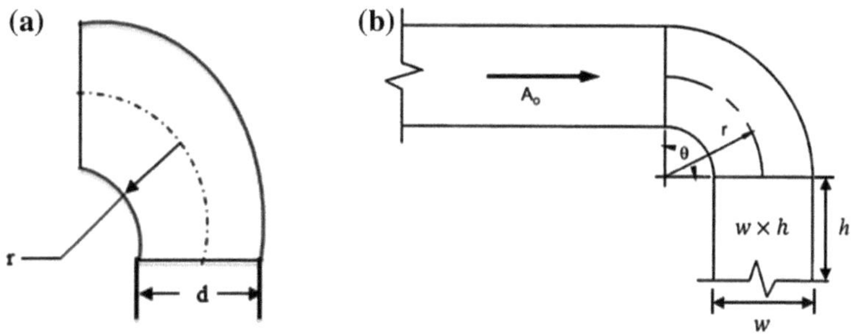

Fig. 10.13 Schematic diagram of (**a**) round elbow duct; (**b**) rectangular elbow

Therefore, a diffuser effect takes place near the outer wall while a bell mouth effect exists at the inner wall. After turning, a reverse effect occurs as the air flows from the curved part to the straight part of the duct.

The magnitude of local resistance coefficient of an elbow is affected by the following factors: the angle of turn of the elbow (45° or 90°); ratio of centerline radius r to diameter d given as r/d for round elbows or ratio of throat radius r to duct width w given as r/w for rectangular elbows; number of pieces making up the elbow in case of piece (joined) elbows; shape of cross-sectional area of the duct; and presence or absence of splitter vanes.

The physical models of the various types of elbows are shown in Fig. 10.13.

The relationships between inputs and outputs of an elbow are as follows:

$$G_{\text{in}} = G_{\text{out}} \tag{10.17}$$

$$P_{\text{out}} = P_{\text{in}} - 8\xi \frac{\rho_{\text{a}} G_{\text{in}}^2}{\pi^2 d_{\text{e}}^4} \tag{10.18}$$

The local resistance coefficients for different shapes of elbow ducts refer to Tables 10.1 and 10.2.

The Simulink block diagram for elbow duct is shown in Fig. 10.14.

10.3.3.3 Reducing/Expanding Duct

When air flows through a contraction or enlargement, there is always creation of flow separation which results into eddies and large-scale turbulences after the enlargement. The same occurs before and after contraction. These cause loss in total pressure. The velocity of air attains its maximum value at vena contracta region.

Table 10.1 Local resistance coefficient of round 90° elbow ducts [49]

r/d	0.5	0.75	1.0	1.5	2	3	4
ξ	0.9	0.43	0.33	0.24	0.19	0.17	0.15

Table 10.2 Local resistance coefficient for rectangular 90° elbow ducts [49]

	h/w									
r/w	**0.50**	**0.75**	**1.00**	**1.5**	**2.00**	**3.00**	**4.00**	**5.00**	**6.00**	**8.00**
0.50	1.38	1.29	1.18	1.06	1.00	1.00	1.06	1.12	1.16	1.18
0.75	0.52	0.48	0.44	0.40	0.39	0.39	0.40	0.42	0.43	0.44
1.00	0.25	0.23	0.21	0.19	0.18	0.18	0.19	0.20	0.21	0.21
1.50	0.20	0.19	0.17	0.15	0.14	0.14	0.15	0.16	0.17	0.17
2.00	0.18	0.16	0.15	0.14	0.13	0.13	0.14	0.14	0.15	0.15

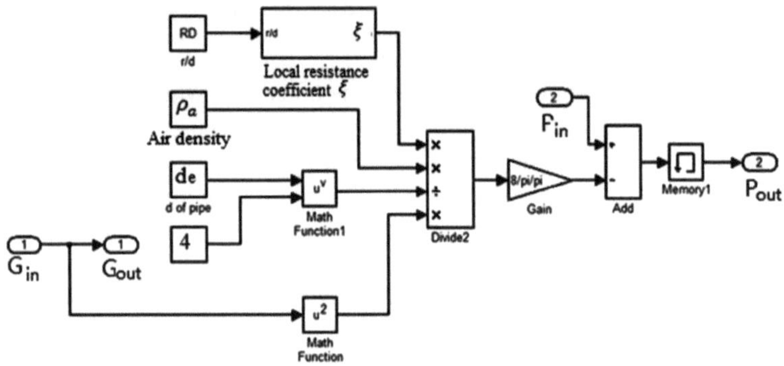

Fig. 10.14 Simulink block of elbow duct sub-system

Vena contracta is the region where the cross section of airflow is minimum and the static pressure is negative. To minimize energy loss, a gradual expansion known as a diffuser or a converging transition is recommended.

For expanding or reducing ducts, their physical models are shown in Fig. 10.15. The input variables are input volumetric flow (G_{in}) and input pressure (P_{in}). The output variables are set to be outlet volumetric flow (G_{out}) and outlet pressure (P_{out}). The relationship between inputs and outputs is as follows:

$$G_{\text{in}} = G_{\text{out}} \tag{10.19}$$

$$P_{\text{out}} = P_{\text{in}} - 8\xi \frac{\rho_{\text{a}} G_{\text{in}}^2}{\pi^2 d_{\text{e}}^4} \tag{10.20}$$

where $\xi = \left(\frac{A_2}{A_1} - 1\right)^2$ for contracting duct and $\xi = \left(1 - \frac{A_2}{A_1}\right)^2$ for expanding duct [50].

The Simulink block diagram for expanding or reducing ducts is similar to that for straight or elbow duct.

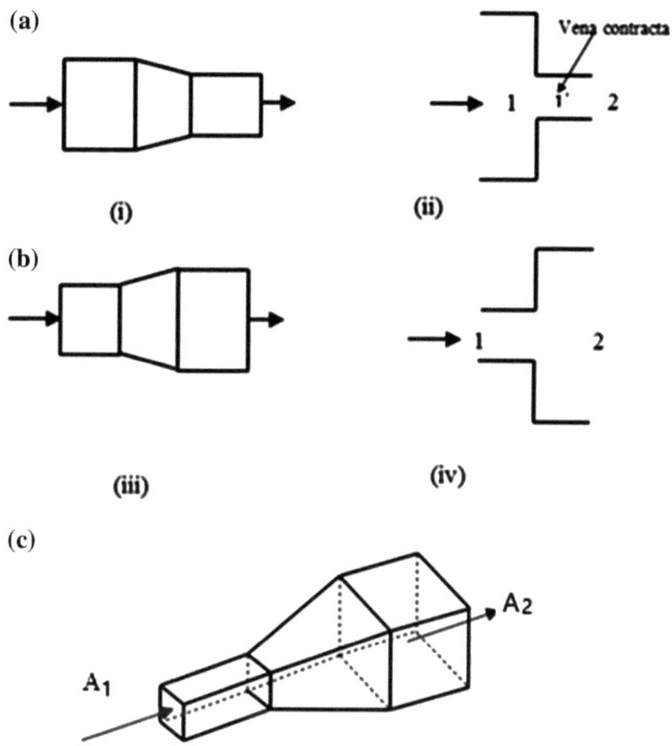

Fig. 10.15 Enlargements and contractions: (**a**) sudden contraction (**b**) sudden expansion (**c**) pyramidal reducer

10.3.3.4 Tee-Duct

A branch duct that diverges or combines with the main duct is referred to as a tee-duct, as shown in Fig. 10.16.

Tee-ducts can be round, flat oval, or rectangular. The tee-pipe comprises converging and diverging sections. The pressure loss in tee-ducts is dependent on branch internal radius, volume flow rate of the fluid entering or leaving from a branch and branch angle. In the tee-duct models, input variables include junction inlet airflow volume ($G_{\rm in}$) and junction inlet air pressure ($P_{\rm in}$) while output

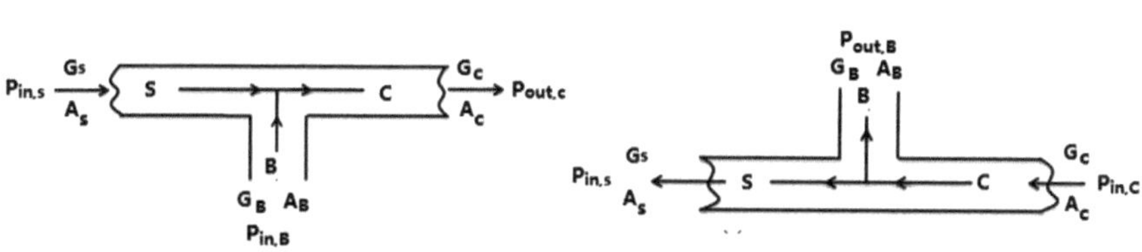

Fig. 10.16 Schematic diagram of (**a**) Converging tee-duct, (**b**) Diverging tee-duct

Table 10.3 Resistance coefficient for converging branch of tee-ducts [51]

A_B/A_C	ξ_{B-C}								
	G_B/G_C								
	0.1	0.2	0.3	0.4	0.5	0.6	0.7	0.8	0.9
0.1	0.40	3.80	9.20	16.3	25.5	36.7	42.9	64.9	82.0
0.2	0.37	0.72	2.27	4.30	6.75	9.70	13.0	16.9	21.2
0.4	0.54	0.03	0.58	1.30	2.06	2.98	3.90	4.92	6.10
0.6	0.59	0.17	0.27	0.75	1.20	1.68	2.20	2.70	1.92
0.8	0.60	0.22	0.15	0.55	0.89	1.25	1.60	3.20	2.25
λ_{S-C}									
$\boldsymbol{G_B/G_C}$	0.1	0.2	0.3	0.4	0.5	0.6	0.7	0.8	0.9
λ_{S-C}	0.16	0.27	0.38	0.46	0.53	0.57	0.59	0.60	0.59

parameters include junction outlet airflow volume (G_{out}) and junction outlet pressure (P_{out}).

For the converging tee-duct, we have the following equations:

$$G_{\text{out,C}} = G_{\text{in,S}} + G_{\text{in,B}} \tag{10.21}$$

$$P_{\text{in,S}} = P_{\text{out,C}} + \frac{8}{\pi^2}\lambda_{\text{S-C}}\frac{l_{\text{d,S-C}}}{d_{\text{e,c}}^5}\rho_{\text{a}}G_{\text{in,S}}^2 \tag{10.22}$$

$$P_{\text{in,B}} = P_{\text{out,C}} + \frac{8}{\pi^2}\xi_{\text{B-C}}\frac{\rho_a}{d_{\text{e,B}}^4}G_{\text{in,B}}^2 \tag{10.23}$$

where $\lambda_{\text{S-C}}$ is the local resistance coefficient from duct S to C; $\xi_{\text{B-C}}$ is the local resistance coefficient from duct B to C. These coefficients can refer to Table 10.3. However, they should be adjusted according to real situations.

The Simulink block for the converging tee-duct is presented in Fig. 10.17.

And for diverging tee-duct,

$$G_{\text{out,S}} = G_{\text{in,C}} - G_{\text{out,B}} \tag{10.24}$$

$$P_{\text{out,S}} = P_{\text{in,C}} - \frac{8}{\pi^2}\lambda_{\text{C-S}}\frac{l_{\text{d,C-S}}}{d_{\text{e,S}}^5}\rho_{\text{a}}G_{\text{out,S}}^2 \tag{10.25}$$

$$P_{\text{out,B}} = P_{\text{in,C}} - \frac{8}{\pi^2}\xi_{\text{C-B}}\frac{\rho_{\text{a}}}{d_{\text{e,B}}^4}G_{\text{out,B}}^2 \tag{10.26}$$

where $\lambda_{\text{C-S}}$ is the local resistance coefficient from duct C to S; $\xi_{\text{C-B}}$ is the local resistance coefficient from duct C to B. These coefficients can refer to Table 10.4.

The Simulink block diagram for diverging tee-duct is shown in Fig. 10.18.

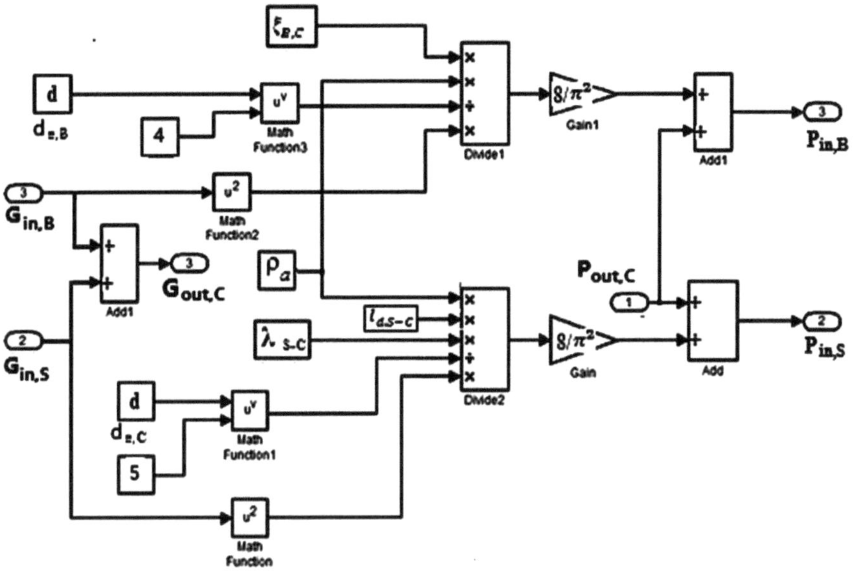

(a) Detailed block

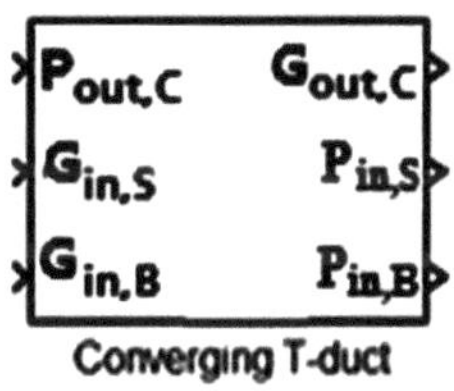

(b) Subsystem package

Fig. 10.17 Simulink block of converging tee-duct

Table 10.4 Local resistance coefficients for diverging branch of tee-ducts [51]

A_B/A_C	$\xi_{C\ B}$								
	G_B/G								
	0.1	0.2	0.3	0.4	0.5	0.6	0.7	0.8	0.9
0.1	2.06	1.20	0.99	0.87	0.88	0.87	0.87	0.86	0.86
0.2	5.15	1.92	1.29	1.03	0.99	0.94	0.92	0.90	0.89
0.4	15.90	4.35	2.24	1.48	1.11	0.88	0.80	0.75	0.72
0.6	34.60	8.70	4.03	2.41	1.65	1.22	1.04	0.94	0.87
0.8	76.67	18.49	8.05	4.49	2.86	1.98	1.59	1.36	1.20
λ_{C-S}									
G_S/G_C	**0.1**	**0.2**	**0.3**	**0.4**	**0.5**	**0.6**	**0.7**	**0.8**	**0.9**
λ_{C-S}	32.40	6.40	2.18	0.90	0.40	0.18	0.07	0.03	0.00

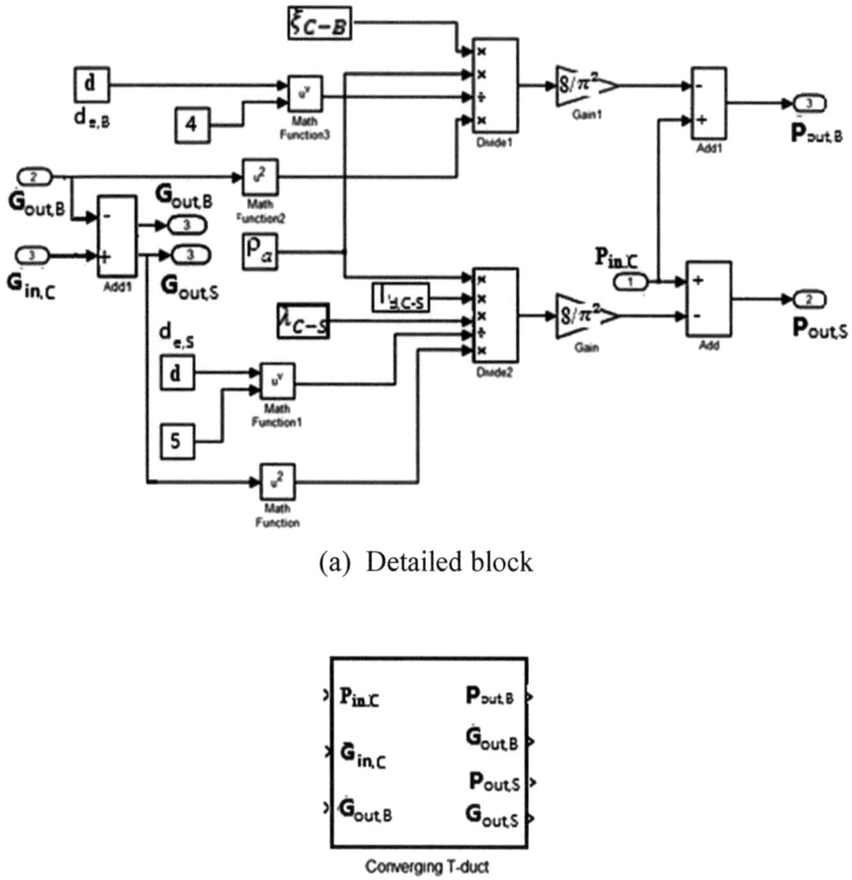

(a) Detailed block

(b) Subsystem package

Fig. 10.18 Simulink block of diverging tee-duct

10.3.4 Other Local Resistance Components

The other local components in VAV system mainly include filters, steam heaters, coolers, and water retainers. Their local resistance model can be uniformly written as:

$$G_{\text{out}} = G_{\text{in}} = G_{\text{a}} \tag{10.27}$$

$$P_{\text{out}} = P_{\text{in}} - \frac{\rho}{2}\xi_x\left(\frac{G_{\text{a}}}{A_x}\right)^2 \tag{10.28}$$

where ξ_x is the local resistance coefficient of a local component; A_x is the cross-sectional area of a local component. The Simulink block diagrams for these local components are similar to that for straight or elbow duct.

10.3.5 Application of Component Model Library: Case Study

In the previous section, the component model library of VAV system has been developed by using the Simulink platform. In this section, a real VAV system will be modeled based on the model library. The schematic diagram of the duct system is presented in Fig. 10.19. All the ducts are of circular pipes, and the duct system consists of four sub-systems, as shown in Fig. 10.20. Table 10.5 is a summary of the duct diameters.

The sub-systems can be efficiently modeled with the component model library, which are presented in Fig. 10.21. The Simulink blocks all start from the supply fan through ducts of different types mainly including straight ducts, elbow ducts, and tee-ducts. These blocks are connected in turn according to the real system. Simulations can be performed easily with the Simulink tool after all the model parameters have been reasonably set.

The summary of pressure loss coefficients is given in Table 10.6. The pressure loss coefficient for all straight pipes (λ) can be calculated by Eq. (10.12). For all curved or elbow pipes, the local pressure loss coefficient (ξ) is taken to be 0.33 according to Table 10.1. For the supply duct system, the local pressure loss coefficients of the tee-pipe can be obtained according to Table 10.4.

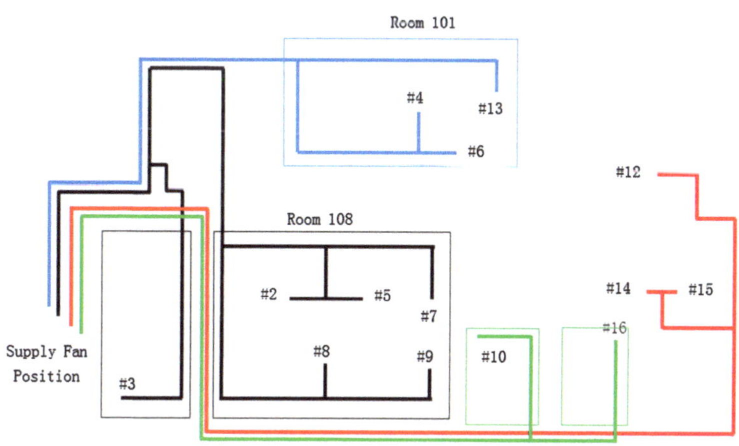

Fig. 10.19 Schematic diagram of the VAV duct system

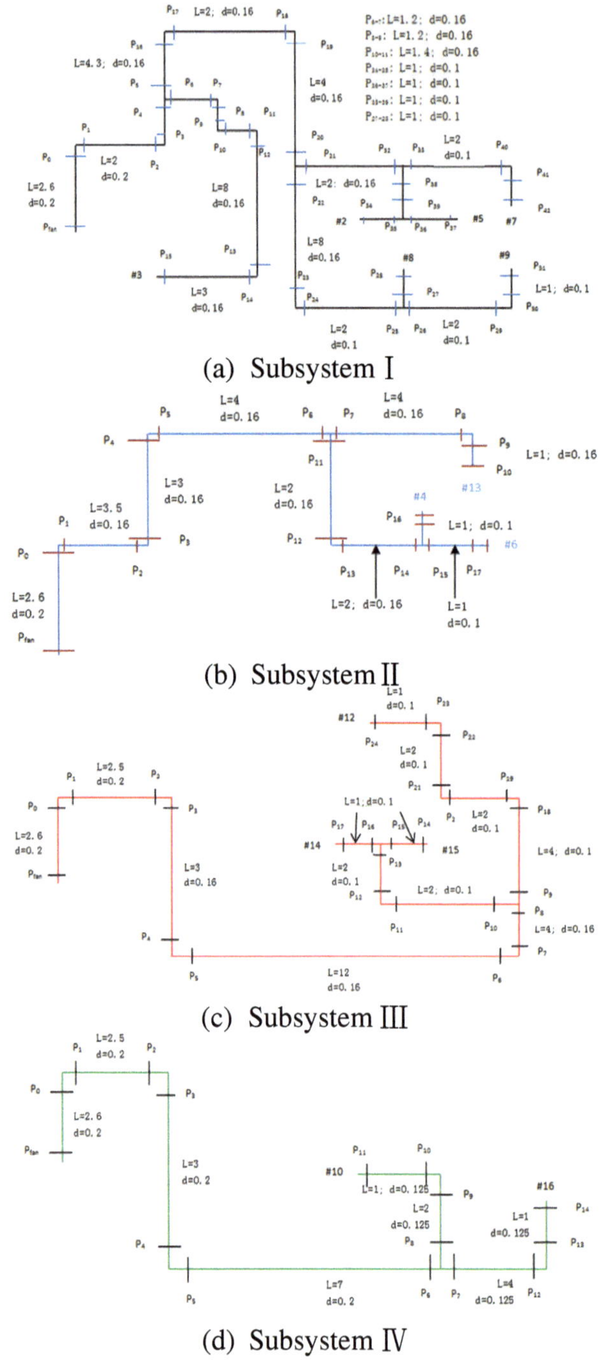

Fig. 10.20 Detailed dimensions of the sub-systems (unit: m)

Table 10.5 Summary of description of duct dimensions

LOOP	Sectional description	Diameter (m)
1	$P_{fan} - P_0$	0.2
	$P_0 - P_1$; $P_1 - P_2$; $P_2 - P_3$; $P_3 - P_4$; $P_4 - P_5$; $P_5 - P_6$; $P_6 - P_7$; $P_7 - P_8$; $P_8 - P_9$; $P_9 - P_{10}$; $P_{11} - P_{12}$; $P_{12} - P_{13}$; $P_{13} - P_{14}$;	0.16
	$P_{15} - P_{17}$; $P_{14} - P_{16}$	0.1
2	$P_{fan} - P_0$; $P_0 - P_1$; $P_1 - P_2$; $P_2 - P_3$	0.2
	$P_3 - P_4$; $P_6 - P_7$; $P_7 - P_8$; $P_8 - P_9$; $P_9 - P_{10}$; $P_{10} - P_{11}$; $P_{11} - P_{12}$; $P_{12} - P_{13}$; $P_{13} - P_{14}$; $P_{14} - P_{15}$; $P_{15} - P_{16}$; $P_{16} - P_{17}$; $P_{17} - P_{18}$; $P_{18} - P_{19}$; $P_{19} - P_{20}$; $P_{20} - P_{21}$; $P_{21} - P_{22}$; $P_{22} - P_{23}$; $P_{23} - P_{24}$	0.16
	$P_{24} - P_{25}$; $P_{27} - P_{28}$; $P_{26} - P_{29}$; $P_{29} - P_{30}$; $P_{38} - P_{39}$; $P_{34} - P_{35}$; $P_{36} - P_{37}$; $P_{33} - P_{40}$; $P_{40} - P_{41}$; $P_{41} - P_{42}$	0.1
3	$P_{fan} - P_0$; $P_0 - P_1$; $P_1 - P_2$;	0.2
	$P_2 - P_3$; $P_3 - P_4$; $P_4 - P_5$; $P_5 - P_6$; $P_6 - P_7$; $P_7 - P_8$;	0.16
	$P_{10} - P_{11}$; $P_{11} - P_{12}$; $P_{12} - P_{13}$; $P_{14} - P_{15}$; $P_{15} - P_{16}$; $P_{16} - P_{17}$; $P_8 - P_9$; $P_9 - P_{18}$; $P_{18} - P_{19}$; $P_{19} - P_{20}$; $P_{20} - P_{21}$; $P_{21} - P_{22}$; $P_{22} - P_{23}$; $P_{23} - P_{24}$	0.1
4	$P_{fan} - P_0$; $P_0 - P_1$; $P_1 - P_2$; $P_2 - P_3$; $P_3 - P_4$; $P_4 - P_5$; $P_5 - P_6$	0.2
	$P_6 - P_7$	0.16
	$P_8 - P_9$; $P_9 - P_{10}$; $P_{10} - P_{11}$; $P_7 - P_{12}$; $P_{12} - P_{13}$; $P_{13} - P_{14}$	0.125

The summary of constants 'a_1' and 'a_2' that appear in pressure resistance model of VAV terminal unit [i.e., Eq. (10.3)] is obtained by fitting experimental data, as shown in Table 10.7. The other parameters include the dynamic viscosity of air (μ) which is set as $1.82 \times 10^{-5}\,\text{kg/m s}$ and the density of air (ρ) which is set as $1.12\ \text{kg/m}^3$.

The performance curve of the fan used in this VAV system is shown in Fig. 10.22. The coefficients a_4, a_3, a_2, a_1 and a_0, in Eq. (10.7) are determined by fitting the data provide in Fig. 10.22, which are 3438.2, −13753, 19381, −12122, and 4303.4, respectively. The regression coefficient $R^2 = 0.9997$.

Table 10.8 gives six cases in which all the terminal units have different valve opening degrees. Table 10.9 gives airflow rates of terminal devices required for the six cases. The calculated airflow distributions in the duct system for the sic cases are listed in Table 10.10. The calculated pressure drops due to the duct system for the six cases are compared to the experimental ones. As shown in Fig. 10.23, the relative errors of model results compared with the experimental ones in the six cases are all within 15 %, which means that the system model established based on the modular component library is valid.

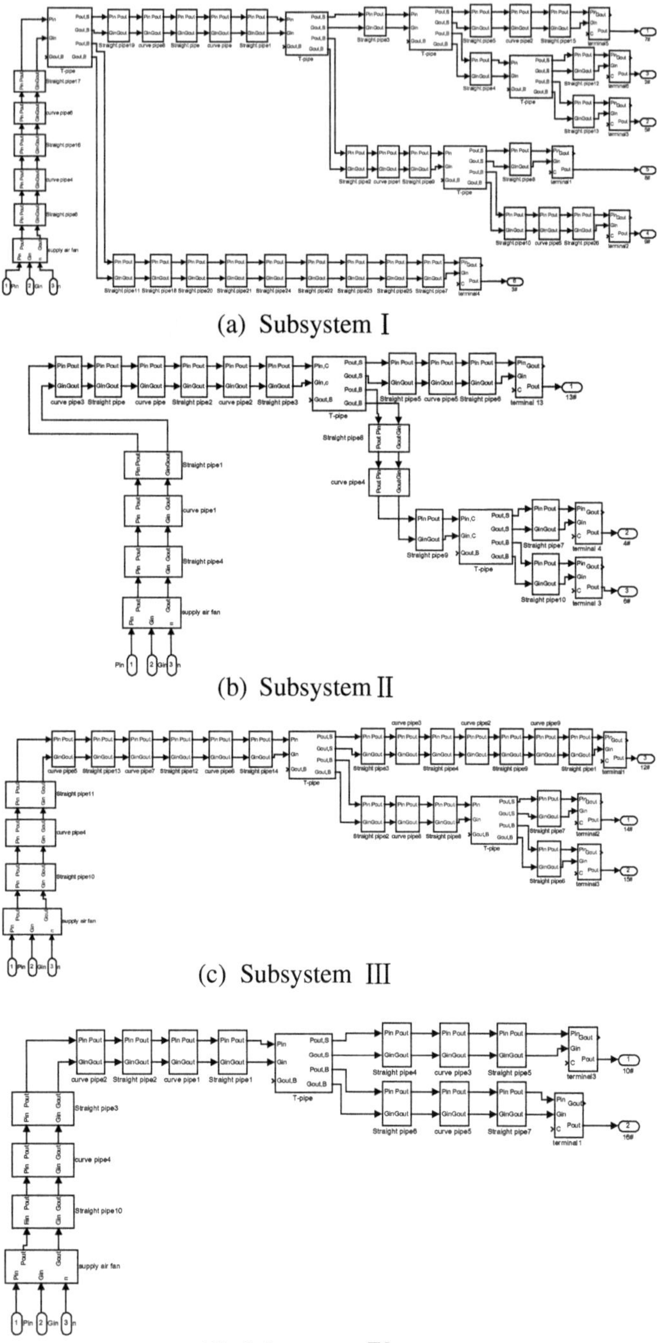

(a) Subsystem Ⅰ

(b) Subsystem Ⅱ

(c) Subsystem Ⅲ

(d) Subsystem Ⅳ

Fig. 10.21 Simulink model for the sub-systems

Table 10.6 Summary of pressure loss coefficients

Duct type	Local pressure loss coefficient (ξ)	Friction coefficient (λ)
Straight ducts	N/A	By Eq. (10.12)
Elbow ducts (1:1)	0.33 (Refer to Table 10.1)	N/A
Tee-ducts	Obtained by Table 10.4	By Eq. (10.12)

Table 10.7 Constants 'a_1' and 'a_2' in Eq. (10.3) for different terminal units

No. of terminal unit	a_1	a_2
2#	3.21899E7	−1.08004
3#	3.73482E7	−1.11162
4#	9.85272E7	−1.71578
5#	5.60809E7	−1.24104
6#	2.55427E8	−1.623
7#	3.09864E8	−1.65655
8#	7.12895E8	−1.70995
9#	5.79444E7	−1.19956
10#	5.28444E10	−2.92827
12#	6.75643E11	−3.53853
13#	9.81743E12	−3.88766
14#	3.94437E7	−1.74502
15#	3.56572E6	−0.84164
16#	3.99346E6	−0.84285

10.4 Control Strategies for VAV System

VAV air-conditioning system uses the method of varying supply air volume to match the room thermal load (i.e., a heat rate which must be removed from or added to the conditioned space in order to maintain the design indoor condition) and minimum airflow requirements,adjusting the room air temperature and humidity. Terminal devices regulate the opening degree of the valve to achieve the purpose of adjusting the supply air volumc bascd on the room's heat load. For VAV systems, the airflow rate can bc rcduccd according to the load variations to conserve fan power and possible reheat energy. However, a minimum airflow rate (MAR) is usually maintained for indoor ventilation purposes during low cooling and heating load conditions. The MAR is typically 30 % for most buildings specified by ASHRAE [52]. When the load ratio is higher than the MAR, the airflow ratio follows the load ratio, and when the load ratio is lower than the MAR or becomes negative in heating seasons, the airflow ratio is usually kept at the MAR.

Total system supply air volume is equal to the sum of supply air volume demand at each of the terminal devices. The fan rotational speed is adjusted through some specific control strategies to achieve the objective of regulating the amount of air volume delivered to terminal devices. Basically, control methods of system supply

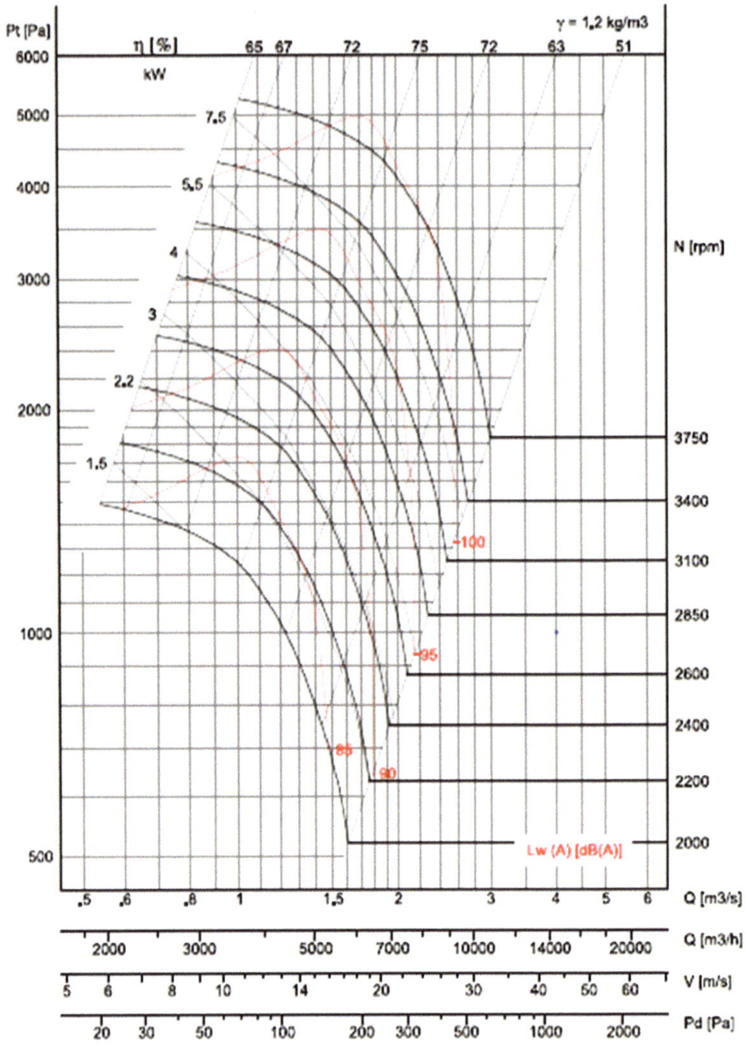

Fig. 10.22 Performance curve of fan in the VAV system

air volume include constant static pressure method, total air volume method, and variable static pressure method.

10.4.1 Constant Static Pressure Method

Constant static pressure control is considered as a base case where the static pressure is maintained at a constant throughout different load conditions. This is the

Table 10.8 Valve opening degree (%) of all terminal units for different cases

No. of terminal unit	Valve opening degree (%)					
	Case 1	Case 2	Case 3	Case 4	Case 5	Case 6
#2	25.1	43.8	43.9	49.1	56.7	61.2
#3	31.0	43.0	55.2	46.1	37.1	37.2
#4	33.2	54.6	65.4	54.8	43.5	44.1
#5	18.2	33.1	29.4	41.2	59.1	61.1
#6	39.7	70.6	74.3	86.2	66.1	65.1
#7	35.0	49.0	50.1	56.2	62.3	68.1
#8	25.0	46.0	44.1	59.1	83.2	86.3
#9	40.0	67.1	67.1	72.3	86.3	86.2
#10	52.0	72.1	71.1	70.2	68.2	69.1
#12	49.5	79.6	79.1	99.9	72.6	72.5
#13	41.0	60.1	63.1	71.1	46.1	48.1
#14	17.0	39.3	38.8	46.3	33.9	33.9
#15	17.0	45.1	45.1	54.2	44.1	47.2
#16	14.7	34.3	33.2	38.2	45.9	50.3

Table 10.9 Required airflow rates of terminal devices for the six cases

No. of terminal unit	Required air volume flow rate (m^3/h)					
	Case 1	Case 2	Case 3	Case 4	Case 5	Case 6
#2	140	300	300	300	300	300
#3	150	269	330	264	202	199
#4	275	550	600	524	424	416
#5	140	300	300	300	300	300
#6	150	330	330	330	298	289
#7	140	300	300	300	300	295
#8	140	300	300	300	300	300
#9	140	300	300	300	300	300
#10	225	495	495	461	424	418
#12	150	330	330	330	285	279
#13	75	165	165	165	122	120
#14	225	488	495	495	414	403
#15	225	495	495	495	440	428
#16	225	495	495	494	495	495

most common strategy in use especially in China [53]. It boasts of the following merits: simple control strategy, stable system operation, less control loops, and works independently from the building automation system (BAS).

For the constant static pressure method, the maximum duct static pressure required by design load is maintained constantly under partial load conditions. Reduced airflow rate due to part-load conditions causes the static pressure loss

Table 10.10 Airflow distributions in the duct system for the six cases

Duct item No.	Calculated air volume flow rate (m^3/h)					
	Case 1	Case 2	Case 3	Case 4	Case 5	Case 6
$P_{fan} - P_0$	2400	5118	5252	5028	4604	4542
$P_1 - P_2$	2400	5118	5252	5028	4604	4542
Loop leading to valves 4, 6, and 13	500	1045	1095	1019	844	825
Loop leading to valves 4, 6	425	880	930	854	722	705
Loop leading to valves 2, 5, 7, 8, and 9	700	1500	1500	1500	1798	1495
Loop leading to valves 2, 5, and 7	420	900	900	900	900	895
Loop leading to valves 2 and 5	280	600	600	600	600	600
Loop leading to valves 8 and 9	280	600	600	600	600	600
Loop leading to valves 10 and 16	450	990	990	955	919	913
Loop leading to valves 12, 14 and 15	600	1148	1320	1320	1139	1110
Loop leading to valves 14 and 15	450	983	990	990	854	831

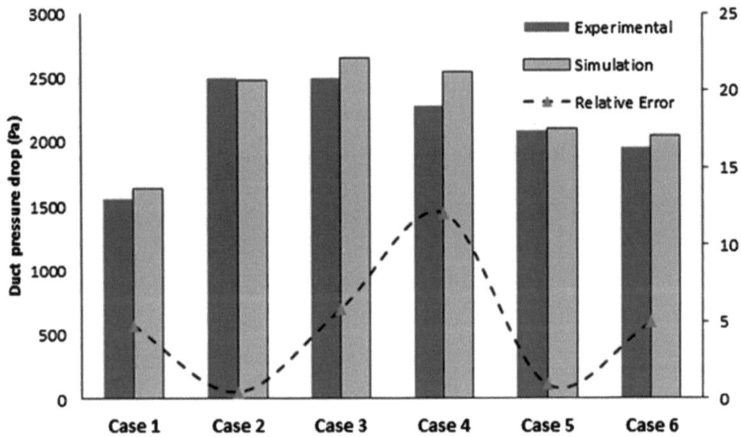

Fig. 10.23 Experimental versus calculated results on the duct pressure drop for the six cases

between the sensor location and terminal unit to be much less than that under full-load conditions. The decrease in lost pressure causes an increase in static pressure in front of the VAV terminal unit whenever a constant static pressure setpoint value is used, hence increasing the pressure drop across the VAV terminal unit damper. In order to maintain the space temperature, we increase the flow coefficient by gradually closing the terminal unit damper in efforts to cut down airflow rate. When the VAV box damper opening degree is small, the duct static pressure becomes higher than the setpoint value; the AHU supply air volume becomes excess, and the DDC output value to the frequency converter is reduced to decrease air volume and static pressure so that VAV box damper opening degree is not too small. A low static pressure setpoint value may fail to meet all rooms' air volume requirements; on the other hand, a high static pressure setpoint value will increase VAV box noise and energy consumption. A lot of fan power is wasted through throttling at the VAV terminal unit. As a result, it does not run in the best fan energy-saving mode at part-load conditions. This is contrary to the main control objective.

10.4.2 Total Air Volume Method

The total air volume method just happens to evade/avoid the static pressure setpoint as used in the static pressure control methods. It is essentially model-based control whose effectiveness largely depends on the effectiveness of fan model and pressure resistance model of ducts as well as other local resistance parts. In practice, we can use the law of proportionality when we have a good selection of fan performance curve and the duct resistance curve. The setpoint of the total air volume can be used to control speed of the fan in order to achieve the actual desired total air volume.

10.4.2.1 Optimal Control Process

The system total supply air volume is given by the sum of supply air volume demand of individual rooms. The volumetric flow quantity of the fan serves as system's total supply air volume. The volumetric flow of the fan is related to fan speed as well as the exit air pressure of the fan. The fan's exit air pressure is used to overcome duct network resistance of air in the VAV system. The damper positions of the regulating terminal unit and the duct network resistance undergo change, thus adjusting the supply air volume in each room. To obtain the best operating status in a VAV system, the fan speed should be as low as possible, and the damper opening degree of each terminal unit should be as large as possible, but no greater than 95 %. The status of damper valve position in the VAV terminal device is defined as listed in Table 10.11.

During the operation of VAV system, position of damper valve is expected to range from 85 to 95 %. When damper valve position is too small, it indicates that the fan rotational speed is too large, and vise versa. The status of fan rotational speed based on the status of damper valve position of VAV terminal devices is given in Table 10.12.

Table 10.11 Definition of status of valve position of VAV box

Status of damper valve position	Definitions
Too small	Valve position is less than 85 %
Moderate	The valve position is between 85 and ~95 %
Too large	Valve position is greater than 95 %

Table 10.12 Definition of status of fan speed

Status of fan rotational speed	Definition
Fan speed is too low	At least one of the VAV terminal device damper valve positions is too large in the VAV system
Fan speed is moderate	There is no terminal device whose valve opening is not too large, and there is at least one terminal device whose damper valve position is moderate
Fan speed is too large	Damper valve position of each terminal unit is too small

As shown in Table 10.12, the optimum rotational speed of fan fulfills the condition that the statuses of all the VAV terminal unit damper valve positions are not too large as well as causing the statuses of VAV terminal unit damper valve position to be in moderate conditions. Fan speed control logic can be developed in accordance with these evaluation indicators as shown in Fig. 10.24.

The basic process of optimization control process for fan speed is as follows:

1. Read the number of opening VAV terminal units (denoted by X_z) as well as supply air volume demand of each VAV terminal unit to calculate total supply air volume of the system;
2. Calculate preliminary fan rotational speed from the fan characteristic curve and duct network resistance characteristics of the air duct system;
3. Adjust each damper valve position so that the primary supply air volume of each VAV terminal device meets the room requirements;
4. Record the number (denoted by X_H) of the terminal units whose status is too large and the number (denoted by X_L) whose status is too small.
5. If the status of VAV terminal device damper valve position is too large, increase the fan rotational speed. The rate of increase can be calculated by Eq. (10.29). After increasing the fan rotational speed, return to step (4);

$$\Delta n_v = \frac{10X_H}{X_z T} \tag{10.29}$$

   where Δn_v is rate of increase of fan rotational speed (%/min); T is integration time (min).
6. If VAV terminal device damper valve position is in a moderate state, stop adjusting the fan speed;
7. If the status of VAV terminal device damper valve position is too small, decrease the fan rotational speed. The rate of decrease can be calculated by Eq. (10.30). After the reduction of fan speed, then go back to step (4).

$$\Delta n_v = \frac{10X_L}{X_z T} \tag{10.30}$$

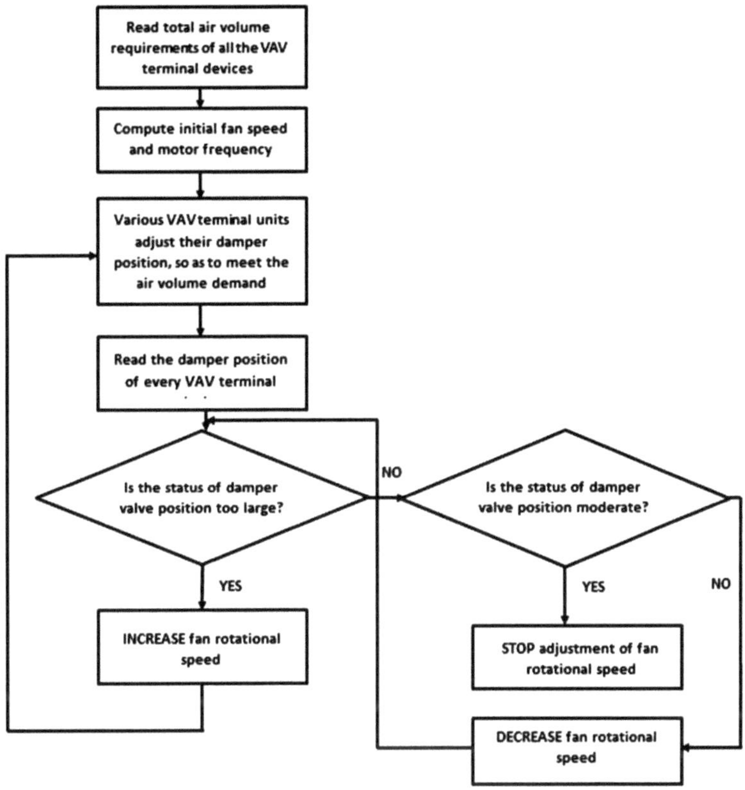

Fig. 10.24 Optimization control process for fan speed

10.4.2.2 Case Study

A real VAV system is taken as the case study. The Simulink diagram for the total air volume control scheme with optimizer is shown in Fig. 10.25, which mainly include air-supply fan, duct sections, valves opening degrees, and optimization controller or optimizer. The detailed information for block 'Sub-system I,' 'Sub-system II,' 'Sub-system III,' and 'Sub-system IV' refers to Fig. 10.21, and the fan's characteristic performance refers to Fig. 10.22. The symbols from 'C#2' to 'C#16' stand for opening degrees of all the terminal units. The optimizer block is carried out in accordance with the control logic as shown in Fig. 10.24.

All the valve opening degrees are fed into the controller which carries out the optimization scheme and outputs the optimal fan rotational speed. During the optimal control process, no valve is allowed to open above 95 %, which ensures that at any given time no damper will be excessively open.

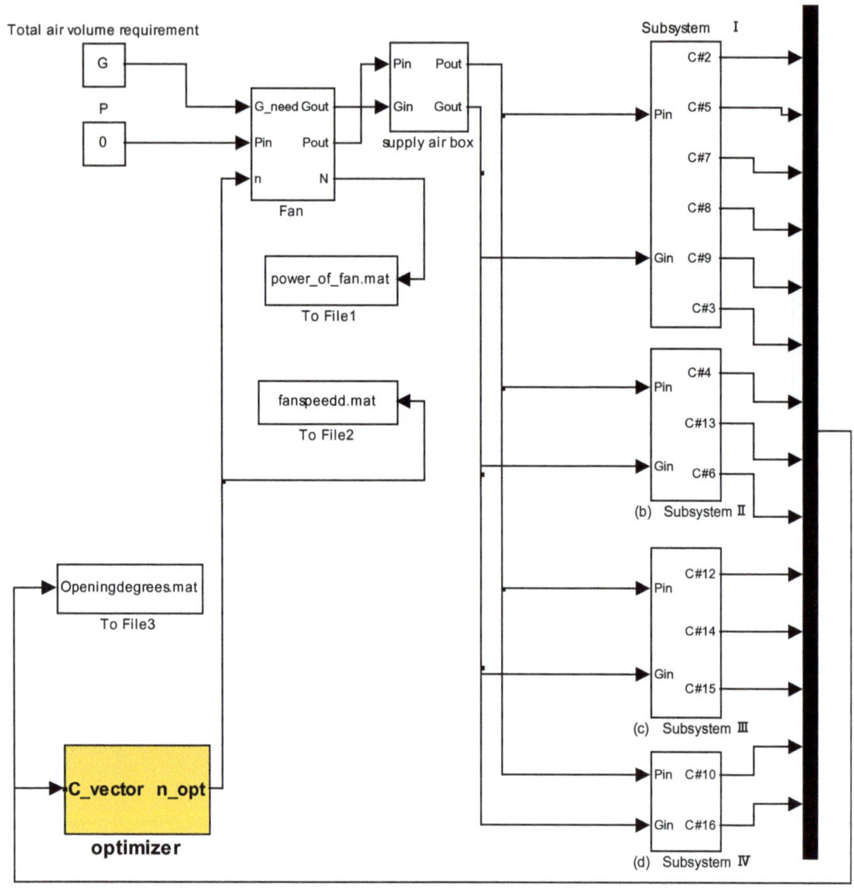

Fig. 10.25 Simulink diagram for the total air volume control scheme with optimizer

Assuming the following operating conditions are given: Airflow rate requirement of room #2 is 0.0694 m^3/s, room #5, #7, #8, and #9 each has 0.0833 m^3/s, rooms #3, #6, and #12 each has 0.0917 m^3/s, rooms #10, #16, #14, and #15 each has 0.1375 m^3/s, room #13 has 0.1667 m^3/s and room #4 has 0.0458 m^3/s. The integration time of the fan rotation speed changes is 1 min (60 s). The initial fan speed is set at 3418 rpm based on the fan characteristic curve and the duct network resistance characteristics, and the simulated control processes are shown in Figs. 10.26 and 10.27.

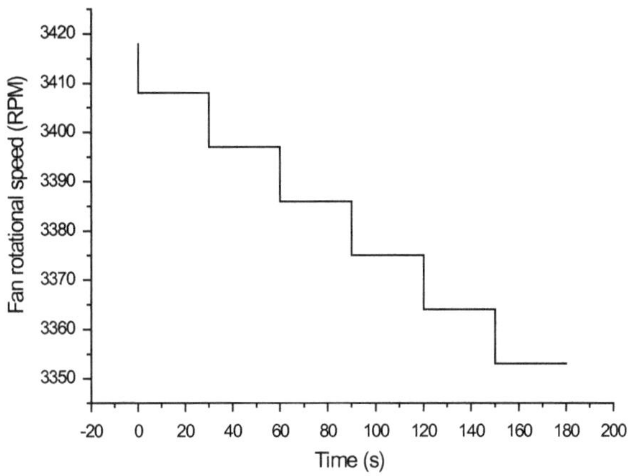

Fig. 10.26 Variations of fan speed with time during the optimal control process

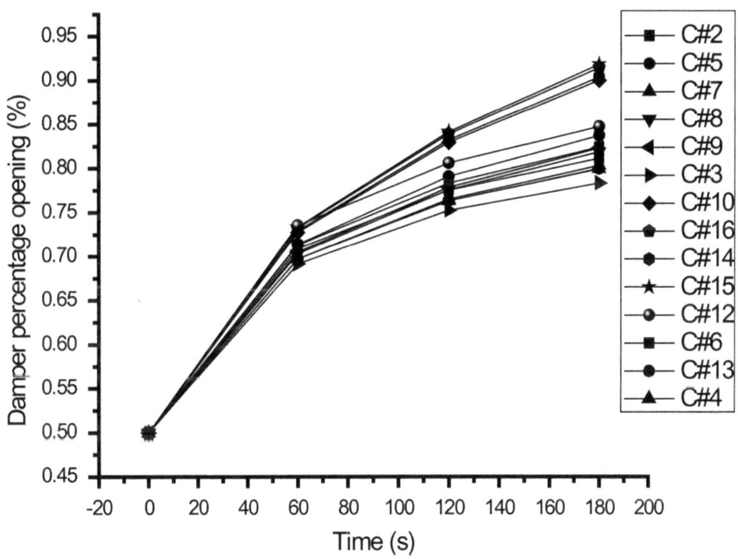

Fig. 10.27 Variations of opening degrees for all the VAV terminal units during the optimal control process

As shown in Fig. 10.26, adjustment of the fan speed is represented by a step-like decline. At 180 s, the opening degree of No. #15 VAV terminal device is already in a moderate state, and the optimization process accomplishes. The corresponding fan speed is 3358 rpm, and the smallest opening degree among the terminal devices is about 75 % which occurs on the No. #3 terminal device (see Fig. 10.27).

10.4.3 Variable Static Pressure Method Based on Trim-and-Respond Logic

Trim-and-respond (T&R) method is a static pressure reset control strategy that monitors the dampers' status. For systems installed with direct digital controllers in every zone, ASHRAE Standard 90.1 and Title 24 requires that the static pressure setpoint reset should be dynamic such that one damper is nearly wide open [16]. The damper position is classified as low or high. If the value of the position of damper is high, the setpoint of the controlled static pressure may be reduced or trimmed by the controller. Similarly, if the observed status of damper position is low, the static pressure setpoint will be increased by the controller, thus making a response. This control strategy is flexible and can be realized in different ways. In addition, this control strategy allows the user to set different trim-and-respond rates which can be of advantage to the system performance.

The T&R controller can be implemented in three different ways. First and foremost is the case in which if the maximum damper position exceeds a certain value, the setpoint of duct static pressure will slightly increase. Likewise, if the maximum damper position is below a certain value, the setpoint of the duct static pressure will slightly decrease. The second approach utilizes system requests to increase or decrease static pressure setpoint. In such circumstance, the system works by making a request for pressure increase for each damper whose value is above the desired setpoint [54]. The controller is also capable of making multiple requests for recorded values that deviate greatly from the desired setpoint. These requests then act as a multiplying factor which can effectively increase the rate of response for greater numbers of requests. The absence of any requests makes the controller to trim the setpoint of static pressure by a small amount. That way, the controller is always seeking to minimize the static pressure while still supplying the needed airflow to each zone. Finally, trim and respond can be implemented using a stepped response technique. In this approach, if the maximum damper position is within the prescribed range of values, no changes are made by the controller; if the damper position deviates from the prescribed range by a small value, the controller creates a small response in the duct static pressure setpoint; and if the damper position deviates from the prescribed range by a large value, the controller creates a large response in the duct static pressure setpoint. The steps are set up at small, moderate, and large deviations from the desired setpoint of the damper position, which results into creation of small, moderate, and large responses in the duct pressure setpoint. This tiered response technique can be used for both decreasing and increasing the static pressure setpoint.

10.4.3.1 Implementation of Trim-and-Respond Logic

In the T&R control strategy, before any increase in the amount of static pressure there is always a decrease (trimming) thus called trim-and-respond control. Demand is measured in terms of 'requests' generated by zone controller based on the position of the damper. Trimming of the static pressure setpoint is systematic through zone pressure requests which suggest need for static pressure adjustment. When the damper is above 95 % open, one or more requests are generated. The T&R algorithm polls all terminal units within an interval of two minutes in order to decide whether to increase or decrease the static pressure setpoint based on number of pressure alarms. Each terminal unit is programmed to send an alarm whenever it is unable to meet its airflow demand. Once the number of alarms is greater than the number of zones that can be ignored (i.e., two zones), the controller increases static pressure setpoint to the required value. Otherwise, the static pressure setpoint is trimmed. When the damper opening position falls below 80 %, no request is generated. The controller constantly decreases the static pressure setpoint at a fixed rate when there are no requests from zones or when two or less zones are generating requests. This strategy resets the static pressure setpoint within the range of minimum static pressure (SP_{min}) to maximum static pressure (SP_{max}). The maximum setpoint is the constant static pressure value needed to satisfy duct system pressure requirements at full-load conditions [55]. The minimum setpoint is static pressure value which can ensure satisfactory operation of all dampers at minimum conditions. This is in most cases assumed to be one-third of the static pressure at full-load conditions [16]. When the fan is off, the setpoint shall be equal to initial static pressure (SP_o) which is gotten by trial and error. The reset logic shall be activated whenever the fan is turned on, starting with delay time (T_d) equivalent to five minutes after initial fan start command. The system is considered to be at full flow conditions any time the actual flow of a VAV terminal box reaches 95 % of the designed flow. Therefore, a pressure request alarm will be triggered when the opening position of the damper is greater than 95 % until it drops to 80 % open [56]. When the logic is active, trim the setpoint by 100 Pa every two minutes in case of one or two zone pressure requests. If there are more than ignorable (I) zone requests, respond by changing the setpoint by an amount ($SP_{res} * (R - I)$) but making sure that this value does not exceed the maximum static pressure response value ($SP_{res-max}$). The maximum static pressure response value ($SP_{res-max}$) is chosen to be equal to the minimum static pressure setpoint value. The trim-and-response values must have opposite signs as well as absolute values so that at any given point there is a change in setpoint value. The reset logic may be stuck at a value whenever the response value is equivalent to trim value. The trim-and-respond control logic parameters, which are summarized in Table 10.13, are explained as below:

(i) **Minimum static pressure setpoint (SP_{min})**—The minimum setpoint is static pressure value which can ensure satisfactory operation of all dampers at minimum conditions. This is in most cases assumed to be one-third of the static pressure at full-load conditions [16].

(ii) **Maximum static pressure setpoint (SP_{max})**—The maximum static pressure setpoint is the constant static pressure value needed to satisfy duct system pressure requirements at full-load conditions [55]. This ensures sufficient supply of air in the event that all terminal units require maximum air volume supply. The system performance will not be affected by choosing a high setpoint limit if the reset logic works perfectly. This is because the rest logic will automatically limit duct pressure setpoint. However, under real practical applications, it is recommended to use conventional method to limit the setpoint since this improves performance in case rogue zones exist. Therefore, the static pressure sensor should be as far out into the duct system as possible in order to lower the maximum static pressure setpoint. This promotes efficient operation even in case the reset logic fails due to any problem.

(iii) **Delay time** ($\mathbf{T_d}$)—This is the period between fan start time and reset logic start time, which ensures that the system is stable before the reset logic is initiated. When the reset logic is started simultaneously with fan, system inaccuracies may arise.

(iv) **Ignored zones (I)**—The control logic allows a small number of pressure requests to be ignored before it responds by increasing the setpoint. Ignoring rogue zones improves energy efficiency. However, this may result in discomfort in those zones that have been ignored. For a successful implementation of T&R strategy, rogues zones must be addressed. Rogue zones demand high static pressure setpoint almost constantly, thus eliminating chances of having low static pressure setpoints at reduced load.

(v) **Number of requests from zones (R)**—A 'request' is a call to reset static pressure generated by downstream zones. Requests are made whenever a downstream zone is no longer satisfied with the volume of air supplied. Requests are usually sent to the air-handling unit controller. The setpoint is increased whenever a sufficient number of requests exist and decreases at a fixed rate in the absence of requests. The number of pressure requests made is dependent on the damper position such that different damper positions call for a specific number of requests. Some zones may make a single request while others might make multiple requests. Multiple requests eliminate zone starvation at the expense of saving energy. This therefore means that the rate of response is increased whenever the numbers of requests increase when the system in operation.

(vi) **Trim amount** ($\mathbf{SP_{trim}}$)—A very slow trim rate improves stability.

(vii) **Respond amount** ($\mathbf{SP_{res}}$)—Fast respond rate ensures timely address of zone demands for increased airflow.

(viii) **Maximum respond per time interval** ($\mathbf{SP_{res-max}}$)—The maximum static pressure respond value is chosen to be equal to the minimum static pressure setpoint value. This is to ensure that in the event that the static pressure value is far below the minimum setpoint value, one single response can eliminate system starvation.

The trim-and-respond control method may have the following advantages:

1. Easy tuning. The trim-and-respond logic is easier to tune compared to the traditional PID control logic. Trim-and-respond logic comprises several fairly intuitive parameters while PID logic includes only non-intuitive parameters.
2. Faster response than trimming. This means that we can always adjust the rate of trimming to be much slower than the rate of response. That way, stability can be promoted by a slow trim rate, whereas fast rate of response ensures timely address of the zone demand for increased airflow. On the other hand, PID loops only have one speed: A slow decrease of the setpoint means a corresponding slow decrease of the setpoint. A slow increase may lead to starving the zones when demand for airflow rises quickly.
3. Knowledge of damper position is not mandatory as requests for setpoint changes are generated at zone level which is sent to the controller for controlling the fan when needed.
4. Rogue zones can be easily ignored or avoided.
5. Allows preferential treatment of zones depending on their importance. This means that critical zones can be weighted more heavily than non-critical zones in how the reset logic responds. This is not easy to accomplish with PID logic.

10.4.3.2 Case Study

The previous VAV system is taken as the case study. For the present case, the minimum and the maximum airflow setpoint is 1.32 and 2.1 m^3/h, respectively, with temperature ranging between 20 and 24 °C. At maximum cooling load, nearly all terminal units are fully open and the fan is running at almost optimum speed.

The trim-and-respond control logic is employed for the control of the VAV system, and the corresponding parameters are suggested in Table 10.13. As shown in Fig. 10.28, the sequence starts five minutes after the supply fan status is deemed

Table 10.13 Trim-and respond control logic parameters

Parameter	Definition	Suggested value in the following studying
SP_o	Initial static pressure setpoint	1300 Pa
SP_{min}	Minimum static pressure setpoint	1000 Pa
SP_{max}	Maximum static pressure setpoint	3000 Pa
T_d	Delay time	5 min
T	Time interval	2 min
I	Number of ignorable zone requests	2
R	Number of request from zones	Dependent on current state
SP_{trim}	Trim amount	−100 Pa
SP_{res}	Respond amount	+130 Pa
$SP_{res-max}$	Maximum respond per time interval	1000 Pa

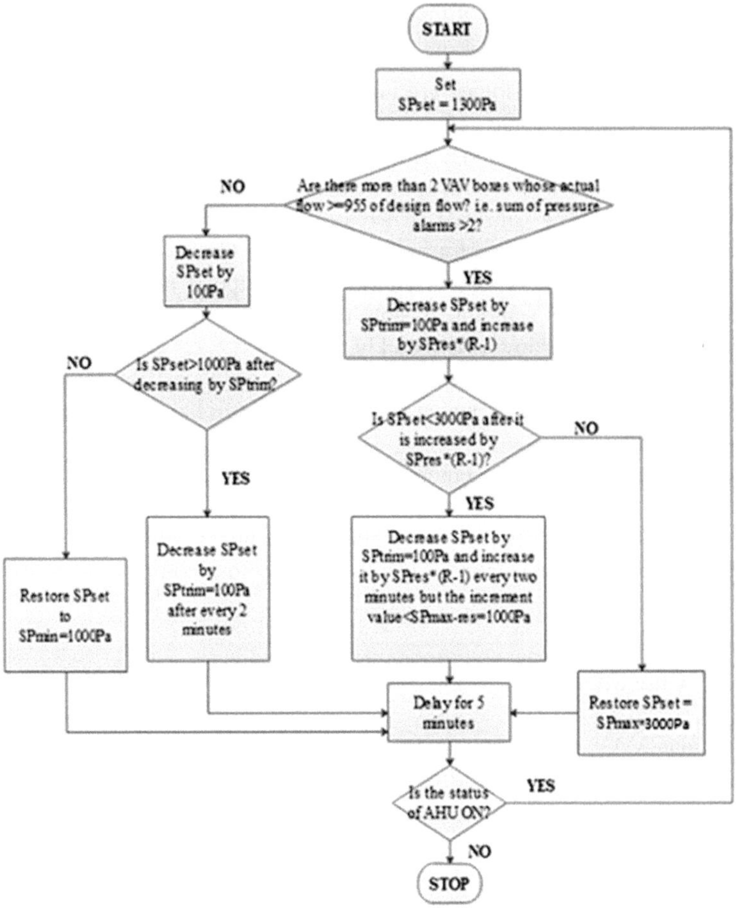

Fig. 10.28 Flowchart of trim-and-respond control strategy

on. The algorithm will gradually reduce the static pressure setpoint of the AHU by 100 Pa for every two minutes. As the static pressure setpoint decreases, downstream VAV terminal units' dampers will open further for a specific load. When the combination of two factors (i.e., reduced static pressure and load changes) cause more than two VAV terminal boxes to be fully open, the system will respond by increasing the static pressure setpoint by 130 Pa for every request but not more than 1000 Pa irrespective of the number of requests. The setpoint will continue to increase every two minutes until only two or less VAV terminal unit boxes are dissatisfied. Afterward, the setpoint will continue to decrease by 100 Pa in every two minutes.

Figure 10.29 shows simulated results of the T&R control strategy for the VAV system. As shown in Fig. 10.29, the VAV system operates under low setpoints for a larger part of the time. As demand for airflow increases, there is insufficiency of

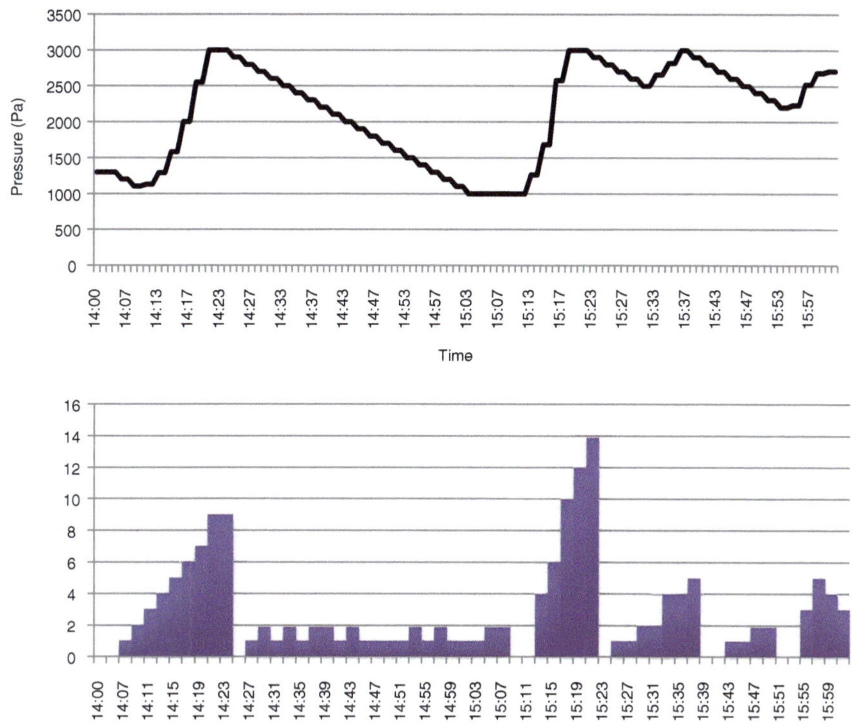

Fig. 10.29 Variations of static pressure setpoints and system requests with time

static pressure required to provide increased airflows. As soon as the airflow demand reaches its maximum (2.1 m^3/h), most dampers are near fully open positions. The static pressure reset loop then increases the fan operating point as most dampers are more than 85 % open. This causes alarms in most zones.

The static pressure reset controller begins to trim the static pressure setpoint as soon as two or less zones generate alarms, and at this moment the opening degree of most dampers is less than 80 %. The controller constantly decreases the static pressure setpoint at a fixed rate when there are no requests from zones or when two or less zones are generating requests. Static pressure setpoint gradually decreases to attain minimum setpoint value, thus resulting in energy saving. On the other hand, increase in demand causes the setpoint to have a rapid increase. It can also be seen that the reset is fast. Static pressure setpoint is reset from the minimum of 1000 Pa to the maximum of 3000 Pa in quarter an hour.

It is possible to ignore rogue zones here since trim and respond makes pressure adjustments based on pressure alarms generated. It means that a low static pressure setpoint is maintained in the VAV system even if one damper keeps on demanding for increased static pressure, which can bring about energy saving at the expense of comfort.

10.5 Control Sequences for VAV System with Different Terminal Units

10.5.1 For Cooling-Only Terminal Unit

The control logic for VAV cooling-only terminal unit under occupied mode operation is depicted schematically in Fig. 10.30. When the zone is in cooling, the cooling loop output shall correspond to the active airflow setpoint from the minimum to the cooling maximum airflow setpoints. Otherwise, the active airflow setpoint shall be the minimum airflow setpoint. The VAV damper shall be modulated by a control loop to maintain the measured airflow at the active setpoint.

It is noted that during the cooling loop, if supply air temperature from air handler is greater than room temperature, cooling shall be locked out.

We should provide software points that interlock to a system-level point to

(a) Force zone airflow setpoint to zero
(b) Force zone airflow setpoint to $V_{\text{cool-max}}$
(c) Force zone airflow setpoint to $V_{\min}$
(d) Force damper full closed/open
(e) Reset request-time accumulator point to zero (provide one point for each reset-type listed below)

If the airflow adjustment cannot satisfy the cooling load requirement, the supply air temperature shall be reset to a lower value through adjusting the chilled water valve of cooling coil. The cooling supply air temperature (SAT) reset requests are listed as below:

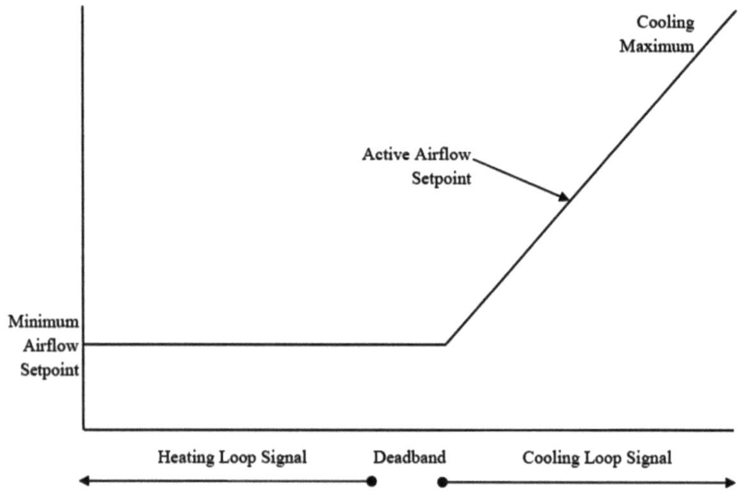

Fig. 10.30 Control logic for VAV cooling-only terminal unit (occupied mode)

(a) If the cooling loop is less than 85 %, send 0 requests (i.e., no action).
(b) If the cooling loop is greater than 95 %, send 1 request (decrease the SAT to the third lowest level).
(c) If the zone temperature exceeds the zone's cooling setpoint by 3 °C for 2 min, send 2 requests (decrease the SAT to the second lowest level).
(d) If the zone temperature exceeds the zone's cooling setpoint by 5 °C for 2 min, send 3 requests (decrease the SAT to the lowest level).

The static pressure will be reset according to the valve opening degree of the VAV box, and the reset requests are given as below:

(a) If the valve opening degree of the VAV box is less than 85 %, send 0 requests (i.e., no action).
(b) If the valve opening degree of the VAV box is greater than 95 %, send 1 request (reduce the static pressure setpoint).
(c) If the measured airflow is less than 70 % of setpoint for 1 min, send 2 requests (increase the static pressure setpoint to a higher level).
(d) If the measured airflow is less than 50 % of setpoint for 1 min, send 3 requests (increase the static pressure setpoint to the highest level).

10.5.2 For Reheat Terminal Unit

The control logic for VAV reheat terminal unit under occupied mode operation is depicted schematically in Fig. 10.31. The logic is also called 'Dual Maximum Logic.' The term 'dual maximum logic' comes from the fact that there are now two maximum airflow setpoints: one for heating, and the other for cooling.

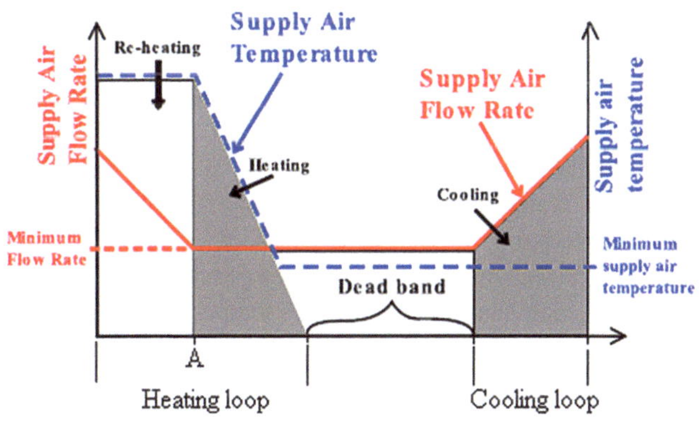

Fig. 10.31 Control logic for VAV reheat terminal unit (occupied mode)

With dual maximum logic, the minimum airflow setpoint is determined as follows:

1. No less than the larger of:
   (a) The zone minimum outdoor air rate that has been discussed in the earlier text.
   (b) The lowest setpoint allowed by the VAV box controls. With modern DDC, the controllable minimum for most DDC manufacturers is usually not an issue because it is normally below the ventilation requirement.
2. No more than 20 % of the cooling maximum airflow setpoint.

The heating maximum airflow setpoint must be:

1. No less than the larger of:
   (a) The MAR determined above.
   (b) The amount of air required to heat the space at a supply air temperature appropriate for the application or as limited by Title 24/90.1, typically 85 to 95 °F (29 to 35.2 °C).
2. No more than 50 % of the cooling maximum airflow setpoint. This is a larger percentage than the 30 % allowed using conventional logic which allows VAV reheat terminals to be used in more applications. This does result in more reheat at high heating loads, but the lower minimum airflow setpoint (a maximum of 20 % vs. 30 % with the conventional control logic) offsets this by reducing reheat at lower heating loads. Since most zones spend far more time at low heating loads than they do at high heating loads, the dual maximums approach will result in overall reheat savings.

The dual maximum logic for VAV reheat terminal unit under occupied mode operation is illustrated as below:

When the zone is in cooling, the cooling loop output shall be mapped the airflow setpoint from the minimum to the cooling maximum airflow setpoints. Hot water valve is closed unless the supply air temperature is below the minimum setpoint. Please note that if supply air temperature from air handler is greater than room temperature, cooling shall be locked out.

When the zone is in deadband, the active airflow setpoint shall be the minimum airflow setpoint. Hot water valve is closed unless the supply air temperature is below the minimum setpoint.

When the zone is in Heating (i.e., the zone temperature drops below the heating setpoint), the heating loop shall maintain space temperature at the heating setpoint as follows:

1. From 0–50 %, the heating loop output shall reset the supply air temperature from the current AHU SAT setpoint to the lesser of 90 °F or 20 °F above space temperature.

2. From 51–100 %, if heating is available, the heating loop output shall reset the active airflow setpoint from the minimum airflow setpoint to the maximum heating airflow setpoint.
3. The hot water valve (or modulating electric heating coil) shall be modulated to maintain the supply air temperature at setpoint no lower than 50 °F.

We should provide software points that interlock to a system-level point to

(a) Force zone airflow setpoint to zero
(b) Force zone airflow setpoint to $V_{\text{cool-max}}$
(c) Force zone airflow setpoint to V_{min}
(d) Force zone airflow setpoint to $V_{\text{heat-max}}$
(e) Force damper full closed/open
(f) Force heating to off/closed
(g) Reset request-hours accumulator point to zero (provide one point for each reset type listed below)

The system requests include cooling supply air temperature (SAT) reset requests and static pressure requests, and they are the same as that for VAV cooling-only terminal unit.

If there is a hot water coil, heating water supply temperature (HWST), reset requests

(a) If the hot water valve is less than 85 %, send 0 requests (no action).
(b) If the hot water valve is greater than 95 %, send 1 request (increase the HWST to the third highest level).
(c) If the supply air temperature is 15 °F less than setpoint for 5 min, send 2 requests (increase the HWST to the second highest level).
(d) If the supply air temperature is 30 °F less than setpoint for 5 min, send 3 requests (increase the HWST to the highest level).

10.5.3 For Series Fan-Powered Terminal Unit

The control logic is for series fan-powered terminal unit depicted schematically in Fig. 10.32.

When the zone is in cooling, the cooling loop output shall be mapped the primary airflow setpoint from the cooling maximum to the minimum airflow setpoints. Heating coil is off.

When the zone is in deadband, the primary airflow setpoint shall be the minimum airflow setpoint. Heating coil is off.

When zone is in heating, the heating loop shall reset the discharge temperature from the current AHU SAT setpoint to 90 °F. Higher temperatures tend to cause air to stratify and bypass into the return air. The hot water valve (or modulating electric heating coil) shall be modulated to maintain the discharge temperature at setpoint.

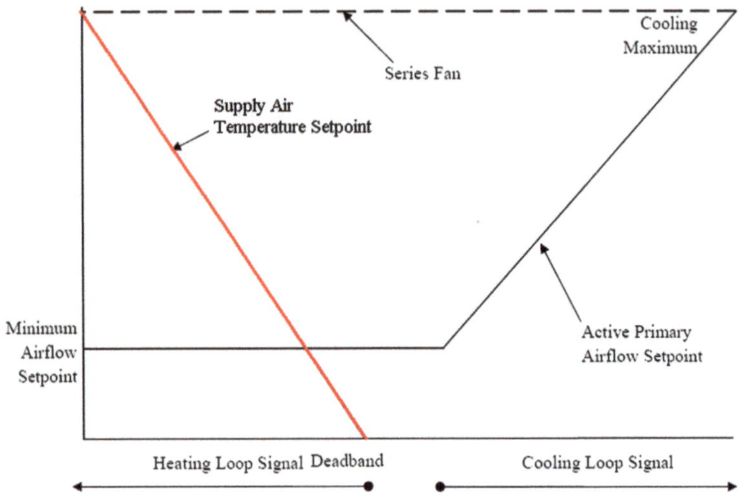

Fig. 10.32 Control logic for series fan-powered terminal unit (occupied mode)

The VAV damper is modulated to maintain the measured airflow at setpoint. The terminal fan shall run whenever zone is in occupied mode. Prior to starting the fan, the damper is first driven fully closed to ensure that the fan is not rotating backward. Once the fan is proven on for a fixed time delay (15 s), the damper override is released.

The system requests are the same as that for the above two types of terminal units.

10.6 Test Script for VAV Control Study

10.6.1 Preparation

1. **Design a test form**. Designing a test form that will be used to collect data and help in evaluating data. Thinking through and documentation of the test procedure in advance makes the testing process easy.
2. **Determine the criteria of acceptance**. The setpoints of both the discharge temperature and duct static pressure should be reset between their minimum and maximum setpoint values. The resetting procedure is based on the respective control variable (i.e., cooling load, damper position, and outdoor air temperature) and the system. They should meet these setpoints without excessive hunting. In addition, trending should be used to evaluate the interaction between the two reset strategies. This will result into modified sequences if the control strategies begin to oppose each other.
   Note: 'Lack of excessive hunting' signifies that after a moderate disturbance the controlled variable (duct static pressure) comes into the control deadband and

remains there for a period of 4 cycles (1 cycle is made up of one overshoot and one undershoot).

3. **Outline instructions/precautions**. If performing the tests while the building is partially occupied or the systems are in temporary operation modes to provide space conditioning, bear in mind that zone comfort may be influenced by the test procedures. In the event of a test failure, the source of the failure should be identified and acted upon appropriately. The system will be retested once the fault is corrected.
4. **Specify participants, their roles and responsibilities**. The guideline of testing outlined in this paper can assist in verifying proper system performance in both new and existing building applications. The testing process may require the participation of the following individuals. A detailed description of the general roles and responsibilities of the participants will be given. These roles and responsibilities will be customized based on actual project requirements.

10.6.2 General Inspection of Air-Handling and Distribution System

1. **Review all prefunctional checklists to ascertain completeness.**

Complete the commissioning prestart, start-up, and prefunctional checklists, as well as applicable manufacturer's prestart and start-up recommendations before performing any functional tests. Some of the prefunctional checklists include the following:

(a) Ensuring that supply fan rotates in the right direction and is free of unusual noise and vibration.
(b) Belt tension, alignment, and condition of supply fan are okay (where applicable).
(c) Ensuring that duct installation associated with the air-handling system being tested is complete. If interior duct liner is used, seal the exposed ends properly.
(d) Ensuring that all terminal units associated with the air-handling system being tested have been functionally tested and are capable of serving normal operating loads.
(e) Ensuring that the respective cooling, heating, and pumping systems associated with the air-handling system being tested have been functionally tested and are capable of serving normal operating loads.
(f) Ensuring that both the water and air systems have been balanced per design.
(g) Ensuring that all control sensors (i.e., for pressure temperature) have been installed and calibrated.
(h) Ensuring that all safeties and interlocks have been tested and are operational.
(i) Ensuring that all sequences of operations have been programmed per design.

2. **Duct Static Pressure Reset Test Procedures**

(2.1) **Verifying the maximum static pressure setpoint**. Disable the discharge air temperature reset prior to testing the static pressure reset strategy in order to avoid any interaction between the two control strategies (handle interaction separately). Regardless of the static pressure reset control strategy employed (whether duct or discharge), modulate all of the VAV box dampers fully open by either commanding each damper to full cooling or by lowering zone temperature setpoint to 15 °C below existing space conditions. Both approaches will simulate a cooling load on the space and result in the VAV boxes being modulated to full cooling positions. As the primary air dampers open, duct static pressure should drop and the controlling static pressure setpoint should reset incrementally upward until the maximum limit is reached.
Check the following:

(a) Static pressure setpoint does not exceed maximum value.
(b) Supply fan ramps up close to full speed (or the speed reported by the balancer as necessary to meet design flow).
(c) The static pressure reported in the BAS meets the setpoint without excessive hunting.

(2.2) **Verifying the minimum static pressure setpoint**. Continuing from above, modulate all of the VAV box dampers to minimum flow position by either commanding each damper to minimum damper position or by raising zone temperature setpoint to 15 °C above the prevailing space conditions. Note that some VAV boxes may be programmed to increase airflow above the minimum flow setting in order to satisfy a maximum heating load. When raising space temperature setpoint to modulate the VAV boxes to minimum flow position, first verify that maximum heating flow setpoint is the same as the minimum flow value. Adjust system setpoints as necessary. As the primary air dampers close, duct static pressure should increase and the controlling static pressure setpoint should reset incrementally downward until the minimum static pressure setpoint is reached, or all zones are satisfied.
Check out the following that:

(a) Static pressure setpoint does not exceed the minimum value.
(b) Supply fan should ramp down.
(c) Supply fan VFD speed should not decrease below the minimum speed recommended by VFD, fan, or motor manufacturers. Minimum speed can range between 10 and 20 Hz, depending on the type of fan installed. Centrifugal fans can typically operate at the lower end of the range, whereas axial fans may require higher minimum speeds compared to centrifugal fans.
(d) Measured static pressure meets setpoint without excessive hunting.

(2.3) **Restoring system to normal operation.** Restore the system to normal operating conditions by removing all overrides and adjusting all set-points to values specified in the design sequence of operations.

3. **Test Procedures for Discharge Air Temperature Reset**

(3.1) **Verifying maximum discharge air temperature setpoint.** Disable the static pressure reset prior to testing the discharge air temperature reset strategy in order to avoid any interaction between the two control strategies (handle interaction separately). The specific control strategy employed determines the way to simulate a low load condition in order to trigger the discharge air temperature reset control strategy. Test procedures for each control strategy are as follows:

Outdoor Air (OA)—The outdoor air temperature value is overwritten to be at the low-end of the reset schedule.

Damper Position—Either instruct VAV box dampers to be at a minimum position or adjust zone temperature setpoint to be 5 °F above actual zone temperature.

Deviation from Setpoint—Modulate the zone temperature setpoint as necessary in order to ensure that the deviation from setpoint is sufficient to trigger the control sequence.

Cooling Demand—Modulate the zone temperature setpoint to be 5 °F above actual zone temperature.

These steps will initiate a decrease in zone cooling loads, which should result into resetting the discharge air temperature incrementally upward until the maximum is reached.

Check out the following that:

(a) Discharge air temperature setpoint does not exceed the maximum value.
(b) Discharge air temperature meets setpoint without excessive hunting.

(3.2) **Verifying the minimum discharge air temperature setpoint.** Continuing from above, the procedure to simulate a high load condition for each control strategy is as follows.

Outdoor Air (OA)—The outdoor air temperature value is overwritten to be on the higher side of the reset schedule.

Damper Position—Either instruct VAV box dampers to a maximum position or adjust zone temperature setpoint to be at 5 °F below actual zone temperature.

Deviation from Setpoint—Modulate the zone temperature setpoint as necessary in order to ensure that the deviation from setpoint is sufficient to trigger the control sequence.

Cooling Demand—Modulate the zone temperature setpoint to be at 5 ° F below the actual zone temperature.

These steps will initiate an increase in zone cooling loads, which should result in resetting the discharge air temperature incrementally downward until the minimum limit is reached.
Check out the following that:

(a) Discharge air temperature setpoint does not exceed the minimum limit value.
(b) Measured discharge air temperature meets setpoint without excessive hunting.

(3.3) **Restoring the system to normal operation**. Restore the system to normal operating conditions by eliminating all overrides and adjusting all setpoints to specifications in the design sequence of operations.

4. **Identifying Interactions of Control Strategies**

(4.1) **Trend reset control strategies**. As pointed out in the section on Reset Strategy Interactions, it is normal for these two reset strategies to counteract one another and result in unstable control. Trend analysis of all polled terminal boxes for a given air handler is the best way to (i) investigate if individual reset control strategies (i.e., discharge temperature reset and static pressure reset) are working properly under normal operating conditions and (ii) to identify possible negative interactions between the static pressure and discharge air temperature resets. In order to detect operational problems such as hunting and excessive cycling, a trend sample rate of no more than 5 min is necessary. Typically, plot the independent variable (for instance, VAV box damper position for static pressure reset or zone temperature control loop output for discharge air temperature reset) along the X-axis and the control variable (static pressure or discharge air temperature setpoint) on the Y-axis.
Another trend plot that can help in identifying reset interactions is the actual speed of the supply fan (assuming the supply fan is controlled by a VFD) over time. If the supply fan speed remains fairly constant (especially if zone loads are known to vary) or seems to be constantly fluctuating, then we can conclude that the two reset strategies may be fighting each other.

(4.2) **Considering modifying the control strategy to minimize negative interaction**. One approach that can be employed to minimize negative interaction between the discharge air temperature reset and static pressure reset strategies is to enable the discharge air temperature reset strategy and disable the static pressure reset strategy (i.e., maintain fixed static pressure setpoint) during colder atmospheric conditions, and enable the static pressure reset strategy and disable the discharge air temperature reset strategy (i.e., maintain fixed discharge air temperature setpoint) during warmer atmospheric conditions.

Colder Conditions: The objective is to minimize as much as possible the amount of simultaneous heating and cooling during winter, avoid over-cooling internal zones that may not have reheat capability, and maximize energy savings in both the chilled and hot water systems. While it is true that supply fan energy savings will be sacrificed, the energy saved at the central plant should far outweigh the fan energy losses. Note that fan energy usage can be minimized if the fixed static pressure setpoint is as low as possible to satisfy all zones. Experimentation may be required to determine the optimum static pressure setpoint.

Warmer Conditions: Maintaining a constant discharge air temperature during warmer weather will improve space temperature and humidity control and allow for maximum supply fan energy savings as zone loads vary and VAV dampers modulate.

A method for implementing such a control strategy is to have two independent control loops that become enabled and disabled based on outside air temperature. A reasonable outdoor air temperature for a changeover between control strategies may range between 70 and 75 ° F. In this range, the economizer will be able to offset more chiller load at the higher discharge air temperature setpoint during cooler ambient conditions. Above this range, the economizer provides minimal benefit, most zones will probably not require any reheat, and humidity control will be improved by maintaining a fixed discharge air temperature setpoint. It may take some experimentation to determine the optimum outside air temperature changeover point.

10.6.3 Trend Data Review

Typical trends showing the room temperature over time (over at least one week) will be recoded and analyzed to evaluate the validity of all control sequences. Figure 10.33 gives data flow for trend analysis. To begin with, trend setup on DDC server is made, and data are extracted to CD or directly transferred through Web. Then, the data should be converted into a readable format for some analysis/graphing tools. Afterward, the performance of a VAV control system is analyzed based on the data. To analyze a VAV control system performance, we will first want to get a bird's-eye view of the entire data set. This will highlight areas that are likely to be problems, and we can then focus in on these. This is so-called statistical analysis as shown in the first part of Fig. 10.33. Most packages feature statistical analysis such as number of data points, average, standard deviation, min/max, start and end time, and sample interval. This will let us cull trends from the database and allow us to find inactive trends. A simple minimum/maximum/average value will show trends that do not change over time, or that are exceptionally high or low. In addition, extreme

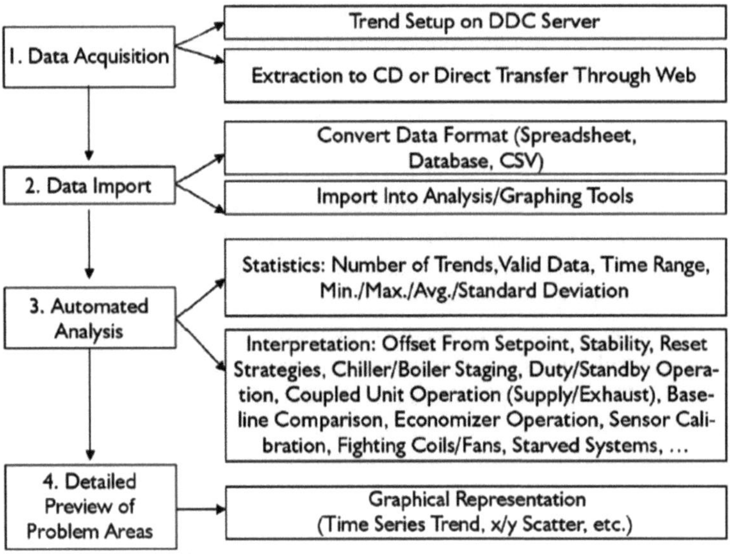

Fig. 10.33 Data flow for trend analysis

values can indicate sensor errors for room temperature, and the number of points in a trend can indicate that data are missing.

1. **Offset from setpoint**

The 'offset from setpoint' is one of the most important tasks for data analysis. Its objective is to take the different trend series and find the worst performers. It mathematically calculates how much a trend point deviates from its setpoint (s), and for how long. We can take all values above the setpoint and calculate the area between the actual value and the setpoint. The same is then done for values below the setpoint. The larger these two areas are, the worse the value adheres to its setpoint.

The formulas for such an analysis are as follows:

$$Offset_{C,n} = Max(0, Temp_n - SP_{cooling,n}) \quad \text{for cooling} \tag{10.31}$$

$$Offset_{H,n} = Max(0, SP_{heating,n} - Temp_n) \quad \text{for heating} \tag{10.32}$$

$$Offset_C = \sum_{n=1}^{N} Offset_{C,n} \times Time\_step_n \quad \text{for cooling} \tag{10.33}$$

$$Offset_H = \sum_{n=1}^{N} Offset_{H,n} \times Time\_step_n \quad \text{for heating} \tag{10.34}$$

where $Offset_{C,n}$ is the difference between the actual value $Temp_n$ and the cooling setpoint $SP_{coolig,\ n}$ at trend Point n. By taking the maximum of 0 and the difference, all negative values are ignored. These negative values are discarded because they indicate that the actual trend value is not above the cooling setpoint and, thus, do not indicate a problem in control. Similar logic applies for the heating offset.

Time_step is simply the time difference between one trend point and the next. In most cases, this is a fixed interval, for example five minutes, and is determined by the initial trend setup at the EMCS. However, in many cases trend data may be 'dropped' or lost, so that actual trend data may not occur in regular intervals. The trending interval also may be changed at the EMCS during the trended period. For this reason, Time_step should be calculated for every trend point. By multiplying $Offset_n$ and $Time\_step_n$ for all trend points and summing up all of the resulting values, the area between setpoint and temperature outside of setpoint is calculated.

The following step is to divide the resulting area by the total time of the trend interval over which calculation takes place, so that trends measured over different time intervals can be compared. In this way, the result for a point trended and calculated over three weeks will not be larger than the result of a point trended and calculated over one week, provided their control performance is similar.

$$Offset_{C,\text{norm}} = \frac{Offset_C}{Time_N - Time_1} \tag{10.35}$$

where $Time_N$ is the time stamp of the last trend point, and $Time_1$ is the time stamp of the first trend point. Each of the Eqs. (10.31)–(10.35) can be easily implemented in a spreadsheet or a graphing package. The result for each trend are two numbers, one offset for how badly overheated and one offset for how undercooled an HVAC zone was during the trend period. This offset takes into account both the amount of time the zone was outside of setpoint values, and the magnitude of this deviation.

By copying the minimum and maximum trend values next to the offset for cooling and heating, a clear picture emerges about how well every zone performed.

Analyzing several hundred trends can be done in one of two ways. In the simpler version, the calculations outlined previously are input into a spreadsheet or graphing program as a template. Trend values then are manually copied/pasted into the template, calculated, and results copied into a table by hand. In this manner, creating an overview of a whole building will take several hours. An alternative is to create a visual basic macro, or a macro in the language of the graphing software, that imports all trends within a directory and then copies the results into a table automatically.

2. **Lead/Lag Operation**

Lead/lag operations also can be easily analyzed. The sum of two status or start/stop trends must equal 1 in a typical duplex setup. Similarly, if a supply fan's operation automatically triggers an exhaust fan's, then the sum of status or start/stop trends

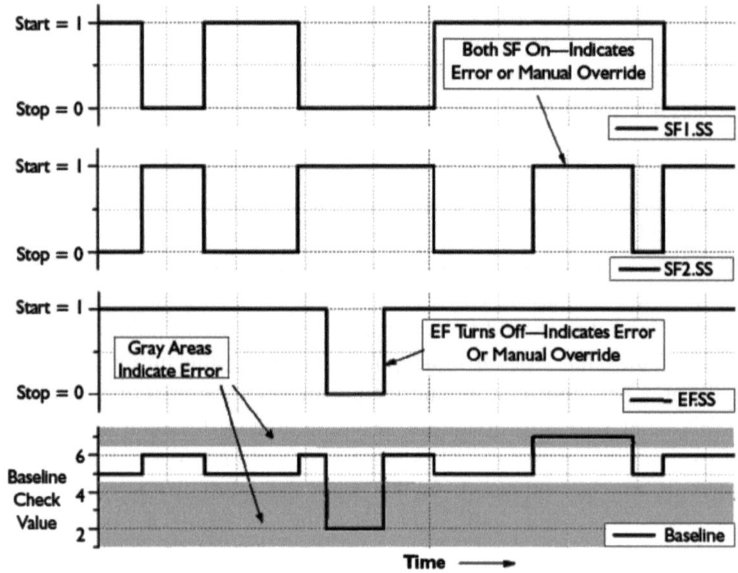

Fig. 10.34 Lead/lag analysis. The *gray* areas show where errors occur

must be either 0 or 2, but never 1. Note that a certain time delay may exist, so that care must be taken not to generate errors if a one-minute delay exists between supply fan and exhaust fan start, and one-minute trends show supply fan on but exhaust fan off right at that moment.

The following example shows this: Dual fan coil units are serving a space, together with an exhaust fan. The sequence calls for either Supply Fan 1 or Supply Fan 2 to run (but not both). If either supply fan is running, the exhaust fan (EF) should also run. We can numerically add the fan start–stop (SS) commands as follows:

$$\text{Baseline} = \text{SF1.SS} + \text{SF2.SS} \times 2 + \text{EF.SS} \times 4 \tag{10.36}$$

This means that baseline values above 6 or below 5 should not occur during normal operation. The gray areas in Fig. 10.34 show where errors occur. By taking a minimum and maximum of all baseline values, we can immediately see if the fans operated incorrectly at any time. By taking a maximum setpoint of 6 and a minimum setpoint of 5, we also can run an offset from baseline analysis and get an idea of how often errors occur, or whether we are looking at very incidental user-overrides.

References

1. Urban, R.A.: Design considerations and operating characteristics of variable volume systems. ASHRAE J. **12**(2), 77–84 (1969)
2. Shepherd, K.: VAV Air Conditioning Systems. Blackwell Science Ltd, United Kingdom (1999)
3. ASHRAE Handbook. HVAC Systems and Equipment [Chapter 4]. In Air Handling and Distribution. American Society of Heating, Refrigerating and Air conditioning Engineers Inc, Atlanta (2008) (4.11)
4. Dodd, M.: Comparing Energy Savings of Different VAV Systems, pp. 1–9. EnergySoft LLC (2012)
5. Bearg, D.W.: Indoor Air Quality and HVAC Systems. Lewis Publishers (1993)
6. Engdahl, F., Johansson, D.: Optimal supply air temperature with respect to energy use in a variable air volume system. J. Energy Build. **36**, 205–218 (2004)
7. Tung, D., Deng, S.: Variable air volume system under reduced static pressure control. Build. Serv. Eng. Res. Technol. **18**(2), 77–83 (1997)
8. Inoue, U., Matsumoto, T.: A study on energy savings with variable air volume systems by simulation and field measurement. J. Energy Build. **2**, 27–36 (1979)
9. Kloostra, L.: VAV systems save 38 % of energy use. Heating Pip. Air Conditioning **51**(12), 61–63 (1979)
10. Mull, T.E.: Energy conservation measures for air distribution and HVAC systems. J. Plant Eng. **58**(10), 60–64 (2004)
11. Mysen, M., Rydock, I.P., Tjelflaat, P.O.: Demand controlled ventilation for office cubicles-can it be profitable? J. Energy Build. **35**, 657–662 (2003)
12. Norford, L.K., Rabl, A., Socolow, R.H.: Control of supply air temperature and outdoor airflow and its effect on energy use in a variable air volume system. ASHRAE Trans. **92**, 30–45 (1986)
13. Wang, S., Burnett, J.: Variable-air-volume air-conditioning systems: optimized static pressure setpoint. J. Build. Serv. Eng. Res. Technol. **19**(4), 219–231 (1998)
14. Khoo, I., Levermore, G.J., Letherman, K.M.: Variable-air-volume terminal units I: steady state models. J. Build. Serv. Eng. Res. Technol. **19**(3), 155–162 (1998)
15. Parameshwaran, R., Karunakaran, R., Iniyan, S., Anand, A.S.: Optimization of energy conservation potential for VAV air conditioning system using Fuzzy based genetic algorithm. Int. J. Mech. Aerosp. Ind. Mechatron. Manuf. Eng. **2**(1), 67–74 (2008)
16. ANSI/ASHRAE/IESNA. Standard 90.1-2010 Energy standard for buildings except low-rise residential buildings. International Code Council, Inc., Washington, D.C. and American Society of Heating, Refrigerating and Air-conditioning Engineers, Inc., Atlanta, GA (2010)
17. Yu, Y., Liu, M., Cho, Y., Xu, K.: Integrated demand controlled ventilation for single duct VAV system with conference rooms. In: Proceedings of ICEBO, California (2007)
18. Yu, Y., Xu, K., Cho, Y.H., Liu, M.: A smart logic for conference room terminal box of single duct VAV system. In: Proceedings of the 7th International Conference for Enhanced Building Operations. San Francisco, CA (2007)
19. Cho, Y.H.: Development of a terminal control system with variable minimum airflow rate. Energies **5**, 4643–4664 (2012)
20. Hartman, T.: TRAV—A new HVAC concept. Heating/Piping/Air Conditioning Eng. HPAC **61**(7), 69–73 (1989)
21. Hartman, T.: Terminal regulated air volume (TRAV) systems. ASHRAE Trans. **99**(1), 791–800 (1993)
22. Hartman, T.: Global optimization strategies for high-performance controls. ASHRAE Trans. **101**(2), 679–687 (1995)
23. Englander S.: Ventilation Control for Energy Conservation: Digitally Controlled Terminal Boxes and Variable Speed Drives. Princeton University (1990)

24. Englander, S.L., Norford, L.K.: Saving fan energy in VAV systems—Part 2: supply fan control for static pressure minimization using DDC zone feedback. ASHRAE Trans. **98**(1), 19–32 (1992)
25. Warren, M., Norford, L.K.: Integrating VAV zone requirements with supply fan operation. ASHRAE J. **35**(4), 43–46 (1993)
26. Wei, G.H., Liu, M.S., Claridge, D.E., Sakurai, Y.: Improved air volume control logic for VAV systems. In: Proceedings of the Twelfth Symposium on Improving Building Systems in Hot and Humid Climates. San Antonio, TX, May 15–17 (2000)
27. Nassif, N., Moujaes, S.: A new operating strategy for economizer dampers of VAV system. J. Energy Build. **40**, 289–299 (2008)
28. Haasl, T., Potter, A., Irvine, L., Luskay, L.: Retro-commissioning's greatest hits. In: Proceedings of the 1st International Conference for Enhanced Building Operations, Austin, TX (2001). http://esl.eslwin.tamu.edu/digital-library.html
29. Song, L.I., Joo, D., Dong, M., Liu, J., Wang, K., Hansen, L.Q., Swiatek, A.: Optimizing HVAC control to improve building comfort and energy performance. In: Proceedings of the 3rd International Conference for Enhanced Building Operations. Berkeley, CA (2003). http://esl.eslwin.tamu.edu/digital-library.html
30. Pang, X., Zheng, B., Liu, M.: Case study of continuous commissioning in an office building. In: Proceedings of the 6th International Conference for Enhanced Building Operations. Shenzhen, China (2006). http://esl.eslwin.tamu.edu/digital-library.html
31. Liu, M.: Variable speed drive volumetric tracking (VSDVT) for airflow control in variable air volume (VAV) systems. J. Sol. Energy Eng. **125**, 318–323 (2003)
32. Liu, G., Liu, M.: Supply fan control methods for VAV systems using a fan airflow station. ASHRAE Trans. **114**(2), 451–457 (2008)
33. Wu, L., Liu, M., Wang, G., Pang, X.: Integrated static pressure reset with fan airflow station in dual-duct VAV system control. In: Proceedings of the Energy Sustainability Conference 2007, pp. 441–49. Long Beach, CA (2007)
34. Liu, M., Claridge, D.E., Turner, W.D.: Continuous commissioning guidebook for federal energy managers. Federal Energy Management Program, U.S. Department of Energy, Washington, D.C. (2002). www1.eere.energy.gov/femp/operations_maintenance/om_ccguide.html
35. Zheng, K., Li, H., Yang, H.: Application of wireless sensor network (WSN) technologies in optimal static pressure reset in variable air volume (VAV) system. In: Proceedings of the 7th International Conference for Enhanced Building Operations. SanFrancisco, CA (2007). http://esl.eslwin.tamu.edu/digital-library.html
36. Murphy, J.: Ventilation control in terminal units with variable speed fan control. ASHRAE J., 12–19 (2013)
37. Meng, Q.L., Yan, X.Y., Ren, Q.C.: Global optimal control of VAV air-conditioning system with iterative learning: an experimental case study. J. Zhejiang Univ.-Sci. A (ApplPhys&Eng), 1–21 (2014)
38. Koulani, C.S.: Optimized damper control of pressure and airflow in ventilation systems. Section of building physics and services department of civil engineering, Technical University of Denmark. Master thesis (2013)
39. The MathWorks. Simulation and Model-Based Design. Simulink Manual Version 6 (2005). www.mathworks.com
40. Karris, S.T.: Introduction to Simulink with Engineering Applications, Orchard Publications (2006). www.orchardpublications.com
41. Zheng, G.R.: Dynamic modeling and global optimal operation of multi-zone variable air volume HVAC system. PHD Thesis, The Center for Building Studies, Concordia University, Montreal, Canada (1997)
42. White, F.M.: Fluid Mechanics, 3rd edn, p. 736. Mc-Graw Hill, New York (1994)
43. Brandemuel, M.J., Gabell, S., Andersen, I.: A Toolkit for Secondary HVAC System Energy Calculation. ASHRAE, Joint Center for Energy Management, University of Colorado, Boulder (1993)

44. Nassif, N.S., Kajl, S., Sabourin, R.: Modeling and validation of existing VAV system components. In: Proceedings of Esim, Canadian Conference on Building Simulation, Vancouver, Canada (2004)
45. Fans and Pumps. Energy Management Series 13 for Industry and Commercial Institutions, Canada
46. Colebrook, C.F.: Turbulent flow in the pipe, with particular reference to the transition region between the smooth and rough pipe laws. J. Inst. Civ. Eng. **11**, 133–156 (1938–1939) (London)
47. Haaland, S.: Simple and explicit formulas for the friction factor in turbulent flow. Trans. ASME J. Fluids Eng. **103**, 89–90 (1983)
48. McQuiston, F.C., Parker, J.D.: Heating, Ventilating and Air-Conditioning, 3rd edn. Wiley, Hoboken (1988)
49. Idelchik, I.E.: Handbook of Hydraulic Resistance. Hemisphere Publishing Corporation (1986)
50. Archer, W.H.: Loss of head due to enlargements in pipes. Trans. Am. Soc. Civ. Eng. **76**, 999–1026 (1913)
51. ASHRAE. HVAC Fundamentals. ASHRAE Handbook (2011)
52. ANSI/ASHRAE/IESNA Standard 90.1-2010. Energy standard for buildings except low-rise residential buildings. American Society of Heating, Refrigerating and Air-conditioning Engineers, Inc, Atlanta (2010)
53. Yuying, S., Zheng, Z., Hou, X., Tian, P.: AHU control strategies in the VAV system. In: Fourth International Conference on Innovative Computing, Information and Control, pp. 119–123 (2009)
54. Nelson, R.M., Householder, B.: A study on static pressure reset and instability in variable air volume HVAC systems. Final report, Iowa Energy Center (2011)
55. Taylor, S.: Resetting set points using trim and respond logic. ASHRAE J., 52–57 (2015)
56. Taylor, S.: Increasing efficiency with VAV system static pressure setpoint reset. ASHRAE J. **49**(6), 24–32 (2007)

MIX
Papier aus verantwortungsvollen Quellen
Paper from responsible sources
FSC® C105338

If you have any concerns about our products,
you can contact us on
ProductSafety@springernature.com

In case Publisher is established outside the EU,
the EU authorized representative is:
Springer Nature Customer Service Center GmbH
Europaplatz 3, 69115 Heidelberg, Germany

Printed by Libri Plureos GmbH
in Hamburg, Germany